Chemical Hazards
in the Workplace

Chemical Hazards
in the Workplace

Chemical Hazards in the Workplace

Measurement and Control

Gangadhar Choudhary, EDITOR

National Institute for Occupational Safety and Health

Based on a symposium

sponsored by the Division of

Chemical Health and Safety at the

Second Chemical Congress of the

North American Continent

(180th ACS National Meeting),

Las Vegas, Nevada,

August 25–28, 1980.

ACS SYMPOSIUM SERIES 149

AMERICAN CHEMICAL SOCIETY

WASHINGTON, D. C.   1981

Library of Congress CIP Data

Chemical hazards in the workplace.
  (ACS symposium series; 149)
  "Based on a symposium sponsored by the Division
of Chemical Health and Safety at the Second Chemical
Congress of the North American Continent (180th
ACS National Meeting), Las Vegas, Nevada, August
25-28, 1980."
  Includes bibliographies and index.

  1. Industrial toxicology—Congresses. 2. Industrial
hygiene—Congresses. 3. Environmental monitoring—
Congresses. 4. Work environment—Congresses.
  I. Choudhary, Gangadhar, 1935–   . II. American
Chemical Society. Division of Chemical Health and
Safety. III. Chemical Congress of the North American
Continent (2nd: 1980: Las Vegas, Nev.) IV. Series.

RA1229.C46   615.9'02   81-130
ISBN 0–8412–0608–2   ACSMC8 149 1-628 1981
                                          AACR2

FOREWORD

The ACS SYMPOSIUM SERIES was founded in 1974 to provide a medium for publishing symposia quickly in book form. The format of the Series parallels that of the continuing ADVANCES IN CHEMISTRY SERIES except that in order to save time the papers are not typeset but are reproduced as they are submitted by the authors in camera-ready form. Papers are reviewed under the supervision of the Editors with the assistance of the Series Advisory Board and are selected to maintain the integrity of the symposia; however, verbatim reproductions of previously published papers are not accepted. Both reviews and reports of research are acceptable since symposia may embrace both types of presentation.

CONTENTS

Preface xi

METHODOLOGY

1. Sampling and Analytical Methodology for Workplace Chemical
   Hazards: State of the Art and Future Trends 3
   C. C. Anderson, E. C. Gunderson, and D. M. Coulson

2. Development of an Analytical Method for Benzidine-Based Dyes .. 21
   E. R. Kennedy and M. J. Seymour

3. An Infrared Analysis Method for the Determination of
   Hydrocarbons Collected on Charcoal Tubes 37
   T. C. Thomas and A. Richardson III

4. Development of Personal Sampling and Analytical Methods for
   Organochlorine Compounds 49
   K. W. Boyd, M. B. Emory, and H. K. Dillon

5. Measurement, Analysis, and Control of Cotton Dust 65
   J. G. Montalvo, Jr., D. P. Thibodeaux, and A. Baril, Jr.

6. Estimation of Airborne Sodium Hydroxide 87
   E. Reid

7. Development of a Method for Sampling and Analysis of Metal Fumes 95
   W. F. Gutknecht, M. B. Ranade, P. M. Grohse, A. S. Damle,
   and P. M. P. Eller

8. Sampling and Breakthrough Studies with Plictran 109
   C. C. Houk and H. J. Beaulieu

9. Sampling and Analysis of Chlorinated Isocyanuric Acids 123
   J. Palassis and J. R. Kominsky

10. Monitoring for Airborne Inorganic Acids 137
    M. E. Cassinelli and D. G. Taylor

MONITORING AND CONTROL

11. Specialized Sorbents, Derivatization, and Desorption Techniques for
    the Collection and Determination of Trace Chemicals in the
    Workplace Atmosphere 155
    R. G. Melcher, P. W. Langvardt, M. L. Langhorst,
    and S. A. Bouyoucos

vii

12. Solid Sorbents for Workplace Sampling 179
    E. C. Gunderson and E. L. Fernandez

13. Diffusional Monitoring: A New Approach to Personal Sampling 195
    D. W. Gosselink, D. L. Braun, H. E. Mullins, S. T. Rodriguez,
    and F. W. Snowden

14. An Evaluation of Organic Vapor Passive Dosimeters Under
    Field Use Conditions 209
    R. S. Stricoff and C. Summers

15. The Role of Biological Monitoring in Medical and Environmental
    Surveillance 223
    C. B. Monroe

16. Permeation of Protective Garment Materials by Liquid Halogenated
    Ethanes and a Polychlorinated Biphenyl 235
    R. W. Weeks, Jr. and M. J. McLeod

17. The Use of a Fiberoptics Skin Contamination Monitor in the
    Workplace 269
    T. Vo-Dinh and R. B. Gammage

18. A Health Hazard Evaluation of Nitrosamines in a Tire
    Manufacturing Plant 283
    J. D. McGlothlin, T. C. Wilcox, J. M. Fajen, and G. S. Edwards

19. Sampling Methods for Airborne Pesticides 301
    E. C. Gunderson

SPECIAL TOXICANTS

20. Occupational Exposure to Polychlorinated Dioxins and
    Dibenzofurans 319
    C. Rappe and H. R. Buser

21. Occurrence of Nitrosamines in Industrial Atmospheres 343
    D. P. Rounbehler, J. W. Reisch, J. R. Coombs, D. H. Fine,
    and J. M. Fajen

22. Gas Chromatography–Mass Spectrometric Characterization of
    Polynuclear Aromatic Hydrocarbons in Particulate Diesel Emissions 357
    D. R. Choudhury and B. Bush

23. Application of Glass Capillary Gas Chromatography for Determina-
    tion of Potential Hazardous Compounds in Workplace Environments 369
    G. Becher, A. Bjørseth, and B. Olufsen

24. Suitability of Various Filtering Media for the Collection and
    Determination of Organoarsenicals in Air 383
    G. Ricci, G. Colovos, N. Hester, L. S. Shepard, and J. C. Haartz

25. Determination of Aromatic Diamines and Other Compounds in Hair
    Dyes Using Liquid Chromatography 401
    K. Johansson, C. Rappe, W. Lindberg, and M. Nygren

26. High Performance Liquid Chromatographic Determination of
    Aromatic Amines in Body Fluids and Commercial Dyes 413
    P. J. M. VanTulder, C. C. Howard, and R. M. Riggin

viii

QUALITY ASSURANCE

27. An Evaluation of Statistical Schemes for Air Sampling 431
    S. M. Rappaport, S. Selvin, R. C. Spear, and C. Keil

28. Industrial Hygiene Logistics 457
    G. Stough and A. Salazar

29. The NIOSH Action Level: A Closer Look 471
    J. C. Rock

30. Industrial Hygiene Air Sampling with Constant Flow Pumps 491
    W. B. Baker, D. G. Clark, and W. J. Lautenberger

31. Statistical Protocol for the NIOSH Validation Tests 503
    K. A. Busch and D. G. Taylor

NEW TECHNOLOGIES

32. The Introduction of Microprocessor-Based Instrumentation for the
    Measurement of Occupational Exposures to Toxic Substances 521
    R. Kriesel, H. Brouwers, and K. Jansky

33. A Versatile Test Atmosphere Generation and Sampling System 533
    S. Kapila, R. K. Malhotra, and C. R. Vogt

34. Development of Workplace Guidelines for Emerging Energy
    Technologies 543
    O. White, Jr. and D. Lillian

35. Recent Developments in Electrochemical SPE Sensor Cells for
    Measuring Carbon Monoxide and Oxides of Nitrogen 551
    A. B. LaConti, M. E. Nolan, J. A. Kosek, and J. M. Sedlak

36. A New Passive Organic Vapor Badge with Backup Capability 575
    W. J. Lautenberger, E. V. Kring, and J. A. Morello

37. New Technology for Personal Sampling of NO_2 and NO_x in the
    Workplace 587
    R. McMahon, T. Klingner, B. Ferber, and G. Schnakenberg

38. Ion Chromatographic Analysis of Formic Acid in Diesel Exhaust
    and Mine Air 599
    I. Bodek and K. T. Menzies

Index 615

PREFACE

The workplace environment is a significant part of the total ecological system. Since it can be measured, some control over it can be achieved, and improvements in the control technologies in the workplace can be made. Because of the rapidly growing production of complex chemical substances and the use of these in modern living during the past three decades, the existence of chemical hazards in workplaces in relation to worker health and safety has become the subject of great concern.

Although the industrial hygiene considerations and the federal government involvement in worker health and safety in the United States began a long time ago, concerted effort and increased attention toward this work-related problem—either by government or by industry—did not become possible until about ten years ago when two sister agencies of the government were created, namely, the Occupational Safety and Health Administration (OSHA), which is part of the Department of Labor, and the National Institute for Occupational Safety and Health (NIOSH), which is part of the Department of Health and Human Services. The functions of these agencies are to clean up the workplace environment and to protect workers' health through worker and industry participation, recognition of potential hazards, and conduction of on-site evaluations. The fulfillment of these responsibilities requires the development of new measurement and control methods as well as improving the existing technology.

To achieve meaningful health hazard evaluations and control technologies for the workplace environment, knowledge of correct measurement and monitoring techniques is necessary. The increased number and complexity of chemical species in workplaces has made the occupational environment intricate in nature. Careful measurements are required for any meaningful controls. Therefore, analytical chemists and industrial hygienists working in the occupational health field face a great challenge in measuring and evaluating the workplace environment. New problems are encountered and solutions are sought on a continuing basis. For instance, the work atmosphere could contain various gases and vapors, aerosols, particulates, vapor–particulate mixtures at various temperatures, humidities, and concentrations to which workers may be exposed. Sampling and analytical methods for the substances must be available before any measurement and control efforts are made to meet their health and

xi

safety threatening challenges. In addition, quality control and compliance statistics must be maintained for any meaningful efforts in this regard.

The symposium upon which this book is based was designed to present a current perspective on the measurement and control of chemical hazards in the workplace and to encourage an exchange of ideas among specialists in related areas. This symposium presented both the state-of-the-art and future directions of monitoring and measurement procedures for the occupational environment. Specific topics included: new analytical techniques and methods development; occupational environmental monitoring and control technology (including medical monitoring and analysis); and quality assurance and requirements of compliance statistics.

The authors represent an excellent cross section of the current knowledge in the field of the measurement and control of the occupational environment. The chapters are organized into sections (based on the logical categorization developed for the symposium) on methodology, monitoring and control, special toxicants, quality assurance, and new technologies. I hope that this book will be a source of useful information to those working in the field, and also a valuable contribution to the literature.

I wish to acknowledge, with sincere appreciation, the contributions of the authors and reviewers; without their time-consuming efforts this work would not have been possible. I would also like to thank the ACS Division of Chemical Health and Safety for inviting me to organize the symposium, and the National Institute for Occupational Safety and Health for supporting my participation in this activity.

GANGADHAR CHOUDHARY
National Institute for
  Occupational Safety and Health
Cincinnati, Ohio

October 2, 1980

METHODOLOGY

Sampling and Analytical Methodology for Workplace Chemical Hazards

State of the Art and Future Trends

C. CLARINE ANDERSON, ELLEN C. GUNDERSON, and DALE M. COULSON

SRI International, Menlo Park, CA 94025

Industrial hygiene sampling and analysis is a rapidly expanding activity in government and industry. Exposures of individuals to toxic substances in workplace environments requires accurate sampling and measurement of gases, liquids and solids. Acceptable methods are now available for at least 400 substances as a result of the NIOSH Standards Completion Program. Miniature impingers and bubblers have been long used for workplace sampling. They are inconvenient to use. Solid sorbent tubes are easier to use and are finding wide applicability.

The conversion to solid sorption media from liquid absorption solutions for collection of gases and vapors is a continuing process. The solid sorbent sampling tube is a small device, easily manipulated, and not prone to lose its contents when being used under awkward sampling conditions or during shipping. These physical factors can improve the accuracy of the final result.

Filter collection media have become available in a wide variety of materials including glass fibers and many synthetic plastic films. This allows for selection of a filter that is compatible with the analytical method in addition to not altering the physical and chemical characteristics of the particles collected. The field of aerosol technology has grown significantly and with the increased knowledge of particle characterization methodology the effects of specific ranges of particulate matter can be analyzed.

Analytical techniques have gone through considerable changes in the past 20 years. With the development of more sensitive and selective analytical instrumentation the analyst has been able to detect and identify minute quantities of materials never before seen. This has brought about a keen awareness of the widespread distribution of toxic hazards and also the need to study the long term effects of low level exposures. The development of new methodology is a dynamic process. However, new methods should always be thoroughly tested to demonstrate the precision and accuracy of the results obtained.

SRI International and Arthur D. Little, Inc. carried out an extensive development and validation study between 1974 and 1979

0097-6156/81/0149-0003$05.00/0

for NIOSH in which approximately 400 methods were studied. (1,2)
The study was carried out in two phases. In the first phase the
major emphasis was on laboratory validation of existing methods.
In the second phase more emphasis was placed on methods development
and the substances that were studied were selected from those for
which validation methods were not available from the first phase.
The results of these studies were presented as individual reports
for each substance. They are a sampling and analytical method
(SAM), a sampling data sheet (SDS), and a backup data report (BUD).
The reports on methods have been published by NIOSH and are available through the U.S. Government Printing Office, Washington, D.C.

Protocol for Methods Validation

A detailed protocol for laboratory validation of sampling and
analytical methods for toxic substances in workplace environments
is given in Figure 1. The literature was searched and a method of
sampling and analysis was selected. The next step was to evaluate
and, if necessary, develop an analytical method that was compatible
with the sampling medium. If a satisfactory analytical method became available only then did we undertake generation of a test atmosphere. Then samples were collected with the appropriate collection
medium. Both capacity and collection efficiency were evaluated.

For each method 18 samples were collected and analyzed—6 samples at each of the 1/2, 1, and 2 times the OSHA Standard level.
If the results to this point indicated a successful method, then
storage stability was evaluated. If all requirements of the protocol were met, the method was considered laboratory-validated and
appropriate reports were prepared. At various stages of the protocol, we evaluated the probability of success within the budget for
each method. If at any time it became apparent that the method
study could not be successfully completed within budget, laboratory
work was discontinued and a failure report was prepared.

The basic criterion for successful validation was that a
method should come within 25% of the "true value" at the 95% confidence level. To meet this criterion, the protocol for experimental
testing and method validation was established with a firm statistical basis. A statistical protocol provided methods of data analysis
that allowed the accuracy criterion to be evaluated with statistical parameters estimated from the laboratory test data. It also
gave a means to evaluate precision and bias, independently and in
combination, to determine the accuracy of sampling and analytical
methods. The substances studied in the second phase of the study
are summarized in Table I.

Selection of Methods of Sampling and Analysis

A literature search was usually the first step that resulted
in the selection of an analytical method consistent with one of the
common sampling methods. The objective of these methods is to

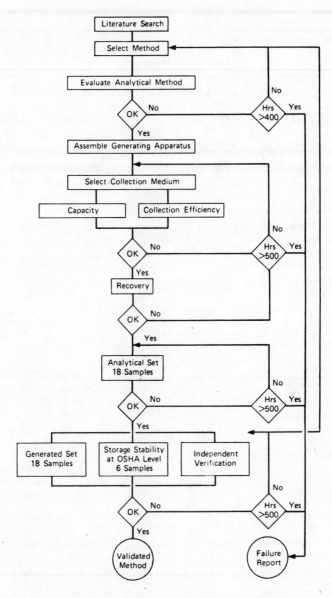

Figure 1. Protocol for method development and validation

Table I   Validated Methods

Code	Compound No.	Analyte	OSHA standard (mg/cu m)	Collection medium	Sample treatment *	Analytical method	Range (mg/cu m)
S	S345	Acetaldehyde	360	Bubbler (Girard T)	-	HPLC	170-670
S	S169	Acetic acid	25	Charcoal	Formic acid	GC/FID	12.5-50
S	S342	Alkyl mercury compounds	0.01 (TWA) 0.04 (C)	Carbosieve B	Thermal desorption	Flame-less AA	.004-.017 (TWA) .02-.08 (C)
A	S346	Allyl glycidyl ether	45	Tenax GC	Diethyl ether	GC/FID	19-87
S	S158	2-Aminopyridine	2	Tenax GC	Thermal desorption	GC/FID	0.91-3.60
S	S347	Ammonia	35	H_2SO_4 treated silica gel	0.1 N H_2SO_4	ISE	17-68
A	S348	Ammonium sulfamate	15	MCEF	Water	IC/ECond	6.4-27.3
S	S163	Anisidine (ortho & para isomers)	0.5	XAD-2	Methanol	HPLC	0.25-1.16
A	S2	Antimony & compounds	0.5	MCEF	HCl	AA	0.258-1.08
S	S276	ANTU	0.3	PTFE filter	Methanol	HPLC	0.128-0.76
S	S253	Benzoyl peroxide	5	MCEF	Diethyl ether	HPLC	3.12-19.10
A	S138	n-Butylamine	15 (C)	H_2SO_4 treated silica gel	50% aq. methanol	GC/FID	8.1-35.5
S	S350	n-Butyl mercaptan	35	Chromosorb 104	Acetone	GC/FPD-S	16.8-74
A	S313	Cadmium fume	0.1 (TWA) 0.3 (C)	MCEF	HNO_3/HCl	AA	0.04-0.175 (TWA) 0.122-0.57 (C)
A	S249	Carbon dioxide	5000 ppm	Bag	-	GC/TCD	2270-10,000 ppm
A	S340	Carbon monoxide	50 ppm	Bag	-	Electro-chemical	24.7-115.4 ppm
A	S278	Chlordane	0.5	MCEF/Chromosorb 102	Toluene	GC/ECD	0.156-1.17
S	S11	Chloroacetaldehyde	3 (C)	Silica gel	50% aq. methanol	GC/ECD	1.8-6.4
S	S120	Chlorodiphenyl, 42% Cl	1	Glass fiber filter/ Bubbler (isooctane)	Isooctane	GC/ECond	0.51-2.7
A	S211	1-Chloro-1-nitropropane	100	Chromosorb 108	Ethyl acetate	GC/FID	51-206
S	S112	Chloroprene	90	Charcoal	Carbon disulfide	GC/FID	44-174
A	S203	Cobalt metal fume & dust	0.1	MCEF	Aqua regia	AA	0.031-0.22 (f) 0.040-0.26 (d)
S	S354	Copper fume	0.1	MCEF	HNO_3	AA	0.05-0.37
S	S356	Crag herbicide	15	MCEF	Water/methylene blue complex	Color.	5-27
S	S279	2,4-D	10	Glass fiber filter	Methanol	HPLC	5.1-20.3
S	S280	Demeton	0.1	MCEF/XAD-2	Toluene	GC/FPD-P	0.06-0.33
S	S111	Dichlorodifluoromethane	4950	Charcoal	Methylene chloride	GC/FID	2940-10,500
A	S109	Dichloromonofluoromethane	4200	Charcoal	Carbon disulfide	GC/FID	1730-7600
S	S108	Dichlorotetrafluoroethane	7000	Charcoal	Methylene chloride	GC/FID	3500-14,100
S	S140	Diethylaminoethanol	50	Silica gel	Acidify with HCl, desorb w/MeOH-H_2O basify w/NaOH-MeOH	GC/FID	25-113

Table I (continued)

Code	Compound No.	Analyte	OSHA standard (mg/cu m)	Collection medium	Sample treatment *	Analytical method	Range (mg/cu m)
A	S141	Diisopropylamine	20	Impinger/ 0.1 N H_2SO_4	0.3 N KOH	GC/FID	8.5-37.4
S	S214	Dinitrobenzene (all isomers)	1	MCEF/Bubbler (Ethylene glycol)	–	HPLC	0.42-2.4
S	S166	Dinitro-o-cresol	0.2	"	–	HPLC	0.070-0.62
S	S215	Dinitrotoluene	1.5	"	–	HPLC	0.90-5.0
S	S24	Diphenyl	1	Tenax GC	CCl_4	GC/FID	0.64-2.4
S	S284	Endrin	0.1	MCEF/Chromosorb 102	Toluene	GC/ECD	0.06-0.31
A	S361	2-Ethoxyethanol	740	Charcoal	Methanol/methylene chloride	GC/FID	340-1460
A	S105	Ethyl chloride	2600	Charcoal	Carbon disulfide	GC/FID	1590-6500
A	S102	Fluorotrichloromethane	5600	Charcoal	Carbon disulfide	GC/FID	2390-10,500
S	S327	Formaldehyde	3 ppm (TWA) 10 ppm (Peak) 5 ppm (C)	Bubbler (Girard T)	–	Polar.	1.4-6.2 ppm (TWA) 4.6-19.8 ppm (Peak)
S	S173	Formic acid	9	Chromosorb 103	Water	IC/ECond	4.4-21.6
S	S17	Furfural	20	Bubbler (Girard T)	–	HPLC	10.1-40
S	S365	Furfuryl alcohol	200	Porapak Q	Acetone	GC/FID	120-470
A	S194	Hafnium	0.5	MCEF	HNO_3, perchloric acid, HF	Plasma emission spectro.	0.26-1.05
S	S287	Heptachlor	0.5	Chromosorb 102	Toluene	GC/EC	0.23-1
A	S288	Hydrogen cyanide	11	MCEF/Bubbler (0.1 N KOH)	–	ISE	5.2-21.0
A	S57	Hydroquinone	2	MCEF	Acetic acid	HPLC	0.84-4.05
S	S366	Iron oxide fume	10	MCEF	HCl/HNO_3	AA	3.9-18.2
S	S370	Malathion	15	Glass fiber filter	Isooctane	GC/FPD-P	8-35
S	S85	MAPP (methyl acetylene/propadiene mix)	1000 ppm	Bag	–	FID	480-1990 ppm
S	S199	Mercury	0.1 (C)	Ag coated Chromosorb P	Thermal desorption	Flameless AA	0.046-0.18
S	S371	Methoxychlor	15	Glass fiber filter	Isooctane	GC/ECond	7.7-31
S	S84	Methyl acetylene	1000 ppm	Bag	–	GC/FID	520-1880 ppm
S	S148	Methylamine	12	Silica gel	Acidify with HCl, desorb with H_2O	IC/ECond	6.24-28.1
S	S99	Methyl chloride	100 ppm (TWA) 200 ppm (C) 300 ppm (Peak)	Charcoal	Methylene chloride	GC/FID	59-220 ppm (TWA) 143-580 ppm (Peak)
S	S374	Methylcyclohexanol	470	Charcoal	Methylene chloride	GC/FID	215-920
S	S375	Methylcyclohexanone	460	Porapak Q	Acetone	GC/FID	210-850
S	S291	Methyl formate	250	Carbosieve B	Ethyl acetate	GC/FID	108-542

Table I (continued)

Code	Compound No.	Analyte	OSHA standard (mg/cu m)	Collection medium	Sample treatment *	Analytical method	Range (mg/cu m)
A	S43	Methyl methacrylate	410	XAD-2	CS_2	GC/FID	193-725
A	S319	Nitric acid	5	Impinger (water)	-	ISE	2.6-10.8
A	S321	Nitric oxide	25 ppm	Draeger oxidizer, TEA coated Molecular Sieve	TEA	Color.	11.1-48 ppm
A	S7	p-Nitroaniline	6	MCEF	Isopropanol	HPLC	3.9-12.9
A	S219	Nitroethane	310	XAD-2	Ethyl acetate	GC/FID	147-600
A	S320	Nitrogen dioxide	5 ppm	TEA coated Molecular Sieve	TEA	Color.	3.1-11.5 ppm
A	S220	Nitromethane	250	Chromosorb 106	Ethyl acetate	GC/AFID	123-500
A	S294	Paraquat	0.5	PTFE filter	Water	HPLC	0.256-1.03
A	S297	Pentachlorophenol	0.5	MCEF	Ethylene glycol	HPLC	0.265-1.130
A	S296	Phosdrin	0.1	Chromosorb 102	Toluene	GC/FPD-P	0.027-0.145
A	S332	Phosphine	0.4	HgCN coated silica gel	Acidic permanganate	Color.	0.195-0.877
A	S257	Phosphorus pentachloride	1	PVC filter/bubbler (water)	Sodium molybdate/ hydrazine sulfate complexation	Color.	0.55-2.0
A	S334	Phosphorus yellow	0.1	Tenax GC	Xylene	GC/FPD-P	0.056-0.244
A	S228	Picric acid	0.1	MCEF	70% aq. methanol	HPLC	0.036-0.189
A	S298	Pyrethrum	5	Glass fiber filter	Acetonitrile	HPLC	1.41-8.5
A	S181	Quinone	0.4	XAD-2	Ethanol/hexane	HPLC	0.17-0.75
A	S299	Ronnel	10	MCEF/Chromosorb 102	Toluene	GC/FPD-P	2.82-17.1
A	S300	Rotenone	5	PTFE filter	Acetonitrile	HPLC	1.16-11.1
A	S182	Silver, metal and sol. compds.	0.01	MCEF	HNO_3	Plasma emission spectro.	0.0036-0.0181
S	S301	Sodium fluoroacetate	0.05	MCEF	Water	IC/ECond	0.020-0.137
A	S381	Sodium hydroxide	2	PTFE filter	HCl	NaOH back-titration	0.76-3.9
A	S243	Stibine	0.5	$HgCl_2$ coated silica gel	Conc. HCl	Color. Rhodamine B	0.119-1.008
A	S302	Strychnine	0.15	Glass fiber filter	HPLC mobile phase	HPLC	0.073-0.34
S	S308	Sulfur dioxide	13	Bubbler (H_2O_2)	Add isopropyl alcohol, adjust pH w/perchloric acid	Barium perchlorate titration	6.6-26.8
S	S244	Sulfur hexafluoride	1000 ppm	Bag	-	GC/TCD	500-2010 ppm
S	S245	Sulfuryl fluoride	5 ppm	Bag	-	GC/FPD-S	2.54 - 10.29 ppm

Table I (concluded)

Code	Compound No.	Analyte	OSHA standard (mg/cu m)	Collection medium	Sample treatment *	Analytical method	Range (mg/cu m)
S	S303	2,4,5-T	10	Glass fiber filter	Methanol	HPLC	4.9-21.4
A	S201	Tantalum	5	MCEF	HNO_3, perchloric acid and HF	Plasma emission spectro.	2.5-10.0
A	S383	Tetraethyl lead	0.075	XAD-2	Pentane	GC/PID	0.045-0.20
A	S384	Tetramethyl lead	0.07	XAD-2	Pentane	GC/PID	0.040-0.18
S	S256	Thiram	5	PTFE filter	Acetonitrile	HPLC	3.0-12.2
A	S391	Vanadium, V_2O_5 dust	0.5	MCEF	NaOH	AA/HGA	0.24-0.90
A	S388	Vanadium fume	0.1 (C)	MCEF	NaOH	flameless AA/HGA	0.060-0.29
A	S316	Zinc oxide fume	5	PVC filter	-	X-ray diffraction	2.4-9.9

Proposed (Class E) Methods

Code	Compound No.	Analyte	OSHA standard (mg/cu m)	Collection medium	Sample treatment *	Analytical method	Range (mg/cu m)
A	P&CAM 304 (S154)	o-Chrorobenzylidene malononitrile (OCBM)	0.4	PTFE filter/ Tenax GC	20% methylene chloride in hexane	HPLC	0.147-0.82
S	P&CAM 291 (S9)	alpha-Chloroacetophenone	0.3	Tenax GC	Thermal desorption	GC/FID	0.18-0.62
S	P&CAM 285 (S177)	Crotonaldehyde	6	Bubbler (hydroxylamine)	-	Polar.	2.9-23.4
S	P&CAM 294 S83)	Cyclopentadiene	200	Chromosorb 104 coated with maleic anhydride	Ethyl acetate	GC/FID	76-380
A	P&CAM 297 (S390)	Dibutyl phosphate	5	PTFE filter	Acetonitrile, trimethylsilyl reagent	GC/FPD-P	2.3-10.0
A	P&CAM 295 (S282)	Dichlorvos (DDVP)	1	XAD-2	Toluene	GC/FPD-P	0.38-1.71
S	P&CAM 302 (S180)	Maleic anhydride	1	Bubbler (H_2O)	-	HPLC	0.50-2.14
A	P&CAM 315 (S305)	TEPP	0.05	Chromosorb 102	Toluene	GC/FPD-P	0.025-0.124
A	P&CAM 313 (S307)	Warfarin	0.1	PTFE filter	Methanol	HPLC	0.054-0.24

Codes for Analytical Methods

AA	=	Atomic absorption spectrometry		
AA/HGA	=	"/high-temperature graphite analyzer		
Color	=	Colorimetric spectroscopy		
GC/AFID	=	Gas chromatography/alkali flame ionization detector		
GC/ECD	=	"/electron capture detector		
GC/ECond	=	"/electrolytic conductivity detector		
GC/FID	=	"/flame ionization detector		
GC/FPD-P	=	"/flame photometric detector-phosphorous mode		

GC/FPD-S	=	"/flame photometric detector-sulfur mode
GC/PID	=	"/photoionization detector
GC/TCD	=	"/thermal conductivity detector
HPLC	=	High-performance liquid chromatography
IC/ECond	=	Ion chromatography/electrolytic conductivity detector
ISE	=	Ion-specific electrode
Polar.	=	Polarography

*Unless otherwise noted, samples are desorbed/extracted in the solvent indicated.

evaluate the exposure of a particular person in a workplace environment. This is most commonly done with a personal sampling pump and a collection device in the breathing zone. The common collection devices are sorbent tubes, filters, midget bubblers and impingers, and gas bags. No passive monitors were included in this study. Figure 2 shows the general setup for personal monitoring. The pump and the sampler (or collection device) are attached to the person. The sampler is in the breathing zone. Figure 3 shows examples of the collection devices. The sorbent tubes used to collect gaseous compounds usually contain activated charcoal according to NIOSH design. (3) In addition to charcoal, porous organic polymers and silica gel have been used extensively.

In the selection of sampling media, an effort was made to use media that would be compatible with a limited number of inexpensive analytical methods. A conscious effort was made to develop methods that would be useful for a class of chemicals with limited success.

The most common analytical methods used were gas chromatography, HPLC, AA spectrophotometry, polarography, colorimetry, and potentiometry with ion-selective electrodes. In this study GC/MS and other more expensive instrumentation were avoided. If sorbent tubes could not be used for gaseous substances, then the less desirable miniature bubblers or impingers were considered. Although these devices are inconvenient they were often used because no better alternatives were available. Bags were used in a few cases where the analyte could not be retained on a sorbent because of volatility and a small tendency to sorb. Filters were used for particulates. Combinations of collection devices were used if we felt that both particulates and vapor might be present in the analyte.

When the analyte is present as particulate and vapor, collection of a representative sample is complicated. Sampling must be done in such a manner that none of the vapor escapes. It requires a filter followed by a vapor collector during sampling, as shown in Figure 4. After the sample is collected, there can be serious losses due to volatilization of the material collected on the filter. This can be avoided by placing the filter in a solvent immediately. The solvent should be the one used in the analytical method. If particulates are left on the filter and the filter is left in the cassette for an extended time, significant amounts of the particulates may volatilize from the filter and deposit on the walls of the cassette. Losses may also occur through the seals of the cassette. Should it be necessary to store the filter in the cassette, washings from the inside of the cassette may be required in addition to the extract of the filter in the analytical method.

A rough guideline was used in this program for determining if a substance exists as a vapor or particulate (or both) at levels near the OSHA standard. If the ratio of the equilibrium vapor concentration at 25°C to the standard was between 0.05 and 50, then contribution from both vapor and particulate was considered. A ratio below 0.05 indicated it would probably be vapor. Frequently,

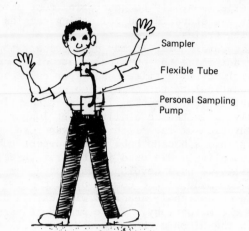

Sampler

Flexible Tube

Personal Sampling
Pump

Figure 2. Personal sampler

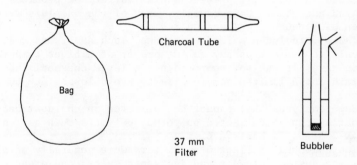

Bag

Charcoal Tube

37 mm
Filter

Bubbler

Figure 3. Collection devices

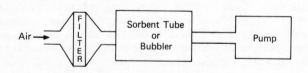

Air → FILTER | Sorbent Tube or Bubbler | Pump

Figure 4. Particulate/vapor sampler

this determination was made experimentally be generating test atmospheres of the substance near the standard and collecting samples
with particulate/vapor sampling trains.

When insufficient data are available, it may be necessary to
generate test atmospheres to determine the physical state of a compound and the collection efficiency of a filter and vapor collector.
The estimation of the vapor/particulate ratio may also depend on
concentration and sample loading. For example, in a short sampling
time, vapor may be efficiently collected on a filter, but longer
term sampling may reveal saturation of the filter with vapor and
eventual passage of the vapor into a backup bubbler or sorbent tube.

Table I summarizes the sampling media used in the last three
years of the study. Previously, we have developed validated sampling and analytical methods for many of the common organic solvents that could be collected on charcoal, desorbed with carbon disulfide, and analyzed by gas chromatography. These procedures are
usually nearly identical with the NIOSH method P & CAM 127. Likewise, methods for substances that give well-behaved particulates
both in collection and analysis had been validated. The substances
summarized in Table I represent a wide variety of problems in sampling and analysis. Consequently, many of the samplers were
charged with unusual collection media.

Table I includes a few substances, such as Freons that are
difficult to collect on any solid sorbent at ambient temperature.
In some cases it was necessary to use a large charcoal tube, a low
flow rate and a limited sample size to avoid losses due to breakthrough.

Some analytes that cannot be collected on charcoal because of
degradation can be collected on porous polymers such as Tenax GC
and XAD-2 or on Carbosieve B.

The compounds collected on filters were generally metals
(fumes and dusts); organic and inorganic salts; and organic compounds that have very low vapor pressures. The procedure most commonly used for collection of particulates involves sample collection with a closed-face 37-mm filter cassette containing a filter
supported by a cellulose backup filter support. The most frequently used filter materials include mixed cellulose ester (MCE), polyvinyl chloride (PVC), glass fiber, silver membrane, and recently,
polytetrafluoroethylene (PTFE). Any high-efficiency filter usually
functions well for collection purposes; therefore, selection was
usually based on the final analytical requirements. However, in
this program, poor recoveries were observed for some substances
(such as thiram) when glass fiber filters were used, and the filter
medium was found to be at fault. The PTFE filters, being inert,
were very useful for collecting reactive substances. In the analytical method for sodium fluoroacetate by ion chromatography, surfactants used as wetting agents on some filters were found to interfere significantly.

MCE filters were generally used when the sample was extracted,
digested, or otherwise treated to create a solution for final

analysis. PTFE filters were not the first choice because of their
expense, but they were useful for reactive substances and in cases
where the analytical procedure required a solvent that was not compatible with other filter media. For example, when HPLC was the
preferred analytical technique, the choice of sampling device,
especially the filter, was dependent upon the solvents used in HPLC
analysis. The material was usually extracted from the filter in
the same organic solvent such as methanol and acetonitrile, which
are commonly used HPLC eluents.

Generally, any of these filters satisfied the requirements for
collection of most of the particulates to be analyzed. However,
certain analytes and their associated analytical methods gave better results with alternative filters. The filter cassette was also
tested as part of the total sampling device. Some substances were
partially collected on the cassette, and the substance had to be
rinsed from the cassette with solvent to avoid loss.

A few substances were collected in bags. These analytes are
generally very volatile and weakly sorbed, even on charcoal. They
included carbon monoxide, carbon dioxide, sulfur hexafluoride, sulfuryl fluoride, methyl acetylene, and methyl acetylene/propadiene
mixture.

Midget bubblers were used as a last resort where sorbent methods had failed. Bubblers were generally charged with a derivatizing agent that would stabilize the analyte. Aldehydes were collected as Girard "T" or hydroxylamine derivatives that could be analyzed by polarography or HPLC with uv detection.

General Methods

A concerted effort was made to develop general methods that
could be applied to several substances within the same compound
class. This approach was reasonably successful for pesticides,
aldehydes, and organolead compounds. The outcome was satisfactory
for substances with similar properties; however, special treatment
was necessary for single compounds within a class that were unusu-
ally reactive or volatile.

Pesticides. An objective of high priority in this program was
to develop a universal filter/sorbent sampling train for pesticides.
Where possible, we used a 37-mm MCE filter followed by a sorbent
tube containing 20/40 mesh Chromosorb 102. The filter and sorbent
were combined and the pesticide was dissolved in toluene. The resulting solution was analyzed by GC with an appropriate detector.
This method was validated for endrin, chlordane, and ronnel. Methods were also validated for phosdrin, TEPP, and heptachlor using a
Chromosorb 102 sorbent tube only. Chromosorb 102 was not a satisfactory sorbent for some substances. A method was validated for
demeton using an MCE filter/XAD-2 sampling train.

When HPLC was the preferred analytical method for a pesticide,
the choice of sampling device and sample treatment procedure was

affected by the HPLC eluent. PTFE or glass fiber filters were used
in methods where solvents such as methanol and acetonitrile were
the appropriate extraction solvents or when the MCE filter was not
satisfactory. Validated methods include ANTU, 2,4-D, paraquat,
pyrethrum, rotenone, 2,4,5-T, thiram, and warfarin.

Aldehydes. The aldehydes studied in this program include acetaldehyde, chloroacetaldehyde, formaldehyde, furfural, crotonaldehyde, and acrolein. Because of the instability of the aldehydes,
they are best collected by derivatizing during sample collection.
The Girard-T derivatives of the aldehydes were studied extensively
and methods were validated based upon formation of the Girard-T
derivative of acetaldehyde, formaldehyde, and furfural. A bubbler
containing Girard-T reagent in an aqueous solution at a controlled
pH was used to collect samples. The Girard-T derivatives are positively charged species that exhibit uv absorptivity, so ion exchange separations were feasible.
Polarographic methods of analysis of the derivatives of formaldehyde and crotonaldehyde without chromatography were developed.
Analysis by HPLC was later considered so as to achieve greater
resolution between the individual aldehydes and potential interferences. Methods validated using HPLC analysis include acetaldehyde and furfural. Tests indicated that HPLC analysis may be applicable to the Girard-T derivatives of other aldehydes, such as formaldehyde, propionaldehyde, and benzaldehyde.
Girard-T derivatives of chloroacetaldehyde, crotonaldehyde,
and acrolein were not stable. Alternative methods were developed
based upon the derivative formed by reaction of crotonaldehyde with
hydroxylamine, and the formation of the hydrate of chloroacetaldehyde.

Organolead Compounds. Methods for sampling and analysis of
tetraethyl lead and tetramethyl lead were developed based on collection on XAD-2, desorption with pentane, and analysis by gas
chromatography with a photoionization detector.
A larger amount of XAD-2 was required for tetramethyl lead because of its higher volatility. It is likely that both tetraethyl
and tetramethyl lead could be collected simultaneously with the
larger sorbent tube.

Others. Attempts were made to develop general methods for mercaptans, amines, alcoholamines, and nitroalkanes. However, results
were not satisfactory. We did not use a single collection medium
for amines. The media used for various amines included silica gel
with and without H_2SO_4, a porous polymer with thermal desorption,
and a bubbler. The preferred method is collection of silica gel
followed by acidification of the sample with dilute HCl immediately
after collection. This method should be successful for ammonia and
most aliphatic amines. Ion chromatography was used successfully
for the analysis of methyl amine. This analytical method may be

generally applicable for aliphatic amines. Additional work may also demonstrate that collection on a porous polymer followed by thermal desorption into a gas chromatograph would be generally useful for aliphatic amines. A method was developed and validated for n-butyl mercaptan based upon collection with a sorbent tube containing Chromosorb 104, and analysis by gas chromatography. The sorbent did not have adequate capacity to collect methyl or ethyl mercaptan. The nitroalkanes gave inconsistent storage stability results on various solid sorbents.

Generation of Test Atmospheres

Test atmosphere generation was accomplished in a three-stage dynamic system. It consisted of a source generator, which introduces test air containing a high analyte concentration into a dilution system, where three concentration levels are automatically prepared and delivered to sampling chambers. Figure 5 gives a simplified sketch of a dynamic generation system. The actual apparatus used at SRI is shown in Figure 6.

In addition to the main generation system, special systems were designed and constructed for use with reactive compounds. Separate systems constructed exclusively of Teflon and/or glass were used with reactive substances including fluorine, mercury, and several amines.

Aerosols. Test atmospheres containing aerosols may be conveniently generated by a number of techniques including:
- Spray drying
- Atomization
- Condensation
- Thermal decomposition
- Dust dispersion.

Spray drying--Spray drying is applicable to substances that are soluble in a convenient solvent. This method involves atomizing a solution to form a mist. Larger particles are removed by impaction. The remaining flow is mixed with solvent-free air to permit evaporation of solvent from the droplets. Residue particles of the solute are left behind as the solvent evaporates.

Atomization--Liquid aerosols of readily melted solids may be generated by direct atomization using, for example, a pneumatic nebulizer such as the Collision. When a pneumatic atomizer is used, the aerosol concentration may be varied by changing the pressure of the atomizer air supply. The mass median diameter of the aerosols produced by the small pneumatic nebulizer is typically in the micrometer range. The direct atomization method was applied to dimethylphthalate, a liquid, and dibutyl phosphate, a low-melting solid.

Condensation--The condensation method requires the formation of a vapor-air mixture that is supersaturated with respect to the vapor. Condensation from the supersaturated vapor may then

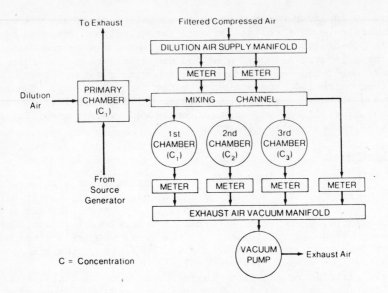

Figure 5.   Schematic of a dynamic generation system

Figure 6.   Generation of apparatus for test atmospheres

occur either on foreign particles present in the gas (heterogeneous nucleation) or on small clusters of vapor molecules that are formed at random (homogeneous nucleation). Particles ranging from about 0.01 to more than 10 μm have been prepared by this method. Although it is best known as a method for generating mono-disperse aerosols, condensation may also be used to produce polydisperse aerosols. Dinitro-o-cresol was generated using this method.

   Thermal decomposition--Thermal decomposition methods may be used to prepare metal oxide fumes. An aerosol of a precursor to the metal oxide (i.e., a substance that is readily decomposed, thermally, to yield the oxide) is first generated and then is heated by passing it through a heated tube to decompose it to the oxide. Metal formates, oxalates, and the like, which readily yield the oxides and do not produce objectionable side products, are commonly used precursors. In this program, fumes of iron oxide, vanadium oxide, and copper oxide were generated using this method.

   Dust dispersion--Aerosols of mineral particles and a few other materials may be prepared by dust dispersal techniques. A number of dust dispersal devices have been constructed. A commonly used device is the Wright Dust Feeder. In this device the dust is compressed into a cylindrical compact, which is placed in an electronically driven mechanism that rotates it and drives it against a fixed knife blade at a fixed rate. Material is scraped off the compact by the blade, entrained in an air stream, and blown against a metal plate to break up particle aggregates.

   Gases and Vapors. Test atmospheres of gases and vapors are generally prepared dynamically by the following methods:
- Dilution
- Syringe injection
- Vaporization
- Permeation
- Chemical reaction.

   Dilution--A cylinder of the compressed gas is fitted with a metering apparatus (regulator and/or critical flow orifice) and the effluent stream is injected directly into the dilution system. The flow rates of the analyte can be calibrated and the concentration of each test atmosphere can be calculated from the known dilution air flows.

   Syringe injection--Injection of a volatile liquid at a known rate into an air stream of known flow rate is a convenient method for producing vapor mixtures of known composition. Since the rate of delivery of the syringe drive may be accurately determined by weighing the amount of liquid delivered over a known period of time and by measuring the dilution air flow rate, very good accuracy may be obtained with this method.

   Vaporization--A saturated air stream is produced by contacting the air with the analyte and then passing the stream through a condenser at a lower temperature. The saturated air stream is then diluted to the desired concentration with clean air.

    Permeation--Permeation tubes are often used to prepare
analyte concentrations in the ppb to hundreds of ppm levels.

Directions for the Future

    The determination of worker exposure to toxic substances has
been a challenging and continuously changing task for many years.
Sampling and analytical methodologies have been adapted to meet the
needs for specific contaminant identification and quantification as
well as for determining compliance relative to a legal exposure
limit. Since passage of the Occupational Health and Safety Act of
1970, which set standards for over 400 substances, much greater emphasis has been placed on achieving highly accurate results from
sampling and analytical methods. In this discussion we have focused on personal sampling methodology where samples are collected
near the breathing zone of the worker.

    Sampling methodology has steadily improved over the years,
especially in the area of integrated sampling. Personal sampling
pumps with electronic flow controllers can significantly increase
the accuracy of resulting sample volumes.

    Philip West (4) reviewed the status of passive monitors for
toxins such as chlorine, SO_2, vinyl chloride, alkyl lead, benzene,
H_2S and HCN. Dr. West predicts a bright future for passive monitors. He points out the disadvantages of active monitors and the
merits of passive devices. In my opinion, we need more information
on precision and accuracy in sophisticated validation studies before we let our hopes rise too high.

    In the future there will be more emphasis on passive monitors,
including those that are exposed in the field but analyzed in the
laboratory and those that can be read in the field. At present, a
number of methods require midget bubblers for sampling. These will
be largely replaced by collectors containing solid sorbents; the
sorbents may be very specialized. Surface coatings of nonvolatile
derivatization reagents and reagents bonded to inert surfaces will
likely be developed. Samplers of this kind are needed for the collection and stabilization of some of the more reactive compounds
such as conjugated diolefins, aldehydes, and amines. Finally, there
will be more emphasis on class methods that will allow measurement
of several compounds at once so as to simplify the work of the
industrial hygienists and the analytical chemist.

    We used a rather simple protocol in our laboratory validation
studies. Future studies should include other factors such as temperature effects, the effects of potential interfering substances
on recovery and actual field trials. These field trials should be
monitored by an independent method in which we have a high degree
of confidence.

    As we move toward passive monitors, validation protocols must
also include extensive examination of such factors as face velocity
and variable concentration, especially where short spikes occur.

Someday we will have small portable devices that sample and analyze during exposure to the workplace atmosphere. Devices small enough to be worn comfortably by an active person, devices that will give warnings in real-time when danger of hazardous exposure occurs. These devices may be in the form of integrated circuits with sensors covered with permi-selective membranes, the combination of which will lead to both sensitive and selective measurements. Ten years from now current methods and devices may be, for the most part, but a memory.

In conclusion, let us make a few suggestions. Those of us who are primarily chemists need to become familiar with the needs of the industrial hygienist and with conditions that exist in workplaces. Industrial hygienists need to understand the problems of the chemist in analyzing the samples. Knowledge of the sampling and analytical method is important in selection of sampling media. Both the analyst and the hygienist must appreciate the strengths and weaknesses of the methods. And, of course, quality assurance programs must always be used.

Acknowledgements

This work was carried out under NIOSH Contracts CDC-99-47-45 and 210-76-0123. We are especially indebted to Drs. Laurence J. Doemeny and David G. Taylor of NIOSH for their guidance as project officers and to Dr. Philip L. Levins, Ms. Alegria B. Caragay, Mr. Richard H. Smith, and Ms. Cristine M. Freitas of ADL for their contributions to the work. In addition to the authors, special recognition must be given to Ms. Ellen L. Fernandez and Ms. Jolene Y. Louie at SRI International.

Literature Cited

1. Gunderson, E. C. and Anderson, C. C., "Development and Validation of Methods for Sampling and Analysis of Workplace Toxic Substances," November, 1979, Final Report on NIOSH Contract 210-76-0123 by SRI International.

2. Taylor, D. G. (Manual coordinator), "NIOSH Manual of Analytical Methods," 2nd ed., vols. 1-6, Rockville, MD., 1980.

3. White, L. D.; Taylor, D. G.; Mauer, P. A.; and Kugel, R. E., "A Convenient Optimized Method for the Analysis of Selected Organic Vapors in the Industrial Atmosphere," Am. Ind. Hyg. Assoc., 1970, 31, 225.

4. West, P. W., "Passive Monitoring of Personal Exposures to Gaseous Toxins," Amer. Lab., July 1980, p. 35.

RECEIVED October 14, 1980.

Development of an Analytical Method for Benzidine-Based Dyes

EUGENE R. KENNEDY and MARTHA J. SEYMOUR

National Institute for Occupational Safety and Health, Robert A. Taft Laboratories, 4676 Columbia Parkway, Cincinnati, OH 45226

Benzidine has been an important dye industry intermediate since 1890 (1,2). Over 200 dyes based on benzidine are listed in the Colour Index or are in commercial use. Although the potential of benzidine to cause bladder cancer has been well documented (3,4,5), it was originally believed that when chemically incorporated into a dye, the carcinogenic hazard was removed (6). Recent evidence (7,8,9) has shown that benzidine-based dyes can be reduced to benzidine in living systems and eliminated by the usual benzidine metabolic pathways. Because of this fact, the National Institute for Occupational Safety and Health has recommended that certain benzidine-based dyes be recognized and handled as carcinogens (9).

The only published method available for the determination of personal exposure to benzidine-based dyes utilized the analysis of urine for benzidine and benzidine metabolites (10,11). This method does not allow for quantitation of a daily exposure, since benzidine and its metabolites have been found in the urine of hamsters fed a benzidine-based dye up to 168 hours after a single dosing (12). A method for the determination of personal exposure to azo dyes and diazonium salts has been developed (13), but it is not specific enough to determine an exposure to a benzidine-based dye.

In the development of a sampling and analytical method for benzidine-based dyes, the most important feature was the verification of the benzidine moiety in the dye molecule. Since reduction of some of these dyes in vivo was known to release benzidine, a human carcinogen, the determination of benzidine released by chemical reduction is the most logical approach.

Specificity for a particular dye was not reasonable due to the large number of benzidine-based dyes and the possibility of dye substitution. The method should provide quantitative collection and recovery at the microgram level and be free

from interferences, especially those arising from benzidine
congeners, such as o-tolidine (3,3'-dimethylbenzidine) and
o-dianisidine (3,3'-dimethyoxybenzidine). Various conditions
for reduction of the azo linkages in these dyes were known
(14), so cleavage of the benzidine moiety was possible. Also,
conditions for the analysis of aromatic amines were well
established (12,15,16,17). Preliminary work in our laboratory
had shown that benzidine and its congeners could be separated
by high pressure liquid chromatograph (HPLC). Based on this
reasoning, a method for the determination of benzidine in
benzidine-based dyes was developed.

Experimental

Apparatus. The HPLC system used in this study, assembled
from modular components manufactured by Waters Associates,
consisted of two Model 6000A pumps, a Model 660 solvent
programmer, a Model 440 ultraviolet detector with a 280-nm
filter and a Waters Model 710A Intelligent Sample Processor.
Data were recorded on a Soltec Model B281 dual channel strip
chart recorder and integrated by a Hewlett-Packard Model 3354A
Laboratory Automation System. The columns used for the
analyses were a Waters Associates μ-Bondapak C_{18} and a
Waters Model RCM100 Radial Compression Module with a Radial
Pak A cartridge. The mobile phase was 60% methanol (Burdick
and Jackson) and 40% of an aqueous phosphate buffer. This
phosphate buffer was prepared by dissolving 3.390 g (0.025
mole) of KH_2PO_4 (Fisher Scientific Corp.) and 3.530 g
(0.025 mole) of Na_2HPO_4 (Matheson Coleman Bell) in 1 L of
water (18). A helium purge was maintained in the solvent
reservoirs to eliminate dissolved air. The columns were
maintained at ambient temperature and run at a 2-mL/min flow
rate. Resulting pressure at this flow rate was 500-1900 psi
for the Radial Compression Module and 3000 psi for the
μ-Bondapak column. A precolumn filter could not be used
because the extra dead volume reduced resolution of reduction
products.
Millipore 37 mm Mitex (Teflon) filters (5.0 μm) with
backup pads and three-piece cassettes were used for the spiked
filter studies. A Doerr Model 0272X vacuum pump with a
10-port sampling manifold and 1-L/min critical orifice
(Millipore Corporation) was used to pull air through the
dye-spiked filter cassettes for the stability studies. Two
methods of exposing the cassettes to humidified air were
employed. For the 28-day storage study, air was pulled
directly through midget impingers filled with a saturated
sodium chloride solution. For the 7-day storage study, air
was blown through a large saturated sodium chloride-filled
impinger and then supplied to a 10-port manifold to which the
cassettes were attached. Air was pulled through the cassettes

using the previously described vacuum pump. With this system
the cassettes were always maintained at atmospheric pressure.
Relative humidity generated by either of these techniques was
approximately 75% (19).

Visible-spectrum studies of the course of the dye
reduction were performed on a Beckmann Model 25
Ultraviolet-Visible Spectrophotometer scanning the region of
750-350 nm.

Reagents. The dyes used in this study (see Figure 1) were
obtained from the following sources: Congo Red (Colour Index
(C.I.) Direct Red 28, C.I. No. 22120) A. D. Mackay Inc.;
Direct Black GX (C.I. Direct Black 38, C.I. No. 30235), Direct
F. Blue 2B 250% (C.I. Direct Blue 6, C.I. No. 22610), Direct
Brown BRL 200% (C.I. Direct Brown 95, C.I. No. 30145)
Fabricolor Inc.; Evans Blue (C.I. Direct Blue 53, C.I. No.
23860), Benzo Azurine G (C.I. Direct Blue 8, C.I. No. 24140)
Pfaltz and Bauer Inc. Benzidine was obtained from Sigma
Chemical Co., o-tolidine from Fisher Scientific Corp. and
3,3'-dimethoxybenzidine (o-dianisidine) from Eastman Kodak
Company. Aniline, p-aminophenol, p-phenylenediamine and
p-nitroaniline used in the interference study were obtained
from Chem Service Inc.

The phosphate buffer solution used for reduction of the
dyes was prepared by dissolving 1.179 g (0.0087 mole) of
KH_2PO_4 and 4.300 g (0.0303 mole) of Na_2HPO_4 in water
to make 100 mL of solution. For the reduction of the dyes, a
solution was prepared which contained 10 mg of sodium
hydrosulfite (Fisher Scientific) in 1 mL of the above buffer
solution (0.087 M KH_2PO_4 - 0.303 M Na_2HPO_4). This
sodium hydrosulfite containing solution was prepared
immediately before addition to the desorbed dye to prevent
decomposition of the sodium hydrosulfite. All solutions,
including the HPLC aqueous phase, were filtered through a
0.22-μm cellulose ester membrane filter before use to prevent
plugging of the HPLC system during analysis.

Procedure. A primary standard solution of benzidine in
methanol was prepared. Standards of lower concentrations were
prepared by dilution of the primary standard. Typical liquid
chromatograph calibration curves were linear in the region
from 0.38 to 30.6 μg/mL. The standards were stable for
several months when stored in the dark. Aqueous solutions of
known dye-formulation concentration were prepared.

Filter samples for the stability and repeatability studies
were prepared by spiking the filter with a known volume of the
dye solution using fixed volume pipettes or volumetric
syringes. The filters were then dried in a dessicator filled
with anhydrous calcium sulfate and phosphorous pentoxide. The
dessicator provided more rapid and uniform drying of the

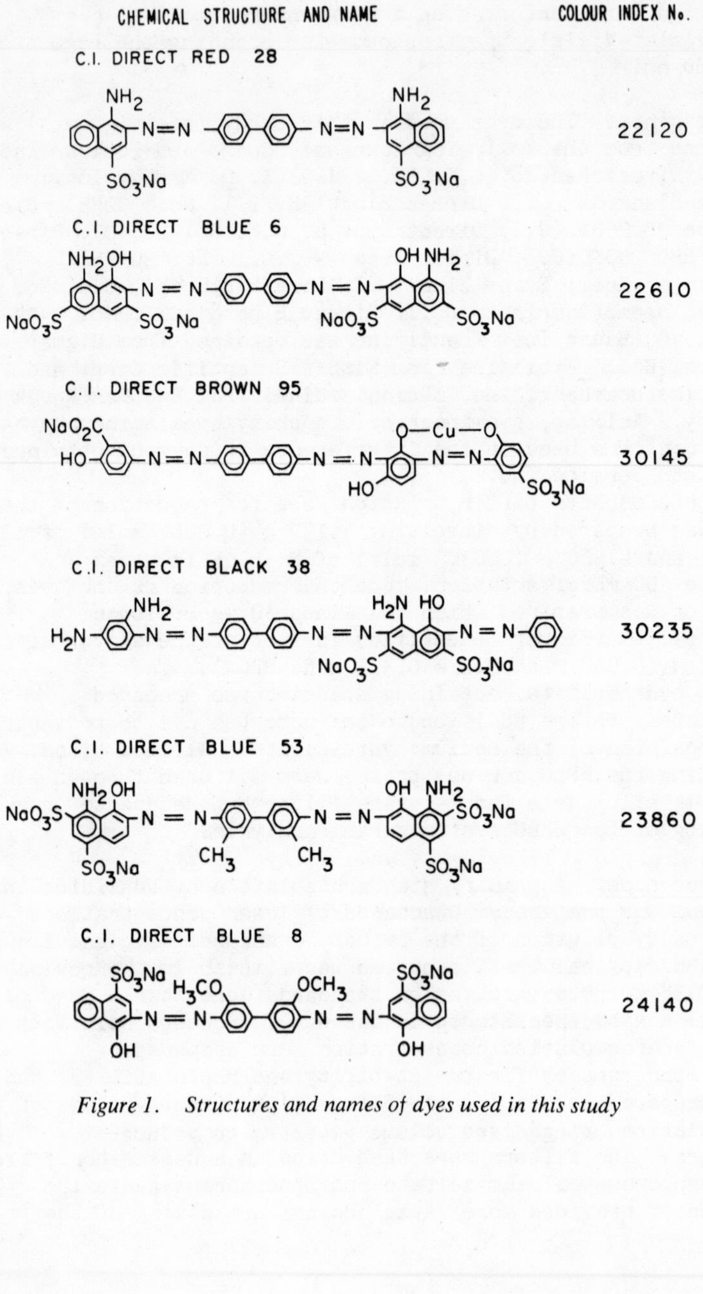

CHEMICAL STRUCTURE AND NAME   COLOUR INDEX No.

C.I. DIRECT RED   28

22120

C.I. DIRECT BLUE 6

22610

C.I. DIRECT BROWN 95

30145

C.I. DIRECT BLACK 38

30235

C.I. DIRECT BLUE 53

23860

C.I. DIRECT BLUE 8

24140

Figure 1.   Structures and names of dyes used in this study

filters than air drying. The filters were then stored at room temperature in a filter shipping case until reduction and analysis.

The spiked filters were placed in a 50-mL beaker with the spiked side up and 1 mL of water was added. The beaker was shaken so that all of the filter area had been washed by the water. One mL of the reduction-buffer solution without the sodium hydrosulfite was added to the filter and water. The beaker was shaken a second time. The filter was then turned over (spiked side down) and the beaker placed in an ultrasonic bath for 15 minutes. At the end of this period, the solution in the beaker was colored. A 1-mL aliquot of this solution was transferred to a 4-mL vial. One mL of a freshly prepared solution of 100-mg sodium hydrosulfite in 10-mL phosphate reduction buffer was then added to the vial. The vial was then capped and shaken several times during the course of an hour. During this time, the original color of the solution disappeared or changed to a different color, depending on the dye present. This solution was then injected into the liquid chromatograph. A 10-μL aliquot was used, giving a measurement limit of 0.38 ng benzidine/μL. The analytical reproducibility at this limit was 10% coefficient of variation (CV).

Results and Discussion

Initial work on the reduction reaction involved a study of the completeness of the reaction and verification of the reduction product, benzidine. The presence of benzidine in the reduced dye sample was confirmed by gas chromatographic/mass spectrometric analysis. In order to determine the completeness of the dye-reduction reaction, the reduction of C.I. Direct Black 38, C.I. Direct Brown 95 and C.I. Direct Blue 6 were studied individually in the visible spectrum. A baseline was recorded using the phosphate reduction buffer in both cells. Subsequent additions of known amounts of dye and scanning allowed absorption maxima and molar absorptivity to be determined. Then 3 mg of sodium hydrosulfite was added to the dye-containing cell. The concentration of dye remaining after reduction was calculated using Beer's Law. The remaining dye varied from 0 to 6% of the original amount of dye added (Table I). Reduction was complete within 30 minutes.

In the initial phases of the analytical method development a Waters Associates C_{18} μ-Bondapak column was used. With certain dyes, such as C.I. Direct Black 38, aniline is used as a terminal group (Figure 1). This column caused the benzidine peak of the reduced dye to obscure a peak due to aniline. If these two compounds were not resolved, the method could not differentiate between benzidine-based and aniline-based dyes. This problem was removed by use of the Waters Radial Compression Module Model RCM 100 with a Radial Pak A cartridge.

Table I
Visible Spectrum Studies of Benzidine-Based Dyes

Dye	Molar Absorptivity	Concentration (μg/mL)	
		Before Reduction	After Reduction
C.I. Direct Blue 6	26.5	9.1	0.1
C.I. Direct Brown 95	23.8	18.1	1.1
C.I. Direct Black 38	19.0	15.9	0.0

This system resolved the aniline peak (retention time (rt) =
2.67 min) from the benzidine peak (rt = 2.27 min) as can be
seen in Figure 2. Other potential interferences were selected
for study by looking at the expected fragments from the
reduction of various dyes. Reduced dye samples were spiked
with aniline (rt = 2.67 min), p-aminophenol (rt = 1.97 min),
p-phenylenediamine (rt = 1.93 min) and p-nitroaniline
(rt = 3.16 min). None of these materials interfered with the
detection of the benzidine peak. To determine if other types
of dyes might interfere with the analysis, two sets of filters
were spiked at low and high levels separately with C.I. Direct
Red 28 (13.7 μg and 137 μg), C.I. Direct Blue 53 formulation
(o-tolidine-based) (21.2 μg and 212 μg) and C.I. Direct Blue 8
formulation (o-dianisidine-based)(23.3 μg and 233 μg).
Results from the analyses showed there were no interferences
from the other dyes present (Figure 3 and Table II). The
coefficient of variation for the average analytical method
recovery (CV$_{AMR}$) has been defined by the following equation
(20) to account for the propagation of error resulting from
the division of the amount of benzidine found on the filter by
the amount in the liquid sample:

$$CV_{AMR} = [(CV_L)^2 + (CV_F)^2)]^{1/2}$$

where: CV$_L$ = coefficient of variation for the liquid
                 samples.
        CV$_F$ = coefficient of variation for the filter
                 samples.

Three of the dyes (C.I. Direct Blue 8, C.I. Direct Black
38 and C.I. Direct Brown 95) had been analyzed for residual
benzidine content during previous work (21). None of the dyes
contained sufficient quantities of residual benzidine to
require a correction to be made to the total amount of
benzidine found in the reduced dye samples.

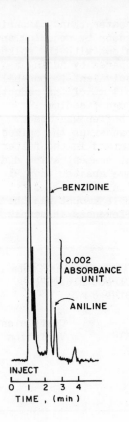

Figure 2.   Chromatogram of a 10-μL injection of reduction products of C.I. Direct Black 38 at the 124.0-μg level showing resolution of benzidine (retention time = 2.27 min) and aniline (retention time = 2.67 min). Chromatographic conditions: Radial Pak A; 2   mL/min; 60% methanol/40% phosphate buffer; ambient temperature.

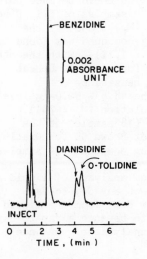

Figure 3.   Chromatogram of a 10-μL injection of reduction products of a mixture of C.I. Direct Red 28 (28.8 μg), C.I. Direct Blue 53 (28.6 μg) and C.I. Direct Blue 8 (34.4 μg) showing resolution of benzidine (retention time = 2.33 min), o-dianisidine (retention time = 4.02 min), and o-tolidine (retention time = 4.27 min). Chromatographic conditions as in Figure 2.

Since no chemically pure dyes (greater than 95% purity) were available, recovery studies were done by reducing and analyzing a liquid sample of the dye along with the spiked filter. These liquid samples were prepared by adding a quantity of dye solution (2-200 µL) equivalent to one-half the amount present on the spiked filter to 1 mL of the reduction buffer solution. To this solution 10 mg of sodium hydrosulfite in 1 mL of the reduction buffer was added. The sample was then analyzed in the same manner as the spiked filter sample. The amount of dye contained in the filter and liquid spiked samples was the same, but concentration did vary. However, volume variations became greater than 9% (200 µL addition to 2 mL of solution) on only the most concentrated samples. Since total sample amount was the same for filter and liquid samples, any differences in sample amounts would be due to recovery losses.

Table II

Recovery of Free Benzidine from Filter Samples Containing
C.I. Direct Red 28, C.I. Direct Blue 53,
and C.I. Direct Blue 8

Loading of C.I. Direct Red 28 (μg)	Average Recovery (%)[a]	Coefficient of Variation (%)
13.7	98.0	4.3
136.7	106.0	2.6

[a]Averages of six samples.

Based on the analyses of these liquid samples an approximation of amount of the benzidine-containing dye compound in a particular dye formulation could be made using molecular weight calculations. For 30 samples of C.I. Direct Red 28 formulation (molecular weight (M_r) = 696) at the 15-µg level, the benzidine-containing compound was found to compose 91.2% of the formulation (CV = 1.1%). With the other dyes the following compositions were found: C.I. Direct Blue 6 formulation (M_r = 932), 30.7% benzidine-containing compound (23 samples at the 48.8 - 52.0-µg level) CV = 8.5%; C.I. Direct Black 38 formulation (M_r = 781), 41.6% benzidine-containing compound (18 samples at the 14.92-µg level) CV = 8.5%; C.I. Direct Brown 95 formulation (M_r = 759), 39.4% benzidine-containing compound (24 samples at the 30.8-µg level) CV = 16.7%.

An examination of the side-by-side analyses of liquid and filter samples should be able to detect any day-to-day variations in analyses. When the raw data from the liquid samples was studied, there appeared to be such a day-to-day variation in the results. However, analysis of variance

indicated no variation at the 0.05 level of significance. To
further investigate any possible sources of variation, the
method was subjected to a ruggedness test as described by
Youden (22), using benzidine concentration as the response
variable. The HPLC mobile phase buffer and reduction buffer
were prepared with 10% additional disodium hydrogen phosphate
and 10% additional potassium dihydrogen phosphate for the high
levels in the test. Sodium hydrosulfite concentration in the
reduction buffer was also increased by 10%. The mobile phase
makeup was changed to 35% buffer/65% methanol for the high
level. Flow rate was increased to 2.2 mL. An old and new
column were used to evaluate column backpressure. Ten μL of
sample solution were removed before addition of the reductant
solution to simulate volume variation caused by use of out of
calibration pipettes. The major contributions to the
variation resulted from HPLC buffer concentration, column
backpressure and mobile phase composition. According to
instrument specifications, mobile phase makeup was controlled
to within ± 2% of set point so that any variation should have
been randomized during the testing. During later work it was
found that ambient temperature variations (± 5° C) caused
changes in volumes delivered by the pumping system. This
volume change amounted to as much as 0.2 mL/min. This problem
was solved by premixing the mobile phase. Column backpressure
could also have been a contributing factor to the variation in
volume delivery since solvent compressibility is affected
greatly by temperature at lower pressures (23). Based on this
ruggedness test, HPLC mobile-phase buffer concentration could
be a major source of variation. To minimize this variation in
the method, care should be used when preparing this buffer.

The analysis method was evaluated with four different
benzidine-based dye formulations (Figure 1). The results of a
repeatability study utilizing spiked samples are shown in
Table III. Spiking levels were arbitrarily chosen but do
reflect the lower limit of the method.

All coefficients of variation passed the Bartlett's test
(24) for homogenity at the 1% significance level, except C.I.
Direct Black 38. With the lowest level of C.I. Direct Black
38 excluded, the coefficients of variation for the remaining
two levels were homogeneous. The pooled analytical
coefficients of variation (CV_1) for each dye are as
follows: C.I. Direct Red 28, 0.055; C.I. Direct Blue 6,
0.083; C.I. Direct Brown 95, 0.108; C.I. Direct Black 38,
0.058 (2 levels only).

To investigate the storage stability of these dyes, a
28-day storage study was undertaken. Filters were spiked and
dried by the procedure described in the experimental section.
Sixty liters of humidified air (75% relative humidity) were
drawn through each spiked filter before storage, using the
previously described vacuum pump system. During the

Table III
Recovery of Benzidine from Benzidine-Based Dye Filter
Samples After One-Day Storage

Dye	Loading of Dye (μg)	Average Recovery(%)[a]	Coefficient of Variation (%)
C.I. Direct Red 28	13.7	94.4	7.2
	27.3	101.0	5.6
	273.4	109.0	2.5
C.I. Direct Blue 6	15.0	100.0	6.6
	30.0	98.9	11.7
	300.0	108.0	4.8
C.I. Direct Brown 95	12.1	78.5	9.2
	24.3	96.4	14.8
	242.6	110.0	6.6
C.I. Direct Black 38	6.2	78.1[b]	16.6
	12.4	91.6	7.9
	124.0	102.0	2.2

[a]Based on six samples. [b]Based on five samples.

humidification step utilizing the midget impingers, several of
the samples were acidentally wetted with the sodium chloride
solution. This was not noticed until the samples were
prepared for analysis. These contaminated samples were not
included in the data analysis. Also, occasional plugging of
an impinger with salt was noted. This resulted in a partial
vacuum being created in the cassette. When the impinger was
unplugged, the rush of air might possibly have dislodged a
portion of the dye from the filter causing greater variation
in the analysis. When several of the cassettes were opened
for analysis, the spiked dye spot seemed to be lying loose on
the filter. These samples also were not included in the data
analysis. A pump malfunction in the HPLC system invalidated
the day-14 analyses for C.I. Direct Blue 6, C.I. Direct Brown
95 and C.I. Direct Black 38. The remaining data from the
storage study is shown in Table IV.

The variability of the recoveries was subjected to
analysis of variance at the 0.05 level of significance.
Significant differences were observed with the recoveries from
C.I. Direct Red 28. Duncan's multiple range test indicated
that the recovery on day 1 was significantly different from
the other three analyses at days 7, 14 and 28. Analyses at
days 7, 14, and 28 were not significantly different. With
C.I. Direct Blue 6, C.I. Direct Black 38 and C.I. Direct Brown
95 there were no differences between recoveries at days 1, 7,
and 28 for each dye. However, the analytical coefficients of
variation became larger on day 28 for these three dyes.

Table IV
Evaluation of the Stability of Filter Samples Containing
Benzidine-Based Dyes Stored Up to Twenty-Eight Days

Dye	Loading (µg)	Average Recovery[a] (%)			
		Day 1	Day 7	Day 14	Day 28
C.I. Direct Red 28	13.7	101.0[b] (± 8.7)	83.8[c] (± 16.0)	86.4[d] (± 9.3)	88.8[c] (± 21.1)
C.I. Direct Blue 6	15.0	103.0[b] (± 12.7)	105.0 (± 7.0)	--[e]	108.0 (± 12.5)
C.I. Direct Brown 95	12.1	71.4 (± 7.0)	73.9 (± 8.6)	--[e]	66.0 (± 18.4)
C.I. Direct Black 38	6.2	78.1[b] (± 16.1)	68.8 (± 8.6)	--[e]	69.5 (± 21.7)

[a]Based on six samples. 95% confidence limit contained in parentheses. [b]Based on five samples. [c]Based on three samples. [d]Based on four samples. [e]Analyses lost due to equipment malfunction.

The repeatability and stability studies were done at concentration levels which preliminary evidence on liquid samples had shown to be reproducible within 10% CV. However, this was not the case for three of the dyes (C.I. Direct Red 28, C.I. Direct Brown 95, and C.I. Direct Black 38). In this situation, the lowest analytically quantifiable limit (LAQL) had to be redefined. From the results of the repeatability study, the second highest level was the logical choice for this redefined LAQL. A second abbreviated stability study was conducted with C.I. Direct Red 28, C.I. Direct Black 38, and C.I. Direct Brown 95 at this new LAQL. Because of the problems encountered with the midget impinger humidification system, changes were made to incorporate one large impinger and supply the humidified air to the filter cassettes at atmospheric and not reduced pressure. Again, 60 L of humidified air were pulled through each cassette. The results of this study are shown in Table V. Recoveries of C.I. Direct Black 38 and C.I. Direct Brown 95 were approximately 90%. There were no significant differences between recoveries of the first day's analyses and the last day's analyses at the 0.05 level of significance.

Although the LAQL varied with each dye, the second stability study indicated that the LAQL should be in the range of 25-30-µg dye per sample for at least two of the dyes. Since the method did not differentiate between benzidine-based dyes, one LAQL was necessary for the method. By utilizing the 30-µg level as the LAQL, reduction and analysis in the working

Table V
Evaluation of the Stability of Filter Samples Containing
Benzidine-Based Dyes Stored Up to Seven Days

Dye	Loading (μg)	Average Recovery[a] (%)	
		Day 1	Day 7
C.I. Direct Red 28	28.8	97.8 ($\pm$ 6.3)	95.9 ($\pm$ 6.6)
C.I. Direct Blue 6	15.0	103.0 ($\pm$ 12.7[b])	105.0 ($\pm$ 7.0)
C.I. Direct Brown 95	26.6	90.4 ($\pm$ 5.7)	91.2 ($\pm$ 6.3)
C.I. Direct Black 38	12.3	88.5 ($\pm$ 10.0)	85.6 ($\pm$ 11.5)

[a]Based on six samples. 95% confidence intervals contained in parentheses. [b]Based on five samples.

range of each of the four dyes evaluated is assured.

To minimize loss by decomposition, samples should be analyzed as soon as possible after collection, preferably within seven days. The samples should be stored in a dark environment while awaiting analysis to prevent photosensitive compounds from degrading and changing sample composition.

Since the method does not distinguish between various benzidine-based dyes and analytically pure dyes are not easily available, recovery correction factors cannot be used. This will cause the results to be equal to or below the true level of dye exposure.

This method has been used to analyze both symmetrical (C.I. Direct Red 28 and C.I. Direct Blue 6) and unsymmetrical (C.I. Direct Black 38 and C.I. Direct Brown 95) benzidine-based dyes. Based on this work, the application of the method to other benzidine-based dyes should be straightforward. When field samples are submitted for benzidine-based dye analysis, bulk samples of the dyes present in the sample also should be submitted. With these bulk samples, the analyst should be able to determine if this method is applicable to the various dyes submitted and if any interferences are present. The method presently has not been tested on field samples. An existing sampling method (13) for azo dyes and diazonium salts should be directly applicable to this method with a change from a cellulose ester to a Teflon filter. This change is necessary to insure quantitative recovery of the sample from the filter.

In summary, a method for the identification and quantification of benzidine-based dye containing samples has been developed. This method could be expanded to include samples taken from media other than air with minor modification, for instance, in assaying benzidine-based dye formulations at the microgram level.

Disclaimer

Mention of company names or products does not constitute endorsement by the National Institute for Occupational Safety and Health.

Abstract

A method for the determination of personal exposure to benzidine-based dyes has been developed. This procedure involved the reduction of benzidine-based dye filter samples to free benzidine with neutral buffered sodium hydrosulfite solution. The benzidine-containing reduction solution was then analyzed by high performance liquid chromatography. The reduction was found to be quantitative by visible-spectrum analysis. This reduction and analysis method was evaluated with four benzidine-based dyes over the range from 12 to 300 micrograms per sample. Precision for the reduction and analysis of the four dyes falls within 11% coefficient of variation. This method can differentiate between benzidine- and benzidine congener-based dyes. Results are reported in terms of free benzidine.

Literature Cited

1. Venkataraman, K. "The Chemistry of Synthetic Dyes"; Academic Press: New York, 1952; pp. 506-514.

2. Lubs, H. A. "The Chemistry of Synthetic Dyes and Pigments"; American Chemical Society Monograph No. 127: Washington, D.C., 1952.

3. Clayson, D. B. Prev. Med., 1976, 5, 228-244.

4. Haley, T. J. Clin. Toxicol., 1975, 8, 13-42.

5. Case, R. A. M. Ann. R. Coll. Surg., 1965, 39, 213-235.

6. Billiard-Duchesne, J. J. Urol. Med. Chir., 1959, 65, 748-791.

7. Rinde, E; Troll, W. J. Nat. Cancer Inst., 1975, 55, 181-182.

8. Hartman, C. P.; Fulk, G. E.; Andrews, A. W. Mutation Research, 1978, 58, 125-132.

9. "NIOSH/NCI Joint Current Intelligence Bulletin 24, Direct Black 38, Direct Blue 6 and Direct Brown 95 Benzidine-Derived Dyes." DHEW (NIOSH) Publication No.

78-148, National Institute for Occupational Safety and
Health, Cincinnati, OH, April 17, 1978.

10. Taylor, D. G., Ed. "NIOSH Manual of Analytical Methods,"
    2nd ed.; DHEW (NIOSH) Publication No. 79-141, Vol. 5,
    P&CAM 315, National Institute for Occupational Safety and
    Health, Cincinnati, OH, August, 1979.

11. Nony, C. R.; Bowman, M. C. Intern. J. Environ. Anal.
    Chem., 1978, 5, 203-220.

12. Nony, C. R.; Bowman, M. C. Jefferson, AR, March 1979,
    National Center for Toxicological Research Technical
    Report for Experiment No. 196.

13. Taylor, D. G., Ed. "NIOSH Manual of Analytical Methods,"
    2nd ed.; DHEW (NIOSH) Publication No. 77-157A, Vol. 1,
    P&CAM 234, National Institute for Occupational Safety and
    Health, Cincinnati, OH, April, 1977.

14. Rounds, R. L. In "Encyclopedia of Industrial Chemical
    Analysis": Snell, F. D.; Hilton, C. L., Ed.; J. Wiley and
    Sons: New York, 1968; Vol. 6, pp. 404-408.

15. Riggin, R. M.; Howard, C. C. Anal. Chem., 1979, 51,
    210-214.

16. Schulze, J.; Ganz, C.; Parkes, D. Anal. Chem., 1978, 50,
    171-174.

17. Morales, R.; Rappaport, S. M.; Weeks, Jr., R. W.;
    Campbell, E. E.; Ettinger, H. J. Los Alamos, New Mexico,
    December 1977, Los Alamos Scientific Laboratory Progress
    Report LA-7058-PR.

18. Dean, J. A., Ed. "Lange's Handbook of Chemistry," 11th
    ed.; McGraw-Hill Book Company, New York, 1967; p. 5-73.

19. Ibid. p.10-79.

20. Meyer, S. L. "Data Analysis for Scientists and
    Engineers." John Wiley and Sons, Inc.: New York, 1975;
    Chapter 10, p. 40.

21. Boeninger, M. "Carcinogenicity and Metabolism of Azo
    Dyes, ..."; DHHS (NIOSH) Publication No. 80-119; National
    Institute for Occupational Safety and Health, Cincinnati,
    Ohio, July, 1980, pp. 94-95.

22. Youden, W. J. "Statistical techniques for collaborative tests,"; The Association of Official Analytical Chemists: Washington, D.C., 1969; pp. 33-36.

23. Perry, John H., Ed. "Chemical Engineer's Handbook," 4th ed.; McGraw-Hill Book Company, New York, 1963; p 3-106.

24. Taylor, D. G.; Kupel, R. E.; Bryant, J. M. "Documentation of the NIOSH Validation Tests"; DHEW (NIOSH) Publication No. 77-185, National Institute for Occupational Safety and Health, Cincinnati, OH, April, 1977, pp. 2-11.

RECEIVED September 19, 1980.

An Infrared Analysis Method for the Determination of Hydrocarbons Collected on Charcoal Tubes

THOMAS C. THOMAS and ANDREW RICHARDSON III

USAF Occupational and Environmental Health Laboratory,
Brooks Air Force Base, San Antonio, TX 78235

In handling worldwide industrial hygiene problems for the Air Force, our Laboratory receives a heavy work load of charcoal tube and vapor monitor air samples. With our work load increasing, and more methods being applied to gas chromatography, any method which would remove some of the work load from our overworked gas chromatographs would be welcomed.

Any alternate methods would have to have similar sensitivities, be relatively specific and able to analyze the samples quickly. We have found that infrared spectroscopy, when applied to hydrocarbon samples collected on charcoal tubes or vapor monitors, meets these requirements.

To analyze charcoal tubes or vapor monitors for hydrocarbons by IR, the desorption solvent would have to be IR inactive in the region 3100 to 2750 cm^{-1}. Two such solvents: Freon ^{R} 113 (1,1,2 trichloro-1,2,2-trifluoroethane) and perchloroethylene meet this requirement.

Freon 113 desorbs hydrocarbons from charcoal at consistent % recoveries, however, does not meet the minimum desorption efficiency of >75% recommended by NIOSH ([1]). However, a mixture of perchloroethylene and Freon 113 does provide a desorption efficiency >80% and can be analyzed by this IR method.

Therefore, to check the possibility of using this IR procedure in our Laboratory, it was decided to evaluate the procedure against the recommended NIOSH GC procedures. We have limited our study to the hydrocarbons: JP-4 aviation fuel and PD-680 cleaning solvent. In addition to evaluating the IR method for charcoal tubes and vapor monitors, we also compared the two methods on actual field samples.

Experimental

Gas chromatography. A Varian Model 1800 gas chromatograph equipped with dual flame ionization detectors was used and

it was coupled to a Hewlett-Packard 3350 Data System. A
10' X 1/8" stainless steel column of 10% OV-101 on 80/100
mesh chromosorb W was used. The column, injector, and
detector were at 85°C, and 225°C, respectively. Nitrogen
carrier gas flow was at 20 ml/min; hydrogen gas flow was at
35 ml/min and air flow at 400 ml/min.

The charcoal tubes were broken open and the charcoal
transferred into a stoppered glass test tube. One milliliter
of carbon disulfide was pipetted into each tube, and 1.5 ml
of carbon disulfide was pipetted into each 3M vapor monitor
badge. After 30 minutes, aliquots of carbon disulfide were
injected into the gas chromatograph and compared versus hydrocarbon standards prepared in carbon disulfide. The total areas
of the sample and standard peaks were measured by the data
system.

Infrared Spectroscopy. A Perkin-Elmer Model 283 Infrared
Spectrometer was used. Normal operating conditions were
used: normal slit, 1X expansion, 12-minute scan speed (4000-200
wave numbers), and normal gain, in accordance with Perkin-Elmer
setup instructions. A Beckman 2 cm path length, near infrared
silica cell (holds 8 ml sample) was used to hold the desorbing
solution in the sample compartment (Figure 1).

The charcoal tubes were broken open and the charcoal
transferred to stoppered glass test tubes. When checking
Freon 113 as a desorbant, 10 ml was pipetted into each tube.
After 30 minutes, the solution was mixed and analyzed. When
checking the perchloroethylene/Freon 113 mixture, 0.5 ml of
perchloroethylene was individually pipetted into each tube.
After 30 minutes, 9.5 ml of Freon 113 was added and each tube
mixed on a vortex stirrer. For the 3M vapor monitors, 0.5 ml
of perchloroethylene was pipetted into each monitor and allowed
30 minutes for desorption. The monitors were then carefully
rinsed with Freon 113 through a glass funnel into a 25 ml
graduated cylinder. The final volume was adjusted to 10 ml with
Freon 113 and mixed. The solutions were individually poured into
the silica cell, the cell placed in the IR sample compartment and
a scan made from 3130 to 2750 wave numbers. A sloping line was
drawn across the resulting spectra from the 3130 to the 2750
wave number (Figure 2). The peak intensities were measured
at 2958, 2926, and 2855 wave numbers. For each standard, the
intensities at these three wave numbers were measured and the
sum plotted versus concentration. The sum from each sample
was then compared to the standard curve (Figure 3). Hydrocarbon standards were prepared by injecting known amounts of
JP-4 and PD-680 into a 5% perchloroethylene - 95% Freon 113
mixture.

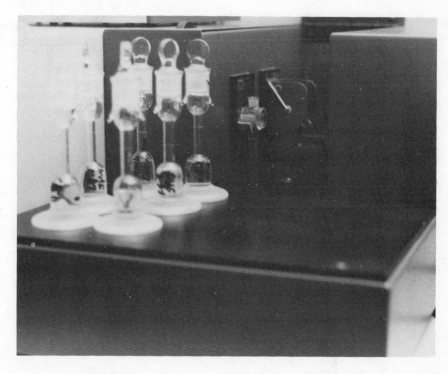

Figure 1.   Infrared cell used to hold extraction solutions

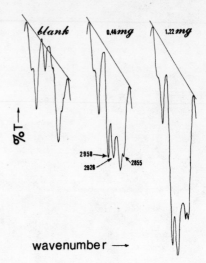

Figure 2.   Method of measuring hydrocarbon peak intensities

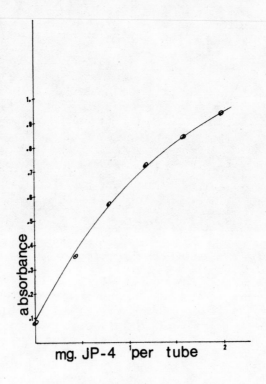

Figure 3.   JP-4 Aviation fuel standard curve

Scope of the Study

Analytical Desorption Recovery. Spiking with known amounts of JP-4 and PD-680 was done in two ways. First, known amounts were injected into the center of the front portion of the charcoal tube (SKC, NIOSH approved charcoal tubes). Secondly, the vapor monitors (3M 3500 organic vapor monitors) were spiked by injecting known amounts through the center elutriation port of the monitor cap, onto a piece of filter paper placed between the elutriation cap and the diffusion plate of the vapor monitor (2). The filter paper was removed after 24 hours and analyzed for any unvaporized portion. The monitor was analyzed as described earlier. The spiking was done at three levels, roughly corresponding to 0.5X, 1.0X, and 2.0X the occupational standard of these hydrocarbons, for a 10 liter air volume.

Recovery was initially checked using only Freon 113 as a desorbant, however, when recoveries were found to be below 75% (Table I), the Freon 113/perchloroethylene mixture was evaluated.

Analytical Precision. Additionally, the analytical method was tested to assure that its precision was acceptable. This test was performed at three different levels as before, except that the spiking was done directly into known amounts of the Freon 113/perchloroethylene mixture. Six samples at each of the three concentrations were used to form the basic statistical set of data. The samples were compared to previously prepared standards.

Field Sampling. An opportunity arose where actual field samples could be analyzed by both the infrared and gas chromatographic methods. At Robins AFB, Georgia, workers were inspecting and repairing the interior and exterior of C-141 aircraft fuel tanks. They were exposed only to JP-4 fuel fumes. Duplicate charcoal tubes or vapor monitors were attached to each worker, one on each lapel. Samples were drawn through the charcoal tubes at 0.20 to 0.26 lpm by portable pumps attached to the worker's belt. Because of slight variations, the total volumes collected for the duplicates were close, but not exactly the same in all cases. Samples were then labeled and shipped to our Laboratory for analysis by both methods.

Results and Discussion

Recovery Studies. As can be seen from Table I, Freon 113 desorbed both hydrocarbons with consistency, however, at levels less than 75% recovery. The desorption efficiency was the same, whether the sample was allowed 30 minutes or 24 hours for desorption. One surprising fact was that the NIOSH approved gas chromatographic technique for PD-680 also showed a desorption recovery at less than 75%; this was rechecked several times

to confirm this finding. The documentation of NIOSH validation
tests for petroleum distillates showed an average recovery of
95.7 ± 7.44% (3). Our study showed a recovery for JP-4 close to
100% by NIOSH approved method.

It was found that initial desorption with 0.5 ml of
perchloroethylene increases the recovery to greater than 80%.
It is theorized that perchloroethylene increases the desorption
of unsaturated and aromatic components that would not desorb into
pure Freon 113. There were several reasons why the initial
desorption was accomplished with 0.5 ml perchloroethylene and
then brought up to 10 ml with Freon 113. Using a 2 cm infrared
cell, perchloroethylene is not completely IR inactive in the
region of interest. It has a weak peak at approximately
2870 cm^{-1}, which would cause problems if pure perchloroethylene
were used. Secondly, use of Freon 113 lowers the inhalation
hazard to the analyst. Table II lists the recoveries from
charcoal tubes at different levels using the Freon 113/perchlo-
roethylene mixture.

Table III lists the recoveries from the 3M vapor monitors using
the Freon 113/perchloroethylene mixture. Poor recovery was also
obtained for PD-680 using the NIOSH GC method.

Statistical Studies. The present suggested standard for
air monitoring accuracy is that the absolute total error
(sampling and analysis) should be less than 25% in at least 95%
of samples analyzed at the level of the standard (1). This
implies that the true coefficient of variation of the total
error should be no greater than 0.128 derived as follows:
CV_t = 0.25/1.96 = 0.128. The number 0.128 is the largest true
precision for a net error at ± 25% at the 95% confidence level.
The number 1.96 is the appropriate t - statistic from the t
distribution at the same confidence level. Since the coefficient of variation of pump error is assumed to be 5%, a method
should have a CV analysis ≤0.102 to meet the CV accuracy
standard. Tables IV and V, shows that the infrared technique
meets this requirement.

Field Studies. As seen in Table VI, with the exception of
three sets of duplicate charcoal tube samples, the results
obtained with each method corresponded fairly well. The
engineer in the field admitted that the duplicate 598, 602 and
610 could have been accidentally contaminated with some of the
liquid aviation fuel. This could possibly explain the
differing results by each method on these samples. It appears
one set of samples was from workers exposed to low levels of
JP-4 vapors, while the other set was from an area with very high
concentrations. The variations in results on the low-level
samples could be explained by several factors. Among these would

TABLE I.   RECOVERY OF PD-680 AND JP-4 USING FREON 113
           A DESORBANT.[a]

Charcoal Tube Spiked with 3.715 mg PD-680

	IR Method	NIOSH Method
Run #1	57%	67%
Run #2	60%	63%
Run #3	57%	71%
	AVG: 58%	AVG: 67%

Charcoal Tube Spiked with 3.80 mg JP-4

	IR Method	NIOSH Method
Run #1	49%	102%
Run #2	48%	94%
	AVG: 48.5%	AVG: 98%

[a]Six tubes spiked for each run and recoveries averaged.

TABLE II.   RECOVERY OF PD-680 AND JP-4 FROM CHARCOAL TUBES USING
            FREON 113/PERCHLOROETHYLENE AS A DESORBANT.[a]

PD-680

	IR Method	NIOSH Method
Run #1 - 1.11 mg spike	83%	--
Run #2 - 3.715 mg spike	86%	67%
Run #3 - 5.95 mg spike	86%	--
	AVG: 86%	

JP-4

	IR Method	NIOSH Method
Run #1 - 1.22 mg spike	82%	--
Run #2 - 3.80 mg spike	88%	98%
Run #3 - 6.08 mg spike	88%	--
	AVG: 86%	

[a]Six tubes spiked for run and recoveries averaged.

TABLE III. RECOVERY OF PD-680 AND JP-4 FROM 3M VAPOR MONITORS
            USING FREON 113/PERCHLOROETHYLENE AS A DESORBANT.[a]

PD-680

	IR Method	NIOSH Method
Run #1 - 0.743 mg spike	89%	57%
Run #2 - 3.715 mg spike	96%	60%
Run #3 - 5.944 mg spike	95%	54%
AVG:	93%	AVG: 57%

JP-4

	IR Method	NIOSH Method
Run #1 - 0.76 mg spike	94%	106%
Run #2 - 3.04 mg spike	85%	101%
Run #3 - 6.08 mg spike	84%	96%
AVG:	88%	AVG: 101%

[a]Three vapor monitors spiked for each run and recoveries
averaged.

TABLE IV. STATISTICAL DATA ON THE IR METHOD FOR JP-4 ANALYSIS.

SPIKE LEVEL	FOUND	
1.29 mg	1.24 mg	
1.29	1.24	Mean: 1.24 mg
1.29	1.24	S.D.: 0.008
1.29	1.23	CV: 0.006
1.29	1.23	
1.29	1.23	
3.00 mg	3.02 mg	
3.00	3.03	Mean: 2.98 mg
3.00	3.02	S.D.: 0.060
3.00	2.98	CV: 0.020
3.00	2.89	
3.00	2.94	
6.08 mg	6.08 mg	
6.08	6.08	Mean: 6.04 mg
6.08	6.02	S.D.: 0.063
6.08	5.97	CV: 0.010
6.08	6.10	
6.08	6.02	

TABLE V.   STATISTICAL DATA ON IR METHOD FOR PD-680 ANALYSIS.

SPIKE LEVEL	FOUND		
1.11 mg	1.11 mg	Mean:	1.18 mg
1.11	1.07	S.D.:	0.073
1.11	1.27	CV:	0.062
1.11	1.22		
1.11	1.19		
1.11	1.19		
2.97 mg	2.97 mg	Mean:	2.94 mg
2.97	2.99	S.D.:	0.041
2.97	2.94	CV:	0.014
2.97	2.92		
2.97	2.97		
2.97	2.88		
6.02 mg	6.02 mg		
6.02	6.05	Mean:	6.05 mg
6.02	6.07	S.D.:	0.019
6.02	6.08	CV:	0.003
6.02	6.06		
6.02	6.05		

be: the distance between each duplicate set on each lapel; the position each worker was occupying while the samples were taken; air currents around each worker; concentration zones within the areas surrounding the workers; and, possible slightly different sampling rates. For the large amounts of JP-4 collected in the high concentration areas, the two-methods compared surprisingly closely. The variations can also be explained by the factors listed above, as well as slightly differing sampling time durations as noted in Table VI.

As can be seen in Table VII, the milligram amounts collected in the duplicate monitors corresponded fairly well. Except for the duplicate sample 649, the variations could be explained by the factors listed above. Again contamination during sampling could have caused the wide difference in the one set.

TABLE VI.   COMPARISON OF GC AND IR METHODS ON DUPLICATE FIELD
            CHARCOAL TUBE SAMPLES.

Sample No.	Method Used	Sample Size in liters	mg JP-4 Per Tube	mg/m^3
586	IR	15.4	0.04	2.6
586	GC	15.4	<0.01	<0.6
588	IR	15.4	<0.01	<0.4
588	GC	15.4	<0.01	<0.4
590	IR	15.2	<0.01	<0.7
590	GC	15.2	<0.01	<0.7
592	IR	15.2	0.12	7.9
592	GC	15.2	0.03	2.0
594	IR	14.6	<0.01	<0.7
594	GC	14.6	<0.01	<0.7
596	IR	28.6	<0.01	<0.4
596	GC	28.6	<0.01	<0.4
598	IR	23.8	0.03	1.3
598	GC	23.8	3.60	151.
600	IR	35.0	0.06	1.7
600	GC	35.0	<0.01	<0.3
602	IR	28.2	<0.01	<0.4
602	GC	28.2	7.20	255.
604	IR	--	<0.01	Blank Tube
606	IR	24.0	<0.01	<0.4
606	GC	24.0	<0.01	<0.4
608	IR	18.4	<0.01	<0.5
608	GC	18.4	<0.01	<0.5
610	IR	24.2	6.10	252.
610	GC	24.2	0.42	17.5
626	IR	31.0	43.0	1386.
626	GC	33.8	30.5	902.
628	IR	33.8	38.6	1142.
628	GC	36.7	48.0	1307.
630	IR	28.8	36.6	1271.
630	GC	36.4	45.0	1236.
632	IR	34.3	30.5	889.
632	GC	33.6	27.0	804.
634	IR	32.8	26.7	814.
634	GC	30.2	34.2	1131.
636	IR	8.36	21.8	2428.
636	GC	9.12	20.3	2226.
638	IR	8.64	30.4	3518.
638	GC	9.36	22.3	3014.
640	IR	7.0	8.02	1146.
640	GC	28.0	31.7	1132.
642	IR	8.58	10.9	1271.
642	GC	8.4	8.0	952.
643	IR	--	0.01	Blank Tube
644	GC	--	0.01	Blank Tube

TABLE VII.   Comparison of GC and IR Methods on Duplicate 3M
Vapor Monitor Field Samples.

Sample No.	Method Used	Exposure Time, Minutes	mg JP-4 Per Monitor	mg/m³[a]
611	GC	235.	0.11	22.
611	IR	235.	0.08	16.
613	GC	127.	0.12	44
613	IR	127.	0.034	12
615	GC	213.	0.52	114.
615	IR	213.	0.44	97
617	GC	269.	0.12	21.
617	IR	269.	0.011	2.
619	GC	236.	0.82	162.
619	IR	236.	0.91	180
621	GC	128.	0.18	66.
621	IR	128.	0.17	62.
623	GC	83.	0.11	62.
623	IR	83.	0.045	25
645	IR	42.	2.84	3162.[b]
645	GC	42.	1.8	2004.
647	IR	44.	2.4	2551.
647	GC	44.	1.9	2019.
649	IR	28.	0.034	57
649	GC	28.	0.76	1269.

[a]Since the monitor sampling rate has not been determined for
JP-4 aviation fuel, the sampling rate (24.3 cm³/min) for
Stoddard solvent was used to calculate mg/m³ (4). Since these
are similar mixtures, this approach was taken to calculate
mg/m³.
[b]It should be noted in Table VI and Table VII that workers wore
supplied air systems when entering these high concentration
areas.

Conclusion

It appears there would be several advantages in using the
proposed infrared method: analyses could be performed much faster
using the IR method. An IR scan takes 45 seconds versus ~16
minutes by the NIOSH GC method. The IR method is adapted to
charcoal tubes sampled by current recommended techniques. The
majority of laboratories have infrared spectrometers and the
simplest of models will do; utilization of the IR method would
relieve pressure of the heavily used gas chromatographs. The

solvents used in the IR method present less of a hazard to the analyst than carbon disulfide. TLV of the Freon 113/perchloroethylene mixture is 5009 mg/m^3 versus 30 mg/m^3 for carbon disulfide. The use of the IR method for hydrocarbons would eliminate occasional problems of variations in FID response by the GC method. The IR method is so simple that technicians can be easily, and quickly, trained to perform the analysis.

There are also several limitations of using the IR method. They are: use of Freon 113/perchloroethylene for hydrocarbon desorption prevents further analysis by gas chromatography for other solvents. Samples requesting hydrocarbons, plus any other solvents, would have to be analyzed solely by the GC method. The IR method is best adapted to pure hydrocarbon exposures such as to JP-4 aviation fuel, gasoline, petroleum distillates, Stoddard solvent, etc. Other organic compounds containing C-H bonds will introduce positive interferences in the IR method; however, the NIOSH GC methods are also very susceptible to extraneous compounds. With less sophisticated infrared spectrometers, the analysts will have to manually measure the peak intensities. However, since a majority of samples are usually negative, a quick scan can eliminate these without any measurement. The working range is somewhat small, up to approximately 2 mg hydrocarbons per tube being on chart. For off-chart samples, dilutions can be made quickly and simply.

We believe it has been shown that this method for infrared analysis of hydrocarbons collected on charcoal tubes and vapor monitors is a valid and acceptable one. Further work is being done to validate the method for other hydrocarbons such as petroleum naphtha, Stoddard solvent, and other JP aviation fuels. Additionally, work is being done to determine the 3M monitor sampling rate for JP-4.

Literature Cited

1. Taylor, David G., Richard E. Kupel, and John M. Bryant, "Documentation of the NIOSH Validation Tests," DHEW (NIOSH) Publication No. 77-185, pages 1-12, April 1977.

2. Rodriguez, S.T., D.W. Gosselink, and H.E. Mullins, "Determination of Desorption Efficiencies in the 3M 3500 Organic Vapor Monitor," presented at American Industrial Hygiene Conference, Houston, Texas, May 20, 1980.

3. Ibid, S380-1 to S380-4.

4. 3M Brand Organic Vapor #3500 Compound Guide, Occupational Health and Safety Products Division, St Paul, Minnesota.

RECEIVED September 19, 1980.

Development of Personal Sampling and Analytical Methods for Organochlorine Compounds

K. W. BOYD, M. B. EMORY, and H. K. DILLON

Southern Research Institute, 2000 Ninth Avenue South, Birmingham, AL 35255

The need for air sampling and analytical methods for toxic contaminants in the workplace arises from provisions of the Occupational Safety and Health Act of 1970 requiring that regulations be prescribed to limit the exposure of employees to substances or physical agents that may endanger their health or safety. To prescribe such regulations and to ensure compliance, it is necessary to have available sampling and analytical methods suitable for use by employers, Governmental personnel, and others interested in analyzing air samples from the workplace.

In 1971 when safety and health standards were established by the U. S. Department of Labor for several hundred chemical substances, there were analytical methods available for some of the compounds, but few were validated to ensure the accurate monitoring of the exposure of workers to these toxic substances (1). Consequently, programs were undertaken by the National Institute for Occupational Safety and Health (NIOSH) to develop and validate sampling and analytical methods. The initial intent was to provide methods that would be useful to industry in measuring the exposures of personnel to potentially toxic materials at concentration levels near the accepted standard levels. Consequently, many earlier methods were developed around the standard levels established by the Occupational Safety and Health Act with validation at, for example, levels ranging from one-half to twice the established standard level (2). Often these methods were not validated at lower concentration levels, say, one-tenth of the original level.

The concern over the workplace hazards of chemical substances has increased with the determination that some compounds are carcinogens or suspect carcinogens. Consequently, it has been recommended by NIOSH that several established standards be lowered (3). It is, therefore, very important to develop methods that can be readily adapted to lower standard levels without the need for additional costly and time-consuming research.

This presentation describes work performed under contract with NIOSH to develop and validate sampling and analytical

0097-6156/81/0149-0049$05.00/0

methods for three potentially toxic chlorinated organic compounds—hexachlorocyclopentadiene (HCCP), hexachlorobutadiene
(HCBD), and 1,2-dichloropropane (1,2-DCP) (4, 5, 6). In anticipation of the possible establishment of relatively low standards
for these compounds, NIOSH requested that the developed methods
be validated over substantially lower ranges of concentration
levels than were previously considered. The ranges of the
methods were based on the determination of the lowest analytically quantifiable level (LAQL), defined as the smallest amount
of a substance that can be determined in a sample with a recovery
greater than 80% and with a relative standard deviation less than
10%. Consequently, sampling and analytical methods were developed that are reliable at levels well below the current OSHA standard of 350 mg/m^3 for 1,2-DCP and well below the TWA (8-h time
weighted average) for HCCP of 0.1 mg/m^3 recommended by the American Conference of Governmental Industrial Hygienists (ACGIH) (7,
8). Standards for HCCP and HCBD have not yet been promulgated.
       The methods developed for HCCP, HCBD, and 1,2-DCP involve
the collection of vapors of the compounds from air with solid
sorbents in tandem with personal sampling pumps, desorption of
the sorbed compounds in appropriate solvents, and analysis of the
extracts by gas chromatography (GC).

Method Development

       To achieve optimal sensitivity and selectivity, it was neces-
sary to develop three totally separate methods, one for each compound. Initially, it was necessary to develop, optimize, and
calibrate a procedure for quantitating each analyte. With these
steps successfully completed, candidate collection media were
screened in tests designed to find a material with three attri-
butes: (1) an acceptable sorption capacity for the appropriate
analyte (ideally, high enough to provide a sampling volume of at
least 12 L with no more than 5% breakthrough), (2) an efficient
desorption (≥80% recovery) of the compound for analysis, and
(3) a stability of the sorbed analyte at room temperature for at
least 7 d without significant degradation (9).

       Optimization and Calibration of Analytical Procedure. The
first step undertaken in the laboratory was the establishment of
an optimum procedure for determining HCCP, HCBD, and 1,2-DCP by
GC. (Other analytical techniques were eliminated on the basis of
a preliminary literature search.) Two tasks were involved:
(1) the choice of the most suitable GC detector, column, and
operating conditions for each method and (2) the calibration of
the resulting procedures.
       Detector selection was relatively straightforward. Because
the electron capture detector (ECD) offered sensitivities for
HCCP and HCBD that could not be equaled by any other GC detection
system, the ECD was employed for the determination of these two

compounds (10). The ECD, the flame ionization detector (FID), and the Hall electrolytic conductivity detector (in the halogen mode) were evaluated for the determination of 1,2-DCP. It was found that the Hall detector offered better sensitivity than the others; this detector was chosen for the development of a method for 1,2-DCP.

In choosing the most suitable GC column, the primary criterion employed was the degree of resolution of each analyte from potential interferents that could be achieved under the optimum operating conditions for that column. The compounds considered as potential interferents for each method are given in Table I. Many of these compounds are halocarbons that are likely to coexist with the analytes in the workplace. Others possess physical and chemical properties similar to the analytes. A number of different GC columns and operating conditions were evaluated for each method before optimum results were obtained.

The optimized operating conditions for each analytical method including the detector system of choice are reported in Table II. The reported columns and operating conditions yield satisfactory peak shapes and resolution of all the potential interferents evaluated for HCCP and HCBD. Two potential interferents—tetrachloro-1,2-difluoroethane and 1,2-dichloroethane—could not be separated from 1,2-DCP with conventional packed columns. Tetrachloro-1,2-difluoroethane, a compound with physical properties similar to 1,2-DCP, is not likely to be found with 1,2-DCP in air samples and, therefore, should seldom cause a problem (11, 12). The other potential interferent, 1,2-dichloroethane, is an impurity in reagent-grade 1,2-DCP but typically represents less than 1% (w/w) of the reagent (13). Thus, this compound should not ordinarily pose an interference problem. (With a Carbowax 20M glass capillary column (30 m by 0.25 mm i.d.) in place of a conventional packed column, 1,2-DCP was resolved from 1,2-dichloroethane.)

The GC response was calibrated for each analyte to determine the reproducibility of injections, the detection limit, and the working range of the method. Both peak height and peak area response measurements were taken.

For HCCP the chromatographic response was found to be a linear and reproducible function of HCCP concentration in the range of about 5.0 to 142 ng/mL (25 to 710 pg injected) with a correlation coefficient of 0.9993 for peak height measurement. A linear response was only obtained, however, if the column was conditioned daily with several 5-μL injections of a relatively concentrated solution of HCCP in hexane. The detection limit was about 5 ng/mL (25 pg injected). At this limit, the precision of peak height measurements corresponded to a relative standard deviation (RSD) of 6% with a ratio of about 7:1 for peak height to background noise. The RSD for peak area measurements with a mechanical integrator was about 33%, corresponding to a much lower precision than that obtained with peak height measurements.

Table I. Compounds Selected for Interference Testing in the Development of Analytical Procedures for the Three Chlorocarbons

HCCP	HCBD	1,2-DCP
Tetrachloroethylene	tetrachloroethylene	1,3-dichloropropane
Hexachloroethane	hexachloroethane	2,2-dichloropropane
1,2,4-Trichlorobenzene	1,2,4-trichlorobenzene	trans- and cis-1,3-dichloro-1-propene
HCBD	HCCP	trichloroethylene
Octachlorocyclopentene	4-chlorobiphenyl	tetrachloroethylene
		dibromomethane
		1,2-dichloroethane
		tetrachloro-1,2-difluoroethane

Table II. Optimized GC Analytical Procedure for Each of the Chlorocarbons[a]

	HCCP	HCBD	1,2-DCP
Detector	electron capture	electron capture	Hall electrolytic conductivity
Column	3% OV-1 on Gas-Chrom Q (100/120 mesh) in glass (4 mm i.d. by 2 m)	3% OV-1 on Gas-Chrom Q (100/120 mesh) in glass (4 mm i.d. by 2 m)	3% Carbowax 1500 on Chromosorb W HP (60/80 mesh) in nickel (2 mm i.d. by 3 m)
Operating conditions			
Carrier gas	5% CH_4, 95% Ar, 20 mL/min	5% CH_4, 95% Ar, 20 mL/min	N_2, 25 mL/min
Temperatures			
Injection port	150 °C	240 °C	150 °C
Column	135 °C	135 °C	50 °C
Detector	250 °C	250 °C	–
Detector parameters	detector purge, 5% CH_4, 95% Ar, 80 mL/min	detector purge, 5% CH_4, 95% Ar, 80 mL/min	furnace temperature, 850 °C; electrolyte, 50% $(CH_3)_2CHOH$ in H_2O, 0.9 mL/min; H_2, 50 mL/min
Solvent for compound[b]	hexane	hexane	15% (v/v) acetone in cyclohexane

a. A Hewlett-Packard 5750A GC was used for the HCCP and HCBD methods and a Perkin-Elmer Sigma 2 GC was used for the 1,2-DCP method.

b. The injection volume was 5 µL of sample and 1 µL of solvent flush.

Detector sensitivity was better for HCBD than for HCCP. The detection limit was 0.8 ng/mL (4 pg injected) with an RSD of 10%. The response, however, to HCCP was nonlinear for both peak height and peak area measurements in the working range of 0.8 to about 170 ng/mL (850 pg injected). (A slight curvature in response versus concentration such as that observed is not uncommon for the response of an ECD (14).) For HCBD as for HCCP, peak height measurements were more precise than peak area measurements with a mechanical integrator.

For 1,2-DCP, the Hall detector response was linear in the range of 6.93 to 347 µg/mL (34.7 to 1735 ng injected). A plot of peak area versus concentration in this range (as determined with the Perkin-Elmer Sigma 10 Data System) yielded a correlation coefficient (r) of 1.0000.

Selection of Collection Media. The methods developed for HCCP, HCBD, and 1,2-DCP involve the collection of the analytes from air on solid sorbent materials in small Pyrex tubes. Each tube is 7 cm long by 6 mm o.d. and 4 mm i.d. The recommended sampling tubes contain two beds of sorbent material—one layer for sorption and a second, smaller, backup layer to monitor breakthrough if the capacity of the sorbing layer is exceeded.

A sampling tube of this type offers several advantages for personal monitoring. The portability of the device allows it to be used for sampling the breathing air of an individual. Thus, the exposure of an individual worker to a chemical substance can be ascertained. Many previously available sampling methods required the use of bubblers with liquid absorbers or other bulky and complicated apparatus that was inconvenient for personal monitoring. The sorbent tube is not only very convenient to use; its compactness is convenient for shipping and handling.

The materials considered as sorbents for HCCP, HCBD, and 1,2-DCP were subjected to two types of preliminary tests— capacity tests and desorption efficiency tests. On the basis of these tests, a sorbent material was tentatively selected for each analyte; the selection was confirmed only after the overall sampling and analysis method was validated.

The sorbent materials evaluated for each analyte are listed in Table III. (It was necessary to clean some of the sorbent materials to remove impurities prior to use. The cleaning procedure consisted of Soxhlet extraction with an 80:20 mixture of acetone and methanol for 4 h followed by extraction with hexane for 4 h.)

To be an acceptable substitute for wet collectors and to satisfy the NIOSH criterion for acceptable methods, a sorbent material must have a demonstrated sorption capacity for the analyte that is adequate for sampling a reasonable volume of workplace air at an established rate. Typically, a sample volume of at least 12 L (1 h at 0.2 L/min) is desirable.

In the capacity tests, the sorbent materials were challenged

Table III. Sorbent Materials Evaluated
for Each Analyte

HCCP	HCBD[a]	1,2-DCP
Amberlite XAD-2 (20/50 mesh)[b]	Tenax-GC (35/60 mesh)[c]	coconut charcoal (20/40 mesh) (SKC Cat. No. 226-01-01)
Porapak R (50/80 mesh)[c]	Amberlite XAD-2 (20/50 mesh)[b]	petroleum charcoal (20/40 mesh) (SKC Cat. No. 226-36-01)
Ambersorb XE-340 (20/50 mesh)[c]		petroleum charcoal (20/40 mesh) (SKC Cat. No. 226-38-01)
Chromosorb 104 (60/80 mesh)[c]		petroleum charcoal (20/40 mesh)[c] (Barnebey-Cheney Type 580-26)
Tenax-GC (35/60 mesh)[c]		Carbosieve B (45/60 mesh)[c]
Porapak T (80/100 mesh)[c]		silica gel (60/80 mesh)[c]
Porapak T (50/80 mesh)[c]		Amberlite XAD-4 (16/50 mesh)[c]
		Porapak R (50/80 mesh)[c]

a. Only two sorbent materials were evaluated for the HCBD method because both performed satisfactorily and it was considered unnecessary to evaluate other materials.

b. Precleaned by the supplier, Applied Sciences, Inc.

c. Soxhlet extracted with acetone/methanol mixture and hexane.

with laboratory test atmospheres of the appropriate analyte until
the capacity of either a 50- or 100-mg quantity of each material
in a Pyrex tube was exceeded and analyte breakthrough was
observed. (In field applications, the recommended sampling device
would contain two sections of sorbent material packed into a
single tube—the sorbing section and a smaller backup section.)

The capacity tests required the construction and evaluation
of a vapor generator, a sampling system, and a generator monitor
for each analyte. The system used for 1,2-DCP is depicted in the
figure. Similar systems were constructed for HCCP and HCBD except
bubbler measurements were used to monitor the generator effluent
for these compounds whereas a total hydrocarbon analyzer was used
to monitor the 1,2-DCP effluent. The systems operated on the
vapor saturation technique with the analyte concentration controlled by the low temperature cooler and the dilution volume. A
midget impinger was employed as a reservoir for each analyte with
both the temperature of the impinger and the flow of nitrogen
through the reservoir closely regulated. The nitrogen laden with
the appropriate analyte was mixed with precleaned dilution air in
a glass splash trap—the mixing chamber—and was passed into a
cylindrical glass sampling chamber with seven sampling ports for
sorbent tubes and other sampling devices. The excess generator
effluent was vented through a bed of charcoal and then into a
hood. To provide clean, dry air for the generator, the airstream
(taken from a laboratory air supply system) was passed sequentially through a bed of charcoal, a felt filter, and finally
through a membrane filter with an average pore diameter of 0.2 μm.
The dilution air was humidified by metering part of the airstream
through a heated Greenburg-Smith impinger containing distilled,
deionized water.

To determine the capacity of a sorbent material, generator
effluent was sampled into sorbent tubes at a known rate and
breakthrough from the tubes was monitored.

In some desorption efficiency tests, the sorbent materials
that had yielded the most promising results in the capacity tests
were spiked with solutions of HCCP in hexane, HCBD in hexane, or
1,2-DCP in 15% (v/v) acetone in cyclohexane. In other tests,
desorption efficiency was determined prior to capacity tests.
The spiking procedure for all tests was as follows:

- Fifty or one hundred milligrams of sorbent was added
  to a glass vial.

- Five or ten microliters of hexane containing a known
  amount of the analyte was injected into the sorbent
  bed in the vial. This amount of solvent evaporated
  rapidly.

- The vial was sealed and stored overnight.

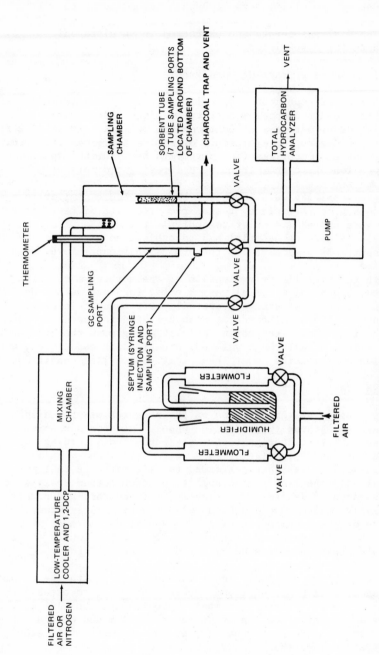

Figure 1. Schematic of vapor generator and sampling system for 1,2-DCP

- One or two milliliters of hexane was added to the vial.

- The vial was capped and extracted in an ultrasonic bath for a specified time.

- The sample extract was analyzed.

A series of experiments was also performed to determine the sample blank and the optimum extraction time. When required, several solvents were evaluated to obtain optimal extraction efficiencies. A desorption efficiency of at least 0.8 was required.

The sorbent materials that performed best in the capacity and desorption efficiency tests were investigated further with respect to the stability of the sorbed analyte. Preliminary tests of analyte stability were conducted by a procedure similar to that in the desorption efficiency tests; the procedure differed in that samples were stored 7 d prior to analysis rather than 1 d. To be acceptable, a sorbent material had to exhibit no statistically significant loss of analyte at the 0.05 significance level by a two-tailed t test.

With the preliminary evaluation of the sorbent materials completed, a sorbent material was selected for each method on the basis of the most satisfactory overall performance. In addition, a solvent or solvent mixture was also selected for each method. The selections are presented in Table IV along with the breakthrough times, breakthrough volumes, breakthrough capacities, and desorption efficiencies under the specified sampling conditions.

Method Validation

The purpose of this portion of the research was to validate the developed methods by generating enough data for a statistical evaluation. The validation tests were conducted according to NIOSH guidelines (9).

Initially, tests were performed to determine the LAQL for each analyte. The long-term stability of each sorbed analyte at its LAQL was also determined. Finally, the accuracy and precision of the analytical method alone and also the overall accuracy and precision of the combined sampling and analytical methods were determined. All tests were performed with the previously developed analytical procedures and sampling devices.

Determination of LAQL. Tests were performed with the recommended analytical procedures and sorbent materials to establish the LAQL for each analyte. As stated previously, the LAQL is the smallest amount of a compound that can be determined with a recovery from the sorbent greater than 80% and a relative standard deviation less than 10% (9).

The procedure for these tests was similar to that described for the desorption efficiency tests. The spiked samples were stored overnight and extracted, and the extracts were analyzed.

Table IV. Collection Media Selected for the Three Analytes

	HCCP	HCBD	1,2-DCP
Sorbent material	Porapak T[a] (80/100 mesh)	Amberlite XAD-2 (20/50 mesh)	petroleum charcoal (20/40 mesh) (SKC Cat. No. 226-36-01)
Breakthrough time,[b] h	>8 (0.2 L/min)	>8 (0.2 L/min)	>1[c] (0.05 L/min)
Breakthrough volume,[b] L	>100	>100	4[c]
Tube capacity,[b] μg	>100	>1000	2000[c]
Average desorption efficiency of indicated quantity of analyte	0.94 (27.4 ng)	1.00 (4 ng)	0.92 (1.04 μg)
Sorbent tube configuration[d]	75 mg sorbing layer, 25 mg backup layer	100 mg sorbing layer, 50 mg backup layer	100 mg sorbing layer, 50 mg backup layer
Extraction solvent	hexane	hexane	15% (v/v) acetone in cyclohexane

a. This material required cleaning by Soxhlet extraction.

b. For these tests the temperature of the generator effluent was maintained at 25 to 28 °C and the relative humidity at greater than 80%. The concentrations of the analytes in the generator effluent was 1 mg/m^3 of HCCP, 10 mg/m^3 of HCBD, and 700 mg/m^3 of 1,2-DCP.

c. Breakthrough time, breakthrough volume, and tube capacity were increased by a factor of about 4 at a relative humidity of about 80%.

d. The sorbent tubes were Pyrex (7 cm long by 6 mm o.d. and 4 mm i.d.). Silanized glass wool plugs separated the sections.

The LAQL was about 25 ng for HCCP, about 20 ng for HCBD, and about 400 ng for 1,2-DCP.

Determination of Long-Term Stability of Sorbed Analyte.

Tests were performed to determine the stability of sorbed analytes in tubes that had been exposed to generator effluent and that were then stored for at least 7 d. For each method, the analyte concentration in the generator effluent was maintained at about 0.3 X LAQL per liter; 3 L of the generator atmosphere was sampled at a rate of 0.2 L/min to yield a tube loading of about the LAQL. Following exposure, the sorbent tubes were sealed with Teflon tape and plastic caps for storage.

The results as summarized in Table V indicated that the stability was satisfactory for all three methods. To be acceptable, the average recovery of each analyte had to be at least 80%, and the difference between the average recovery on the first day and the average recovery after storage (for at least 7 d) had to be statistically insignificant at the 0.05 significance level by a two-tailed t test (9).

Table V. Long-Term Stability of Sorbed Analytes[a]

	HCCP	HCBD	1,2-DCP
Average concentration after 0 d,[b,c] $\mu g/m^3$	9.93 ± 0.27	9.04 ± 0.28	121 ± 4
Storage period, d	28	34	7
Average concentration after indicated storage period,[c] $\mu g/m^3$	9.47 ± 0.37	8.93 ± 0.31	118 ± 5
t	0.67	0.72	1.11
t critical	2.18	2.12	2.26

a. Each value represents at least six samples except the 0-d value for 1,2-DCP; it represents only five samples.

b. Samples were extracted and analyzed immediately after exposure.

c. Ninety-five percent confidence limits are given for these data.

In the HCCP method, some migration of the compound from the sorbing section to the backup section occurred during storage. To minimize this problem in the application of the method in the field, it is recommended that the two sorbent sections be separated prior to storage.

Determination of Accuracy and Precision of the Analytical Procedure. The desorption efficiency was determined for each method at widely separated analyte quantities to establish the average recovery to be expected. The spiking and analysis procedures for these tests were similar to those described earlier for the preliminary desorption efficiency tests. For HCCP and 1,2-DCP, the analyte levels chosen were approximately the LAQL, 10 X LAQL, and 1000 X LAQL. For HCBD, only two levels were tested—the LAQL and 1000 X LAQL.

The experiments gave average desorption efficiencies of 1.004 for HCCP, 0.984 for HCBD, and 0.954 for 1,2-DCP. Because the precision was found to be homogeneous by Bartlett's test for each of the three analytes at the levels tested, the values of the RSD were pooled to obtain an overall RSD for each compound (9). The pooled values were 3.0% for HCCP, 1.1% for HCBD, and 3.1% for 1,2-DCP.

Determination of Accuracy and Precision of the Overall Sampling and Analytical Procedure. The final step in completion of each method was the determination of the accuracy and precision of the combined sampling and analytical steps. In these tests, sorbent tubes were exposed to test atmospheres of the analytes in air at a relative humidity of 80% or greater and a temperature of 25 to 28 °C. The sampling rate was nominally 0.2 L/min, and a total of about 3 L of test gas was sampled into each tube. To evaluate the results by an independent procedure, the test gas was sampled simultaneously with sorbent tubes and (1) with bubblers containing hexane at 0 °C for HCCP or HCBD or (2) with a total hydrocarbon analyzer for 1,2-DCP. After exposure, the sorbent sections in the tubes were transferred along with the glass wool plugs to vials. The contents were then extracted with 1 mL of the appropriate solvent, and the extracts were analyzed.

The minimum requirements of the results were as noted above for other tests: 80% recovery with 10% relative standard deviation. Also, the results could not be biased from the independent procedures by more than ±10% (9).

The HCCP method was evaluated in the concentration range of about 13 to 873 $\mu g/m^3$ in 3-L air samples. The average bias from the independent analytical procedure (bubbler measurements) was +1% and the pooled RSD was 8%. (This estimate of precision includes an assumed RSD of 5% for the precision of air metering with a personal sampling pump. The air sampling in these laboratory tests was performed with critical flow orifices where little variation in air sampling rates was experienced. Thus, a correction was considered necessary to include the variations expected in sampling rates in field measurements.)

The HCBD method was evaluated in the concentration range of about 10 to 2000 $\mu g/m^3$ in 3-L air samples. The average bias from the independent analytical method (bubbler measurements) was -7% and the pooled relative standard deviation was 9%.

The method for 1,2-DCP was evaluated in the concentration range of 0.124 to 128 mg/m^3 in 3-L air samples. The average bias from the independent analytical procedure (total hydrocarbon analyzer) was less than 1% over the range of the method. The pooled relative standard deviation was 6.4% over this range.

A summary of the results of the method validations appears in Table VI.

Table VI. Summary of Results for Method Validations

	HCCP	HCBD	1,2-DCP
LAQL, ng	25	20	400
Desorption efficiency[a]	1.004	0.984	0.954
Range of validation in 3-L air sample,[b] $\mu g/m^3$	13 to 865	10 to 2,000	124 to 128,000
Overall pooled RSD, %	8.0	9.0	6.4

a. The desorption efficiency was averaged for levels ranging from near the LAQL to 1000 X LAQL.

b. Data obtained in this range were used to calculate the pooled RSD.

Application of Methods

Methods for all three compounds have been approved by NIOSH. The methods for HCCP and HCBD were published in the "NIOSH Manual of Analytical Methods", Vol. 5 (15). The method for 1,2-DCP will be included in Volume 6 of the Manual to be published soon.

The method for HCCP has been used routinely by industry over the past year to determine employee exposures to the compound in air (16). The method has been reported to be reliable. Its application has also facilitated planning for engineering controls. Plans are underway to employ the method developed for HCCP in personal monitoring and perimeter sampling during the cleanup of waste disposal sites (17, 18). In this endeavor, the method will also be employed to determine other volatile chlorinated compounds that are likely to be present including tetrachloroethylene, trichloroethylene, hexachlorobenzene, hexachloroethane, and HCBD.

Acknowledgments

    Ms. Debra Y. Harton, Assistant Chemist, assisted in the
laboratory work. Overall supervision of the project was the
responsibility of Dr. William J. Barrett, Director, Applied
Sciences Research, and Dr. Herbert C. Miller, Head, Analytical
and Physical Chemistry Division. Other personnel of Southern
Research Institute provided valuable advice. These include
Ms. Ruby H. James, Head, Environmental Analytical Chemistry Section; Dr. Thomas P. Johnston, Head, Pharmaceutical Chemistry Division; and Dr. Edward B. Dismukes, Senior Research Adviser. This
work was conducted under contract with NIOSH (210-78-0012);
Dr. Robert H. Hill, Jr., Project Officer, Mr. Robert A. Glaser,
Project Officer, and Dr. Alexander W. Teass, Head, Organic Methods
Section of the Measurements Research Branch, provided effective
guidance and encouragement.

Literature Cited

1.   Taylor, D. G. "NIOSH Manual of Analytical Methods", Vol. 1,
     National Institute for Occupational Safety and Health,
     Cincinnati, OH, 1979, p v.

2.   Taylor, D. G.; Kupel, R. E.; Bryant, J. M. "Documentation of
     the NIOSH Validation Tests", National Institute for Occu-
     pational Safety and Health, Cincinnati, OH, 1977.

3.   Mackison, F. W.; Stricoff, R. S.; Partridge, L. J. "Pocket
     Guide to Chemical Hazards", National Institute for Occu-
     pational Safety and Health, Cincinnati, OH, 1978; pp 1-3.

4.   "Development of Air Sampling and Analytical Methods for
     Toxic Chlorinated Organic Compounds—Research Report for
     Hexachlorocyclopentadiene", Southern Research Institute,
     Birmingham, AL, NIOSH Contract No. 210-78-0012, National
     Institute for Occupational Safety and Health, Cincinnati,
     OH, 1980.

5.   "Development of Air Sampling and Analytical Methods for
     Toxic Chlorinated Organic Compounds—Research Report for
     Hexachlorobutadiene", Southern Research Institute,
     Birmingham, AL, NIOSH Contract No. 210-78-0012, National
     Institute for Occupational Safety and Health, Cincinnati,
     OH, 1980.

6.   "Development of Air Sampling and Analytical Methods for
     Toxic Chlorinated Organic Compounds—Research Report for
     1,2-Dichloropropane", Southern Research Institute,
     Birmingham, AL, NIOSH Contract No. 210-78-0012, National
     Institute for Occupational Safety and Health, Cincinnati,
     OH, 1980.

7.   "Threshold Limit Values of Airborne Contaminants", Federal
     Register, 1979, 44, 8855.

8.   Kelley, W. D. National Safety News, October 1979, 83.

9.   "Development of Air Sampling and Analytical Methods for
     Toxic Chlorinated Organic Compounds", Southern Research
     Institute, Birmingham, AL, NIOSH Contract No. 210-78-0012,
     National Institute for Occupational Safety and Health,
     Cincinnati, OH, 1978.

10. Krejci, M.; Dressler, M. Chromatog. Rev., 1970, 13, 1-59.

11. McBee, E. T.; Haas, H. B.; Chao, T. H.; Welch, Z. D.; Thomas,
     L. E. Ind. Eng. Chem., 1941, 33, 176.

12. Standen, A. "Kirk-Othmer Encyclopedia of Chemical Tech-
     nology", Vol. 9, Interscience:New York, 1966, pp 743-750.

13. Aldrich Chemical Company, personal communication, 1980.

14. Lillian, D.; Bir Singh, H. Anal. Chem., 1974, 46, 1060-3.

15. Taylor, D. G. "NIOSH Manual of Analytical Methods", Vol. 5,
     National Institute for Occupational Safety and Health,
     Cincinnati, OH, 1979, Methods Nos. 307 and 308.

16. Nagle, G. Velsicol Chemical Corporation, personal communica-
     tion, 1980.

17. D'Appolonia, K. D'Appolonia Consulting Engineers, Inc.,
     personal communication, 1980.

18. Eimutis, E. Monsanto Research Corporation, personal communi-
     cation, 1980.

RECEIVED September 19, 1980.

Measurement, Analysis, and Control of Cotton Dust

JOSEPH G. MONTALVO, JR., DEVRON P. THIBODEAUX, and ALBERT BARIL, JR.

Southern Regional Research Center, New Orleans, LA 70179

The OSHA standard on occupational exposure to cotton dust defines "cotton dust" as "the dust present during the handling or processing of cotton, which may contain a mixture of substances including ground-up plant matter, fiber, bacteria, fungi, soil, pesticides, non-cotton plant matter and other contaminants which may have accumulated during the growing harvesting, and subsequent processing or storage periods" (1).

The standard presents OSHA's determination that exposure to cotton dust presents a significant health hazard to employees and establishes permissible exposure limits for selected processes in the cotton industry and for non-textile industries where there is exposure to cotton dust. The cotton dust standard also provides for employee exposure monitoring, engineering controls and work practices, respirators, employee training, medical surveillance, signs and record keeping.

In order to provide a healthier, safer work environment for cotton textile workers, the Cotton Textile Processing Laboratory at the Southern Regional Research Center, New Orleans, LA, is conducting research on agricultural particulates, including the measurement, analysis and control of cotton dust generated in the handling and processing of cotton. This paper presents a review of research that includes determining the effectiveness of the standard vertical elutriator for measuring airborne cotton dust, an impaction precutter dust sampler, and in situ cotton particulate assays. Mathematical modeling includes predicting the internal flow patterns and elutriator errors associated with magnitude and direction of ambient air currents, and property relationships and errors associated with establishing particulate burden in cotton particulate reference materials. Research and development of devices for removal and capture of dust include a wet wall electroinertial precipitator and a fluid electrode precipitator.

Measurement and Analysis of Airborne Cotton Dust

The current criteria for limiting airborne dust concentrations in textile mills are based on the epidemiological studies of Merchant et al. (2). The sampling device used for studying the dose-response characteristics of textile mill workers for cotton dust in that study was the Lumsden-Lynch vertical elutriator (VE) as described by Lynch (3). The VE was selected as the device for measuring airborne cotton fibers because of the widely-held belief that byssinosis is associated with the non-lint, respirable fraction of cotton dust in the air of cotton textile mills. NIOSH (4) ultimately recommended the VE because it is designed to collect cotton dust in an upward laminar flow of air with an average velocity equal to the settling velocity of a unit density 15 μm diameter sphere. The cotton dust standards promulgated by OSHA for cotton dust concentration in the textile industry are as follows (1):

1. 200 μg/m^3 or less in yarn manufacturing
2. 750 μg/m^3 or less in slashing and weaving
3. 500 μg/m^3 or less elsewhere in the cotton industry.

Since its adoption as a cotton dust sampler, the VE has been recognized as a device which sampled other than merely lint-free respirable particles. Bethea and Morey (5) reported that the VE operating under standard conditions (7.4 Lpm) did collect a portion of lint. Several other researchers including Neefus (6), Matlock and Parnell (7), and Claassen (8) have also reported collections of lint and other particulates greater than 15 μm diameter.

All of these problems have led researchers at the SRRC to concentrate on three areas of airborne cotton dust measurements. These include: a) analytical (theoretical) modeling of the performance characteristics of the VE; b) experimental measurements of the VE's aerodynamic flow characteristics; and c) development of an alternate air sampling method capable of accurately measuring lint-free, airborne, respirable cotton dust.

Analytical Models. A realistic semi-empirical model (coded VELUT) of the isokinetic sampling efficiency of the VE has been developed at SRRC by Robert (9). Robert deduced that the air flow in the VE is not laminar as assumed by its originators, but is characterized by flow separation at its inlet accompanied by the shedding of turbulent vortices, reverse flow, and local upward flow velocities significantly larger than the average velocity calculated assuming laminar or parabolic flow. The model calculates the sampling efficiency (E) of the VE for collecting airborne cotton dust as a function of the size distribution characteristics of the particles.

The modeling algorithm is based upon the assumptions of two physical processes acting in series to capture dust. These are an aerodynamic separation of larger particles based on elutriation (settling) in the lower VE chamber and the capture of the particles upon a semi-permeable membrane filter.

For a given particle diameter (s) the efficiency, $\varepsilon(s)$ can be expressed in this model as:

$$\varepsilon(s) = p_a \cdot r_f = p_a \cdot (1 - p_f) \qquad (1)$$

where p_a and p_f are the penetration probabilities of the aerodynamic collector and the filter, respectively, and where r_f is the retention probability of the fiber. The average efficiency E may be computed by integrating $\varepsilon(s)$ as follows:

$$E = \frac{1}{C} \int_0^\infty \varepsilon(s) \cdot F(s,MMAD,GSD,C) \cdot ds \qquad (2)$$

where the function F is the log-normal distribution function, MMAD the mass median aerodynamic diameter, GSD the geometric standard deviation, and C is the true total mass concentration in sampled air given by

$$C = M/V = m/E \cdot V \qquad (3)$$

where m is the measured and M the actual mass of dust contained in volume V. The evaluation of Eq. 2 requires either knowledge of or assumption of the distribution F and the evalaution of $\varepsilon(s)$ as a function of p_a and r_f. The determination of p_a is made difficult because analysis of the continuum mechanics involved indicates that the air will separate from the walls of the VE close to its inlet as shown in Figure 1. This is accompanied by regions of backflow and recirculation within the sampler, which reduces the effective diameter of the VE and concurrently increases its centerline velocity. This phenomenon will certainly promote the transport of particles larger than the nominal 15 μm cut-off to be captured by the filter.

In order to test the validity of the VELUT model it was used to generate predicted isokinetic values of VE differential sampling efficiency for values of MMAD and GSD corresponding to experimental values reported in the literature. A comparison of the experimental data obtained by Carson and Lynch (10) using mono-disperse aerosols of dioctylphthalate, with the predictions of VELUT for the same conditions is given in Table I. This limited data is in excellent agreement with the predictions of VELUT.

TABLE I

Comparison of Experimental and Theoretical Values
of Isokinetic Sampling Efficiency

MMAD (μm)	GSD	Efficiency	
		Experiment Carson and Lynch(10)	Best-Fit Theory Program VELUT
6.7	1.13	0.95 ± 0.08	0.923
15.4	1.05	0.72 ± 0.07	0.653
19.2	1.10	0.49 ± 0.035	0.507
24.5	1.16	0.34 ± 0.05	0.320

The model has been used to predict the sampling efficiency
of the VE for a wide range of MMAD and GSD values typical of
what might be encountered in cotton textile processing. These
parameter values are for the actual size distribution of particles in the sampled air, and not for those collected on the
membrane filter. These results are summarized in Table II. A
remarkable feature of this model is that it predicts that the VE
will collect significant amounts of particles with aerodynamic
diameters greater than 30 μm.

TABLE II

Absolute Values of Isokinetic Sampling Efficiency for Various
Cotton Dust Distributions

GSD	MMAD (Micrometers)									
	05	1	3	6	10	15	20	25	40	60
1.5	0.82	0.94	0.98	0.92	0.80	0.62	0.46	0.32	0.10	0.02
2.0	0.81	0.91	0.95	0.88	0.74	0.58	0.46	0.36	0.18	0.08
2.5	0.79	0.89	0.92	0.83	0.70	0.56	0.46	0.38	0.22	0.12
3.0	0.78	0.87	0.89	0.80	0.67	0.55	0.46	0.39	0.25	0.16
5.0	0.74	0.80	0.78	0.70	0.61	0.53	0.46	0.41	0.31	0.23
10.0	0.69	0.70	0.67	0.61	0.55	0.49	0.45	0.42	0.35	0.29

Most recently the VELUT model has been expanded to handle
the problem of simulating the particle collection efficiency of
a VE sampling dust-laden air under non-isokinetic conditions
(11). These cases include sampling from quiescent air as well
as from air moving with both horizontal (crossflow) and vertical

(upflow or downflow) velocity components with respect to the VE axis. Robert's conclusions from this study were that unless the conditions of local airflow in the vicinity of the VE inlet are carefully monitored, the experimental results obtained can not be considered to have an accuracy of better than a factor of two relative to the true dust concentrations.

   Experimental Measurements. Besides theoretical modeling of the performance of the VE as a cotton dust sampler, the SRRC has also conducted experimental research on the device. Claassen (8) has carried out extensive studies of the airflow character- istics of the VE and, in general, his results are in qualitative agreement with the predictions of the VELUT model. Observations of a vapor cloud as it entered the transparent elutriator revealed that the airflow separated from the VE wall within about 2.5 centimeters of the entrance, confirming the predictions from VELUT. As the flow proceeded upward through the VE, turbulent vortices of the order of one-inch diameter were formed in the lower cone and propagated upward. Also in verification of the prediction of VELUT, regions of recirculation and backflow were identified in the vicinity of the walls with no visible damping of the turbulence in the straight section.
   The second experimental technique involved the introduction of flow-tracing smoke injected radially through small holes drilled into an axially inserted smoke injector tube. Studies of multiple exposure photographs lead to several conclusions concerning the flow characteristics of the VE. Included among these are the following: a) large variations in both the magnitude and direction of the velocity was observed with strong indications of downflow at locations near the centerline; b) definite indications of non-axisymmetric flow conditions with upflow at some radial locations and downflow at other radial positions; c) evidence of a high velocity jet of air along the centerline of the VE; and d) observations of high radial diffusion of the cloud at the higher portions of the VE in the vicinity of the upward converging cone. These qualitative observations are consistent with the predictions of VELUT.

   Alternate Sampling Methods. The third area of SRRC's research on airborne cotton dust has concentrated on developing an alternate air sampling method for accurately measuring lint-free respirable cotton dust. The device chosen for further study was originally developed by Battelle Memorial Institute while under contract to USDA, SRRC (12). This instrument, called a "precutter," utilized the impaction of dust on a moving strip of tape as a method of discriminating against larger particles in an airstream. A sharp 15 μm aerodynamic cutoff was observed. As shown in Figure 2, the sampled air flows through a U-shaped geometry with particles larger than a fixed size being

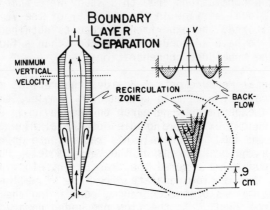

Figure 1.   Boundary layer separation in inlet section of the vertical elutriator

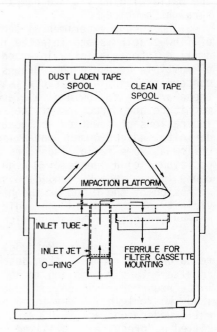

*Figure 2.   Schematic illustration of the
SRRC precutter sampler*

collected on the tape and the smaller particles being collected
by a filter mounted in a cassette attached to the sampler's
outlet.

An evaluation of the precutter was performed by a coopera-
tive research program between SRRC and the North Carolina State
University School of Textiles. The chief objective of this
research was to determine if the precutter had dust collection
characteristics which were equivalent to the VE. The approach
followed was to test a prototype of the precutter along side of
a VE in the N.C.S.U. model card room where cottons having a wide
variety of properties were being processed. Results from this
study (13) showed that when the precutter was operated under
conditions corresponding to 50% collection efficiency at
15 μm, it sampled considerably less mass of material than the VE
operated in close proximity (18 in. away). The chief
differences between the two samplers was that the VE collected
an appreciable amount of lint and lint fragments while the
precutter sampled essentially fine respirable particles. For
this reason the precutter gave readings which were of the order
of 30% to 40% less than the corresponding VE measurements. It
was found that in order to make the two samplers statistically
equivalent it was necessary to enlarge the exit jet diameter of
the precutter. Under these conditions the precutter has a
nominal 29 μm aerodynamic cutoff diameter and will allow the
collection of lint and lint fragments on the sampling filter.
These results further enforce the predictions of VELUT and raise
the question of the suitability of the VE as a lint-free air
sampler. The standard, however, is based on environmental data
obtained by a VE. At this writing, it is not known if
modification of the standard will be recommended by OSHA.

Measurement and Analysis of Trash and Dust in Cotton

Measurement and analysis of trash and dust in cotton is
important in the prediction of dust generation potential, in
relation to byssinosis (14) and rotor spinning performance (15).
By definition, trash is the nonlint particles including dust
which settle under the influence of gravity in the processing of
cotton. Dust is the finer material capable of being dispersed
in air. Particulate burden is the combined trash and dust level
in cotton.

Techniques are available to measure particulate and dust
burden in cotton. Unfortunately, the accuracy associated with
any given method is unknown. Cotton particulate and dust
reference materials, needed to asses accuracy and compatability
of measurement systems between concerned laboratories, are also
not available.

To overcome this dilemma, theory and approaches are being
developed to verify particulate and dust burdens in bulk quantities of cotton stock designated in advance as reference

materials. The cotton reference materials will, in turn, be used to assess accuracy of the state-of-the-art measurement systems, and facilitate material balances and mathematical modeling of ginning and textile studies. Particulate burden verification is discussed in the following section. Theory and operation of dust verification techniques will be reported elsewhere.

In Situ Native Standard Method. A fundamental approach to verification of particulate burden in cotton reference materials is under evaluation (16) based on a null hypothesis. The hypothesis states that upon rendering a cotton free of foreign material, the recoverable particulates-lint ith property constant λ_i (for example, color) of the synthesized mixture is equal to that for the in situ particulate constant, ψ_i. The experimental scheme to test the hypothesis is as follows.

Cotton is mechanically cleaned by cycling the lint through a mechanical cleaner such as a Shirley analyzer (17) or SRRC nonlint tester (18). Particulates recoverable from cleaning are added back to the cleaned lint and the ith property measured to facilitate computation of λ_i. Cleaned lint spiked with recoverable particulates constitutes a native standard. Assuming that $\lambda_i = \psi_i$, an apparent particulate concentration in the stock material is computed by measuring the in situ property value. The spiking process is repeated, and different properties measured. Means of group apparent in situ particulate levels are compared to decide which, if any, differ because of rejection of the null hypothesis. Particulate burden in the reference material is the average of the accepted group means. The method is demonstrated on ginned cotton.

Two equations were derived to predict $\tau_i^\%$, the in situ particulate burden error anticipated for a specific ith property measurement under alternative hypothesis conditions, $\lambda_i \neq \psi_i$. This error or bias may be expressed as a function of

$$\tau_i^\% = \left[\frac{\lambda_i}{\psi_i} - 1 \right] 10^2 \tag{4}$$

or
$$\tau_i^\% = \left[100 - \%_{nt} \right] \cdot \left[\frac{\lambda_i}{\theta_i} - 1 \right] \tag{5}$$

λ_i and ψ_i, Equation 4 or as a function of $\%_{nt}$, λ_i, and θ_i, Equation 5 where

$\%_{nt}$ = the percent nonlint trash (relative to the in situ particulate burden) recovered from mechanical cleaning for spiking purposes and

θ_i = other particulates ith property constant (proportionality constant between in situ nonrecoverable particulates and the ith property value)

The two error equations quantify the model predictions, Figure 3, where

τ_i = _in situ_ particulate burden based on the ith property measurement and

ε_i = ith property value for the uncleaned stock cotton

Table III gives predicted τ_i error for operational limiting $\%_{nt}$ and λ_i/θ_i values. The projected upper limit of recoverable nonlint trash for spiking purposes is about 70-90% of the _in situ_ particulate burden. Assuming then that $\%_{nt}$ = 90% and

$$\theta_i = \lambda_i \pm 0.5\,\lambda_i \tag{6}$$

then λ_i/θ_i varies within the arbitrarily chosen limits

$$0.67 \leq \lambda_i/\theta_i \leq 2.0 \tag{7}$$

so that $\tau_i^\%$ will be within the range -3.33 to 10.0%. For a large number of mutually independent properties measured the projected average error would be 3.34%.

TABLE III

Dependence of $\tau_i^\%$ on Operational Limiting λ_i/θ_i and $\%_{nt}$ Values

λ_i	θ_i	λ_i/θ_i	$\%_{nt}$	ψ_i	$\tau_i^\%$
1.000	1.750	0.571	90.0	1.044	-4.29
			80.0	1.093	-8.58
			70.0	1.147	-12.87
1.000	1.500	0.667	90.0	1.034	-3.33
			80.0	1.071	-6.67
			70.0	1.111	-9.99
1.000	1.000	1.000	variable	1.000	0
1.000	0.500	2.000	90.0	0.909	+10.0
			80.0	0.833	+20.0
			70.0	0.769	+30.0
1.000	0.250	4.000	90.0	0.769	+30.0
			80.0	0.625	+60.0
			70.0	0.526	+90.0

Deviation from λ_i/θ_i = 1 may be due to changes in the relative amounts of leaf, bract, stem, etc. found in the recoverable trash as opposed to the nonrecoverable particulates. It is doubtful that the spread of λ_i and θ_i values would approach the maximum range suggested by Equations 6 and 7 unless $\%_{nt}$ is quite small.

Predicted τ_i errors may be computed from Equations 4 or 5 or the graphic analysis presented below. All three methods give equivalent error values, which lends credence to the derivations. Figure 4 shows a graphic analysis of the ith property values under the constraints indicated. Curve A

$$\lambda_i = \theta_i = \psi_i = 1.000$$

would result in a 0% error for $\tau_i^\%$, regardless of percent nonlint trash recoverable for spiking. Thus, $\Delta_i = \tau_i$ at point (a), see also the model in Figure 3. For $\lambda_i/\theta_i = 2$, curves B and C, point (c) on the curve corresponds to $\Delta_i > \tau_i$. The error in $\tau_i^\%$ is equal to the line segment ac. Note that curve C, slope = $\psi_i = 0.090$, is the true _in situ_ slope for $\lambda_i/\theta_i = 2$ and $\%_{nt} = 90$. For $\lambda_i/\theta_i = 0.67$, curves D and E, point (e) on the curve corresponds to $\Delta_i < \tau_i$. The error is equal to the line segment ae. The actual _in situ_ slope would be $\psi_i = 1.034$.

To demonstrate feasibility, the _in situ_ native standard method is illustrated on a ginned cotton by measuring three mutually independent properties: color, conductance, and pH. Color was measured on the solid samples by reflectance spectroscopy, conductance and pH on aqueous suspensions. An example of the calculation protocol based on color measurements is shown in Table IV. $X_{j=1}$ through $X_{j=3}$ denote $j = 3$ replications of meter readings on each sample. $\overline{X}_{color}$ is the average value for the j replications. Nonlint trash added values are regressed on $\overline{X}_{color}$. $A\tau_{color}$ is the apparent particulate value based on color measurement.

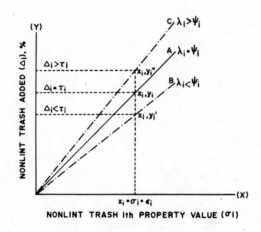

Figure 3.   Model of the in situ native standard method

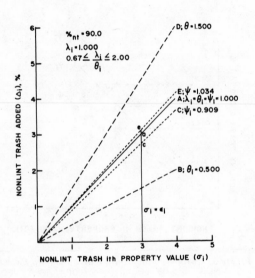

Figure 4.   Graphic analysis of the ith property constants

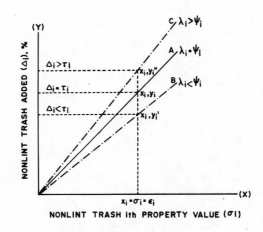

Figure 3.   Model of the in situ native standard method

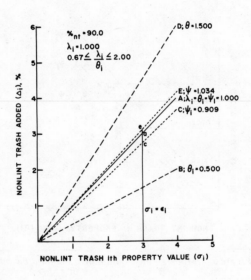

Figure 4.   Graphic analysis of the ith property constants

TABLE IV

Apparent In Situ Particulate Burden Based on Colorimetric Measurement

Nonlint Trash(nt) Added, Percent Y	Meter Reading			
	$X_{j=1}$	$X_{j=2}$	$X_{j=3}$	$\overline{X}_{color}$
1.06	62.9	63.3	63.9	63.4
2.06	58.7	60.0	59.6	59.4
3.00	51.0	51.7	52.2	51.6
4.00	46.2	47.0	47.3	46.8
5.00	43.2	44.1	44.0	43.8
Stock	50.0	50.2	49.9	50.0

$$\Delta_{color} = \lambda_{color} \cdot \overline{\sigma}_{color} + \beta_{color}$$

correlation coefficient, $r = -0.988$: slope, $\lambda = -0.1853$; intercept, $\beta = 12.841$; $\overline{\epsilon} = 50.0$; assume $\lambda = \psi$, evaluate $A\tau_{color}$ at $\overline{\sigma} = \overline{\epsilon}$

$$A\tau_{color} = \psi_{color} \cdot \overline{\epsilon}_{color} + \beta_{color}$$

$$A\tau_{color} = (-.1853) \cdot (50) + 12.84 = 3.58\%$$

Results in Table V are tabulated for the three properties studied. One hundred eighty-four property measurements were performed in obtaining the data. The apparent particulate ($A\tau_i$) value calculated by measuring pH, set 4, is an obvious outlier and is reported but not included in the statistical calculations. Group mean apparent particulate ($\overline{A\tau_i}$) value ranged from a low of 3.46 to a high of 3.51%. Set standard deviation (s) ranged from 0.150 to 0.259%. Coefficient of variation (CV) ranged from 4.27 to 7.47%. Variation of $\overline{A\tau_i}$ values was in the second decimal place. The pooled $\overline{A\tau_i}$ variance was only 0.04%.

TABLE V

Results

Set Number	Percent Apparent In Situ Particulates (A^τ_i)		
	Color	Conductance	pH
1	3.58	3.50	3.39
2	3.31	3.66	3.47
3	3.21	3.33	3.68
4	3.78	3.34	(4.74)
$\overline{A^\tau}_i$	3.47	3.46	3.51
s,%	0.259	0.156	0.150
CV,%	7.47	4.50	4.27

PERCENT IN SITU PARTICULATES (τ)

Υ = 0.01 level of significance

τ = (3.47 + 3.46 + 3.51)/3 = 3.48%

Based on a level of significance, Υ, of 0.01, there is no reason to believe the calculated $\overline{A^\tau}_i$ values differ (statistical decision is based on the following degrees of freedom: total, 11; property groups, 3; and $\overline{A^\tau}_i$ set observations, 11 - 3 = 8). Therefore, the in situ particulate burden in the stock is the average of all three $\overline{A^\tau}_i$ values, 3.48% based on the available data. The data suggests the absence of a significant alternative hypothesis error, $\tau^\%_i$, in this trial run and is in excellent agreement with the model predictions. Refinement of measurement procedures to reduce variances and development of additional systems to increase degree of confidence in the null hypothesis is on-going.

Control of Cotton Dust

A wide variety of systems are currently being used to clean the air in textile mills processing lint cotton. Most of these are custom designed, and use proprietory equipment such as bag filters, rotary drum filters, condensers, V cells, etc. (19), but many systems have evolved as a result of needed improvements. The better systems have several stages, including a waste separator that can reduce dust content of the air by some form of filtration, and a fine dust separator capable of removing respirable dust (less than 15 μm). The effective control and removal of dust in this size range is quite difficult and

expensive. Presently, there are few air handling systems in the
textile industry capable of economically and efficiently collecting and removing respirable dust. For the past several
years, our laboratory has sponsored and carried out research on
the systems outlined below.

Wet Wall Electroinertial Precipitator. The wet wall
electroinertial precipitator (WWEP) air cleaner was developed to
meet the requirements of high operating efficiency, low maintenance, low operating cost, and simplicity of design. To
achieve high efficiency with low pressure loss, an electrostatic
precipitation was the best candidate method. A wet wall was
desirable to flush the precipitate away and to minimize maintenance. Inertial effects were added to assist in moving
material toward the wall and to improve the cleaning action of
the water on the wall (20).

These features were best incorporated in the design shown
in Figure 5. The unit consists of a concentric wire-in-tube
precipitator. The charging wire is located axially in a
stainless steel tube in a vertical orientation. Air enters
tangentially at the upper end of the tube (through a duct with a
flattened cross section) and acquires a rotating movement. When
high voltage is applied to the wire, an ionizing coronal discharge charges the dust in the air. As the air flows through
the tube, the charged dust is driven to the wall by the radial
electrical field. Water from the upper water inlet flows down
the wall and flushes the precipitated dust into the water outlet
at the lower end of the tube. The rotational movement of the
air also induces rotational flow of the water, which assists in
uniformly wetting the surface of the tube. Clean air is
expelled from the tube's lower end.

Tests of the WWEP were conducted with 4 in, 8 in and 16 in
diameter units at flow rates ranging from 100 to 4000 ft^3/min.
Particle collection efficiency of these units was measured with
atmospheric dust, AC fine test dust (mass mean diameter (MMD) 12
μm), artificial cotton dust (MMD = 4.0 μm), and cotton dust
drawn from the processing area of a card (MMD = 3.0 μm).

Tests conducted with the 4 in diameter WWEP provided guidelines to determine the effects of the tube length, air flow
rate, and voltage polarity on efficiency of particle collection.
Tube lengths of 1, 2, and 4 ft were evaluated. Shearing of
water droplets from the wall at higher velocities made the
practical upper limit on flow rate about 200 ft^3/min. Results
of the efficiency tests with the 2 ft unit are included in Table
VI (water flow at 0.25 gal/min). Runs made at zero potential
and zero current with the rather course AC dust show
efficiencies in the 80% range due merely to centrifugal forces.
Increasing the voltage to -30 kV raises the efficiency to 99%,
which diminishes to 98.5% when +33 kV is applied. Tests with

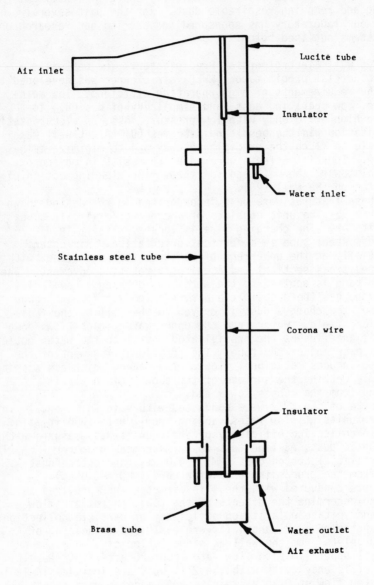

Figure 5. Wet wall electroinertial precipitator

the fine artificial cotton dust indicate a somewhat lower
efficiency (98.1 and 96.1% respectively) for 100 and 200
ft^3/min.

TABLE VI

Total Mass Efficiency of the 4 in Wet
Wall Electroinertial Precipitator

Test Dust	Circulation Rate (ft^3/min)	Potential (kV)	Current (mA)	Dustfeed Rate (g/min)	Pressure Drop (in H_2O)	Efficiency (%)
AC fine	100	0	0	1.0	1.25	81.2
AC fine	200	0	0	2.0	3.75	87.6
AC fine	100	-30	2.4	0.2	1.25	99.0
AC fine	200	-30	2.4	0.2	1.25	99.0
AC fine	100	+33	---	1.0	1.25	98.5
Cotton dust	100	-30	2.2	0.83	1.25	98.1
Card trash	200	-30	2.2	0.67	3.75	99.7
Cotton dust	200	-30	2.1	0.62	3.75	96.1

Under a memorandum of understanding relative to the
research and development of the wet wall electroinertial
precipitator as a commercial product, a 16 in diameter, 60 in
long unit capable of handling up to 4000 CFM was installed at
SRRC for test and evaluation. Runs made at zero potential with
dust from operating cards, prefiltered by a V-cell, showed
efficiencies in the range of 70-80% due to the centrifugal
forces. Applying a potential of 100 kV raised the efficiency up
to 95%, with an air flow ranging from 2900 to 3700 CFM. Under
another agreement for the commercialization of the precipitator,
a 23 in diameter, 72" long unit was developed capable of
handling up to 12,000 CFM. This unit was powered by a pulsed
power supply to meet allowed ozone levels. Several of these
models were built and are currently undergoing evaluation. It
is anticipated that these units will be commercially available
by mid 1980.

Fluid Electrode Precipitator. In general, the best way to
remove fine dust from an air stream is to use conventional
electrostatic precipitators. Although the efficiencies of these
devices are quite high, there are inherent drawbacks that do not
make them acceptable in continuously operating air systems. The
collection plates eventually become saturated with the deposited

dust and the unit has to be cleaned. Most of the cleaning methods require that the unit be shut down in order for the cleaning cycle to occur.

The development of the fluid electrode precipitator is an attempt to alleviate this problem (21). The fluid electrode precipitator illustrated in Figure 6 consists of an array of electrodes in a casing to charge, direct, and collect fine dust. Tungsten wires serve as discharge electrodes for all units tested, auxiliary driving electrodes are charged metal plates or screens that can be inserted between the fluid electrodes, and the collection electrodes are formed by vertical tubes that release a laminar flow of water falling through the height of the casing. The precipitator operation is quite simple. The falling columns of grounded water act as cylinders in cross flow and create vorticies for enhanced particle collection. Auxiliary electrodes and discharge electrodes can be positioned in arrays to direct charged dust into the grounded fluid.

Two series of tests, one at low capacity and one at high, were performed with cotton dust from different sources. For low capacity tests, air was drawn through a model-size cotton carding machine running at 2 lb/hr. The air was prefiltered to remove lint and large particles and was then passed through the test tunnel (9 in x 9 in straight tunnel 9 ft long containing traversable sampling probes located upstream or downstream of the precipitator test section). In the high capacity tests, fine dust released by mechanically tapping a V-cell nonwoven filter was used. The filter was previously loaded with lint and dust from two production cotton carding machines. The air drawn through the filter and passed through the test tunnel contained fine dust that was similar to the prefiltered air from the model card.

Samples of dust from each of the two dust sources were collected for testing. They were dispersed in an electrolyte and analyzed for size on a volume basis. Model card dust had a mass mean diameter of 3.6 μm with a geometric standard deviation of 1.6. The cotton dust released by tapping the loaded filter varied from 4.5 to 6.7 μm mass mean diameter with a geometric standard deviation of about 2. All test concentrations were in the range of 0.5 to 1.5 mg/m^3, which are typical of cotton mill fine dust concentration.

Table VII contains the operational data and collection efficiency for a typical configuration tested. Collection efficiency is expressed in terms of particle and size efficiency.

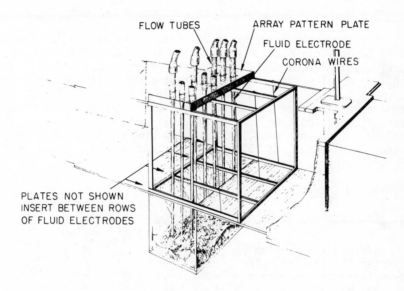

Figure 6. Fluid electrode precipitator

TABLE VII
Efficiency of Typical Fluid Electrode
Precipitator Configuration

Velocity Efficiency (ft/min)	Field Strength (kV/cm)	Dust Supply Concentration (mg/m3)	Collection Efficiency Particle Size (1.5 μm- 10 μm) (%)	Mass (%)
250	6.2	0.8	75	86
350	5.9	0.5	81	93
500	4.9	0.8	60	78

Once properly adjusted an array of fluid electrodes (up to 45 in length) can operate with little maintenance to continuously remove dust. Dust entrained in the fluid can continue to circulate through the system until its accumulation warrants a fluid cleaning cycle. Fluid cleaning would not require percipitator shutdown. Applied to textile mills, these precipitators could be incorporated via parallel or series configurations into existing duct work where continuous fine cotton dust removal of 85% or better would greatly reduce levels of fine airborne dust.

Literature Cited

1. Occupational Health and Safety Administration, U.S. Department of Labor: "Occupational Exposure to Cotton Dust, Final Mandatory Occuaptional Safety and Health Standards." Federal Register, Part III, June 23, 1978, 27350-27463.
2. Merchant, J. A.; Lumsden, J. C.; Kilburn, K. H.; O'Fallon, W. M.; Ujda, J. R.; Germino, V. H.; Hamilton, J. D. J. Occup. Med., 1973, 15, 22-230.
3. Lynch, J. R. Trans. Nat. Conf. Cotton Dust Health, Charlotte, N.C., 1970, 33-43.
4. National Institute for Occupational Safety and Health. U.S. Dept. Health Educ. Welf. Pub. 75-118, 1974, 117.
5. Bethea, R. M.; Morey, P. R. Am. Conf. Gov. Ind. Hyg. Nat. Cotton Dust Symp. 1974, Atlanta, Ga., 285-327.
6. Neefus, J. D. Am. Ind. Hyg. Assoc. J., 1975, 37, 475.
7. Matlock, S. W.; Parnell, C. B. Am. Soc. Mech. Engrs. 76-Tex-10, 1976.
8. Claassen, B. J., Jr. Am. Ind. Hyg. Assoc. J., 1979, 40, 993-941.
9. Robert, K. Q., Jr. Am. Ind. Hyg. Assoc. J., 1979, 40, 535-542.

10. Carson, G. A.; Lynch, J. R. "Calibration of the Vertical Elutriator Cotton Dust Sampler." Presented at the Annual Meeting, A.I.H.A., Boston, Mass. 1973. (Also: personal communication to K. Q. Robert).
11. Robert, K. Q., Jr. Relationship between Crossflow, Isokinetic, and Calm-Air Sampling with a Vertical Elutriator, Proc. 1980 Beltwide Cotton Prod. Res. Conf. (Special Session on Cotton Dust), 86-96.
12. Reif, R. B.; Albrechtson, L. R.; Neville, F. E.; Thompson, W. B.; Hanks, C. L.; McCrady, P. E.; Gieseke, J. A.; Schmidt, E. W.; Miga, L. W. Final Report. 1977, Battelle Memorial Institute, Contract No. 12-14-7001-365.
13. Batra, S. K., Shang, P. P.; Hersh, S. P.; Robert, K. Q., Jr. Proc. 1980 Beltwide Cotton Prod. Res. Conf. (Special Session on Cotton Dust) 97-102 .
14. Ayer, H. E. and Kilburn, K. H, CRC Crit. Rev. Environ. Control, 1971, 2, 207-241.
15. Langley, E. D., Text. Res. J., 1979, 49, 455-457.
16. Montalvo, J. G., Jr.; DeLuca, L. B.; Segal, L.; Proc. 1980 Beltwide Cotton Prod. Res. Conf. (Special Session on Cotton Dust) 70-71.
17. American Society for Testing and Materials, Annu. Book ASTM Stand. 1977, Part 33, 576-583.
18. Rusca, R. A.; Little, H. W.; Gray, W. H., Text. Res. J., 1964, 34, 61-68.
19. Barr, H. S.; Hocutt, R. H.; Smith, J. B., U.S. Dept. Health Educ. Welf. Rep. HSM 99-72-44, March 1974.
20. Thibodeaux, D. P.; Baril, A., Jr.; Reif, R., IEEE Trans. Ind. Applic. January/February 1980, Vol 1A-16, No. 1, 80-86.
21. Claassen, B. J., Jr., Conf. Proc. 70th Annu. Meet. Air Pollut. Control Assoc. Toronto, June 20-24, 1977, 1-15.

RECEIVED October 27, 1980.

Estimation of Airborne Sodium Hydroxide

E. REID

Wolfson Bioanalytical Unit, Robens Institute of Industrial and Environmental
Health and Safety, University of Surrey, Guildford GU2 5XH, United Kingdom

The NIOSH method for determining sodium hydroxide in air (1)
entails collection by bubbler into standard hydrochloric acid, and
electrometric titration of the remaining acid with standard sodium
hydroxide solution. Even with the stipulated long collection time,
the method has to rely on establishing a titration difference which
is rather small and which would be vitiated by any inadvertent
spillage during or after collection. The sensitive method now advocated entails trapping of the alkali aerosol into boric acid, in
a quantity which need not be exact, and final colorimetric titration with standard acid. The nature of the two methods is evident
from Figure 1.

The indicator (2) is present at the outset; hence arrival of
traces of alkaline aerosol in the collecting liquid is signalled
by a color change. The chosen indicator in conjunction with boric
acid as trapping agent has long been in use for determining the
ammonia formed in micro-Kjeldahl digestions, more simply than if
standard mineral acid is the trapping agent. This use of boric
acid dates back to 1913 (3).

The method description now given is followed by a comparison
with the NIOSH method, mainly on a simulated basis entailing
spiking with alkali rather than on an aerosol-collection basis such
as ought to be included in hoped-for verification in other laboratories. Late in the study, a draft (lacking the Figures) of the
still unpublished NIOSH validation by D.V. Sweet (4) was kindly
furnished by Dr. G. Choudhary; it is reassuring in respects such
as the air-collection mode and rate (bubbler) in the method now
advocated, and is candid about problems with the NIOSH method.
Neither the latter nor the present method are specific for NaOH as
distinct from alkali in general.

Proposed Method

Materials and Apparatus. In general the chemicals are of
analytical reagent quality. Water is distilled from glass, with
no precautions to exclude CO_2 thereafter.

0097-6156/81/0149-0087$05.00/0

Boric acid: this is of 1% w/v strength (or 2% if preferred for the sake of very high trapping capacity).

Bromcresol green, 0.1%, and methyl red, 0.01%, each in 96% ethanol: mix in a volume ratio of 1 : 2. [A stronger methyl red stock solution (2) is difficult to prepare.] The working solution, which is stable indefinitely, is made by adding 12 ml of the mixed indicator per 100 ml of the boric acid. Only exceptionally need adjustment to the desired reddish-purple color be performed, by adding drops of acid or alkali.

Standard acid for the titration: 3.75 mM sulfuric acid (hydrochloric acid of twice this molarity would be a suitable alternative), which can be made by dilution from exactly 1 M sulfuric acid as furnished by an ampule obtained from a reliable commercial source (e.g. BDH Chemicals Ltd., Poole, Dorset, U.K.).

Standard NaOH, as used for spiking in method testing, appropriately 0.50 M: this is made up from NaOH pellets and standardized electrometrically against standard acid.

Colorimetric titrations are done with a 10 ml buret of conventional type.

A suitable trap for aerosol collection is the U.S.-made portable Midget Impinger (Model PMI-A), made of polyethylene, as supplied in the U.K. by MDA Scientific (UK) Ltd. (Wimborne, Dorset). It can be used repeatedly as long as the fabric disperser remains intact, provided that our recommended soaking in methanol after use is successful in substantially removing any indicator that has become adsorbed onto the fabric. A minimum volume of 11 ml of boric acid is advisable.

Procedure. Place approximately 12 ml of the boric-acid/indicator solution in the trap. With a pump which can be of a personally worn type, and must allow of volume assessment, draw air through at a rate which can suitably be 0.8 L/min, but should not exceed 1.0 L/min lest the bubbling be rather violent. For method-checking, a second trap in series should be tried in an aerosol-collection exercise, and a single-trap blank should be done with clean air: in neither case should any color change occur.

Collection should be for at least 1 h, or say 10 times as long as is needed for the color to change from reddish-purple to blue, which occurs after about 30 μg of NaOH has been collected. Collection may have to be for as long as 3 h if the concentration of NaOH is well below the 'Threshold Limit Value'/'Time Weighted Average' (TLV-TWA), viz. 2 mg/m^3 of air.

Finally tip the contents into a titration vessel, and add water rinsings (2-4 ml). Titrate with the standard acid back to the initial color. Use standard alkali to correct any inadvertent overshoot as indicated by a red color.

Calculation: 1 ml of 3.75 mM sulfuric acid corresponds to 30 μg of NaOH.

Validation by Spiking to Simulate Aerosol Collection

In comparisons of the NIOSH approach with the present approach, different amounts of standard NaOH were spiked into boric acid in an Erlenmeyer flask, or into 0.075M HCl in a tube (130 × 25 mm; immaterial whether flat-bottomed or round-bottomed) which was then sealed with 'Parafilm'. Electrometric titrations in the NIOSH method were done by pH meter, essentially as in the NIOSH description (1), and the titrant was added by means of a syringe-type microburet ('Agla'; Wellcome Foundation, Beckenham, Kent, U.K.), mounted so as to discharge, just above the solution in the near-sealed tube, via fine-bore polypropylene tubing; a stream of nitrogen was conducted via similar tubing into the vessel, and magnetic stirring was performed as well as hand-swirling.

Comparative results. Figure 2 in conjunction with its legend documents the excellent accuracy and reproducibility obtained over a wide spiking range, even well below the high spike quantities that have to be used with the NIOSH method in the interests of an accurate titration difference. The latter was only 0.036 ml (0.425 ml subtracted from 0.461 ml) with the 0.49 mg spike.

Illustrative Titrations. Figure 3(a) shows a representative electrometric titration as laboriously performed with the NIOSH approach to obtain values for Figure 2. Electrometric checking of the simple colorimetric titrations performed with the boric acid approach is not worthy of illustration. However, an electrometric titration with alkali of the boric acid as put into the trap, shown in Figure 3(b), is informative in showing the lack of any marked inflexion even up to pH 10, and hence the high capacity of the trapping liquid (at least 40 mg of NaOH) in the proposed method. The initial pH value, namely 4.57, is in good accord with that reported by Ma and Zuazaga (2), namely 4.52; these authors stated that the alkaline bluish color had already been attained at pH 4.90 (and, conversely, the acid red color, relevant to any overshoot of acid titrant, at pH 4.26).

Attempted Validation by Aerosol Generation and Collection

The comparisons were based on introducing, and finally estimating, amounts of NaOH which were not ascertainable at the time but which were comparable for the two methods in a given dual test where final titration showed how much had been collected.

Apparatus. Generation of aerosols containing NaOH was achieved by applying a gentle stream of nitrogen to an atomizer (as used for spraying chromatograms) with 50 ml of 20% NaOH in the reservoir, the outflow from the jet being directed into a horizontal glass chamber (approx. 1 L) with free ingress of air. From the exit the air-diluted outflow was pulled through a single trap

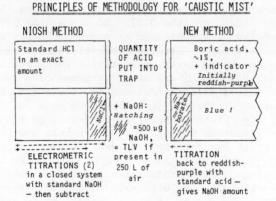

Figure 1.   Principles of the two methods for estimating airborne sodium hydroxide

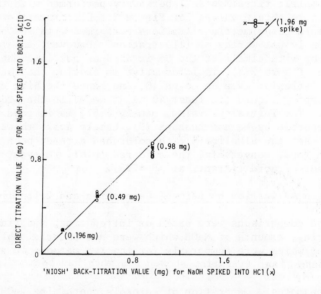

Figure 2.   NaOH values by the two methods, following spiking of the amounts shown parenthetically into boric acid or HCl. The plotting of the replicate points for the boric acid method exaggerates the variability: the coefficient of variation was only 0.2% for the 0.98-mg spikes and 0.5% for the 0.49-mg spikes (6 and 4 observations, respectively). Boric value = (1.054 × NIOSH HCl value) + 0.0024; r = 0.999.

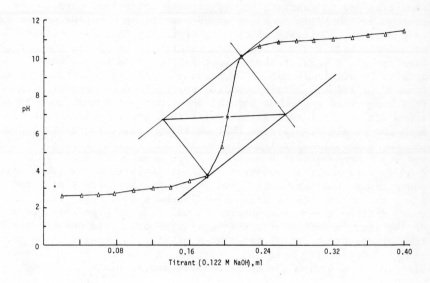

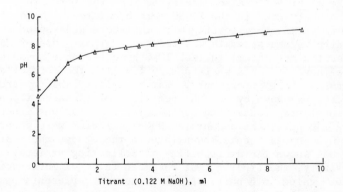

Figure 3. Electrometric titration curves showing the characteristic shapes: (a) NIOSH-type method: titration of 7.5 mL of HCl (originally 7.5mM but one-fifth neutralized for a spiking experiment) with 0.489M NaOH, by a syringe-type micro-buret (Agla); the intersection of the diagonals was taken as the neutralization point. (b) Titration of 12 mL of 1% boric acid with 0.122M NaOH; note the wide buffering range and consequent high capacity for trapping NaOH (40 mg could be coped with).

which was alternated with a partner during each test, or through a
T-piece to two parallel traps that allowed the two methods to be
tested side by side. The final unit in the system was an exhaust
pump that gave a constant but adjustable total flow rate in the
range 0.5 - 1.0 L/min as ascertained exactly by a rotameter for each
trap outflow. Delivery of NaOH sufficiently fast for the NIOSH
method was barely attainable, but with suitable positioning of the
atomizer jet the color change in the boric acid method was
attainable within 15 min. The total air volumes collected ranged
from 50 to 250 L. Notwithstanding apparent smooth functioning of
this improvised system during the 3-6 h period of each dual test,
a precaution was taken against drift that could imperil results
with the single-trap system that entailed collection serially for
the two methods: switching between the respective traps (facilitated by cone-and-socket connections) was done every 30-60 min.

   In early trials, before the Midget Impinger became available,
conventional bubblers were tried, made of glass with a sintered
frit: the design was as described elsewhere (5). Sometimes, however, difficulty was encountered in maintaining pre-test constancy
of the starting color with the sensitive boric acid method: the
bubbler was prone to release traces of alkali, possibly from the
sintered dispersing unit, as evidenced on occasion by an immediate
blue color on adding the boric acid/indicator mixture.

   Results. It was soon realized that the strength of trapping
acid normally used in the NIOSH method was too high to offer any
hope of satisfactory testing with the above generation system.
Accordingly, the strength was reduced at least three-fold. Nevertheless the net volume in the NIOSH by-difference approach could
not be increased to a satisfactory level in the aerosol as distinct
from the spiking experiments. Whilst collections appeared to be
complete, as shown by the absence of a color change if an
additional trap containing boric acid/indicator was inserted just
before the pump, the amounts of NaOH collected were generally
below 0.5 mg; as became known late in the study, the tests done at
NIOSH had shown the desirability of collecting at least 1 mg (4).
Accordingly, the finding that the mean ratio of boric-acid value
to NIOSH value in 6 pairs of tests was 1.4 represents passable
agreement, although not satisfactory validation such as a more
powerful aerosol generator could have allowed.

Concluding Comments

   The putative boric acid method, which hopefully will be tried
in laboratories properly equipped for aerosol generation, offers
notable advantages over the established NIOSH method.
(1) The collected sodium hydroxide is determined by direct titration, rather than by establishing a difference, typically small,
between two titrations. Accordingly, (a) the trapping acid need
not be of exact strength and dispensed exactly, nor assigned a

a limited shelf life (the boric acid/indicator mixture is stable
indefinitely at room temperature, as shown by inspection for color,
and does not encourage microbial growth); (b) any small spill
during collection or the final transfer of the trapping acid from
the bubbler to the titration vessel is merely regrettable rather
than analytically disastrous.
(2) The sensitivity is so good that the air sample can be
relatively small, as little as 50 L collected during 1 h.
(3) The presence of NaOH in the air is manifest by a color change
during the actual sample collection.
(4) The final titration is simple, being colorimetric with a
conventional buret rather than electrometric.

If, improbably, trapping capacity were in question, the boric
acid could be set at 2%, not 1%. It is striking that air lacking
NaOH can be passed through the trap contents for hours without
producing any change in color. In common with the NIOSH method,
the specificity is merely for alkali as distinct from NaOH; the
customary use of boric acid has in fact hitherto been for estimating ammonia in Kjeldahl distillates.

One feature of NIOSH method descriptions in general is an
elaborate exercise to prepare a primary standard. This disincentive
to setting up particular methods for occasional needs can be
obviated if, as may have to be verified, bought-in ampules of
standard acid are trustworthy.

Summary

For estimating NaOH aerosol a direct-titration method is
proposed. The trapping liquid is a boric acid/indicator mixture,
rather than standard acid in exact amount as in the NIOSH method;
finally there is a colorimetric titration with standard acid,
rather than an electrometric titration with standard alkali to give
a by-difference estimate. Although the method is, in common with
the NIOSH method, specific only for alkali rather than NaOH, it
offers notable precision, sensitivity and simplicity, and would be
advantageous for field use especially since the presence of airborne NaOH is manifested during collection by a color change. It
has agreed well with the NIOSH method in spiking tests. Complemen-
tary tests that entailed actual aerosol collection need to be
repeated in a laboratory equipped with an aerosol generator
sufficiently powerful to give a meaningful titration difference
with the NIOSH method.

Acknowledgements

Especial thanks are due to Mrs. Linda L. Basarab for
meticulous help. Dr. J.P. Leppard and Dr. I.R. Tench made useful
suggestions, and Mr. T. McDonald delivered the paper at the actual
ACS Symposium.

Literature Cited

1. NIOSH Manual of Analytical Methods, 2nd edn., Vol. 1, U.S. Department of Health, Education and Welfare (NIOSH), Cincinnati, 1977, #241.

2. Ma, T.S.; Zuazaga, G.; Ind. Eng. Chem. (Anal.), 1942, 14, 280–282.

3. Winkler, L.W.Z.; Agnew. Chem., 1913, 26, 231–232.

4. Sweet, D.V.W.; Haartz, J.C.; Hawkins, M.S. "Determination of Sodium Hydroxide Aerosol in Industrial Hygiene Samples", unpublished work cited in ref. 1: personal communication (draft Report) by D.V. Sweet via G. Choudhary.

5. Perry, R.; Young, R.J. "Air Pollution Analysis", 1977, Chapman and Hall, London, 506 pp. (vide p. 265).

RECEIVED October 14, 1980.

Development of a Method for Sampling and Analysis of Metal Fumes

W. F. GUTKNECHT, M. B. RANADE, P. M. GROHSE, and A. S. DAMLE

Research Triangle Institute, Research Triangle Park, NC 27709

P. M. P. ELLER

National Institute for Occupational Safety and Health, Cincinnati, OH 45226

Fine dusts and fumes of metallic compounds are produced as a result of material handling and thermal processing operations involving metals and ores. The mechanism of dust production is primarily by comminution and the resulting particles range from one to several micrometers in size. Thermal operations involving vaporization produce much finer fumes in the 0.01 to 1 micrometer range. These dust and fume particles remain suspended in the work-place atmosphere for long periods of time and expose workers to several health risks arising through inhalation of these particles.

Inhaled particles can deposit in the respiratory system. Larger particles (several micrometers in size) are deposited in the ciliated portion and are cleared from the respiratory system by muco-ciliary action into the gastronomical tract, but may produce systemic toxic effects by absorption in body fluids. Finer particles reach the lower non-ciliated portion of the lungs, are cleared very slowly, and are responsible for diseases such as pneumoconiosis and lung cancer. Metallic lead (Pb), tellurium (Te), selenium (Se), and platinum (Pt) are known to cause both systemic and respiratory toxicity in laboratory animals and several cases of acute and chronic poisoning among metal workers have also been documented.

A method for sampling and analysis of the metallic dusts and fumes is necessary to assess the exposure of workers to these dusts. Personal sampling devices are used to collect samples from the work-place atmosphere in a representative manner. The samples are then analyzed by convenient analytical techniques such as atomic absorption spectroscopy.

The purpose of the present study is to develop and validate a method for sampling and analysing metal fumes. The techniques under examination are several of those presently prescribed by the National Institute for Occupational Safety and Health (NIOSH); they include collection of particles of the substance from the air by filters, the acid digestion of the filters, and measurement of the residue using atomic absorption spectroscopy. The several components of the study are:

0097-6156/81/0149-0095$05.00/0

- Determination of the instrumental detection limit of atomic absorption spectrometry and also precision as a function of concentration for the four elements listed.

- Verification that presently prescribed dissolution procedures will result in greater than 90 percent recovery of the elements deposited on the filters in the form of pure metals or compounds of the metal.

- Development of modified or new dissolution procedures to replace those not providing satisfactory recovery.

- Generation and collection of metal fumes and determination of collection efficiency.

- Characterization of the generated fumes, i.e., particles.

- Analysis of the collected fumes and determination of the overall accuracy and precision of sampling and analysis for each fume.

Experimental effort has been devoted in parallel to evaluate the sampling operation as well as the analytic procedure.

The sampling operation involves collection of an aerosol sample that is representative of the particle size distribution and concentration of the sampled atmosphere. The efficiency of particle transport and collection operations are dependent on the particle size, sampling velocity, the geometry of the sampling apparatus and the properties of the collection medium. In the present work, a 37 mm diameter membrane filter (0.8 μm pore size) is the primary collection medium under evaluation. The filter is housed in a standard filter cassette and effects of filter-holder inlet geometry are also being investigated.

The dissolution and measurement experiments start with attempts to dissolve a sample (300 μg or less) of a metal or metal compound species using the prescribed NIOSH procedure, followed by measurement using the NIOSH atomic absorption spectrometric (AAS) procedure. If 90 percent recovery of the metal is not achieved, the dissolution procedure is modified, or changed completely, to achieve 90 percent recovery. From previous studies, it is expected that the metal oxides, and selenium generally, would pose problems. The NIOSH-AAS procedures are to be evaluated also, especially when the dissolution matrix is changed. The AAS detection limits using standards are determined through the measurement of blanks.

Experimental

1. Aerosol Sampling Studies. Metallic fumes can be formed by atomization of molten metals, by condensation of metal vapors, and by electric discharge (1). To realistically simulate the metal fumes expected from thermal processing and to cover worst-collection properties, a size range of 0.01 to 1 micrometers is most desirable as a target range in the selection of an appropriate generation technique. A condensation technique was selected in this study to generate fumes of Pb, Te and Se. Platinum was excluded from fume

studies due to high costs. A flow chamber was used to create a
uniform aerosol atmosphere from which samples were taken for collection efficiency studies and to produce filters loaded with
prescribed amounts of the aerosols for subsequent analytical
studies. A schematic diagram of the experimental facility is shown
in Figure 1. Details of the aerosol generation and the collection
chamber are discussed below.

(a) <u>Aerosol Generation</u>. The condensation technique for
aerosol generation has been used by several workers to produce
spherical metallic particles in sizes ranging from 0.01 to several
micrometers (<u>2</u>, <u>3</u>, <u>4</u>). The apparatus shown schematically in Figure
2 is similar to one used by Ranade, <u>et al.</u> (<u>4</u>) for generation of
$PbCl_2$ aerosols. The material to be aerosolized is contained in a
ceramic boat placed in a quartz tube heated by an electric furnace.
A chromel-alumel thermocouple is used to monitor the temperature of
the ceramic boat and also acts as a sensor for the furnace temperature controller. A constant flow of clean, dry N_2 gas is maintained
through the quartz tube. The metal vapor is carried out of the
furnace with N_2 and condensed to form a polydisperse aerosol. The
primary aerosol is diluted by mixing with a filtered dry air stream.
The diluted aerosol is then routed to the chamber as required and
the excess aerosol is vented out through a glass fiber filter.

Effects of carrier gas flow rate, dilution flow rate, and the
combustion boat temperature were studied by sampling the aerosol
stream with an electrical aerosol analyzer to obtain the particle
size distribution. Filter samples were taken for chemical analysis
to determine mass concentrations. Aerosol samples were also
collected in an electrostatic sampler for electron microscopic
examination.

(b) <u>Aerosol Characteristics</u>. Initial tests were made with
potassium chloride as a surrogate material and showed little influence of air flow rate on the size distribution in the flow rate
range of 1 to 5 liters/min (Lpm). The aerosol output, however,
depended on the flow rate. Temperature showed a major effect on
the aerosol size distribution. In the aerosol generation experiments with Pb, Te, and Se, the N_2 carrier gas flow rate was therefore kept constant at 3 Lpm. A dilution air flow was maintained at
8 Lpm.

Effect of temperature was studied by stepwise variation of the
temperature accompanied by a size analysis using the electrical
aerosol analyzer. The size distribution data for selected temperatures are shown in Figures 3 through 5 for Pb, Te, and Se,
respectively. The particle size increases as the generation
temperature is increased. An aging chamber was used with the Te
aerosol in some tests and increase in size by coagulation in the
aging chamber is also evident in Figure 4.

Stable aerosol generation is important to maintain a stable
atmosphere in the chamber. Good stability of the aerosol size
distribution and mass output was observed for Pb aerosol over

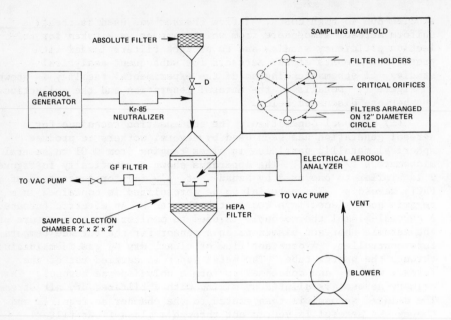

Figure 1.   A schematic of the flow chamber for filter sample collection

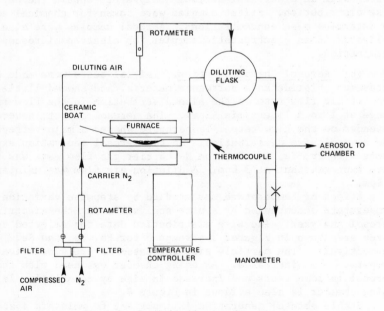

Figure 2.   Lead chloride aerosol generator

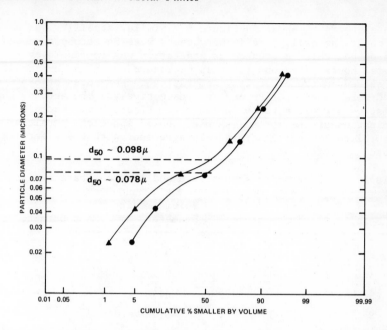

Figure 3. Particle-size distribution for lead aerosols: furnace temperature (●) —840°C, (▲) —895°C; carrier N₂ flow rate ∼ 3 Lpm

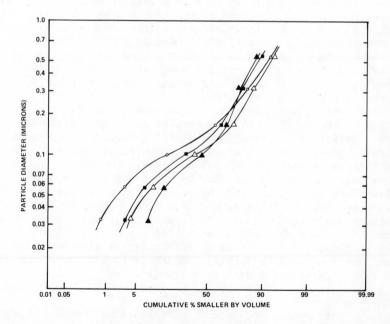

Figure 4. Particle-size distribution of tellurium aerosols at various conditions: (○) 500°C. (□) 455°C, (△) 500°C, (▲) 455°C; (○, □) with coagulation, (△, ▲) without coagulation

several hours as shown in Figure 6. Similar stability was observed
for the Te aerosol. Preliminary tests with Se aerosols showed low
mass output which decreased as generation was continued over
several hours. Reasons for this occurance are being explored.

Based on these results, generation temperatures of 895°C and
500°C were chosen for Pb and Te, respectively. The aerosol characteristics are given below in Table I. Electron microscopic
examination of the aerosol samples showed spherical particles in
the 0.01 to 1.0 μm size range--in agreement with the electrical
aerosol analyzer data.

Table I. Generation Temperature and Aerosol Characteristics

Metal	Temperature	Volume Median Diameter, μm	Mass Output, μg/min
Pb	895°C	0.10	47
Te	500°C	0.12	62
Se	375°C	0.40	–
Se	400°C	0.40	–

(c) Test Chamber and Collection Studies. A plexiglass chamber
(2'x2'x2', active volume) was used as a test chamber (see Figure 1).
The test aerosol is introduced at the top in a filtered air stream.
The dilute aerosol is distributed across the chamber cross section
via screens at inlet and outlet followed by a HEPA filter. A
centrifugal blower is used to maintain a constant flow through the
chamber as monitored by an orifice meter in the inlet section. The
chamber is operated at a slightly negative pressure ($\approx$ 2" H_2O) to
prevent leaks into the work area.

The sampling manifold shown in the insert in Figure 1 is
designed to obtain six replicate samples on 37 mm filters in cassettes. Individual critical flow orifices are used to regulate the
flow rate through each cassette. The chamber atmosphere is monitored using an electrical aerosol analyzer. Uniformity of the
aerosol concentration was checked by obtaining six replicate
samples of Pb aerosol followed by atomic absorption analysis. The
results are shown in Table II and show good reproducibility.

The chamber is being used to produce replicate samples of Pb,
Te, and Se at several loading levels for the analytical studies.
Filter sampling and collection efficiencies are also being explored
using Pb aerosols and by varying the sample flow rate through the
37 mm filter cassettes from 0.1 to 4.0 Lpm.

2. Dissolution Studies

(a) Experimental Apparatus. A Perkin Elmer Model 603 Atomic
Absorption Spectrophotometer equipped with an HGA-2100 heated

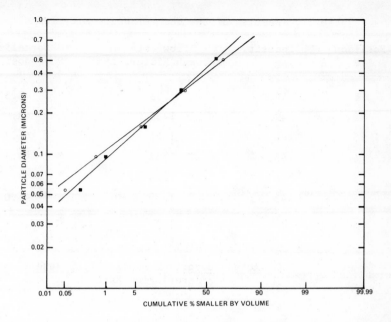

Figure 5.   Particle-size distribution of selenium aerosols: (■) 400°C, (○) 375°C

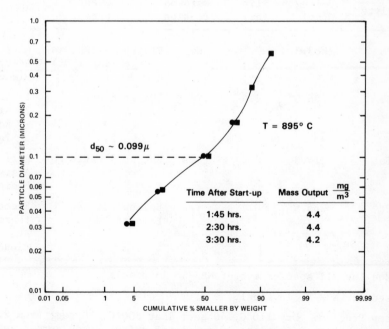

Figure 6.   Particle-size distribution of lead aerosols at different times from start-up

Table II.   Results of the Chamber Uniformity Tests

Position #	Pb Deposit, μg	Flow Rate Through Filter, Lpm	Normalized* Pb Deposit, μg
1	44	1.38	43.3
2	42	1.30	43.9
3	48	1.41	46.3
4	45	1.38	44.3
5	44	1.38	43.3
6	43	1.30	45.0
		1.36 avg.	44.4±1.1 avg., std. dev.

*Normalized Deposit = $\dfrac{\text{Actual Deposit x Average Flow Rate}}{\text{Actual Flow Rate}}$.

Table III.   Instrument Settings

Analyte	Wavelength (nm)	Slit (nm)	Background Correction	Element Source	Atomization
Pb	283.3	0.7	No	Electrodeless Discharge Lamp (EDL) at 10 watts	Air/ Acetylene Flame
Se	196.0	2.0	Yes	EDL at 6 watts	1) Ar/H$_2$[a] 2) Air/ Acety-lene[b]
Te	214.3	0.2	Yes	EDL at 9 watts	Air/ Acetylene Flame
Pt	265.9	0.7	Yes	Hollow Cathode Lamp at 30 ma	Electro-thermal Atomiza-tion[c]

[a]Used for all species except NBS-Se in steel.

[b]Used for NBS-Se in steel.

[c]Temp Program: Dry 110°C, 40 sec; Char 350°C, 20 sec; Atom 2750°C, 7 sec. 50 μL injection; 10 sec integration; argon purge gas.

graphite atomizer, a conventional flame atomizer including single slot and three-slot burner heads, a deuterium arc, background correction system, hollow cathode (HCL) and electrodeless discharge lamps (EDL) and a Model 56 recorder were used during the study. The graphite furnace was loaded manually using MLA microliter pipets. The instrument settings used for the vaious elements are shown in Table III.

   (b) <u>Sample Preparation</u>. Initially, dissolution tests were to be made using test samples simulating real-world samples. Attempts were made to deposit species upon cellulose ester filters by spraying a slurry composed of the species in water. It was found difficult, however, to achieve a homogeneous slurry, even with the addition of surfactants. The mass of sample deposited on the filter could not be reliably related to the volume of slurry deposited. Due to the difficulty in obtaining a constant weight for the cellulose ester filters at the μg level, weighing could not be used to quantify the deposit and this work was discontinued. Depositions were then performed by weighing out the powder onto a 1-cm Teflon disk which was placed in the digestion flask with a filter. A minimum of six replicates were prepared in this manner for each species.

   (c) <u>Dissolution and Measurement</u>. The prescribed dissolution and measurement procedures for the Pb, Te, Se and Pt species were tested. Modifications were made when recoveries greater than 90 percent were not achieved. These modifications are described below and Table IV shows the recoveries obtained with the NIOSH procedures or modifications of these procedures.

   <u>Lead</u>. The existing NIOSH analytical procedure for Pb in air, S-341, involving dissolution in nitric acid was developed using $Pb(NO_3)_2$ (5). This procedure was to be evaluated using Pb metal, PbO, PbO_2, PbS_2 and NBS-SRM 1579 (Pb in paint). The procedure yielded adequate recoveries and precision for all species except PbO_2. With the addition of 1 mL, 30 percent H_2O_2, the oxidation state of lead was converted from +4 to +2 and greater than 90 percent yields were then obtained for PbO_2.

   <u>Selenium</u>. The NIOSH method for Se in air, S-190, describes an extraction procedure which utilizes 0.1 N HNO_3 (6). This procedure was developed for K_2SeO_3. The species Na_2SeO_4, Se metal, SeS_2, SeO_2 and NBS-SRM 339 (Se in steel) were to be tested in this study.

   The NIOSH procedure was originally developed in order to avoid the matrix effects which arise from acid-digested, cellulose ester filters; these effects suppress the selenium signal approximately 20 percent when using the argon (Ar)/H_2 flame (7). In the extraction procedure, the filter remains complete and has no effect upon the analysis. However, while the extraction procedure gives excellent recoveries for a soluble selenium salt such as K_2SeO_3, it does not solubilize Se metal or NBS-SRM 339 (Se in steel). Therefore, a digestion procedure was reconsidered. Since nitric acid dissolves

Table IV.   Recoveries Obtained using NIOSH or
Modified NIOSH Procedures

Species	Digestion Procedure	% Recovery	% Coefficient of Variance
Pb	S-341	94.5±5.8	6.1
Pb	Modified*	103.3±3.4	3.3
PbO	S-341	92.9±3.5	3.8
PbS	S-341	92.6±4.9	5.3
PbO_2	S-341	81.6±2.9	3.5
PbO_2	Modified	99.8±1.3	1.2
Pb in Paint	S-341	94.5±2.7	2.9
Pb in Paint	Modified	93.9±5.2	5.5
Se	Modified	92.6±7.0	7.6
Na_2SeO_4	Modified	97.2±4.4	4.5
SeS_2	Modified	90.8±5.3	5.9
SeO_2	Modified	91.3±7.1	7.8
Se in Steel	Modified	90.6±1.5	1.6
Te	S-204	97.2±5.3	5.4
Te	Modified	99.9±5.4	5.4
Pt	Modified	94.2±9.2	9.7
$(NH_4)_2PtCl_4$	Modified	96.7±7.9	8.2
PtO_2	Modified	98.0±9.5	9.8

*Modified procedures were developed and used when one or more
species of a particular element did not yield >90 percent recovery
(see text).

only one of three forms of selenium metal, an aqua regia digestion
was examined. It was decided that the filter interference could
be dealt with by employing one of three possible procedures:

1. employing standard additions and Ar/H_2 flame;

2. digesting standards with filters added and using Ar/H_2
    flame; or

3. utilizing the slightly less sensitive but hotter air/
    acetylene flame.

Since maximum sensitivity was desired, the first two procedures were considered. Standard additions proved slightly more
time consuming in terms of both sample and data manipulations,
thereby leaving the second option as the procedure of choice. This
procedure yielded >90 percent recoveries for Se metal, Na_2SeO_4,
SeS_2 and SeO_2. Recoveries for NBS-SRM 339 were <10 percent. It
was quickly realized that the high iron matrix in the sample was
playing a major role in signal depression, as the sample otherwise
appeared completely solubilized. Although addition of iron to the
standards improved the recovery to >90 percent, the sensitivity was
still poor, ~5-10 percent of that obtained in an iron-free matrix.
To reduce this effect, a hotter air/acetylene flame was chosen.
Sensitivity proved to be only a factor of two lower than that
obtained with the argon/hydrogen flame (in absence of interferences). Recoveries were still >90 percent. Consequently, should
there be any questions regarding interferences in the sample, one
could select the air/acetylene flame for any selenium-in-air
analysis without sensitivity suffering significantly.

   Tellurium. The NIOSH Method S-204 for analysis of Te in
air was originally developed using $Te(OH)_2$ (8). RTI applied this
procedure to the analysis of the pure metal. The procedure involves
the use of both nitric and perchloric acids during the digestion.
Recoveries were excellent, ~97 percent. An additional set of
samples were subjected to a digestion involving no perchloric acid,
i.e., nitric acid only. This digestion yielded recoveries
averaging 99.9 percent. Since the analyst may wish to avoid the
use of perchloric acid wherever possible, he might opt for the latter procedure.

   Platinum. The NIOSH Method S-191 was developed for
soluble Pt salts in air (9). The digestion involves the use of
nitric and perchloric acids which are ineffective with Pt metal.
Therefore, it was decided to use a nitric-hydrochloric acid
digestion procedure. Pt and $(NH_4)_2PtCl_4$ were recovered satisfactorily but PtO_2 appeared inert to the dissolution mixture and
recoveries for PtO_2 were less than 5 percent. However, PtO_2 is
converted to a mixture of Pt metal and PtO at temperatures greater
than 380°C (10). Therefore, a series of six replicates were
prepared in quartz crucibles. The samples (with filters included)
were first wet ashed in the crucibles which were then placed in an
oven and heated to 500°C. The residue was then treated with aqua
regia as before. Recoveries subsequently averaged >90 percent with
a precision of about ±10 percent.

The electrothermal AAS measurement of Pt is linear to approximately 2 µg/mL using 50 µL injections, above which a high-point-density calibration curve is absolutely necessary. Poorer precision for platinum species is probably related to some degree to the graphite furnace analysis. More specifically, the graphite tube degradation affects the sensitivity, thereby necessitating frequent standard injection—on the order of one standard every three or four samples. Generally, pyrolytic graphite tubes have greater lifetimes than uncoated tubes.

(d) <u>Detection Limits</u>. Instrumental detection limits were determined for each of the elements using standard solutions prepared and measured according to the modified procedures described above. The detection limits were calculated as corresponding to twice the standard deviation of the blanks, and are presented in Table V.

Table V.   Instrumental Detection Limits Based on Twice the Standard Deviation of the Blank

Element	Standard Curve Concentration Range, µg/mL	Detection Limit Determined with Standard Solutions, µg/mL
Pb	0–80	0.27
Pt	0–0.20	0.0030
Se	0–40	0.20
Te	0–40	0.29
Se (in steel)	0–64	1.8

Summary and Conclusions

Sampling and analysis of several metal fumes were investigated in order to develop a reliable method for monitoring work place atmospheres. Significant results are listed below.

1.   Lead, tellurium, and selenium aerosols in 0.01 to 1 micrometer size range could be generated using a condensation aerosol technique. The particle size is dependent on the generation temperature. The aerosol output is dependent on both the generation temperature and the carrier gas flow rate.

2.   The stability and reproducibility were excellent for lead and tellurium aerosols. Selenium aerosol output decreased with time at a given temperature.

3.   The uniformity of aerosol concentration in the sample collection chamber was excellent.

4. A nitric acid/hydrogen peroxide digestion was found to satisfactorily solubilize all lead species in the study.

5. In order to dissolve adequately all selenium species including selenium metal and selenium in steel samples, an aqua regia digestion procedure was used in lieu of the existing NIOSH extraction method.

6. For platinum compounds, the nitric/perchloric acid digestion mixture used in NIOSH procedure S-191 was replaced with aqua regia to solubilize platinum metal. Platinum dioxide was solubilized by first heating the compound to >380°C to convert the PtO_2 to the aqua regia-soluble forms of platinum metal and platinum monoxide.

7. Tellurium metal was found to be recovered adequately not only with the NIOSH procedure using nitric acid/perchloric acid but also with a simple nitric acid procedure.

Literature Cited

1. Zimmerman, N.; Drehmel, D.C.; Abbott, J.H. "Characterization and Generation of Metal Aerosols"; U.S. Environmental Protection Agency; Report No. EPA-600/7-78-013; February, 1978.

2. Spurny, K.; Lodge, J.P. Coll. Czech. Chem. Comm., 1971, 36, 3358.

3. Homma, K.; Kawai, K.; Nozaki, K. "Metal-Fume Generation and Its Application to Inhalation Experiments"; National Institute of Industrial Health, personal communication.

4. Ranade, M.B.; Wasan, D.T.; Davies, R. AIChE Journal, 1974, 20, 273.

5. Taylor, D.G.; Man. Coord. "NIOSH Manual of Analytical Methods"; Second Edition; Vol. 3; DHEW (NIOSH) Publication No. 77-157C; April, 1977; p.S341-1.

6. Taylor, D.G.; Man. Coord. "NIOSH Manual of Analytical Methods"; Second Edition; Vol. 3; DHEW (NIOSH) Publication No. 77-157C; April, 1977; p.S190-1.

7. Taylor, D.G.; Man. Coord. "Documentation of the NIOSH Validation Tests"; DHEW (NIOSH) Publication No. 77-185; 1977; p. S190-1.

8. Taylor, D.G.; Man. Coord. "NIOSH Manual of Analytical Methods"; Second Edition; Vol. 3; DHEW (NIOSH) Publication No. 77-157C; April, 1977; p.S204-1.

9. Taylor, D.G.: Man. Coord. "NIOSH Manual of Analytical Methods"; Second Edition; Vol. 3; DHEW (NIOSH) Publication No. 77-157C; April, 1977; p.S191-1.

10. Grandadam, P. Ann. Chim., 1935, 4 (II Ser), 84.

RECEIVED October 27, 1980.

Sampling and Breakthrough Studies with Plictran

CLIFFORD C. HOUK

Department of Chemistry, Ohio University, Athens, OH 45701

HARRY J. BEAULIEU

110 Veterinary Science Building, Colorado State University, Fort Collins, CO 80523

Organotin compounds are used in three major types of applications: biocides, catalysts and stabilizers in polymers (1). As stabilizers, particularly dialkyltin compounds, they prevent degradation of halogen-containing polymers and polyamides, and nonhalogenated substances such as lubricating oils, hydrogen peroxide, polyolefins and other plastics. The largest use of organotin compounds is photostabilizers in polyvinyl chloride. Diorganotin compounds are also used as heat stabilizers for plastics, catalysts in the production of polyurethane foams, cold curing of silicone rubber, and corrosion inhibitors in chlorinated heat exchange fluids (2). Triorganotin compounds are used mainly in biocidal applications, preservatives for wood, textile, paper, leather and glass, rodent repellents, molluscicides, fungicides, and insecticides.

Plictran (tricyclohexyltin hydroxide), (TCHH), is an organotin most generally used as a miticide for fruit trees and ornamental trees and shrubs. It was developed and is produced by Dow Chemical Company and sold as a 50% wettable powder acaracide, or 50% TCHH-50% inert diatomaceous earth. It is light tan in color and has a negligible vapor pressure (3).

In general, triorganotin compounds show greater toxicity than diorganotin compounds (4). The major concern in occupational exposure evaluations of organotin compounds, in general, is the potential for liver, kidney, pulmonary, and central nervous system effects. NIOSH, in 1976 and the American Conference of Governmental Industrial Hygienists (ACGIH), in 1971 both recommended a threshold limit value (TLV) of 0.1 mg/M³ for all organotin compounds (5,6). ACGIH established a TLV specifically for TCHH of 5.0 mg/M³ (1.2 mg/M³, measured as tin) in 1973 which was then adopted in 1975 (7).

Common usage of TCHH and other organotin compounds has caused a great deal of interest in the development of personal sampling techniques for organotin compounds in general. Although the vapor pressure of TCHH is theoretically negligible, significant breakthrough of TCHH through glass fiber filters and off

activated charcoal has been reported (8). Current concensus
among organotin researchers is that both particulate (nonvolatile) and volatile organotins can be collected with high efficiency by impinger methods using concentrated nitric acid (9).

Graphite furnace atomic absorption spectrophotometry (AAS)
has been shown to be a versatile technique for the detection of
low levels of tin, reproducibly over a wide linear working range
and was the method of analysis used in this study (10,11).

Current methods of collecting personal samples of organotin
compounds produce erratic, inconsistent results. Use of the most
efficient collection scheme, concentrated HNO_3 in a midget
impinger, is certainly unsafe.

This study tested combinations of glass fiber, cellulose
ester filters and HNO_3 impingers for breakthrough of Plictran .
Aerosol samples of Plictran were generated under controlled
laboratory conditions. In most cases, the new systems were compared with the concentrations measured by HNO_3 impinger collection.

Experimental

Reagents. The chemicals used in this study were: concentrated HCl, concentrated H_2SO_4, concentrated HNO_3, concentrated
NH_4OH, 10% v/v HNO_3, 3% v/v HCl, $CuCl_2 \cdot 2H_2O$, 1000 ppm Fisher
Certified AAS Tin Reference Solution, and Plictran .

Standard solutions of inorganic tin in 3% v/v HCl were used.
Serial dilutions of the reference solution were made to obtain
0.01 µg/mL-1.0 µg/mL working standards. Working standards were
prepared fresh prior to analysis of each set of samples.

Materials. Gelman 37 mm, type A/E glass fiber, GN-4 0.8 µ
metricel, and DM-800 0.8 µ membrane filters were used in this
study. Spectral quality, pyrolytic coated, graphite furnaces and
rods were used in the study. The furnace capacity was 5 µL of
solution. Manual sample injections were made with a 5 µL Eppen-
dorf pipet 4700 using disposable tips.

Sample Collection. Samples were collected using a 4 foot
cubic chamber of 1/4 inch plywood. A special pump and cassette
rack was constructed to assure that all cassette and impinger
openings were at the same level, Figure 1.

All samples were collected using the closed face method.
Two schemes were used to mount the filters. One scheme used two
filters in direct contact, without a back-up pad, mounted in a
standard 2 piece 37 mm field cassette. The second scheme separated the two filters in a 3 piece cassette. One filter was
mounted in the center section of the cassette. The second filter
was mounted in the last section both without a back-up pad.

The stationary sampling train consisted of: a sample cassette
containing the filters as described, a midget impinger containing
10 mL of concentrated HNO_3 to trap any breakthrough or volatile

components of Plictran , which in turn was followed by a midget impinger containing 10 mL deionized water to trap any acid vapors from the preceding impinger and finally an empty 2-piece cassette to collect water vapor before air entered the pump.

Some filters were coated with tetraamine copper(II) chloride to determine if a coated filter would improve the collection efficiency for TCHH. The coating was applied by suction filtration of a suspension of the complex through the filters in question. The filters were dried in a desiccator over Drierite, placed in the cassettes and stored until used.

Aerosols were drawn to the filters by Model G, MSA sampling pumps at flow rates of 1.0-2.0 L/min. Pumps were pre- and post calibrated using standard bubble burette methods with an identical sampling train in line.

Laboratory aerosols were generated with a small, 35 psi, 750 mL capacity home paint spray device, Figure 2. A concentration equal to 150 ppm Plictran or 46.2 ppm Sn(IV), was used as the aerosol. During each collection period, midget impingers containing concentrated HNO_3 were used to provide comparison samples since concentrated HNO_3 is considered the most efficient means of collecting organotin compounds. Back-up impingers containing concentrated HNO_3 were used to determine if any breakthrough of the Plictran occurred during sample collection.

Sample Digestion. Filter samples were wet-ashed in 50 mL beakers at 120°C for one hour in 6 mL of a 2:1 mixture of concentrated HNO_3 and concentrated H_2SO_4. After the low temperature digestion, the temperature was increased to 230°C for another hour. Two mL aliquots of the acid mixture were added to prevent evaporation to dryness. At the end of the second hour the volume was reduced to 0.5 mL or less. The remaining solution was quantitatively transferred to a 50 mL volumetric flask with 3% HCl solution and brought to volume. These solutions were stored in nitric acid washed Nalgene bottles until analysis by graphite furnace AAS. All impinger samples were treated and analyzed in the same manner.

Instrumentation. A Varian Model 63 atomic absorption spectrophotometer, equipped with a graphite atomizer head was used to analyze all samples. Instrument parameters appear in Table I. A tin, hollow cathode lamp was used to generate the desired wavelength. A deuterium arc lamp was used to correct for non-atomic absorption signals.

Figure 1.   Pump and cassette rack

Figure 2.   Chamber and spray device

Table I

INSTRUMENT PARAMETERS (12)

HCL - 286 nm, 8 ma D_2 - continuum, 30 ma

Reading mode - peak height, single pen recorder with variable 1-10 mv input

Inert Gas - N_2 Slit Width - 0.2 nm

Furnace	- dry	110°C	30 sec.
	char	500°C	20 sec.
	atomize	2300°C	4 sec.

Standard Curve. The standard curve established during this study is shown in Figure 3. Working standard solutions of 1.0 µg/mL, 0.5 µg/mL, 0.1 µg/mL, 0.05 µg/mL, 0.03 µg/mL were prepared by serial dilution of 1,000 µg/mL certified tin reference solution. Mean values of replicate, three or more, 5 µL injections of each concentration were used to establish the curve by standard regression methods.

Data Treatment. The experimental design of sample collection did not permit statistical comparison of the collection efficiencies of filter combinations versus concentrated HNO_3 impingement. The laboratory chamber was designed to simulate field conditions and not provide a "closed" system in which the concentrations of Plictran could be controlled.

Results and Discussion

To determine the most efficient combination of filters several samples with filters in contact were collected concurrently with concentrated HNO_3 impinger samples. The data, Table II, suggested that concentrated HNO_3 was a better collector of Sn(IV) than any of the filter combinations and that uncoated glass fiber filters were better collectors than other filter combinations. The large breakthrough of Sn to the back-up impingers in all cases but one was a concern. Each impinger in this case contained only 10 mL of acid. With filters in contact in a cassette and both filters ashed at the same time in a beaker, it could not be determined if any Plictran was breaking through the filters. To answer that question, the filters were separated as described. Each filter was separately analyzed. In addition, the volume of acid in the first impinger was increased to 20 mL, the H_2O trap solution was analyzed and the sampling time decreased. The data

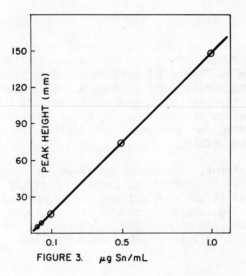

FIGURE 3.   μg Sn/mL

Figure 3.   The standard curve

Table II

LABORATORY CHAMBER RESULTS (12)

Filter[a]	Sn (ppm)[b]	Sn (total)[c]	Vol. Air Sampled (L)	Conc. In Air (µg/L)
GFF-GFF	0.006	0.30	94.2	0.0032
" "	nd	--	--	--
" "	nd	--	--	--
" "	0.015	0.75	68.25	0.011
" "	0.02	1.0	61.75	0.016
" "	0.01	0.50	59.8	0.0084
" "	0.005	0.25	61.1	0.0042
GFF-GN-4	nd	--	--	--
" "	nd	--	--	--
GN-4-GN-4	nd	--	--	--
" "	nd	--	--	--
GFF*-GN-4*	nd	--	--	--
" "	nd	--	--	--
GFF-GFF*	nd	--	--	--
HNO$_3$ imp A	0.148	7.4	60.45	0.122
" " A+	0.126	6.3	60.45	0.104
" " B	0.421	21.05	64.03	0.329
" " B+	nd	--	--	--
" " C	0.153	7.65	63.38	0.121
" " C+	0.045	2.25	63.38	0.036

a = filters in contact, both filters digested together as one,
    filter to left was first in the sampling train
b = concentration of tin
c = total µg of Sn collected
* = coated with $CuCl_2 \cdot 4NH_3$
nd = none detected
+ = back-up impinger

in Table III indicated that increasing the volume of acid and decreasing the sampling time virtually eliminated breakthrough of Sn to the back-up impinger and that HNO_3 was a better collector than the filters. No detectable level of Sn reached the H_2O traps. It also appeared, however, that there was no breakthrough from the first filter to the second filter. To further test this apparent lack of breakthrough, additional samples were collected. In this instance no concurrent acid collection was used. Instead, acid impingers were employed as back-up impingers to the cassettes. If a volatile component of Plictran would break through the filters, the acid would trap it. The data in Table IV indicated that the first filter effectively collected the Plictran since no detectable levels of Sn were found on the second filter. It also appeared that Plictran did not volatilize, and pass through the filters as evidenced by no detectable levels of Sn in the acid. To further test the preceding results, two additional sets of samples were collected. All conditions were the same except sampling times were increased. The data in Tables V and VI support the earlier data and indicate that the first filter effectively collected the Plictran and that there was no breakthrough or volatilization.

Summary and Conclusions

This study tested dual filter combinations compared to concentrated HNO_3 impingement for sampling an organotin compound, Plictran (tricyclohexyltin hydroxide). Glass fiber and cellulose ester filters were studied in two configurations, in direct contact in 2 piece 37 mm field cassettes and separated in 3 piece 37 mm field cassettes. Regular filters and filters coated with $CuCl_2 \cdot 4NH_3$ were tested in both configurations. The data seemed to indicate that Plictran could be safely collected as a personal sample with a filter without relying upon HNO_3. It appeared that a single, uncoated glass fiber filter was sufficient to collect Plictran since breakthrough from filters appeared not to be a problem as in earlier reports. Coated filters appeared not to be more efficient than uncoated but more data was needed to complete such a comparison. Coated filters increased air flow resistance which reduced the sampling flow rate that could be used.

Abstract

An air sampling technique for an organotin miticide, Plictran (tricyclohexyltin hydroxide) has been studied. Different types of filtration sampling were compared to impingement of the sample into concentrated HNO_3. Laboratory sampling consisted of two and three piece field cassettes containing plain and coated glass fiber and cellulose ester filters. The coating material was $CuCl_2 \cdot 4NH_3$. The filters were wet-ashed in 2:1 concentrated HNO_3/concentrated H_2SO_4 and taken up in 3% HCl as working samples. All

Table III

LABORATORY CHAMBER ([12])
(43 minute sample time)

Filter[a]		Sn (ppm)[b]	Sn (total)[c]	Vol. Air Sampled (L)	Conc. In Air (µg/L)
GN-4*	4 pair	nd	--	--	--
GN-4		nd	--	--	--
GN-4*		0.002	0.10	39.99	0.0025
GN-4		nd	--	--	--
GFF	4 pair	nd	--	--	--
GN-4		nd	--	--	--
GFF		0.002	0.10	38.80	0.0026
GN-4		nd	--	--	--
HNO_3 imp A		0.075	3.75	39.99	0.094
" " A+		0.004	0.20	39.99	0.005
H_2O " A"		nd	--	--	--
HNO_3 " B		0.326	16.3	38.92	0.419
" " B+		nd	--	--	--
H_2O " B"		nd	--	--	--
HNO_3 " C		0.153	7.65	38.92	0.196
" " C+		nd	--	--	--
H_2O " C"		nd	--	--	--

a = filters separated as described, first filter typed was first
    in the sampling train.
b = concentration of tin
c = total µg of tin collected
+ = back-up HNO_3 impinger
* = coated
" = H_2O impinger before pump

Table IV

LABORATORY CHAMBER (12)
(32 minute sample time)

Filter[a]		Sn (ppm)[b]	Sn (total)[c]	Vol. Air Sampled (L)	Conc. In Air (µg/L)
GFF	2 sets	nd	--	--	--
GN-4		nd	--	--	--
HNO₃ imp		nd	--	--	--
GFF		0.028	1.25	30.08	0.042
GN-4		nd	--	--	--
HNO₃ imp		nd	--	--	--
GFF		nd	--	--	--
GFF*	2 sets	nd	--	--	--
HNO₃ imp		nd	--	--	--
GFF		0.005	0.25	30.24	0.0083
GFF*		nd	--	--	--
HNO₃ imp		nd	--	--	--
GFF		0.035	1.75	30.56	0.057
GFF*		nd	--	--	--
HNO₃ imp		nd	--	--	--
GFF		0.002	0.10	30.08	0.0033
GFF*		nd	--	--	--
HNO₃ imp		nd	--	--	--

Legend same as Table III

Table V

LABORATORY CHAMBER ($\underline{12}$)
(54 minute sample time)

Filter[a]	Sn (ppm)[b]	Sn (total)[c]	Vol. Air Sampled (L)	Conc. In Air (μg/L)
GFF	0.022	1.1	50.92	0.022
GN-4	nd	--	--	--
HNO$_3$ imp	nd	--	--	--
GFF	0.062	3.1	51.46	0.060
GN-4	nd	--	--	--
HNO$_3$ imp	nd	--	--	--
GFF	0.048	2.4	49.31	0.049
GN-4	nd	--	--	--
HNO$_3$ imp	nd	--	--	--
GFF	0.040	2.0	50.38	0.040
GN-4	nd	--	--	--
HNO$_3$ imp	nd	--	--	--
GFF	0.010	0.5	49.85	0.010
GN-4	nd	--	--	--
HNO$_3$ imp	nd	--	--	--
GFF	0.026	1.3	51.99	0.025
GN-4	nd	--	--	--
HNO$_3$ imp	nd	--	--	--
GFF	0.017	0.85	51.19	0.017
GN-4	nd	--	--	--
HNO$_3$ imp	nd	--	--	--
GFF	0.064	3.2	49.04	0.065
GN-4	0.197†	9.8	49.04	0.201
HNO$_3$ imp	nd	--	--	--

Legend same as Table III

† Glass fiber filter split during sampling.

Table VI

LABORATORY CHAMBER ([12])
(73 minute sample time)

Filter[a]	Sn (ppm)[b]	Sn (total)[c]	Vol. Air Sampled (L)	Conc. In Air (μg/L)
GFF	0.052	2.6	70.1	0.037
GN-4	nd	--	--	--
HNO₃ imp	nd	--	--	--
GFF	0.041	2.0	65.3	0.032
GN-4	nd	--	--	--
HNO₃ imp	nd	--	--	--
GFF	0.039	2.0	69.7	0.028
GN-4	nd	--	--	--
HNO₃ imp	nd	--	--	--
GFF	0.046	2.3	65.0	0.035
GN-4	nd	--	--	--
HNO₃ imp	nd	--	--	--
GFF	0.057	2.85	70.1	0.041
GN-4	nd	--	--	--
HNO₃ imp	nd	--	--	--
GFF	0.043	2.2	73.0	0.029
GN-4	nd	--	--	--
HNO₃ imp	nd	--	--	--
GFF	0.046	2.3	66.8	0.034
GN-4	nd	--	--	--
HNO₃ imp	nd	--	--	--
GFF	0.030	1.5	64.2	0.023
GN-4	nd	--	--	--
HNO₃ imp	nd	--	--	--
GFF	0.039	2.0	68.6	0.028
GN-4	nd	--	--	--
HNO₃ imp	nd	--	--	--

Legend same as Table III

samples were analyzed by graphite furnace AAS with manual injections of 5 μL samples.

Data gathered during the study indicated Plictran could be safely collected as a personal sample with a single glass fiber filter without breakthrough of the Plictran . Coated filters did not appear to improve collection efficiency of the organotin compound. Concentrated HNO_3 appeared to be a more efficient collector of Plictran .

Literature Cited

1. Ross, A. Ann. N.Y. Acad. Sci., 1965, 125, 107-123.
2. Piver, W.T., Environ. Health Perspec., 1973, 4, 61-79.
3. Dow Chemical Company, "Material Safety Data Sheet, Plictran 50 W Miticide"; Midland, Michigan, 1976.
4. Neumann, W.P., "The Organic Chemistry of Tin"; Interscience Pub., London, 1970; pp. 1-282.
5. National Institute for Occupational Safety and Health, "Criteria for a Recommended Standard . . . Occupational Exposure to Organotin Compounds"; DHEW, NIOSH, Cincinnati, Ohio, 1976; No. 77-115.
6. American Council of Governmental Industrial Hygienists, "Documentation of the Threshold Limit Values for Substances in the Workwoom Air . . . Organotins"; ACGIH, Cincinnati, Ohio, 1971.
7. American Council of Governmental Industrial Hygienists, "Documentation of the Threshold Limit Values for Substances in the Workroom Air . . . Tricyclohexyltin Hydroxide, "Plictran)"; ACGIH, Cincinnati, Ohio, 1975.
8. Geissert, J.: Occupational Health and Safety Section, Institute of Rural Environmental Health, Colorado State University, Fort Collins, Colorado, 1978, Personal Communcation.
9. Sinnon, I.: Director of Analytical Services, M&T Chemical Co., Rahway, N.J., 1979, Personal Communication.
10. Frachman, H., Tyberg, A., Branigan, P. Anal. Chem., 1977, 49, 1090-1093.
11. Merganger, J. J. Assoc. Off. Anal. Chem. 1975, 58, 1143-1146.
12. Houk, C.C. "Sampling and Analysis for an Organotin, Plictran (tricyclohexyltin hydroxide)", Master of Science Thesis, Colorado State University, Fort Collins, Colorado, 1980.

RECEIVED November 13, 1980.

Sampling and Analysis of Chlorinated Isocyanuric Acids

JOHN PALASSIS and JOHN R. KOMINSKY

National Institute for Occupational Safety and Health, 4676 Columbia Parkway, Cincinnati, OH 45226

Chlorinated isocyanurates have applications for both swimming pools and as commercial sanitizers (1, 2, 3, 4). Sodium dichloro-isocyanurate dihydrate (NaDCC) and trichloroisocyanuric acid (TCCA) are the most common derivatives of chlorinated isocyanurates. Generally they are used as bactericides, algicides, and sanitizers. They are also used as active ingredients in dry bleaches, dish-washing compounds, scouring powders, water and sewage treatment and generally as replacement of calcium hypochlorite.

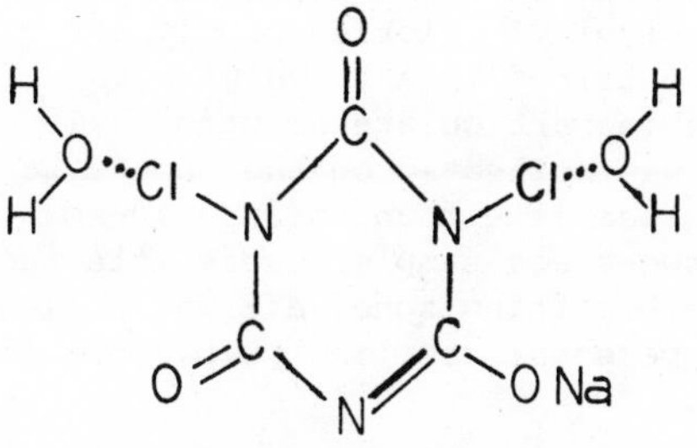

Sodium Dichloroisocyanurate
Dihydrate
(NaDCC)

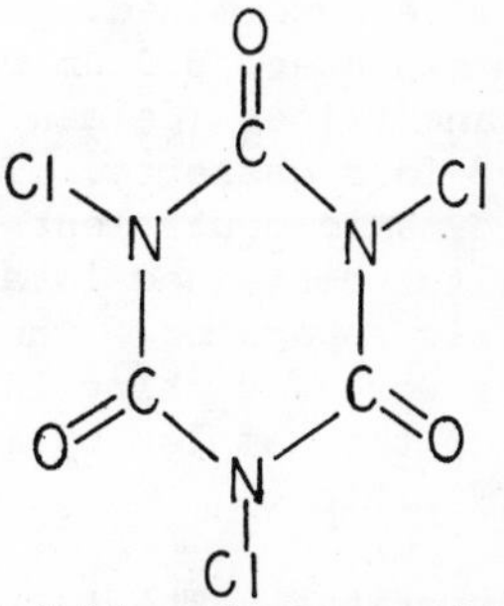

Trichloroisocyanuric
Acid
(TCCA)

Despite apparent widespread use, the toxicological data on these compounds are insufficient (5). Because of the strong oxidation properties, low concentrations of NaDCC and TCCA particulates are extremely irritating to the respiratory tract and mucous membranes of the eyes (6). They can also cause irritation

and erythema of exposed skin areas. The systemic toxicological
characteristics are not known. Currently there is no TLV or
Federal Standard on either compound. The National Institute for
Occupational Safety and Health (NIOSH) conducted a Health Hazard
Evaluation (HHE) at a chlorine dry bleach plant to determine the
extent and effects of exposure to these chlorinated isocyanuric
acids (6). The work reported in this paper is a result of the HHE
request to develop a method for the analysis of air samples for
NaDCC and TCCA.

The preferred method for analyzing chlorinated isocyanuric
acids and their salts is the iodometric titration (7). The main
limitation of the iodometric titration method is sample size;
generally one would need a minimum of 100 milligrams for titrating
either compound. For industrial hygiene samples, this is a problem because sample weights are usually smaller by a factor of 20,
i.e. less than 5 milligrams. Therefore, modifications of the
iodometric titration method were needed to make the method
applicable to industrial hygiene samples.

Experimental

Air Sampling. Since NaDCC and TCCA were manufactured in
different buildings, the compounds were collected separately, thus
eliminating problems with interferences in analysis. Personal exposures to both total and respirable particulates of NaDCC and
TCCA were determined. The total particulates were collected on a
37-mm diameter, 5.0 μm pore size polyvinyl chloride copolymer
membrane filter (Gelman DM-5000) contained in a three-piece
closed-face cassette. The respirable particulate fraction (<10 μm
aero-dynamic equivalent diameter) was collected on the same type
of filter contained in a two-piece cassette mounted in a 10-mm
cyclonic separator. The one-and-two-stage samplers were attached
to the worker's shirt lapel at the breathing zone; air was pulled
through the sampler by means of a personal sampling pump operating
at 1.7 L/min.

Apparatus. The titration assembly consisted of a pH meter
with a millivolt display and capable of supplying 10 microamperes
constant current to the electrodes, two platinum electrodes, a
5-mL microburette, nitrogen gas supply, a magnetic stirrer and
bar, a 250-mL beaker for titration, and a 50-mL burette.

Reagents and Stock Solutions. Deoxygenated distilled water,
prepared by bubbling nitrogen gas into distilled water for 10-15
minutes using flexible tubing connected to a glass Pasteur Pi-
pette, was used.

Sodium Dichloroisocyanurate Dihydrate (NaDCC). A 10^{-3}
M solution was prepared by dissolving 25.59 mg into a 100-mL volumetric flask with distilled water.

Trichloroisocyanuric Acid (TCCA). A 10^{-3} M solution
was prepared by dissolving 23.24 mg into a 100-mL volumetric flask
with distilled water.

Potassium Iodide (KI). A 1 M solution was prepared by
dissolving 16.60 g into a 100-mL volumetric flask with distilled
water.

Potassium Iodate (KIO$_3$). A 0.1000 N solution was prepared by weighing 3.567 g and dissolving into a 1000-mL volumetric
flask with distilled water.

Sodium Thiosulfate (Na$_2$S$_2$O$_3$). 0.005 N and 0.001 N
solutions were prepared by dissolving 1.25 g and 0.250 g, respectively, into 1000-mL volumetric flasks with distilled water.
These solutions were standardized against potassium iodate (see
standardization section).

Sulfuric Acid. A 6 N solution was prepared by slowly
adding 84 mL of 95% concentrated sulfuric acid into 400 mL distilled water in a 500 mL volumetric flask and bringing up the
volume to the mark with distilled water.

Hydrochloric Acid. A 1 N solution was prepared by slowly adding 43.1 mL of concentrated (11.6 M) hydrochloric acid into
a 500-mL volumetric flask containing 400 mL distilled water. The
total volume was brought to the mark with distilled water.

Standardization. A 50-mL aliquot of deoxygenated distilled
water was measured with a graduated cylinder and placed in the
titrating beaker. When the 0.005 N sodium thiosulfate solution
was standardized, 2 mL of 0.1000 N potassium iodate was placed in
the titrating beaker; for the standardization of 0.001 N sodium
thiosulfate solution only 0.2 mL of 0.1000 N potassium iodate
solution were placed in the titrating beaker. A 6-mL aliquot of
the 1 M potassium iodide solution, plus a 15-mL aliquot of the 1 N
hydrochloric acid solution, were added into the titrating beaker.
The solution in the beaker was stirred and the nitrogen purge was
over the surface in the beaker during titration. The potassium
iodate was then titrated with proper sodium thiosulfate
solution. For each new addition of sodium thiosulfate, a stabilized millivolt reading was recorded. The titration was continued
well beyond the end-point until no change in the millivolts was
observed when excess sodium thiosulfate was added. A plot of
millivolts versus volume of sodium thiosulfate used was made and
the end-point was determined from the plot. The normality, N_S,
of sodium thiosulfate was calculated from the equation:

$$N_s = \frac{M_p \cdot N_p}{M_s}$$

where M_p = milliliters of potassium iodate titrated (2 or 0.2 mL)
     N_p = normality of potassium iodate
     M_s = milliliters of sodium thiosulfate at the end point.

Preparation of Standards. To account for the errors which may
be introduced in the presence of residual oxygen in the titration,
particularly at low analyte concentrations, greater accuracy was
obtained by using a calibration curve.

Sodium Dichloroisocyanurate Standards. Using the 10^{-3} M
solution, each volume of 0.5, 1, 2, 4, and 8 mL yielded,
respectively, 0.128, 0.256, 0.512, 1.02, and 2.05 mg of sodium
dichloroisocyanurate dihydrate.

Trichloroisocyanuric Acid Standards. Using the 10^{-3} M solution, each volume of 0.5, 1, 2, 4, and 9 mL yielded, respectively,
0.116, 0.232, 0.464, 0.928 and 2.09 mg of trichloroisocyanuric
acid.
Replicates of six samples per each level of concentration were
prepared for each compound. Each standard was prepared and titrated immediately. The data sets of end-point volumes and
milligrams of the standards were treated statistically, where the
standard deviations and regression analysis were performed.

Titration. Each standard or extract solution along with a
50-mL aliquot of distilled deoxygenated water were added into the
titrating beaker. Three milliliters of 1 M potassium iodide followed by 5 mL of 6 N sulfuric acid solution were also added in the
titrating beaker. The two platinum electrodes were lowered into
the sample solution and the nitrogen purge started over the solution. The sample solution was stirred vigorously. For samples
with filter loading of more than 0.5 milligram, the 0.005 N sodium
thiosulfate solution was used for titration. For samples with
filter loading of less than 0.5 milligram, the 0.001 N sodium
thiosulfate solution was used for titration. For each addition of
sodium thiosulfate, a new stabilized millivolt reading was taken.
The titration was continued beyond the end-point until no change
in the millivolts was observed when excess of sodium thiosulfate
was added. The end-point was determined from a plot of millivolts
versus the volume of sodium thiosulfate used for titrating that
sample.

For consistency, all the 0.001 N sodium thiosulfate end-point
volumes were converted to 0.005 N end-point volumes as follows:

$$M_1 = M_2 \cdot \frac{N_2}{N_1}$$

where M_1= end-point volume of 0.005 N solution
      M_2= end-point volume of 0.001 N solution
      N_1= 0.005
      N_2= 0.001

Filter Extraction. Each spiked or field filter sample was
placed in the titration beaker containing 25 mL deoxygenated distilled water at room temperature. After five minutes, the filter
was lifted with tweezers above the water level and was rinsed
slowly on both sides with 25 mL of deoxygenated distilled water.
Then the filter was rolled with the tweezers and squeezed against
the inside wall of the beaker to remove excess water. Then the
filter was discarded. The procedure described in the "Titration"
section was followed to analyze the extract.

Recovery Study. Spiking solution. A 10 mg/mL spiking
solution was prepared by dissolving 0.500 g of sodium dichloroisocyanurate dihydrate in 50 mL distilled water. (It may take up
to 2 hours to dissolve all the material.)
     Six filters for each level of concentration were spiked using
variable volume pipettes. The following volumes 13, 25, 50, 100,
and 200 μL of the 10 mg/mL solution yielded 0.13, 0.25, 0.50, 1.00
and 2.00 mg of sodium dichloroisocyanurate dihydrate on a filter,
respectively. When the filter was dry, the "filter
extraction" section was followed. The end-point volumes were
converted to milligrams from the calibration curve and the percent recoveries were calculated.
     Filter extraction efficiency study could not be performed for
trichloroisocyanuric acid. When trichloroisocyanuric acid was
dissolved in water or methanol and then was spiked on a filter and
dried, it was not recovered as trichloroisocyanuric acid because it had reacted with the solvent and its chemical structure
had changed. It is recommended that aerosol generating equipment
be used in this case that generate known concentrations of dry
trichloroisocyanuric acid aerosols which are deposited on filters.

Results and Discussion

     A computer literature search revealed no direct analytical
method specific for sodium dichloroisocyanurate dihydrate (NaDCC)
or trichloroisocyanuric acid (TCCA). Each compound dissolved in
water released chlorine in the positive oxidation state and formed
complex equilibria reactions dependent on the pH of the solutions.
NaDCC and TCCA are very strong oxidants and very reactive compounds, therefore, incompatible for chromatographic analysis. The
only method that is used for analysis of compounds containing

chlorine in the positive oxidation state is the classical iodometric titration. The two drawbacks of this method are, first,
that one cannot distinguish two compounds in the same solution
because titration is not a separating technique. Second, the
iodometric method is recommended for samples of 100 milligrams or
more; thus, it is incompatible for small samples (less than 5 mg).

Sampling Test. This experiment was conducted before any other
experiments or titrations were conducted. The reason for doing
this test was to determine if chlorine may be released and lost
during air sampling of these two compounds.

The methyl-orange test for free chlorine in air from NIOSH
method P&CAM 209 was performed. The lower limit of detection of
chlorine by this method was 0.05 μg/mL and, for 20 mL that were
used, it was 1 μg. Normally, the orange color of the solution
turns clear with chlorine. An amount of 3.55 mg of TCCA was
placed on a DM-5000 filter in a three-piece cassette. The humidity of the room was checked with a Bendix Psychron (wet and dry
bulb) hygrometer that indicated 50% humidity. The air was drawn
through the cassette to a bubbler containing 20 mL of methyl
orange reagent solution and through a sampling pump operating at
1.7 L/min. The flowrate of the pump was checked every 10 minutes.
After four hours and thirty-five minutes, the test was stopped.
No color change was observed in the solution; therefore, chlorine
was not detected in less than 0.028%, concluding that no sample
loss occurred in that sampling interval. To check the validity of
the solution, a small particle was taken from the filter,
introduced in the bubbler, and the color immediately turned clear.

The sampling test was repeated with NaDCC. A 4.26 mg amount
was tested with a fresh reagent. After four hours and forty-five
minutes of sampling no color change was noted. The reagent did
turn color after a small particle was introduced in the bubbler.
Again, the conclusion was that no sample loss occurred during the
sampling period.

Method Development. Solid state specific ion electrodes for
chlorine and chloride were investigated and both showed
interferences from iodide ions. Since no other techniques were
available, we decided to modify the iodometric method for smaller
samples. Samples of 2 mg were titrated using the visible end-
point change from very faint blue to clear. The detection of the
visible end-point was very difficult to determine, and precision
results of replicate standards were very poor. A 5-mL microburette was used for small additions, but that did not really
improve the precision of end-point detection. It was decided to
determine the end-point potentiometrically; and thus we employed
two platinum electrodes connected to a digital pH meter with
millivolt readout, providing 10 microamperes constant current to
the electrodes. The precision of detecting the end-point for
titrated standards did improve. The starch indicating solution

that gave the faint blue color was eliminated from that point on, since the visible end-point was not needed. Without the starch indicator, the samples had a faint yellow color (from the iodine) which turned clear at the end-point. The problem of poor precision at the levels - less than one milligram - still existed. Also, potassium iodide in different amounts may have contributed in the imprecision, but standards prepared purposely with different amounts of potassium iodide and then titrated did not show any significant difference in the precision.

The solution to the problem was discovered when a titrated sample (clear solution) was left on the bench and, after a period, it started changing back to a faint yellow color. We hypothesized that air oxidation may have caused that effect and, consequently, air may have interfered with analysis. Standard samples prepared and purposely delayed during the analysis showed that end-point volumes were larger, indicating that some of the iodide ions turned into free-iodine by air oxidation which, in turn, required more thiosulfate for titration and, therefore, larger end-point volume. The following chemical equations obtained from the literature (8) show what happens before, during, and after titration. The reaction of a chlorinated isocyanuric acid compound with potassium iodide in acidic pH is:

$$OCl^- + 2I^- + 2H^+ \longrightarrow Cl^- + I_2 + H_2O$$

Because of excess potassium iodide in solution, the iodine turns into the triiodide form:

$$I_2 + I^- \rightleftharpoons I_3^-$$

During titration, still in acidic pH, the triiodide ion is reacted with sodium thiosulfate to form iodide ion:

$$I_3^- + 2S_2O_3^= \rightleftharpoons 3I^- + S_4O_6^=$$

After titration is completed or even during a titration without a nitrogen purge oxidation does occur and the iodide ion goes back to form free iodine (I_3^-):

$$6I^- + O_2 + 4H^+ \rightleftharpoons 2I_3^- + 2H_2O$$

Therefore, it was decided that a nitrogen purge be used during titration and the distilled water to be deoxygenated before use. Consequent experiments that were performed with either NaDCC and TCCA indicated that precision did improve at all levels of concentration.

Calibration Curves. After many experiments the procedure was optimized and the final version is described in (9) and in the "experimental" section in this paper. The NaDCC calibration curve

was determined from 34 standards ranging from 0.128 to 2.05 mg,
indicating a pooled relative standard deviation of 4.5%.   Table I
contains the statistical analysis of the calibration curve data
for NaDCC.   Figure 1 depicts the calibration curve graph for
NaDCC.   A typical end-point titration graph at the 2.05 mg NaDCC
level using constant current amperometry is shown in Figure 2.

TABLE I

Statistical Analysis of NaDCC Calibration Curve Data

Number of Samples	Concentration Level (mg)	Relative Standard Deviation (%)
9	2.05	4.56
7	1.02	2.90
6	0.51	4.72
6	0.256	2.22
6	0.128	6.70

Y-Intercept = −0.024 ± 0.08 at 95% Confidence Limits
Slope = 2.62 ± 0.07 at 95% Confidence Limits
Linear Correlation Coefficient = 0.9992

Since all the method development was performed with NaDCC, the
established procedure was applied to TCCA.   The calibration curve
was determined from 19 TCCA standards ranging from 0.116 to 2.09
mg, indicating a pooled relative standard deviation of 3.9%.
Table II shows the statistical analysis of the calibration curve
data for TCCA.   Figure 3 depicts the calibration curve graph for
TCCA.

TABLE II

Statistical Analysis of TCCA Calibration Curve Data

Number of Samples	Concentration Level (mg)	Relative Standard Deviation (%)
6	2.09	1.25
6	0.93	2.22
7	0.116	5.10

Y-Intercept = 0.234 ± 0.072 at 95% Confidence Limits
Slope = 4.62 ± 0.05
Linear Correlation Coefficient = 0.9997

Recovery Study.   Recovery of NaDCC from DM-5000 filters was
performed according to the "recovery" procedure described in this
paper.   Table III contains the recovery data and variance results
for each NaDCC concentration level.

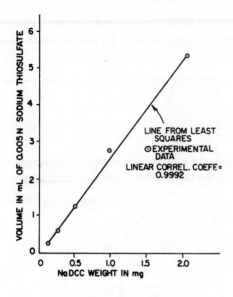

Figure 1.   *Sodium dichloroisocyanurate dihydrate calibration curve*

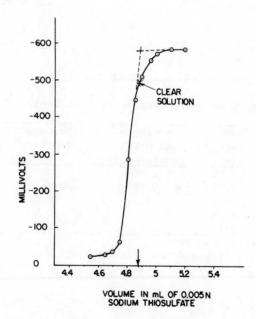

Figure 2.   *A typical constant-current amperometric titration of 2.05 mg sodium dichloroisocyanurate dihydrate*

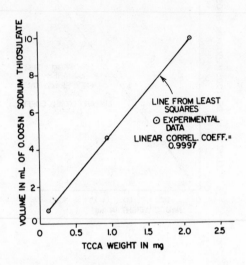

*Figure 3.   Trichloroisocyanuric acid cal-
ibration curve*

TABLE III

NaDCC Recovery Efficiency From DM-5000 Filters

Number of Samples	Amount Spiked (mg)	Average Amount Recovered (mg)	Average Recovery (%)	Relative Standard Deviation
5	2.00	1.96	98.0	1.89
6	1.00	1.06	106.0	1.97
5	0.50	0.49	98.0	3.44
6	0.25	0.232	92.8	5.34
6	0.13	0.131	100.8	4.01

Average Recovery = 99.1%
Pooled Relative Standard Deviation = 3.63%

TABLE IV

Sodium Dichloroisocyanurate Dihydrate Field Sample Results

Sample Number	Job Classification	Air Volume Liters	NaDCC Respirable Dust mg/Filter	mg/m³	NaDCC Total Dust mg/Filter	mg/m³
01, 02	Patrol Operator	433	28.3	65.4	288.8	528.4
03, 04	Patrol Operator	448	0.02	0.04	92.5	206.5
05, 06	Patrol Operator	51	5.81	113.9	68.6	1345.1
07, 08	Patrol Operator	544	0.02	0.04	0.54	0.99
09, 10	Patrol Operator	680	0.20	0.29	14.2	20.9
11, 12	Maintenance	340	0.02	0.06	1.04	3.06
13, 14	Maintenance	340	0.02	0.06	0.88	2.60
15, 16	Maintenance	76	0.02	0.26	1.90	24.8

Field Samples. Field samples were collected using DM-5000
filters. The two producing plants for NaDCC and TCCA were located
in two different areas of the same manufacturing facility; therefore, we anticipated no interferences from either process.
Results showing the respirable fraction and total dust for different employees in the two processes are included in Tables IV and
V. Extracts of samples with filter loadings higher than 2 mg were
diluted and then were titrated.

TABLE V

Trichloroisocyanuric Acid Field Sample Results

Sample Number	Job Classifi- cation	Air Volume Liters	TCCA Respirable Dust		TCCA Total Dust	
			mg/Filter	mg/m^3	mg/Filter	mg/m^3
20,21	Patrol Operator	656	0.02	0.03	0.02	0.03
22,23	Patrol Operator	471	0.02	0.04	0.02	0.04
24,25	Patrol Operator	34	0.02	0.59	14.51	426.8

Conclusions. A sampling and analytical method for two chlorinated isocyanuric acids, NaDCC and TCCA, in air has been
described. Briefly, these acids can be collected from air with
DM-5000 (PVC copolymer) filters. The filter samples are extracted
with water and titrated against sodium thiosulfate using constant-
current potentiometry. The titration method will neither separate
or distinguish NaDCC in the presence of TCCA or the reverse. The
identity of either compound must be known in the workplace
environment along with the identities of any other interfering

substances, i.e., compounds containing chlorine in a positive oxidation state. Concentration results for a compound will be valid only when there are no interferences; otherwise the results will be a combined result of the concentrations of all the compounds containing positive chlorine. Calibration curves in the range of 0.1 to 2 mg of NaDCC or TCCA indicated good linearity (r = 0.999 minimum). Recovery study of NaDCC from filters indicated that NaDCC can be extracted with 99.1% efficiency and 3.6% precision. Experimental results indicate that this method may be used for the analysis of other compounds containing halogens in a positive oxidation state, i.e., hypochlorites, chloramines and chlorinated isocyanuric acids and their salts.

Acknowledgements

We gratefully acknowledge the valuable suggestions and consultations of Dr. J.C. Posner.

Literature Cited

1. Gardiner, J. Water Res., 1973, 7(6), 823-833.

2. Nelson, G.D., "Monsanto Company Special Report No. 6862," 1967.

3. Fitzgerald, G.P.; Der Vartanian, M.E. Appl. Microbiol. 1967, 15, 504-509.

4. Kowalski, X; Hilton, T.B. U.S. Public Health Report, 1966, 81(3), 283-288.

5. Canelli, E. Amer. J. Pub. Health, 1974, 64(2), 155-164.

6. Kominsky, J.R.; Wilcox, T., National Institute for Occupational Occupational Safety and Health, Health Hazard Evaluation Report No. 78-36, 1979, Cincinnati, Ohio 45226.

7. Snell, F.D.; Hilton, C.L., "Encyclopedia of Industrial Chemical Analysis," Interscience Publishers Inc., N.Y., N.Y. (1968) Vol. 7, p. 228.

8. Kolthoff, I.M.; Sandell, E.B. "Textbook of Quantitative Inorganic Analysis," McMillan Co., N.Y., N.Y., 1952; p. 597.

9. Palassis, J., "NIOSH Manual of Analytical Methods," 2nd Ed., Vol. 5, Method No. 314, Cincinnati Ohio, 45226.

RECEIVED October 14, 1980.

Monitoring for Airborne Inorganic Acids

M. E. CASSINELLI and D. G. TAYLOR

National Institute for Occupational Safety and Health, 4676 Columbia Parkway, Cincinnati, OH 45226

Airborne inorganic acids exist in the industrial environment in the form of both vapors and particulates. This study was undertaken to answer a need for a simple sampling and analytical method for monitoring both vaporous and aerosol acid contaminants quantitatively.

Inorganic acids have similar acute toxic properties: corrosive action on the skin, the respiratory tract, and especially the eyes where corneal damage may occur. Severe exposures may cause blindness, pulmonary edema, and even death. The onset of symptoms may be delayed for several hours after exposure. Prolonged exposures to low concentrations produce chronic effects such as tooth erosion, chronic bronchitis, and photosensitization of the skin (1,2,3).

Past sampling methods for acid mists used impingers containing liquids as diverse as the acids being sampled. The analytical methods are equally as varied. (4). The Occupational Safety and Health Administration (OSHA) along with many industrial hygienists have expressed a desire for a non-liquid sampling device which will collect all the common inorganic acids (HCl, H_3PO_4, HBr, HNO_3, and H_2SO_4) and a method for determining these acids collected on a single sampler.

Ion chromatography (IC) offers the analytical tool for the determination of each of the inorganic acids in a single sample. The principle of ion chromatography is the separation and measurement of ions in solution using ion exchange resins, background suppression, and conductimetric detection (5).

This overall study of acid mists began with the development of a sampling and analytical method for hydrogen chloride (6,7). Various solid sorbents and filters, both treated and untreated, were evaluated as collection media and for compatibility with ion chromatography. The sorbent of choice was silica gel which had been washed with deionized water to remove inorganic impurities.

Recently, packed beds have been used for the collection of

aerosols. Lee and Gieseke developed a theoretical approach to
predicting the collection efficiency of packed beds for
aerosol particles (8). In another study using charcoal tubes,
it was reported that approximately 70% of the airborne
particulate mass was deposited on the glass wool plug
preceding the primary bed of charcoal, while 17% was found on
the two beds of charcoal (9). Thus, we hypothesized that the
silica gel tube, which is traditionally a gas and vapor
collector, may have the capability of collecting aerosols as
well as gases and vapors. Sampling tubes, as used in the
earlier HCl study, were packed with 400 mg of silica gel and
evaluated using test atmospheres containing inorganic acids
(6,7). Collection properties were excellent for the
vapor-forming acids, but a greater collection capability was
required for the aerosol-forming acids, phosphoric and
sulfuric acids.

The latest work has focused on the development of a silica
gel sampling tube which will collect aerosol-forming acids as
well as the other acid vapors. This paper describes the
development of the improved sampler. Several variables for
particulate collection were examined to determine the optimum
collection tube geometry. Experiments were performed by
varying the tube diameter which affects the sampling velocity,
the silica gel loading which determines the length of the
sorbent bed, and the mesh size of the silica gel which
determines the airflow properties. All three variables affect
the pressure drop across the tube. Ultimately, a sampling
device was designed which collects both vaporous and
particulate forms of inorganic acids for subsequent analysis
by ion chromatography.

Experimental

Apparatus. Atmospheres containing the inorganic acids
were generated in a system illustrated by the schematic in
Figure 1. Dilute solutions of mixed acids were nebulized with
a Retec medical nebulizer and entered the chamber through a
Teflon tube directed downward through the center of the
chamber to within approximately 12 inches from the bottom.
Dilution air entered the chamber through four ports in the
bottom cover. Flows through the chamber were controlled with
Gilmont flowmeters at 20-30 Lpm for the dilution air and 2-5
Lpm for the nebulizer. A Gast air compressor provided the air
for both nebulization and dilution at approximately 25 psi.
The pressure within the chamber was balanced at atmospheric
pressure, as indicated with a manometer, through the use of an
AADCO vacuum pump. All components exposed to the acids were
fabricated from plastic or glass.

The top cover of the chamber contained an exhaust outlet,

probes for a hygrometer and for a particle sizer, and
initially six sampling ports but later modified to include
twelve sampling ports. Two curved glass ports entered the
sides of the chamber for attaching impingers or filter
cassettes. Samples were drawn by a 1/10 horsepower Millipore
vacuum pump at a nominal flowrate of 0.2 Lpm using critical
orifices. The Retec medical nebulizer delivers aerosols with
a volume median diameter (VMD) of 5.1 µm with a σ_g of 2.0
(13). The concentrations of acids used to generate the
atmospheres were 0.4-0.8 mL of H_3PO_4 per liter of
solution, 0.3-0.5 mL H_2SO_4 per liter, 3-6 mL HNO_3/L, and
10-20 mL of 35% HBr/L. As the aerosols travel through the
chamber evaporation occurs resulting in much smaller
particles. Calculations show that complete evaporation would
result in approximately 0.4 µm VMD particles. The particle
size was considered to be less than 1 µm.

Analyses were done on a Dionex Model 14 Ion Chromatograph
(IC), equipped with a Waters WISP 710A autosampler, Linear
recorder, and interfaced with a Hewlett-Packard 3354
Laboratory Automated System. The principal components of the
IC, shown in Figure 2, are (A) eluent reservoir, (B) pump, (C)
injection valve, (D) separator column, (E) suppressor column,
(F) conductivity cell, and (G) conductance meter with a
recorder (integrator).

Reagents and Standards. The eluent used in the ion
chromatograph was a buffer solution, 0.003 M $NaHCO_3$/0.0024 M
Na_2CO_3, at a pH of 10.4. Stock standards in the
concentration of 1000 µg/mL were prepared in filtered
deionized water from the sodium or potassium salts of each of
the acids studied. From these stock solutions, mixed working
standards were prepared in the eluent solution. These working
standards are stable for at least three days. All chemicals
used were reagent grade or better.

Procedures

Silica Gel Tubes. Glass tubes were packed with silica gel
obtained from Fisher, Grade 01. To remove inorganic
impurities, the silica gel was washed in heated deionized
water for approximately 30 minutes, with occasional stirring,
decanted, and rinsed four to five times with deionized water.
It was then heated again in deionized water for 15-30 minutes
and rinsed thoroughly. The silica gel was then dried
overnight in a 100°C oven until free flowing. If a blank of
the silica gel shows any impurities, particularly sulfate, the
washing procedure is repeated.

Collection tubes (Figure 3) are made from 7-mm O.D./4.8-mm
I.D. glass tubing approximately 13 cm in length, packed with

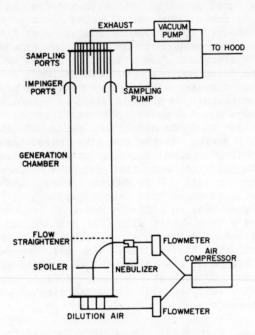

Figure 1.   Generation system for acid mists

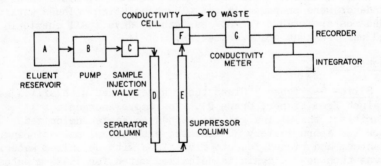

Figure 2.   Schematic of ion chromatograph

700 mg of 20-40 mesh silica gel in the front section and
200 mg in the backup section. Polyurethane foam plugs are
used between the sections and at the end. The front section
of silica gel is held in place with a glass wool plug.
Sampling is done at a nominal flow of 0.2 Lpm. Actual flow
rates averaged 0.17 Lpm with the sampler in line.

Analytical. Samples were desorbed by placing the front
section of silica gel with the glass wool plug and the backup
section in separate 15-mL graduated centrifuge tubes, adding
approximately 6 mL of eluent, and heating in a 100°C
waterbath for 10 minutes. Upon cooling the samples were
brought to 10-mL volume with eluent, covered with Parafilm ,
and shaken thoroughly. The samples were then filtered through
a 12-mL plastic syringe fitted with an Acrodisc in-line
filter into a second syringe for manual injection or into an
autosampler vial.

Samples and standards are injected into the IC in 100-μL
aliquots. The ions are separated by their varying affinites
for the ion exchange resin in the anion separator column, the
oppositely charged ions are stripped away by the suppressor
column leaving only the separated anions and water to be
detected by the conductivity cell. Samples are quantitated by
comparison with a calibration curve.

Since ions are identified only by retention times,
interferences may be difficult to identify. Typical
instrumental operating conditions are listed in Table I.

Table I
IC Conditions

Eluent:	0.003 M $NaHCO_3$/0.0024 M Na_2CO_3
Flow Rate:	138 mL/hr (30% pump capacity)
Columns:	3 X 150 mm Anion precolumn
	3 X 500 mm Anion Separator
	6 X 250 mm Anion Suppressor
Conductivity setting:	10 μmho/cm
Injection volume:	100 μL

A conductivity meter setting of 10 μmhos is appropriate
for most samples. If bromide or nitrate at two times the OSHA
standard levels or chloride are being analyzed, the less
sensitive setting of 30 μmho/cm may be used.

Results and Discussion

Analytical Method. Ion chromatography offers a viable
analytical tool for the determination of each of the inorganic
acids in a single sample. IC is a relatively new technique

developed by Small, Stevens, and Bauman (6). Both anions and
cations may be analyzed by this technique; however, its most
significant contribution is in the determination of anions,
since it separates each of the ionic species from the others
(e.g. NO_2^- , NO_3^- ,SO_3^{-2}, SO_4^{-2}). With proper selection
of eluent, flowrate, and conductance setting, the ions present
in the sample are separated and measured with essentially no
interference.

All samples and eluents must be filtered to prevent
obstruction of flow through the system, especially at inlets
to the columns. For manual injection 1-2 mL of sample are
required to assure good rinse of the injection loop. For
automatic injection, as with the WISP autosampler, the amount
of sample required is relatively less since only the volume
injected is actually used.

Factors governing peak separation and elution time are the
eluent strength, the flow rate, and the length of the
separator column. Buffered eluent solutions are generally
used to maintain a constant pH, thereby keeping constant the
elution order of multivalent ions. Elution times vary
directly with the eluent concentration and inversely with the
flow rate (10). The eluent used for the acids work, 0.003 M
$NaHCO_3$/0.0024 M Na_2CO_3, pumped at a flow rate of 138
mL/hr (30% of pump capacity) gave good separation of the
anions studied. With prolonged use (~one yr) the separator
column begins to lose resolution. This is significant in the
separation of bromide and nitrate ions since they are closely
eluting species.

Calibration curves were prepared by analyzing mixed
standards in the concentration range of 1 to 20 μg/mL, and
plotting peak height vs. concentration. Figure 4 illustrates
typical calibration curves obtained from mixed standards at a
sensitivity setting of 10 μmhos. Variation of points was
approximately ±5%, mainly owing to the fact that the
conductivity cell is sensitive to temperature changes. The
measurement of peak height is recommended over peak area for
IC calculations (10).

Based on a 4-hour sample taken at a nominal flowrate of
0.2 Lpm (48 L) at air concentrations from 0.2 to 2 times the
OSHA standard, the acid mist samples would be expected to
contain: 10-100 μg of H_2SO_4 and H_3PO_4, 100-960 μg
HBr, and 50-500 μg HNO_3. At the 10 μmho conductance
setting, the calibration curves are linear to 20 μg/mL.

Since 10 mL is the dilution volume, larger samples of HBr
and HNO_3, would require further dilution or the use of a
less sensitive conductance setting to fall in the linear range
of the method. Under the instrumental conditions stated in
the experimental section, the presence of 0.3 μg/mL of each of
the acids can be detected with a precision of 10% relative

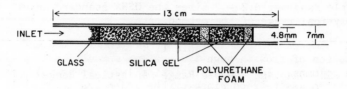

Figure 3.   Silica gel collection tube

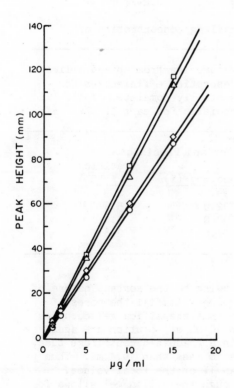

Figure 4.   Calibration curves for inorganic acids: (□) HBr, (△) HNO₃, (◇) H₂SO₄, (○) H₃PO₄

standard deviation, and 0.5 μg/mL at 5% RSD. The RSD for each
of the acids at a concentration of 0.5 μg/mL (n=5) was
determined to be: H_3PO_4 - 3.4%, HBr - 2.9%, HNO_3 -
3.5%, and H_2SO_4 - 2.9%.

Table II lists the OSHA permissable exposure limits ([11]),
the sample range at 0.2 - 2 times the OSHA standard level, and
the analytical range of the method.

Table II

Acid	OSHA Standard (mg/m^3)	Sample Range (μg/sample)	Analytical Range (μg/mL)
H_3PO_4	1	10-100	0.5-20
HBr	10	100-960	0.5-20
HNO	5	50-500	0.5-20
H_2SO_4	1	10-100	0.5-20
HCl	7(C)[a]	20-200	0.2-20

[a]Based on 15-L sample owing to ceiling concentration of
7 mg/m^3 ([6]).

Analytical Recovery. Mean recoveries from spiked silica
gel samples indicated that the desorption efficiencies for
each of the acids studied is essentially complete. Table III
lists the mean recoveries and precision for each of the acids.

Table III
Analytical Recovery and Precision

Acid	N	Mean Recovery (%)	Precision (% RSD)
H_3PO_4	18	99.1	2.9
HBr	8	102.0	5.6
HNO_3	6	102.0	1.8
H_2SO_4	12	97.6	2.8

This complete desorption is owing to the sorbent nature of
the silica gel. An aqueous solution will fill the pores of
the silica gel to such an extent that the silica gel occupies
less than 1% of the space. When 10-mL of solution are added
to a 1-mL volume of silica gel in a graduated centrifuge tube,
the resulting volume in the tube is less than 10.1 mL. Thus
the silica gel occupies less than 1% of the total volume.
This free movement of liquid through the silica gel allows for
complete solution of the absorbed substances.

Aerosol Collection In A Sorbent Tube. Aerosols in the
atmosphere are drawn into the sampling tube by a vacuum
applied to the tube at a constant flow rate. Very large

particles will follow the air stream lines and enter the
sampler. The incoming particle may impact or intercept on the
glass wool plug, or it may penetrate the glass wool by
following the air stream to the sorbent bed. Each granule of
sorbent in the packed bed offers an opportunity for capturing
the aerosol particle by impaction, direct interception or
diffusion. The possibilities for capture may be increased by
lengthening the sorbent bed and/or by using a smaller mesh
sorbent, thereby reducing the void space within the packed
bed. These alterations, however, are limited by the
increasing pressure drop across the tube which may completely
impede the air flow.

   Optimization of Collection Tube Geometry. A solid sorbent
tube, a traditional gas and vapor collection device, has been
adapted to collect particles as well. The physical processes
of collecting aerosols required modification of the sorbent
tube to efficiently retain the particulates as well as the
vapors. Several factors play a role in the collection of an
aerosol by a packed bed: the aerosol particle size, the inlet
velocity, the sorbent mesh size, and the length of the sorbent
bed (12). Initially, silica gel collection tubes developed
for the HCl method mentioned above (7mm O.D./4.8 mm I.D. glass
tube packed with 20-40 mesh silica, 400 mg front and 200 mg
backup) were evaluated in test atmospheres containing the
inorganic acid mists: HCl, H_3PO_4, HNO_3, and H_2SO_4.
Hydrochloric acid was included here for a comparison with
results of previous work where hydrogen chloride gas was used
in sample generation. Collection of the vapor-forming acids
was excellent, 100% for both HCl and HNO_3 (n=6); however,
the two aerosol-forming acids were not completely collected on
the glass wool plug and front section of silica gel.
Independent determinations of generated acid concentrations
were not made in all of the experiments. Initially impingers
were used as the independent sampling method to collect both
the vapors and aerosols. However, they were found to be less
than 60% efficient. Consequently, in the optimization of the
sampling tube for particle collection, a measure of collection
effectiveness was calculated using quantities found on the
backup section of the sampling tube. The mass of sample
collected on the front section (and the initial glass wool
plug) divided by that mass plus the mass on the backup section
is defined as the collection ratio.

$$\text{Collection (\%)} = \frac{\text{mass found on front section and glass wool}}{\text{total mass found}} \times 100$$

Ratio

This collection ratio was used to indicate better

performance. At very high collection ratios (> 90%) this
number is assumed to be equivalent to the collection
efficiency. In other experiments a direct comparison was made
between the silica gel collection and the results of filter
collection for H_3PO_4 and H_2SO_4. The mean collection
ratio (n=6) of the 400-mg silica gel tube for H_3PO_4 was
94.8% with a relative standard deviation (RSD) of 4.8%, and
86.4% for H_2SO_4 with an RSD of 4.6%.

The bar graphs in Figures 5 and 6 illustrate the effect on
the collection ratio of the sampler made by varying the silica
gel mesh size, tube diameters and resultant inlet velocities,
and the length of the sorbent bed. The collection ratios
given are the mean of results from several sample generations
within the air concentration range of 1-2 mg/m^3.

Mesh Size. An interesting point noted by the data in
Figure 5 is that the silica gel with the larger mesh size,
20-40, was a better collector of the acid mists than the
smaller 35-60 mesh. This may be explained by a combination of
the absorbing properties of the silica gel and the fact that
the particles are liquid. It is speculated that the larger
granules may better absorb and hold the liquid droplets of the
aerosol once they are encountered.

Sampling Velocity. The data illustrated in Figure 6 for
20-40 mesh silica gel indicate decreasing retention of the
aerosol with decreasing velocities in the sampling tube. The
sampling velocity is the flow rate through the tube divided by
the cross-sectional area of the tube. Higher velocities
should improve the particle collection by impaction.

The first group of data in Figure 6 shows collection
ratios of three different sampling velocities. The collection
of H_3PO_4 at 25, 15, and 10 cm/s was 97, 95, and 88%
respectively. The collection of H_2SO_4 was 89, 86, and 76%
respectively for the same velocities. The second group of
data shows H_3PO_4 collection at 100% and 88% for velocities
of 15 and 10 cm/s, and H_2SO_4 collection at 98% and 94% at
these velocities. These data indicate that higher velocities
result in better sample collection.

Bed Length. While the first group of data in Figure 6
indicates that a higher sampling velocity results in better
collection, the collection ratios are not at the optimum. A
means of further optimizing the collection ratios is to
increase the length of the sorbent bed. The first group of
data was derived from silica gel tubes with a bed length of
3.5 cm. At this bed length, the best collection ratios
resulted from a sampling velocity of 25 cm/s, which
corresponds to a 6-mm O.D. tube with a 300-mg silica gel

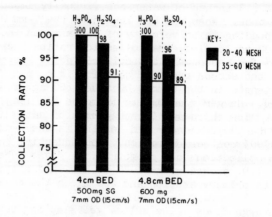

Figure 5.   *Effects of silica gel mesh size on collection ratio*

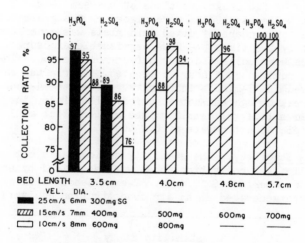

Figure 6.   *Collection of phosphoric acid and sulfuric acid on 20–40 mesh silica gel varying collection tube diameter and bed length*

loading. When the bed length was increased to 4 cm with
400-mg silica gel loading, the pressure drop across the tube
increased to 6.3 inches of water and virtually impeded flow
through the tube. When the bed length of the 7-mm O.D. tube
with corresponding inlet velocity of 15 cm/s was increased to
4-cm by a 500-mg loading, the collection ratios increased to
100% and 98% for phosphoric and sulfuric acids respectively,
as shown in the second group of data in Figure 6. To achieve
a 4-cm bed length in the 8-mm O.D. tube, 800 mg of silica gel
was required. Larger loadings to increase bed length seemed
impractical. The third and fourth groups of data in Figure 6
illustrate the use of longer bed lengths in the 7-mm O.D. tube
to achieve 100% collection ratios. Consequently, the
recommended collection tube geometry is a glass tube, 7-mm
O.D./4.8-mm I.D. packed with 20-40 mesh washed silica gel,
700 mg in the primary section and 200 mg in the backup section.

Breakthrough Study. The silica gel sampling tube has been
optimized for the collection of aerosol-forming acid mists as
well as vapor-forming acids. Now its capability for
collecting at least a 4-hour air sample must be determined.
In the absence of a direct method of monitoring breakthrough
for each of the inorganic acids, breakthrough was determined
by analysis of the backup sections of the samplers.
Twelve collection tubes were placed in the generation
system. An atmosphere of mixed acid mists was generated at
approximately two times the OSHA permissable exposure limits,
and samples were collected over a four-hour period. The first
sample was removed at 1 hour 15 minutes. Thereafter samples
were removed in 15-minute increments. After the 4-hour
period, breakthrough had not occurred. Therefore, the
recommended silica gel sampling tube will collect at least a
4-hour sample at a flowrate of 0.2 Lpm.

Sample Stability. The stability of inorganic acid samples
on the silica gel collection tubes was determined by storing
samples for a period of 10 days. Twelve 3-hour acid mist
samples were generated at an air concentration equivalent to

Table IV
Stability Study

Day Analyzed	N	HBr		HNO$_3$		H$_2$SO$_4$	
		Conc.	Precision	Conc.	Precision	Conc.	Precision
Day 1	6	3.8	10.6	10.1	3.0	1.0	2.8
Day 10	6	4.1	7.4	11.8	2.1	1.2	10

Average concentration in mg/m^3 and the percent relative
standard deviation.

the OSHA standard for each of the acids. Six samples were desorbed on day 1, and six were stored at room temperature for 10 days. The data listed in Table IV indicate that the samples are stable.

Sampling and Analytical Precision and Accuracy. The accuracy of results obtained from a sampling and analytical method are determined by comparison with an independent method. The aerosols of phosphoric and sulfuric acids were collected on mixed cellulose ester filters (14) and analyzed by ion chromatography. Table V shows the collection efficiency of the 7-mm O.D./700-mg silica gel collection tubes with respect to the results obtained from filter samples, and the precision obtained from calculation of the pooled relative standard deviation.

Table V
Sampling and Analytical Precision and Collection
Efficiency Based on Results of Filter Samples

Acid	N	Collection Efficiency(%)	Precision* (RSD)
H_3PO_4	18	88	10.5
H_2SO_4	25	90	8.7

* Pooled value for three concentration levels.

Independent sampling methods for the vapor-forming acids involved the use of impingers or bubblers. Early experiments in this study found that the efficiencies of impingers, containing the buffer solution used as eluent in the analytical method (0.003 M $NaHCO_3$/0.0024 M Na_2CO_3 at a pH of 10.4), averaged approximately 60%. Even when three impingers or bubblers were collected in series, significant amounts of acid were found in the third impinger. For this reason coupled with the variation found in duplicate samples, the use of impingers was discontinued as an independent sampling media.

Silica gel was found to be a better collector of the vapor-forming acids than were impingers or bubblers. Silica gel samplers collected an average of 60% more HCl than parallel impinger samples, and 35% more HNO_3. Although the silica gel tubes used here varied in size and loading, all the HCl and HNO_3 vapors were collected on the primary section. Since the collection ratios were 100% and the samples are stable for at least 10 days, it is concluded that silica gel samplers are superior to impingers for the collection of acid vapors.

The silica gel collection tubes are being field tested in an electroplating operation for H_2SO_4, HNO_3, and HCl, and in metal preparation for painting for H_3PO_4.

Summary

This work has demonstrated that sorbent tubes are viable samplers for inorganic acid mists existing as vapors and aerosols. A silica gel sampling tube was developed which will collect at least a 4-hour sample of inorganic acid at a nominal flow rate of 0.2 Lpm. The optimum sampler geometry was determined to be a 7-mm O.D./4.8-mm I.D. glass tube packed with 20-40 mesh washed silica gel, 700 mg in the primary section and 200 mg in the backup.

The Lee and Gieseke model ($\underline{8}$) for predicting aerosol collection on packed beds assumes that the volume fraction, or solidity, of the packed bed approaches 5/8, and applies to the collection by the packed bed only. Since the silica gel collection tube has a volume fraction of one-half or less, and since the greater percentage of the aerosol is collected on the initial glass wool plug, the model is not applicable to our sampler design.

The analytical method by which all the inorganic acids may be analyzed in a single sample is ion chromatography. Using the stated instrumental conditions the analytical range is 0.5-20 µg/mL for H_3PO_4, HBr, HNO_3, and H_2SO_4, and 0.2-20 µg/mL for HCl (30 µmho conductance). The sample range is 10-100 µg (0.5-2 mg/m^3) for H_3PO_4 and H_2SO_4, 20-200 µg (0.14-14 mg/m^3) for HCl, 50-500 µg (1-10 mg/m^3) for HNO_3, and 100-960 µg (2-20 mg/m^3) for HBr.

Acknowledgement

The comments and suggestions of Drs. Peter M. Eller, Paul A. Baron, and Jerome P. Smith concerning aerosol generation and behavior are gratefully acknowledged.

Disclaimer

Mention of a company name or product does not constitute endorsement by NIOSH.

Abstract

Airborne inorganic acids exist in the workplace environment as both vapors and particulates. To monitor for the common inorganic acids, a single, non-liquid sampling device to collect both vaporous and aerosol contaminants quantitatively, and an analytical method to determine these acids in a single sample was desired.

Ion chromatography offers the analytical tool for the determination of inorganic acids in a single sample. In previous work on hydrogen chloride-hydrogen bromide mixtures, various sample collection media, filters and solid sorbents - treated and untreated, were evaluated. The sorbent of choice was silica gel which had been washed with deionized water. In the present study, collection tubes packed with 400 mg of washed silica gel were evaluated in test atmospheres containing five inorganic acids: HCl, H_3PO_4, HBr, HNO_3, and H_2SO_4. To determine the optimum collection tube geometry, experiments were conducted varying (1) the tube diameter which affects the inlet capture velocity, (2) the silica gel loading which determines the length of the sorbent bed, and (3) the mesh size of the silica gel which determines the air flow properties and the pressure drop of the collection tube.

Ultimately, a single sampling device was designed which collected both the vaporous and particulate forms of inorganic acids with subsequent analysis by ion chromatography.

Literature Cited

1. "Chemical Safety Data Sheets"; Manufacturing Chemists Association: Washington, D. C., 1963.

2. Sax, N. I. "Dangerous Properties of Industrial Materials", 3rd ed.; Van Nostrand Reinhold Co., New York, 1968.

3. "International Labour Office: Encyclopaedia of Occupational Health and Safety"; McGraw-Hill Book Co., New York, 1971.

4. Taylor, D. G., Ed. "NIOSH Manual of Analytical Methods" Vol. 4, DHEW (NIOSH) Publication No. 79-141, 1979.

5. Small, H.; Steven, T. S.; Bauman, W. C. Anal. Chem., 1975, 47, "No. 11", 1801-9.

6. Cassinelli, M. E.; Eller, P. M. "Ion Chromatographic Determination of Hydrogen Chloride"; Paper No. 150, American Industrial Hygienist Conference, Chicago, 1979.

7. Hydrogen Chloride, P&CAM Method No. 310, "NIOSH Manual of Analytical Methods", 2nd ed., Vol. 5; DHEW (NIOSH) Pub. No. 79-141; 1979.

8. Lee, K. W.; Gieseke, J. A. Environmental Science and Technology, 1979, 13, 466-470.

9. Ortiz, L. W.; Fairchild, C. I. "Aerosol Research and Development Related to Health Hazard Analysis", LASL Report LA-6539-PR, USERDA Contract W-7405-Eng. 36, 1976.

10. "Ion Chromatography Training Course"; Dionex Corp:
     Sunnyvale, CA, 1978; 35-7.

11. "Federal Register", 1971, Vol. 36, No. 157, Part 1910.

12. Dennis R., Ed. "Handbook on Aerosols"; Technical
     Information Center, ERDA, 1976.

13. Raabe, O.G. "The Generation of Aerosols of Fine
     Particles", Inhalation Carcinogenesis; Academic Press,
     New York, 1976, 74.

14. "NIOSH Manual of Sampling Data Sheets"; DHEW (NIOSH)
     Pub. No. 77-159, 1977, 33-1, S174-1.

RECEIVED October 14, 1980.

MONITORING AND CONTROL

Specialized Sorbents, Derivatization, and Desorption Techniques for the Collection and Determination of Trace Chemicals in the Workplace Atmosphere

R. G. MELCHER, P. W. LANGVARDT, M. L. LANGHORST, and S. A. BOUYOUCOS

Michigan Division Analytical Laboratories, 574 Building and Health and Environmental Sciences, Dow Chemical U.S.A., Midland, MI 48640

The use of solid sorbents for the collection of industrial hygiene samples has greatly increased in the past few years. This technique has proven valuable, not only to the industrial hygienist because the solid sorbent sampling tubes are easy to use and transport, but also to the analytical chemist who finds the analytical procedures straightforward and adaptable to a wide range of compounds.

In the early development of solid sorbent sampling techniques, a ten minute sampling period was usually adequate to establish the concentration of a compound in air in an area. This technology was quickly followed with emphasis toward portable personal sampling systems which could be used up to eight hours, to determine the "time-weighted-average" (TWA) exposures. Recent emphasis has been in examining and understanding parameters which affect collection and recovery and to develop validation criteria to establish reliability of the methods.

Although methods for many new compounds can be extrapolated by comparison to existing methods, difficulties with collection, recovery, stability or analysis often arise which require further sophistication of the technology. Various techniques have been developed to extend the useful range of available solid sorbents and new specialized sorbents have been developed to solve specific problems. The development of specific and highly sensitive gas chromatographic detectors and other new analytical instrumentation has enabled a greater flexibility in the collection parameters and choice of sorbent/solvent systems. The use of chemical reactions and derivatization will also play an increasingly important part in solid sorbent methods for compounds which are unstable or difficult to analyze.

Some of the recent developments in technique, sorbents and instrumentation which appear to have broad application

are discussed in this paper. A few specific examples are
given in each area to illustrate the concepts; however,
diligent surveillance of the literature will be necessary to
keep pace in this rapidly growing field.

Sorption

In designing and developing methods based on solid
sorbents it is desirable to have some guidelines to predict
the collection (breakthrough) of various compounds.
Although some equations have been developed to relate the
properties of compounds to collection efficiencies (1,2,3),
some of the equations became quite involved and often
required information not readily available for many compounds. Correlations, however, are being developed which
relate a property, such as boiling point, to the breakthrough volume (4,5).
In an approximation, the breakthrough time for organic
compounds on charcoal is dependent strongly on the boiling
point while for polar sorbents such as silica gel and
alumina, the polarity of the collected compound is highly
important. For porous polymers such as Tenax-GC and
Amberlite XAD-2 resin, boiling point and molecular polariz-
ability are important factors. Table I lists a number of
generally used sorbents and the types of compounds collected. One must remember that the choice of a sorbent is
often a compromise between the collection properties and the
desorption properties for a particular chemical. For
example, charcoal is an excellent sorbent for many compounds
but the recovery of many compounds is poor. Various
physical and environmental factors affect collection and
recovery and are discussed in the literature (4, 6,7).
Several techniques for determining the breakthrough are also
discussed (8,9,10). Theoretical treatment of breakthrough
has been suggested (11) in an attempt to accurately determine concentrations in air even though a large amount of the
compound is found on the back-up section. This approach
would not work with highly volatile compounds since compounds such as methyl chloride (12) will elute through a
charcoal collection tube while pumping only pure air.

Desorption and Recovery

Percent recovery is the percent of the chemical collected under actual sampling conditions which is recovered
for analysis. The major contributing factor to low recovery
is often the desorption efficiency, i.e., the partitioning
of a chemical, at equilibrium, between a specific kind and
volume of solvent and a specific batch and amount of solid
sorbent. Recovery and desorption efficiency are not always

TABLE I

SOLID SORBENTS USED FOR THE COLLECTION
OF ORGANIC COMPOUNDS IN AIR

Sorbent	Use
Activated Carbon	Most widely use sorbent for volatile compounds: aliphatic and aromatic solvents, chlorinated aliphatics, acetates, ketones, alcohols, etc.
Silica Gel	Useful for high-boiling compounds and highly polar compounds: alcohols, phenols, chlorobenzenes, aliphatic and aromatic amines and acids, etc.
Activated Alumina	Useful for high-boiling compounds and polar compounds which are difficult to recover from silica gel: alkanolamines, glycols, phenoxyacetic acids, etc.
Porous Polymers Amberlite XAD-2 Tenax-GC Chromosorb 101, 102, 106 Porapak N, Q	Useful for a wide range of compounds: phenols, acidic and basic organics, multi-functional organics, etc.
Carbonized Resins Saran Carbon Carbosive B, S Ambersorb XE-340, XE-347, XE-348	Useful for highly volatile compounds (often better than activated carbon): methyl chloride, vinyl chloride, chloroform, dimethyl ether, etc.
Chemically bonded and other GC packings	Specialized uses: High boiling compounds, pesticides, herbicides, polynuclear aromatics, etc.
Reactant-Coated Sorbent	Reactants coated on a solid support are designed to react with compounds with a specific chemical property or functional group.

equal since other factors such as humidity, temperature,
compound stability to oxidation, etc., may reduce recovery.

Recovery = Desorption Efficiency ± Other Factors

Once the desorption efficiency is determined, the effects
of other factors can be examined by laboratory
experimentation and field validation.

A rough prediction of desorption and recovery can be
made by extrapolation from similar compounds; however,
until we know more about the factors involved, this practice
can lead to difficulties as illustrated in Table II. The
recovery of naphthalene and biphenyl from charcoal are much
lower than one would predict from the recoveries of other
aromatic compounds, and this stronger attraction may be
related to bond angles or steric effects. The order of
recovery, ortho (84%) > meta (81%) > para (76%) methyl
biphenyl, supports this theory. One may predict from this
information that the recovery of benzene would be very
poor, which is not the case.

The most common technique in determining desorption
efficiency is to inject the compound or a solution of the
compound directly into the solid sorbent (13). The mixture
is allowed to stand overnight and then desorbed and analyzed. Gases and highly volatile compounds are usually
introduced as a mixture in air or nitrogen from a SARAN*
film bag or a cylinder.

The percent recovery is determined in the same manner
except a known flow of air is pulled through the sorbent at
the same time to determine the effect of the atmosphere.
The closer the test atmosphere is to the actual air to be
sampled the more accurate the recovery factor will be.

Phase Equilibrium. An equation has been derived (14)
which relates the desorption efficiency to the volume of
solvent and the amount of sorbent. The equation assumes
the system is in equilibrium and can be approached from
either direction. That is, the same desorption efficiency
should be obtained when the compound is initially in the
solvent or the solid phase. This has been shown to apply
to most organic compounds in the concentration range of
interest in Industrial Hygiene analyses. The equations
below can be used to optimize the solid/liquid ratio when
developing an analytical procedure:

$$\frac{1}{D} = K \frac{W_s}{W_1} + 1 \tag{1}$$

where D = desorption efficiency (as a decimal fraction)
      Ws = weight of solid phase

TABLE II

DESORPTION EFFICIENCY OF AROMATIC COMPOUNDS

Sorbent: Coconut Base Carbon 0.5 grams

Solvent: Carbon Disulfide 5 mL

Concentration: 0.2-0.3 mg/5 mL

	Compounds	Desorption Efficiency	
		After 1 hr.	After 16 hrs.
	Benzene	98	98
	Toluene	98	98
	Naphthalene	43	40
	Biphenyl	55	53
	Diphenyl Oxide	93	98
	Diphenyl Methane	93	97
	Para Methyl Biphenyl	76	80
	Meta Methyl Biphenyl	81	86
	Ortho Methyl Biphenyl	84	91

W_1 = weight of liquid phase
K^1 = constant for specific compound and sorbent

Once K is determined, the desorption efficiency can be calculated for any solid and liquid ratio. Another equation (15) can be derived by setting K equal for two different solvent volumes:

$$\frac{1}{D_n} = \frac{1}{n}\left(\frac{1}{D_1} - 1\right)+1 \qquad\qquad (2)$$

where n = the ratio of solvent volumes = $\dfrac{\text{new volume}}{\text{initial volume}}$

D_1 = initial desorption efficiency

D_n = desorption efficiency using new volume

Another useful equation (15) has been derived which can be used to determine the ratio n necessary to obtain a desired desorption efficiency Z:

$$n_Z = \left(\frac{1}{D_1} - 1\right)\frac{Z}{1-Z} \qquad\qquad (3)$$

where n_z is the number of multiples of the original volume of desorbent needed to reach the desired desorption efficiency Z. Although we can increase the desorption efficiency by increasing the volume we also dilute the sample. An optimum system must also consider the dilution factor and the sensitivity of the analytical method.

Improving Recovery

Since charcoal is such a good sorbent and is readily available, the solution to some sampling problems is to find a way to increase the recovery of that compound from charcoal. One way is by increasing the solvent/sorbent ratio as discussed in the phase equilibrium section. Two other approaches are the use of mixed solvents and the two-phase solvent system.

Mixed Solvents. In general, polar compounds usually show low recoveries from charcoal. By adding several percent of a polar solvent to the carbon disulfide desorbent solvent, recovery is often improved by 10 to 20% (16,17,18). Methanol is the most frequently added polar solvent, and it is usually effective as long as it does not interfere with the gas chromatography. If the methanol/carbon disulfide system is used the samples should be run within four hours after desorption since methanol reacts with carbon disulfide in the presence of charcoal (4). In some cases ethanol, butanol, isopropanol or acetone has been added to carbon disulfide to increase desorption efficiency. One or two percent acetone in carbon disulfide has been used to increase the recovery of acrylonitrile from charcoal; however, much larger amounts can be used if needed for other compounds since acetone is completely miscible with carbon disulfide.

Two-Phase Desorption System. The mixed solvent technique has limited use for complex mixtures since it is more difficult to chromatograph, precludes determination of the polar solvent added, and may cause additional interference to other compounds present. A two-phase system has been developed which is capable of measuring both polar and non-polar organic solvents present simultaneously in work environments (19).

The charcoal collection tubes are desorbed with a 50/50 mixture of carbon disulfide and water. After desorption, the water and carbon disulfide layers are analyzed separately. The high recoveries of the polar compounds are attributed to their partitioning into the aqueous phase after desorption from charcoal by carbon disulfide. Table III shows recovery data for 15 common industrial solvents. The chromatograms for each phase are shown in Figure 1A and 1B. Not only does the partitioning eliminate interferences of some polar and non-polar combinations, but the partition coefficients give additional qualitative information. Other modifications using an acidic or basic aqueous phase, or other immiscible solvents such as methanol/carbon disulfide, are presently being studied. Once the compounds are desorbed, a more complex extraction/separation scheme can be used before analysis, if necessary (20).

Other Sorbents

Other solid sorbents have been found more suitable than charcoal for a number of compounds. Silica gel and alumina have been used as a complement to charcoal when sampling polar compounds, but water vapor is strongly adsorbed on these sorbents which leads to deactivation of the sorbent and breakthrough of the compounds by frontal elution. Difficulties also arise with compounds that hydrolyze easily. Alternative sorbents for the collection of polar organic compounds which are sensitive to hydrolysis are porous polymers such as the Chromosorb porous polymer series, Porapak porous polymer series, Tenax-GC and Amberlite XAD sorbent series.

TABLE III

DETERMINATION OF SOME COMMON SOLVENTS IN AIR USING TWO-PHASE DESORPTION

Compound	Molecular Weight	Density* (mg/μL)	Retention Time (minutes)	Partition Coefficient (% in aqueous phase)	Mean Recovery (%)
ethanol	46	0.801	1.83	100	97
acetone	58	0.791	2.41	74	100
2-propanol	60	0.782	3.20	96	95
1-propanol	60	0.804	4.05	95	94
methyl ethyl ketone	72	0.805	5.21	47	83
ethyl acetate	88	0.902	6.10	16	92
2-methoxyethanol	70	0.965	6.38	100	95
1-butanol	74	0.810	7.70	71	91
isopropylacetate	102	0.872	8.65	5	97
2-ethoxyethanol	90	0.930	9.23	100	95
n-hexane	86	0.660	9.33	0	99
n-propyl acetate	102	0.888	10.1	4	92
methyl isobutyl ketone	100	0.801	11.9	3	91
n-heptane	100	0.684	13.6	0	100
toluene	92	0.865	13.7	0	102

*Aldrich Catalog Handbook of Fine Chemicals 1979-1980

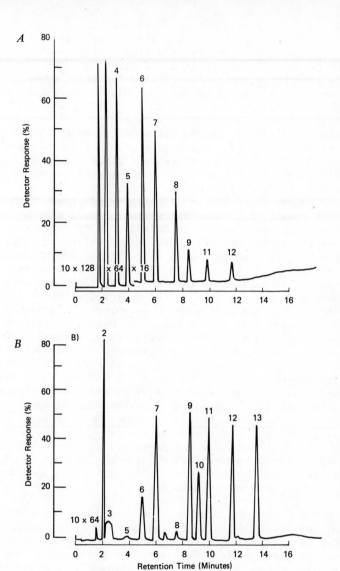

American Industrial Hygiene Association Journal

Figure 1.   Separation of selected solvents spiked into a synthetic sample at TLV levels: (A) aqueous phase, (B) carbon disulfide phase; 1, ethanol, 2, acetone, 3, carbon disulfide, 4, 2-propanol, 5, 1-propanol, 6, methyl ethyl ketone, 7, ethyl acetate, 8, 1-butanol, 9, isopropyl acetate, 10, n-hexane, 11, n-propyl acetate, 12, methyl isobutyl ketone, 13, toluene (19)

The use of various sorbents and desorption systems permits (often necessitates) gas chromatographic detectors other than the most commonly used flame detector or electron capture detector. The use of a selective detector can greatly simplify a difficult analytical problem by reducing interferences and background peaks. Many applications are discussed in the literature (21,22,23,24).

Amberlite XAD-2. XAD-2 polymeric sorbent (25) is composed of single resin beads consisting of an agglomeration of numerous minute microspheres (Figure 2). A clue to the efficiency of collection while minimizing reactivity may be found in the structure of the resin.

The porous structure is of an open-cell type so that water can readily penetrate the pores. During adsorption, the hydrophobic portion of the molecule is preferentially adsorbed on the hydrophobic polystyrene surface of the adsorbent through van der Waals attraction. The compounds being adsorbed do not penetrate substantially into the microsphere but remain adsorbed at the surface thus allowing the adsorbate to be rapidly eluted during the recovery step.

XAD-2 resin has been used for a number of compounds (24,26,27,28,29) and, for example, has been shown valuable for the collection of reactive organo-thiophosphates (30). The reactive compounds, 0,0-diethylphosphorochloridothioate (DEPCT), 0,0-dimethylphosphorochloridothioate (DMPCT) and monoethylphosphorodichloridothioate (MEPCT) were collected on XAD-2 resin, and no loss in recovery was observed after humid air (95% R.H.) was passed through. Hydrolysis is observed for these compounds in atmospheres with high humidity, but once collection has occurred, the resin tends to stabilize the compounds and reproducible recoveries are obtained even after seven days of storage. Figure 3 shows the gas chromatographic determination of these compounds using a flame photometric detector.

Bonded Sorbents. Another type of sorbent which has special applications is the bonded sorbent. The chemically bonded packings developed for gas chromatography were shown to have excellent collection and recovery properties for high boiling compounds (31). These packings, commonly called "brush" packings (32) because of the orientation of the organic moiety on the surface of the support, have been reported (33) to retain the organic compounds through the dominant mechanism of adsorption on the brush. The high liquid desorption recoveries of high boiling compounds can be attributed to the deactivated solid support and to the thin layer and low amount of bonded liquid phase. A bonded packing GC Durapack-Carbowax 400/Porasil F has been used to collect several types of pesticides (Table IV).

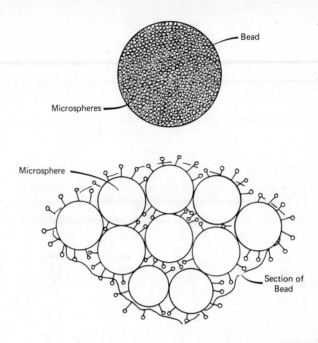

American Industrial Hygiene Association Journal

Figure 2. Schematic of Amberlite XAD-2 resin. The "ball and stick" designation represent molecules containing a hydrophobic ("stick") portion and a hydrophilic ("ball") portion. (30)

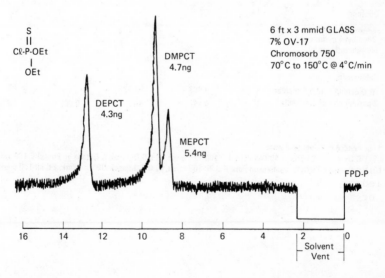

American Industrial Hygiene Association Journal

Figure 3. Chromatogram of organo-thiophosphates using Tracor 550 gas chromatography equipped with flame photomeric detector specific for phosphorous (4)

TABLE IV

**LABORATORY COLLECTION AND RECOVERY EXPERIMENTS
FOR SELECTED PESTICIDES ON BONDED PACKINGS**

Pesticide	Sorbent[a]	Amount Added (μg)	Amount Found (μg)		% Recovery
			Front Section	Back Section	
Ronnel	1	54.5	47.3	0.5	88
[O,O-dimethyl O-	2	51.3	41.1	2.6	85
(2,4,5-trichlorophenyl)	3	50.5	35.7	5.3	81
phosphorothioate]	4	39.0	25.4	1.9	70
Lindane	1	5.93	5.68	0.002	96
(1,2,3,4,5,6-	2	5.21	4.56	0.05	88
hexachlorocyclohexane)	3	5.97	5.74	0.004	96
	4	5.90	5.53	0.02	94
Carbaryl	1	6.4	6.0	Not detected	94
(naphthyl N-methyl	2	11.8	3.6	" "	31
carbamic acid)	3	7.5	4.8	" "	72
	4	9.6	2.4	" "	25
Diazinon	1	1.2	1.0	" "	83
(diethyl 6-isopropyl- 2-methyl-4 pyrimidinyl phosphorothionate)					
Chlorpyrifos					
[O,O-diethyl O-(3,5,6 trichloro-	1	0.083	0.072	" "	87
2-pyridy)-phosphorothioate]	1	0.19	0.17	0.008	93

[a] The bonded packings used were:
1) GC Durapak Carbowax 400 on Porasil F 100-120 mesh, 2) GC Durapak N-Octane on Porasil C 100-120 mesh,
3) GC Durapak Phenylisocyanate on Porasil C 80-100 mesh, 4) GC Durapak OPN on Porasil C 80-100 mesh

Recovery of compounds such as chlorpyrifos is very poor from sorbents such as charcoal, alumina and silica gel. A small tube containing bonded sorbent (Figure 4) can be used for over eight hours for the collection of chlorpyrifos with no breakthrough and 95% recovery. Another advantage of the bonded packing is its low retention of water vapor and volatile solvents. Solvent vapors readily pass through the bonded sorbent without increasing the breakthrough of chlorpyrifos. This property reduces interference and allows collection of the solvent vapors on a subsequent charcoal tube in series.

Carbonized Resins. A special sorbent made by controlled thermal pyrolysis of polyvinylidene chloride (Dow developmental Adsorbent XF-4175L) (34) was shown to be three to five times more effective for the collection of highly volatile compounds, such as vinyl chloride (Figure 5) and methyl chloride, than the best available activated charcoal (31,36,37). Although this sorbent is not commercially available, Carbosive B and Carbosive S show similar collection properties and they are available from gas chromatographic supply houses or may be obtained already packed in small collection tubes (SKC Inc., Eighty Four, PA).

Ambersorb carbonaceous adsorbents (Rohm and Haas Company) are a new class of synthetic adsorbents which show interesting collection properties (37,38). The chemical composition is intermediate between that of activated carbon and polymeric sorbents. Ambersorb sorbents are available in various pore sizes and surface areas.

Ion Chromatography

The new technique of ion chromatography (39,40) is finding increasing application in environmental and industrial hygiene analytical chemistry. (Ion Chromatography is the subject of U.S. Patent No. 3,920,397 and other U.S. patents, exclusively licensed to the Dionex Corporation by The Dow Chemical Company). It combines the separation capabilities of ion exchange with the sensitivity of a conductivity detector for the determination of inorganic cations and anions and for organic compounds which ionize or are convertible to anions or cations through decomposition or other organic reactions. Convenient personal sampling methods based on solid sorbent collection followed by ion chromatographic analysis are now being developed for difficult compounds such as formaldehyde (41) and sulfur dioxide (42).

Methylamines. Methylamines, which are difficult to

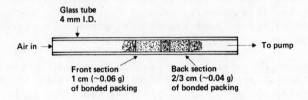

Analytical Chemistry

Figure 4.   Sampling tube containing bonded sorbent used to collect highly boiling compounds and pesticides (31)

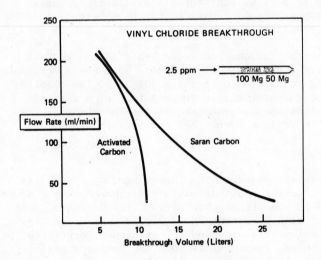

American Industrial Hygiene Association Journal

Figure 5.   A collection efficiency of activated coconut-base carbon and Saran carbon for vinyl chloride (4)

analyze by gas chromatography can conveniently be determined by ion chromatography ($\underline{43}$). A solid sorbent method has recently been developed for the determination of ammonia, monomethylamine, dimethylamine and trimethylamine in air at levels down to 0.3 ppm v/v ($\underline{44}$). Samples are collected on silica gel and desorbed with 0.2N H_2SO_4 in 90-10 (v/v) methanol-water. The extracts are then analyzed using a surface-sulfonated styrene divinylbenzene separating column and an AG1-X10 high capacity stripping column, Figure 6 ($\underline{43}$).

Sulfuryl Fluoride. In the case of sulfuryl fluoride ($SO2F_2$), a colorless, odorless gas used as a structural fumigant, hydrolysis is induced during the desorption step ($\underline{45}$). The charcoal sorbent, used to monitor worker exposures, is desorbed with 0.04 N NaOH and the sulfuryl fluoride undergoes partial hydrolysis:

$$SO_2F_2 + 2\ NaOH \rightarrow NaF + NaSO_3F + H_2O$$

Further hydrolysis of the sodium fluorosulfonate takes place during a heated evaporation step:

$$NaSO_3F + 2NaOH \xrightarrow{\Delta} NaF + Na_2SO_4 + H_2O$$

Ion chromatography is then used (Figure 7) to determine both the F^- and the $SO_4^=$ concentration from which the original amount of sulfuryl fluoride can be calculated.

Derivatization

Derivatizing reagents ($\underline{46}$) have been developed which can be used to stabilize reactive compounds, change their properties so that they can be more easily determined by gas chromatography, or increase their detectability by selective and sensitive detectors which would not be suitable for the original compound ($\underline{47\text{-}54}$).

Derivatization After Desorption. Alkanolamines, highly polar basic compounds, present a difficult analytical problem. Although direct gas chromatographic separations can be achieved, this technique is not applicable to trace analysis due to sorption problems at trace concentrations. A derivatization/gas chromatographic procedure has been developed for the determination of alkanolamines in air as low as 100 ppb ($\underline{54},\underline{55}$). The samples are collected on activated alumina and desorbed with an aqueous solution of 1-octanesulfonic acid. The

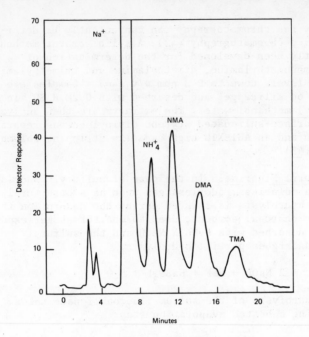

Figure 6.   Typical ion chromatogram of an experimental sample containing ammonia and methylamines

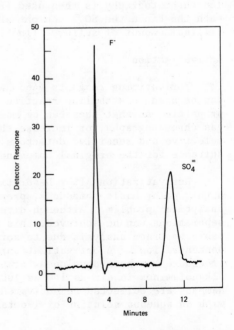

Figure 7.   Ion chromatogram of an experimental sample containing sulfuryl fluoride after complete hydrolysis

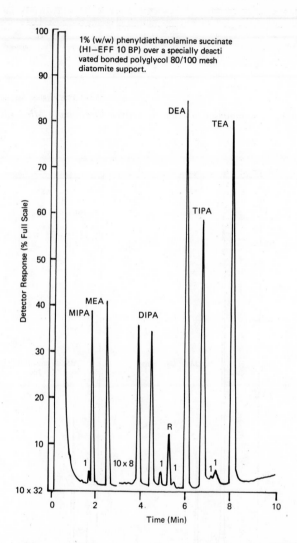

1% (w/w) phenyldiethanolamine succinate (HI—EFF 10 BP) over a specially deacti vated bonded polyglycol 80/100 mesh diatomite support.

Analytical Chemistry

Figure 8. Gas chromatographic separation of selected alkanolamines as their heptafluorobutyrul derivatives using specially prepared deactivated packing: 1, 51 μg/ mL MIPA; 2, 69 μg/mL MEA; 3a, 18 μg/mL DIPA; 3b, 18 μg/mL DIPA₂; 4, 46 μg/mL DEA; 5, 36 μg/mL TIPA; 6, 56 μg/mL TEA; R, reagent artifact; I, impurities in alkanolamine standards (54)

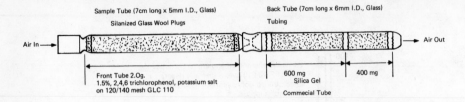

American Industrial Hygiene Association Journal

Figure 9.   Air sampling tube for derivative collection of CMME (57)

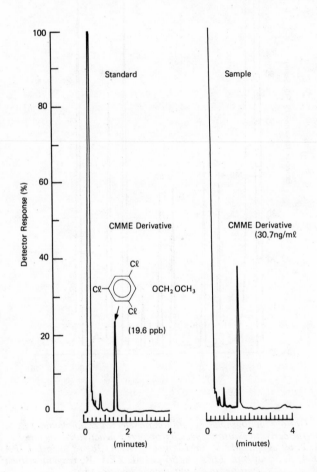

American Industrial Hygiene Association Journal

Figure 10.   GC-EC chromatograms of CMME derivative standard and simulated air sample (57)

acid not only aids in the desorption but also forms a
non-volatile salt which allows complete removal of the
water by lyophilization (freeze drying). The dried salt
is then derivatized using heptafluorobutyryl imidazole,
Figure 8 shows a gas chromatogram of alkanolamines as their
hepta-fluorobutyryl derivatives. Some of the requirements
for a suitable derivatizing reagent are pointed out by this
procedure:

1.   Reaction with both the amino and hydroxyl groups.
2.   Quantitative reaction to form single unique products.
3.   Derivative formed must be stable during removal of
      excess reagent.
4.   Derivative must be suitable for gas
      chromatographic analysis.

Reactant-Coated Solid Sorbent. The most difficult
compound to collect is one which is highly unstable and
whose products cannot be related back to the original
concentration. It is necessary in this case to form a
derivative on contact during the collection step by placing
the derivatizing reagent in the impinger (56) or coating it
on a solid support. A recent solid sorbent method has been
developed for chloromethylmethyl ether (CMME) (57). Air is
sampled with special collection/derivatization sorbent
(Figure 9) prepared by coating 1.5% potassium 2,4,6-tri-
chlorophenate on 120/140 mesh glass beads (GLC-110). A
derivative is formed by the reaction:

$$CH_3-O-CH_2-Cl + Cl\underset{Cl}{\overset{Cl}{\bigcirc}}-OK \rightarrow Cl\underset{Cl}{\overset{Cl}{\bigcirc}}-O-CH_2-OCH_3 + KCl$$

The stable derivative is desorbed from the collection/
derivative tube with methanol. Excess reagent is removed by
diluting with aqueous 1N NaOH and extracting with hexane.
The derivative is then analyzed by gas chromatography using
an electron capture detector. (Figure 10). A back-up tube
containing silica gel is used to collect any of the CMME
derivative which may elute through the front tube.

The most important feature of this procedure is that it
immediately stabilizes CMME for retention on the solid
adsorbent. In addition, the derivative significantly
increases detector sensitivity enabling detection of CMME
(as the derivative) at low parts-per-billion levels in air.
This approach eliminates the need for impingers and the need
for thermal desorption.

Literature Cited

1.   MSA Research Corporation; "Package Sorption Device
     Systems Study"; National Technical Information Service,
     1973, PB-221 138.
2.   Nelson, G. O.; Harder, C. A; Respirator Cartridge
     Efficiency Studies: VI; Effect of Concentration; Am.
     Ind. Hyg. Assoc. J., 1976, 37, 205.
3.   Wood, G. O.; Anderson, R. G.; Personal Sampling for
     Vapors of Aniline Compounds; Am. Ind. Hyg. Assoc. J.,
     1975, 36, 538.
4.   Melcher, R. G.; Langner, R. R.; Kagel, R. O.; Criteria
     for the Evaluation of Methods for the Collection of
     Organic Pollutants in Air Using Solid Sorbents,
     Am. Ind. Hyg. Assoc. J., 1978, 39, 349.
5.   Gallant, R. F.; King, J. W.; Levins, P. L.; Piecewicz,
     J. F.; "Characterization of Sorbent Resins for Use in
     Environmental Sampling", National Technical Information
     Service, 1978, PB-284 347.
6.   Mueller, F. X.; Miller, J. A.; Determination of Organic
     Vapor Mixtures Using Charcoal Tubes; Am. Ind. Hyg.
     Assoc. J., 1979, 40, 380.
7.   Shotwell, H. P.; Caporossi, J. C.; McCollum, R. W.;
     Mellor, J. F.; A Validation Procedure for Air
     Sampling-Analysis Systems; Am. Ind. Hyg. Assoc. J.,
     1979, 40, 737.
8.   Whitman, N. E.; Johnston, A. E.; Sampling and Analysis
     of Aromatic Hydrocarbon Vapors in Air: A Gas-Liquid
     Chromatographic Method. Am. Ind. Hyg. Assoc. J., 1964,
     25, 464.
9.   Severs, L. W.; Melcher, R. G.; Kocsis, M. J.; Dynamic
     U-Tube System for Solid Adsorbent Air Sampling Method
     Development; Am. Ind. Hyg. Assoc. J., 1978, 39, 321.
10. Melcher, R. G.; Severs, L. W., Kocsis, M. J.; Garner,
     W. L.; Specialized Solid Sorbents for Sampling Organic
     Compounds of High and Low Volatility in Air; 172nd ACS
     Natonal Meeting, San Francisco 1976.
11. Peterson, J. E.; Calculation of Vapor Penetration in
     Charcoal Sampling Tubes; Am. Ind. Hyg. Assoc. J.,
     1980, 41, 450.
12. Roush, G. J.; Coyne, L. S.; Charcoal Sampling Method
     for Methyl Chloride, a Gas with Low Charcoal Affinity;
     Am. Ind. Hyg. Conference; Paper 340, 1980.
13. Krajewski, J.; Gromiec, J.; Dobecki, M.; Comparison of
     Methods for Determination of Desorption Efficiencies;
     Am. Ind. Hyg. Assoc. J., 1980, 41, 531.
14. Dommer, R. A.; Melcher, R. G.; Phase Equilibrium Method
     of Determination of Desorption Efficiencies; Am. Ind.
     Hyg. Assoc. J., 1978, 39, 240.
15. Posner, J. C.; Comments on "Phase Euilibrium Method for

Determination of Desorption Efficiencies" and Some
Exensions for Use in Methods Development;
Am. Ind. Hyg. Assoc. J., 1980, 41, 63.

16. NIOSH Manual of Analytical Methods, Vol 1, Second
Edition, April 1977

17. NIOSH Manual of Analytical Methods, Vol 2, Second
Edition, April 1977

18. NIOSH Manual of Analytical Methods, Vol 3, Second
Edition, April 1977

19. Langvardt, P. W.; Melcher, R. G.; Simultaneous Determination of Polar and Non-polar Solvents in Air Using
a Two-Phase Desorption from Charcoal; Am. Ind. Hyg.
Assoc. J., 1979, 40, 1006.

20. Lamb, S.; Analysis of Polar Organic Pollutants, Second
Symposium on Environmental Analytical Chemistry,
Brigham Young University, Paper IIIA; 1980.

21. Ettre, L. D.; Selective Detection in Column Chromatography; J. Chromatogr. Sci. 1978, 16, 396.

22. Gluck, S. J.; The Response of the Model 700A Hall
Electrolytic Conductivity Detector to Sulfur Containing
Molecules; Symposium On Environmental Analytical
Chemistry, Brigham Young University, Paper IF, 1980.

23. Marano, R. S.; Levine, S.P.; Harvey, T. M.; Trace
Determination of Subnanogram Amounts of Acrylonitrile
in Complex Matrices by Gas Chromatography with Nitrogen-
Selective Detector; Anal. Chem. 1978, 50, 948.

24. Langhorst, M. L.; Nestrick. J. J.; Determination of
Chlorobenzenes in Air and Biological Samples by Gas
Chromatography; Anal. Chem.,1980, 51, 2018.

25. Rohm and Haas Company: Amberlite XAD-2, Technical
Bulletin Ion Exchange Department, Revised July 1967.

26. Farwell, S. O.; Bowes, F. W., Adams, D. F.; Evaluation
of XAD-2 as a Collection Sorbent for 2,4-D Herbicides
in Air; J. Environ. Sci. Health, 1977, B12(1), 71-83.

27. Jackson, J. W.; Thomas, T. C.; A Simple and Inexpensive
Method for Sampling 2,4-D and 2,4-T Herbicides in Air;
J. Air Poll. Control Assoc., 1978, 28(11), 1145-1147.

28 Langhorst, M. L.; Solid Adsorbent Collection and Gas
Chromatographic Determination of the Propylene Glycol
Butyl Ether Esters of 2,4,5-T in Air. Am. Ind. Hyg.
Assoc. J., 1980, 41(5) 328-333.

29. Gluck, S. J.; Melcher, R. G.; Concentration and Determination of the Propylene Glycol Butyl Ether Esters of
2,4-Dichlorophenoxyacetic acid (PGBE 2,4-D) in Air;
Am. Ind. Hyg. Assoc. J., in press.

30. Kaminski, F.; Melcher, R. G.; Collection and
Determination of Trace Amounts of Organo-Thiophosphates
in Air Using XAD-2 Resin; Am. Ind. Hyg. Assoc. J., 1978,
39, 678.

31. Melcher, R. G.; Garner, W. L.; Severs, L. W.; Vaccaro,

J. R.; Collection of Chlorpyrifos and Other Pesticides in Air on Chemically Bonded Sorbents; Anal. Chem., 1978, 50, 251.

32. Halasz, I.; Sebestian; Angew. Chem., Int. Ed. Engl., 1969, 8, 453.

33. Pesek, J. J.; Daniels, J. E.; J. Chromatogr. Sci., 1976, 14, 288.

34. Schmidt, D. D.; Jewett, G. L.; Melcher, R. G.; Wessling, R. A.; Method for Quantitatively Analyzing Substances Containing Elements Other Than Carbon, Hydrogen and Oxygen and Volatile Organic Compounds; U.S. Patent 3,967,928, July 6, 1976.

35. Hill, Jr., R. H.; McCammon, C. S.; Saalwaechter, A. T.; Teass, A. W.; Woodfin, W. J.; Gas-Chromatographic Determination of Vinyl Chloride in Air Samples Collected on Charcoal; Anal. Chem., 1976, 48, 1395.

36. Severs, L. W.; Skory, L. K., Monitoring Personnel Exposures to Vinyl Chloride Vinylidene Chloride and Methyl Chloride in an Industrial Work Environment; Am. Ind. Hyg. Assoc. J., 1975, 36 , 669.

37. Rohm and Hass Company; Ambersorb carbonaceous Adsorbents Technical Notes; Company Bulletin, 1977.

38. Neely, J. W.; A Model for the Removal of Trihalomethanes from water by Ambersorb XE-340; ACS Meeting, Miami Beach, Florida, September 1978.

39. Small, H.; Stevens, T. S.; Bauman, W. C.; Novel Ion Exchange Chromatographic Method Using Conductimetric Determination; Anal. Chem., 1975, 47, 1801.

40. Sawicki, E.; Mulik, J. D.; Wittgenstein, E., Eds. "Ion Chromatography Analysis of Environmental Pollutants"; Ann Arbor Science, Ann Arbor, MI, 1978.

41. Kim, W. S.; Geraci, Jr., C. L.; Kupel, R. E.; Solid Sorbent Tube Sampling and Ion Chromatographic Analysis of Formaldehyde; Am. Ind. Hyg. Assoc. J., 1980, 41, 334.

42. Smith, D. L.; Kim, W. S.; Kupel, R. E.; Determination of Sulfur Dioxide by Adsorption on a Solid Sorbent Followed by Ion Chromatography Analysis; Am. Ind. Hyg. Assoc. J., 1980, 41, 485.

43. Bouyoucos, S. A.; Determinaton of Ammonia and Methylamines in Aqueous Solution by Ion Chromatography; Anal. Chem., 1977, 49, 401.

44. Bouyoucos, S. A.; Melcher, R. G.; Collection and Determinaton of Ammonia and Methylamines in Air By Ion Chromatography; Am. Ind. Hyg. Assoc. (Submitted).

45. Bouyoucos, S. A.; Melcher, R. G.; Vaccaro, J. R.; Collection of Sulfuryl Fluoride in Air on Charcoal and Determination by Ion Chromatography; Am. Ind. Hyg. Assoc. J. (Submitted).

46. Blau, K.; King, G. S.; Eds., "Handbook of Derivatives for Chromatography"; Heyden and Son Ltd., London, 1977.

47. Vorbeck, M. L.; Mattick, L. R.; Lee, F. A.; Pederson, C. S.; Preparation of Methyl Esters of Fatty Acids for Gas-Liquid Chromatography; Anal. Chem., 1961, 33, 1513.
48. Kirkland, J. J.; Analysis of Polychlorinated Benzoic Acid by Gas Chromatography; Anal. Chem., 1961, 33, 1521.
49. Stanley, C. W.; Derivatization of Pesticide-Related Acids and Phenols for Gas Chromatographic Determination; J. Agr. Food. Chem., 1966, 14, 321.
50. Argauer, R. J.; Rapid Procedure for the Chloroacetylation of Microgram Quantities of Phenols and Detection by Electron-Capture Gas Chromatography; Anal. Chem., 1968, 40, 122.
51. Kawahara, F. K.; Gas Chromatographic Analysis of Mercaptans, Phenols and Organic Acids in Surface Water with use of Pentafluorobenzyl Derivatives; Env. Sci. Tech., 1971, 5, 235.
52. Yrjanheikki, E.; A New Method for Personnel Şampling and Analyzing of Phenol; 1978, 39, 326.
53. Knarr, R.; Rapporport, S. M.; Determination of Methanethiol at Parts-Per-Million Air Concentrations by Gas Chromatography; Anal. Chem., 1980, 52, 733.
54. Langvardt, P. W.; Melcher, R. G.; Determination of Ethanol- and Isopropanolamines in Air at Parts-Per-Billion Levels; Anal. Chem., 1980, 52, 669.
55. Melcher, R. G.; Langvardt, P. W.; Nauer, L.; Wallace, R. E.; Determination of Alkanolamines in Workplace Atmospheres at Sub Part-Per-Million Levels; Am. Ind. Hyg. Conference, Paper 299, 1980.
56. Langvardt, P. W.; Nestrick, T. J.; Hermann, E. A.; Braun, W. H.; Derivatization Procedure for the Determination of Chloroacetyl Chloride in Air by Electron Capture Gas Chromatography; J. Chromatogr., 1978, 153, 433.
57. Langhorst, M. L.; Melcher, R. G.; Kallos, G. L.; Reactive Adsorbent Derivative Collection and Gas Chromatographic Determination of Chloromethyl Methyl Ether in Air; Am. Ind. Hyg. Assoc. J., In press.

RECEIVED October 2, 1980.

Solid Sorbents for Workplace Sampling

ELLEN C. GUNDERSON and ELLEN L. FERNANDEZ

SRI International, Menlo Park, CA 94025

Worker exposure to toxic vapors is monitored by collecting samples in the breathing zone of the worker. These samples are usually returned to a laboratory for analysis. A suitable sampling pump and collection device are required. Lightweight, battery-operated pumps are available in a variety of flowrates. Collection devices are usually sorbent tubes or bubblers or impingers, but bags and evacuated containers have also been used. This discussion will focus on solid sorbents as collection media.

The examples presented in this paper are based on results of our laboratory method development and validation studies. These studies, performed at both SRI International and Arthur D. Little, Inc., were supported by the National Institute for Occupational Safety and Health (NIOSH) from 1974 to 1979. In an effort to provide validated sampling and analytical methods for determining worker exposure to toxic substances, we validated existing methods when possible and developed and validated new procedures when no methods were available. Evaluation and testing of solid sorbents played a major role throughout this work (1).

Typical of solid sorbent tubes used to sample a workplace environment are the commercially available charcoal tubes. These glass tubes are about 5-6 cm in length (6-mm O.D. by 4-mm I.D.), containing approximately 100 mg of sorbent in a front section and 50 mg of sorbent in a back section separated by a polyurethane plug. Other sorbent tubes vary depending on their application, the most obvious difference being the sorbent contained in the tube. The amount of sorbent may be increased to increase the capacity, or larger I.D. tubing may be used to reduce the pressure drop and allow higher sampling rates. The backup sorbent section may be contained in a separate tube, for instance, if thermal desorption is used or if the analyte has been shown to migrate during storage.

The sorbent tube is placed near the worker's breathing zone, and the outlet of the tube is attached to a calibrated personal sampling pump. A known volume of air is drawn through the tube. Alternatively, several passive charcoal badges are currently commercially available. No sampling pump is required for these devices,

and the diffusion rate of the substance into the charcoal determines the amount collected. Passive badges may soon be available
with sorbents other than charcoal. The same factors may be relevant in selecting a sorbent for a specific compound using these passive sampling methods.

Factors Affecting Sorbent Selection

Solid sorbents are materials with a microporous structure,
whose internal pores and outer surfaces are accessible for sorption.
Typical sorbents used for the collection of air samples have nominal size of 20/40 mesh, with pore diameters less than 50 Å, giving
rise to surface areas up to 1000 m^2/g.

Most solid sorbents rely on vapors being sorbed by a physical
adsorption mechanism: the substance enters the internal pores of
the sorbent and is held there by attractive forces considerably
weaker and less specific than those of chemical bonds. These weakly attractive forces facilitate desorption for subsequent analysis.
The mechanisms for physical adsorption have been studied extensively and are described mathematically by equations such as the Lang-
muir isotherm.

Other sorbent tubes are designed to trap vapors by a chemical
mechanism. Chemical sorption is almost always accomplished by coating a solid sorbent or support with the desired reactant. In this
way, the substance of interest is not only removed from the gas
stream being sampled, but is altered chemically. This type of sorption has the advantages of being more selective and rendering reactive substances stable. Many variables may affect the ability of a
sorbent to collect an analyte. Some of the relevant factors are
discussed below.

Surface Area and Mesh Size. 20/40 mesh sorbent is generally
used to minimize the pressure drop across a sorbent tube. Some sorbents, such as Tenax-GC, are not available in these mesh sizes, but
can still be used by increasing the cross-sectional area of the sampling tube to lower the overall pressure drop for these smaller sorbent particles. Surface areas vary from very low in Tenax-GC to
over 1000 m^2/g in activated charcoal. In some cases, the greater
surface area may increase the capacity of the sorbent.

Pore Size and Distribution. The pore diameter must be sufficient to allow the substances of interest to migrate into the pores
to the adsorbing surface. Sorbents with very small pores, such as
the carbon molecular sieves, are used to collect small molecules
like permanent gases (e.g., methyl formate on Carbosieve B).

Surface Groups. Another factor affecting sorption characteristics is the chemical nature of the adsorbing surfaces. Solid sorbent surfaces may contain hydroxyl or phenyl groups, acid or base
groups, or even reactive groups such as olefins. Polarities also

vary greatly among sorbents. Some sorbents are specially coated to enhance the selectivity for specific substances.

With the possible exception of pore size and distribution, these characteristics can affect both physical and chemical sorption mechanisms. In chemical sorption, the overriding factor is the surface groups on the sorbent. However, the sorbent substrate itself can affect the ability of the coated sorbent to collect a substance, especially if the desired reaction does not occur rapidly.

Table I shows some of the characteristics of the commonly used sorbents. The charcoals are by far the most frequently used solid sorbent for organic vapors. Over 130 methods have been validated in our five-year study using coconut, petroleum, and synthetic charcoal. The other sorbents include silica gel, used primarily for amines, and porous polymers, used for substances not amenable to collection on charcoal or silica gel. Other researchers have used other sorbents, including Florisil, alumina, and molecular sieves.

The characteristics of the substance to be collected also must be considered, including the variables of physical properties, chemical properties, reactivity, and interferences.

Physical Properties. The size, molecular weight, and vapor pressure determine the pore size necessary to trap and release a substance efficiently. Large molecules such as pesticides cannot be collected and recovered on molecular sieves; nor will small molecules like gaseous hydrocarbons be retained well by porous polymers.

Chemical Properties. If a substance is highly polar, it may be readily collected by charcoal, but it may be difficult or even impossible to recover it. Also, some substances may be readily hydrolyzed and it may be best to collect these on hydrophobic sorbents like a porous polymer.

Reactivity. Some substances may decompose or be altered in structure after collection, shipment, and storage of samples on some sorbents. Often decomposition can be minimized by selecting sorbents that are inert or will stabilize the substances of interest. An example is the collection of amines on acid-coated silica gel. The acid forms a salt and prevents the oxidation of the amine groups.

Interferences. Interfering substances may also be present in an industrial environment. Their identities and concentrations may affect not only the stability of the substance of interest on the sorbent, but also the capacity of the sorbent.

A substance collected on a solid sorbent is usually desorbed with a solvent. Different solvents give rise to different recoveries. We have frequently found it necessary to test several sorbent/solvent combinations to determine which best collects and

TABLE I. PROPERTIES OF SOLID SORBENTS

SORBENT	COMPOSITION	SURFACE AREA, m^2/gm	PORE SIZE, Å
Charcoals	Carbon		
Coconut based		800-1000	20
Petroleum based		800-1000	18-22
Synthetic (Carbosieve B)		1000	10-12
Silica gel	SiO_2	300-800	20-40
Porous polymers			
Tenax GC	2,6-Diphenyl-p-phenyleneoxide	19	–
Chromosorb 101	Styrene/divinylbenzene	50	3000-4000
Chromosorb 102/XAD-2	Styrene/divinylbenzene	300-400	85
Chromosorb 103	Cross-linked polystyrene	15-25	3000-4000
Chromosorb 104	Acrylonitrile/divinylbenzene	100-200	600-800
Chromosorb 106	Cross-linked polystyrene	700-800	50
Chromosorb 108	Cross-linked polystyrene	100-200	235
Porapak Q	Ethylvinyl/divinylbenzene	–	75

desorbs the substance of interest. A solvent must meet certain criteria to be acceptable.

The analyte must be efficiently recovered. The usual mechanism for solvent desorption is selective displacement of the analyte. Selective displacement occurs as a more polar solvent displaces a less polar one on charcoal, just as a more active ion displaces a less active one on ion exchange resins. CS_2 is frequently used to recover substances from charcoal, but simple alcohols cannot be displaced from charcoal by CS_2, and it is necessary to add 1%-5% of another alcohol to the CS_2 to facilitate desorption. Frequently, low recoveries can be increased by increasing the quantity of solvent, if analytical sensitivity permits. Prospective solvents may be chosen based on polarity or solubility of the analyte.

It must be compatible with the analytical method. The frequent use of CS_2 as a solvent is favored because CS_2 produces a low response when analysis is performed by gas chromatography with flame ionization detection (GC/FID). Likewise, low UV-absorbing solvents are frequently used in high performance liquid chromatography (HPLC) to minimize solvent interference when using a UV detector.

It must not react with the analyte. For example, CS_2 cannot be used to desorb amines because chemical reactions occur.

It must be compatible with the sorbent. We have found that various solvents, notably CS_2 and CH_2Cl_2, dissolve some porous polymers and are unsuitable for use. Likewise, aqueous bases produce a gel-like substance with silica gel and should be avoided.

The solvent should be nontoxic, where possible. Early work or sorbent/sorbent combinations frequently used benzene or CCl_4 as desorbents. If possible, these should be replaced by less toxic solvents.

Method Validation

Tests must be performed on a prospective sorbent to ensure validity of the sampling and analytical methods. The criteria established by NIOSH for testing a sorbent are summarized below. Details of these criteria are given in Reference 2.

Recovery. The recovery of the substance from the sorbent must be at least 75%; we have preferred recovery to be greater than 90% in our studies. For validated methods, recoveries or desorption efficiencies are given in NIOSH Backup Data Reports (3).

Capacity. Breakthrough volume should be at least 1.5 times the recommended sampling volume at twice the OSHA standard at 80%

relative humidity. This requirement is to ensure that there is sufficient capacity of the sorbent tube, with sufficient margin for interfering substances that may be present. The humidity is also important because high humidity has been shown to reduce the capacity of charcoal and silica gel.

Storage Stability. It must be shown that no greater than 10% loss occurs in samples stored for seven days under ambient conditions. This will ensure ample time to ship the samples from the field to the laboratory. Preliminary tests may be performed on spiked samples, but the final testing must be done with samples collected from test air, since distribution of the analyte on the sorbent is different for spiked samples.

Precision and Accuracy. Statistical requirements for precision and accuracy have been established to ensure that the method is reproducible and free from bias.

Commonly Used Sorbents

The sorbents most commonly used in industrial hygiene sampling are charcoal, silica gel, and the porous polymers. In addition, a number of methods have been developed using coated sorbents. Each sorbent is discussed briefly in this section.

Charcoal. Activated coconut charcoal has gained the status as the almost universal solid sorbent. Petroleum-based charcoal is less active, but is also widely used. Charcoal is a very effective sorbent and is generally used for collection of nonpolar organic solvent vapors. It also collects polar organics, but they frequently cannot be recovered. However, many organic substances that are reactive, polar, or oxygenated (e.g., chloroprene, acetic acid, and acetone) have been successfully collected and recovered from charcoal. Substances for which charcoal tube methods have been validated are listed in Table II.

Many of the charcoal tube methods are based on NIOSH Method P&CAM 127 (4) for organic solvents. In this method, a known volume of air is drawn through a charcoal tube to trap organic vapors, the charcoal is transferred to a vial, and the sample is desorbed with carbon disulfide. The sample is analyzed by gas chromatography (GC) with flame ionization detection (FID). Most methods use CS_2 as the desorption solvent because it yields good recoveries from charcoal and produces a very low flame response.

Charcoal does have limitations that can affect its performance. Humidity may reduce the capacity of the charcoal tube, or interfering substances may displace the substance of interest. Frequent periods of high level exposure may exceed the breakthrough volume for very volatile materials.

The synthetic charcoals or graphitized carbons, such as Carbosieve B, are similar in physical characteristics to the natural

TABLE II.   VALIDATED CHARCOAL METHODS

Acetic acid
Acetone
Acetonitrile
Acrylonitrile
Allyl alcohol
Allyl chloride
n-Amyl acetate
sec-Amyl acetate
Arsine

Benzene
Benzyl chloride
Bromoform
Butadiene
2-Butanone
2-Butoxy ethanol
sec-Butyl acetate
t-Butyl acetate
n-Butyl acetate
sec-Butyl alcohol
t-Butyl alcohol
n-Butyl alcohol
n-Butyl glycidyl ether
p–tert–Butyl toluene

Camphor
Carbon disulfide
Carbon tetrachloride
Chlorobenzene
Chlorobromomethane
Chloroform
Chloroprene
Cumene
Cyclohexane
Cyclohexanol
Cyclohexanone
Cyclohexene

Diacetone alcohol
Dichlorodifluoromethane
Dichloromonofluoromethane
Dichlorotetrafluoroethane
1,1-Dichloro-1-nitroethane
o-Dichlorobenzene
p-Dichlorobenzene
1,1-Dichloroethane
Dichloroethyl ether
1,2-Dichloroethylene
Difluorodibromomethane
Diisobutyl ketone
Dimethylamine
Dioxane
Dipropylene glycol methyl ether

2-Ethoxyethanol
Ethyl chloride
Epichlorhydrin
2-Ethoxyethylacetate
Ethyl acetate
Ethyl acrylate
Ethyl alcohol
Ethyl benzene
Ethyl bromide
Ethyl chloride
Ethyl butyl ketone
Ethyl ether
Ethyl formate
Ethylene chlorhydrin
Ethylene dibromide
Ethylene dichloride
Ethylene oxide

Fluorotrichloromethane

Glycidol

Heptane
Hexachloroethane
Hexane
2-Henanone
Hexone

Isoamyl acetate
Isoamyl alcohol
Isobutyl acetate
Isobutyl alcohol
Isophorone
Isopropyl acetate
Isopropyl alcohol
Isopropyl glycidyl ether
Isopropyl ether

Mesityl oxide
Methyl (n-amyl)ketone
  (2-heptanone)
Methyl chloride
Methylcyclohexanol
5-Methyl-3-heptanone
Methyl acetate
Methyl acrylate
Methyl bromide
Methyl cellosolve
Methyl cellosolve acetate
Methyl chloroform
Methyl iodide
Methyl isoamyl acetate
Methyl isobutyl carbinol

α-methyl styrene
Methylal
Methylcyclohexane
Methylene chloride

Naphthalene
Naphtha

Octane

Pentane
2-Pentanone
Petroleum distillates
Phenyl ether vapor
Phenyl glycidyl ether
n-Propyl acetate
Propyl alcohol
n-Propyl nitrate
Propylene dichloride
Propylene oxide
Pyridine

Stoddard solvent
Stryene

Tellurium hexafluoride
1,1,2,2-Tetrachloro-1,2-
  difluoroethane
1,1,1,2-Tetrachloro-2,2-
  difluoroethane
1,1,2,2-Tetrachloroethane
Tetrachloroethylene
Tetrahydrofuran
Tetramethyl succinonitrile
Toluene
1,1,2-Trichloro-1,2,2-tri-
  fluoroethane
1,1,2-Trichloroethane
Trichloroethylene
1,2,3-Trichloropropane
Trifluorobromomethane
Turpentine

Vinyl toluene

Xylene

Synthetic Charcoal Carbosieve B

Alkyl mercury compounds

Methyl formate

charcoals. However, the surface is not as chemically active perhaps due to impurities in natural charcoals. In some cases this may be used to advantage. For instance, methyl formate could not be recovered from coconut or petroleum charcoal, but satisfactory results were obtained with Carbosieve B.

Silica Gel. Silica gel is a polar adsorbent, its surface containing hydroxyl groups. Thus polar substances have a strong attraction toward silica gel. Since water is highly polar, silica gel retains atmospheric moisture, sometimes preferentially over other substances. This affinity for moisture limits the use of silica gel, although in a dry environment it would be an excellent sorbent. The compounds for which our laboratories have validated methods using silica gel as a collection medium are listed below:

Acetylene tetrabromide	Methylamine
Aniline	Methyl alcohol
Chloroacetaldehyde	Morpholine
Cresol	Nitrobenzene
Diethylaminoethanol	p-Nitrochlorobenzene
Diethylamine	Nitrotoluene
Dimethylacetamide	Phenyl ester-biphenyl
Dimethylformamide	vapor mix
Dimethylamine	o-Toluidine
Ethylamine	Xylidine
n-Ethylmorpholine	

This list includes several aliphatic and aromatic amines that are soluble in water and therefore are not significantly affected by retention of moisture on the silica gel sorbent tube. In some cases, such as with chloroacetaldehyde, collection of moisture may even improve the collector. Chloroacetaldehyde forms a very stable hydrate when collected from humid atmospheres with silica gel, besides being efficiently collected from a dry environment.

Porous Polymers. When charcoal and silica gel are not acceptable--because of poor recovery, low breakthrough volume, or poor storage stability--porous polymers should be considered. These polymers include Tenax-GC and the XAD, Chromosorb, and Porapak series. They are relatively inert, hydrophobic, and usually of high surface area. Most porous polymers do not retain volatiles, including water and solvents, but this may be an advantage when sampling an atmosphere containing high concentrations of water or solvent vapors.

Listed below are the compounds for which methods have been validated in our studies using porous polymers. These methods have been studied more thoroughly that others because they presented more problems than those where charcoal or silica gel could be used. This information is contained in the Backup Data Reports for each method.

Tenax-GC
    Allyl glycidyl ether
    2-Aminopyridine
    Diphenyl
    EGDN and/or nitroglycerin
    Phosphorus yellow

XAD-2
    Anisidine (o- and/or p-)
    Ethyl silicate
    Methyl methacrylate
    Nicotine
    Nitroethane
    Quinone
    Tetraethyl lead
    Tetramethyl lead

Chromosorb 102
    Phosdrin
    Heptachlor

Chromosorb 104
    n-Butyl mercaptan

Chromosorb 108
    1-Chloro-1-nitropropane

Chromosorb 103
    Formic acid

Porapak Q
    Furfural alcohol
    Methylchohexanone

Chromosorb 106
    Nitromethane

Porous polymers are usually most useful for the collection of high molecular weight and nonvolatile substances such as pesticides. They may also be suitable for sampling many of the compounds for which charcoal and silica gel methods have been validated. However, since charcoal and silica gel tubes are less expensive than porous polymer tubes, are commercially available, and are already widely used, they are the favored media.

The most serious limitation to the use of porous polymers is batch-to-batch variation. We have found tremendous variation in recovery and capacity tests with different manufacturer's lots of the same sorbent. Often even an extensive clean-up procedure did not result in reproducible data. For this reason, we recommend testing several manufacturer's lots. This often requires more extensive method development than required with charcoal and silica gel methods.

An example of method development using porous polymers is the work done for n-butyl mercaptan. n-Butyl mercaptan collected on charcoal was found to oxidize readily to the dibutyldisulfide. It was not feasible to analyze for the mercaptan as the disulfide, because the disulfide could also be present in the workplace. Silica gel was an excellent collector for the mercaptan in a dry atmosphere; however, at 80% relative humidity the sorbent collected moisture preferentially, and sorbent capacity was severely reduced.

Porous polymers such as XAD-2, Chromosorbs, and Porapacks were then investigated. Most demonstrated satisfactory capacity and short term recovery, but long term storage stability on certain sorbents was a problem. Chromosorb 104, a porous polymer made from acrylonitrile-divinylbenzene monomers rather than from styrene-divinylbenzene monomers as most others are, demonstrated satisfactory recoveries after a 7-day storage period. Since mercaptans are known to react with olefins and inhibit polymerizations, it was thought that the mercaptan might be irreversibly reacting with unreacted monomers in the polymer sorbent. Prewashing procedures did

not significantly alter the characteristics of the sorbents. The
Chromosorb 104 appeared to have fewer reactive sites for the
mercaptan.

   Coated Sorbents. When collection and recovery of a specific
substance cannot be achieved using charcoal, silica gel, or porous
polymers, chemical sorption with a coated sorbent may be necessary.
Compounds requiring this method of collection are usually too reactive or unstable to be collected and stored by other means. In
this case, a specific stable derivative or unique product characteristic of the compound of interest is desired.
   Examples of some of the products of chemical sorption include
(1) the formation of amine salts when sampling for ammonia and organic amines with acid-coated silica gel, (2) amalgamation of silver with mercury when collecting mercury vapor on silver-coated
Chromosorb P, and (3) formation of the Diels-Alder adduct when collecting cyclopentadiene on Chromosorb 104 coated with maleic
anhydride.
   Validated methods using coated sorbents are listed below.

Sorbent	Analyte
Acid-coated silica gel	Ammonia
	n-Butylamine
Ag-coated Chromosorb P	Mercury
TEA-coated molecular sieve	Nitrogen dioxide
HgCN-coated silica gel	Phosphine
HgCl$_2$-coated silica gel	Stibine
Maleic anhydride-coated Chromosorb 104	Cyclopentadiene

Development and testing of these specialized sorbents is usually a
lengthy process. The reaction must first be shown to proceed on
the sorbent, in addition to recovery, capacity, and storage stability tests. The procedure for preparation of the sorbent must be
shown to be consistent with several batches of prepared sorbent.
The stability of the coated sorbent alone may be a factor and must
be tested. After all the developmental work is completed, the
specialized sorbent is frequently useful for only one substance.

Method Development Using Solid Sorbents

   Figure 1 illustrates the procedure that has been used in our
laboratories to develop sampling and analytical methods for substances that may be collected on solid sorbents. The general
approach will first be discussed, and then an example will be given.
In the general approach, several steps may be isolated, but in
actual method development, these steps are interrelated.

   Literature Search. Method development begins with a literature search, gathering information such as physical and chemcial

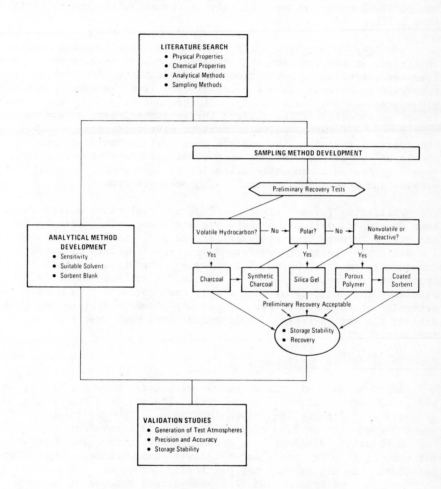

Figure 1.   Solid sorbent method development

properties of the analyte, analytical and sampling methods, toxicological data, and industrial usage and occurrence.

Analytical Method Development. From the point of view of sorbent selection, the important factors to consider in analytical method development are sorbent/solvent compatibility and the detection limit of the analyte.

Sampling Method Development. Selection of a sampling medium is frequently the most lengthy process in method development. Various sorbent/solvent combinations are first tested for preliminary recoveries. Charcoal should be considered first because of its widespread use and availability. If charcoal or synthetic charcoal is not acceptable, silica gel and the porous polymers should be considered. If porous polymers fail to give successful results, then derivatization of the analyte on a surface-coated sorbent should be considered. Whichever sorbent appears most promising should be tested at an early stage for storage stability of spiked samples and breakthrough volumes with test atmospheres.

Validation Tests. After sampling and analytical methods have been developed, validation tests are performed. The criteria for acceptability were discussed earlier and are detailed in Reference 2. Validation tests are performed by collecitng sets of samples from generated atmospheres of a known concentration and then analyzing these samples. Some sets of samples are used to determine precision and accuracy of the method, and another set is used to determine storage stability. If, at any stage of testing and validation, the method fails, developmental work must revert to an earlier stage.

Example of Method Development

As an example of this method development procedure, let us consider the case of diphenyl (biphenyl). The physical and chemical properties of diphenyl are given in Figure 2. We estimated the vapor pressure to be 0.05 mm at 25°C (this is equivalent to 70 ppm).
Chemically, diphenyl is a comparatively nonreactive compound, not subject to hydrolysis, and relatively nonvolatile. These characteristics do not exclude any particular sorbents. The interferences in a workplace cannot be generalized because of the compound's widespread and varied usage. However, because the OSHA standard for diphenyl is only 0.2 ppm, it is likely that interfering materials may coexist at concentrations of greater magnitude. Their presence may affect the capacity of a sorbent and should be considered.
An earlier study at SRI concluded that diphenyl could not be collected on activated coconut charcoal because recovery was poor. The recovery of diphenyl from coconut charcoal is given below.

Chemical Structure

$C_{12}H_{10}$

mw 154.20

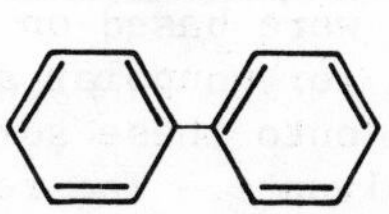

mp   69.2°C
bp   255.2°C
Vapor pressure   1mm at   70.6°C
                    5mm at 101.8°C
                    10mm at 117.0°C

Soluble in ethanol
        ether
        CCl_4
        benzene
        CS_2
        methanol
Insoluble in water
Use: heat transfer fluid
OSHA standard: 0.2 ppm

Figure 2.   Properties of diphenyl

Solvent	μg Taken	Recovery (%)
CS_2	60.2	35.7
	30.1	32.8
	15.1	29.1
Benzene	60.2	52.0
CH_2Cl_2	60.2	<02.0

The report (5) on the study recommended that a less active charcoal be tested. A literature survey did not yield any conclusive evidence for successful collection by recovery from alternative sorbents. On the basis of the little information available, we selected petroleum charcoal and three porous polymers to test (petroleum charcoal is less active than coconut charcoal). All the porous polymers were based on aromatic monomers, and they should be good collectors for nonpolar aromatics. Aliquots of 113 μg diphenyl were spiked onto these sorbents, and the samples were desorbed with various solvents. The results are given below.

Sorbent	Desorbing Solvent	Recovery (%)
Charcoal 104	Methanol	0
(petroleum based)	Carbon disulfide	52
(100 mg)	Benzene	65
	CCl_4	2
	Ethyl ether	0
	Acetone	2
Porapak Q	Methanol	53
(50 mg)	Carbon disulfide	104
	Benzene	98
	CCl_4	93
	Ethyl ether	92
	Acetone	93
Tenax-GC	Methanol	83
(35 mg)	Carbon disulfide	64
	Benzene	79
	CCl_4	99
	Ethyl ether	93
	Acetone	92
XAD-2	CCl_4	94
(30 mg)	Acetone	89

As can be seen, several sorbent/solvent combinations yielded recovery greater than 90%. From these preliminary results, we selected the Tenax-GC/CCl_4 combination because of its high recovery and because previous experience with the Tenax-GC had shown better batch-to-batch consistency and no need for extensive cleanup procedures. The CCl_4 was selected because of its low response and tailing characteristics using the GC analytical method.

Further evaluation proved capacity, storage stability, recovery, precision, and accuracy all to be acceptable, and this method was validated.

Summary

These research efforts have resulted in many sampling and analytical methods for determining workplace exposures to toxic substances. However, there are still many substances for which no suitable methods exist. Much of the information and developmental protocols used in this study can be applied to future studies on these and other substances. In addition, some of the sorbents used for sampling may be directly extendable to passive monitor sampling. There is still a great deal to learn in the area of sorption and sample collection.

Acknowledgments

We would like to express appreciation to staff members at SRI International and at Arthur D. Little, Inc., who performed the experimental work on these programs, and to the National Institute for Occupational Safety and Health for its support. The methods developed in these studies are published in the NIOSH Manual of Analytical Methods (6).

Literature Cited

1. Gunderson, Ellen C.; Anderson, C. Clarine. "Development and Validation of Methods for Sampling and Analysis of Workplace Toxic Substances," U.S. Department of Health, Education, and Welfare (National Institute for Occupational Safety and Health), Cincinnati, Ohio, November 1979, pp. 7–16.

2. Busch, Kenneth A. Memoranda to Deputy Director, DLCD, Dated 1/6/75, 11/8/74, Subject: "Statistical Protocol for Analysis of Data from Contract CDC-99-74-45."

3. Taylor, David G., Ed. "Documentation of the NIOSH Validation Tests," April 1977. This volume contains many Backup Data Reports; others are available from NIOSH, SRI, or ADL.

4. Taylor, David G., Ed. "NIOSH Manual of Analytical Methods," 2nd ed., Vol. 1, April 1977, Method No. P&CAM 127.

5. "Failure Report-Diphenyl," prepared under NIOSH Contract CDC-99-47-45 (Set C).

6. Taylor, David G., Ed. "NIOSH Manual of Analytical Methods," 2nd ed., Vols. 2, April 1977; 3, April 1977; 4, August 1978; 5, August 1979.

RECEIVED October 27, 1980.

Diffusional Monitoring

A New Approach to Personal Sampling

D. W. GOSSELINK, D. L. BRAUN, H. E. MULLINS,
S. T. RODRIGUEZ, and F. W. SNOWDEN

3M, 3M Center, Occupational Health & Safety Products Division, St. Paul, MN 55144

Recent years have seen advances in collection and analytical methods to determine contaminant concentrations in the air. One of these innovations involves the sampling of Organic Vapors without the use of mechanical air pumps; in particular the monitoring of personal exposures.

Philosophically, personal monitoring may perform either of two functions. It may,
1) Define a hazard by generating a record of environmental exposures for a representative worker, or it may be used to
2) Verify the safety of workers in areas of exposure, both extremely important reasons or justifications for the use of personal monitoring.

The conceptual development of diffusion monitoring took place in the late '60's to early '70's. The first paper, published in May of 1972, described diffusion monitoring. Our first product was placed on the market in 1974 and patents have been issued in December of 1975 and April, 1975 to 3M. A second product, the Organic Vapor Monitor, was introduced to the market in January of 1978. An additional patent on personal exposure monitoring was issued in July of 1978.

Of course, on considering personal exposure monitoring by any method, it is helpful to have a point or frame of reference. In this case, that point of reference is the established pump and tube method. It features a mechanical pump whose function is to draw a constant, measured flow of an air-vapor mixture through a tube, normally consisting of two sections, where the vapors are selectively adsorbed. The first section usually contains 100 mg of charcoal and the second, or back-up section, contains 50 mg of charcoal. The pump is powered by a rechargeable battery. Its flow is controlled by setting a dial and is measured by a stroke counter. To obtain valid results all components must be calibrated and function properly. Since it is mechanical, the entire operation is subject to mechanical problems.

In this paper, data is cited which demonstrates that
diffusional monitoring can be used in circumstances in which
pumps and charcoal tubes were previously used. The data
cited emphasizes the comparability of diffusional monitoring
to charcoal tube and pump. However, it must be noted that
the methods are different, and will not in each circumstance
yield comparable results. The most conclusive tests will
be those which employ a third method (e.g., gas chromatography,
infrared analysis) capable of accurately determining the
contaminant, or contaminants concentration.

Theory of Operation

The operation of the 3M Personal Monitoring System is based
on the principle of diffusion. The monitor is comprised of
a velocity barrier, a static air column, and a sorbent layer
at the bottom of the air column.
It is assumed that the contaminant vapor concentration,
at the velocity barrier, is representative of the ambient
concentration or concentration that the worker is exposed to.
Molecules of the contaminant vapor enter the chamber through
the velocity barrier and proceed (by diffusion) at a fixed
rate to the active sorbent layer. The process occurs very
rapidly.
The monitor represents a dynamic, non-mechanical system.
It is not passive. The driving force, its continued operation,
is based on the difference between the contaminant
concentration at the velocity barrier (assumed to be equal
to the ambient concentration) and the sorbent surface (the
principle of diffusion). That difference is maintained by
the continuous adsorption of the contaminant vapors by the
sorbent.
Diffusion controlled processes are measured by Fick's law.
(Equation 1)

$$W = D \frac{A}{L} (C_1 - C_0)t \qquad (1)$$

The weight collected on the sorbent is equal to the product of
the diffusion coefficient, the dimensions of the static
air column (area divided by path length) and the concentration
gradient. C_1 is the concentration at the velocity barrier
which is assumed to closely approximate the ambient concentration. C_0 is the concentration inside the monitor at time t = 0.
The D (A/L) term is the Monitor Sampling Rate and has the
dimensions of flow rate (cm^3/sec.). The exposure time is
measured by t. Therefore, the weight collected, W, is equal
to the flow or sampling rate times the ambient concentration
times the time. This is a dynamic, nature driven type of
monitoring rather than a passive system.

The Average Drift Time or the time it takes the monitor to respond to changes in concentration can be calculated from Equation 2:

$$T = \frac{D^2}{2L} \qquad (2)$$

Where T = Response time (sec.)
   D = The Diffusion Coefficient (cm^2/sec.)
   L = Path length (cm)

For a path length of .65 cm and diffusion rate of 0.12, the response time (Average Drift Time) is less than two seconds. In other works, the monitor responds to changes in concentration of most organics in the atmosphere it is sensing, within two seconds.

How is the process affected by changes in conditions? The diffusion coefficient is an inverse function of pressure, while concentration varies directly with pressure. The result is that weight collected is constant with respect to changes in pressure. Similarly, the diffusion coefficient is a function of absolute temperature to the 3/2's power and concentration is inversely proportional to the square root ($T^{1/2}$) of the absolute temperature. (1)

$$D = f(T^{3/2}) \qquad (3)$$

$$C = f(\frac{1}{T}) \qquad (4)$$

Therefore,

$$W = f(T^{1/2}) \qquad (5)$$

The net result is that W is changes approximately 1% for each increment of 10°F above (increasing) or below (decreasing) 70°F.

By rearranging Equation (1), substituting for the sampling rate, $K = D \frac{A}{L}$ and solving for the environmental concentration, we arrive at Equation (6):

$$C_1 = \frac{W}{KT} \qquad (6)$$

The time-weighted exposure level in mg/m^3 is calculated using equation (6) in the following form:

$$\frac{mg}{m^3} = \frac{\text{corrected weight on monitor (nanograms)}}{\text{sampling rate } (cm^3/min.) \text{ X Sampling Time(min)}}$$

and

$$ppm = \frac{mg}{m^3} \times \frac{24.451}{MW}$$

as detailed in Appendix 1.

A common concern is how to correct for large changes in temperature and pressure as they exist at the sample site. As mentioned above, the rate of mass flux into the monitor is independent of pressure and dependent on the square root of absolute temperature. However, further corrections do exist when mass, by G.C. analysis, is converted to mg/m^3 or PPM. As with rotometers and a pump, the corrections are not the simple application of the gas laws.

The effects of temperature and pressure are given in the Calculation Guide furnished by 3M. These are available upon request. Generally, these effects are not large and can be ignored when temperature is near 25°C and pressure is near 760mm.

Sampling and Analysis

The monitor is a self-contained unit, ready for use, individually packaged. The system contains three parts, the monitor, a closure cap to terminate sampling and a clip to attach the monitor in the breathing zone.

The clip is simply attached, the start time recorded on the designated location on the rear of the monitor, and the Organic Vapor Monitor is placed in the breathing zone. It remains in place for the duration of the exposure period.

At the end of that period the velocity barrier is removed, revealing the static air chamber and the charcoal disc. The closure cap is placed in the body of the monitor, terminating the exposure period and sealing the chamber and sorbent from the atmosphere. At this point, the monitor is placed in its bag, sealed with the tube sealer and sent in for analysis. Information such as the monitor number, date exposed, worker I.D. number, the compounds the OVM was exposed to, temperature and humidity conditions, are all recorded on the package. The 3500 may then be analyzed.

For analysis, the monitor now serves as an analytical vessel for internal desorptions. The Elution Solvent, in most cases Carbon Disulfide, is added to the monitor. At this point, the procedure follows the NIOSH method P & CAM 127. (2) The vessel is shaken or swirled several times over a thirty minute period, after which the Eluent is either decanted into a sampling vial for automatic injection into a G.C. or a sample is extracted manually using a micro-liter syringe.

Subsequently, the sample is injected into the G.C. and the chromatogram is obtained, giving peaks corresponding to each contaminant and the amount present in the sample in milligrams.

Calibration

Monitor calibration requires that the concentration of
the contaminant in the air stream be accurately known.
An apparatus is used to generate known time-weighted average
concentrations of contaminants. The procedure is as follows:
1) Compressed air is passed from the gas cylinder through
   a charcoal bed to remove any organic material in the
   gas stream.
2) An accurately calibrated flow meter is used to determine
   the air flow.
3) The organic liquid (vapor) to be studied is placed in the
   bottom of the diffusion column.
4) The organic vapor diffuses through the porous membrane into
   the air stream at a calculable rate. A wide range of
   contaminant concentrations can be obtained by varying the
   column diameter, length, flow rate, or temperature.
5) The monitors are placed in an exposure manifold.
6) The air stream finally enters a large collection chamber
   where contaminant concentration is verified by a calibrated
   hydrocarbon analyzer and the humidity monitored by a
   hygrometer.
This apparatus was used to generate known concentrations of
Benzene. Separate determinations for each data point were
performed. Agreement between the analyzed monitors and the
known concentration was excellent with low standard deviation.

Through the use of the calibration apparatus, data is
obtained for many organic vapors of industrial hygiene
importance. For compounds with all parameters fully verified,
qualification data sheets are provided. Under general
information, the sampling rate, ($\underline{3}$, $\underline{4}$) recovery coefficient,
and upper exposure limit are given. In addition, physical
constants, sample storage recommendations, STEL sampling
upper exposure limit at 80% relative humidity and concise
equations for calculations of results are provided.

For additional 88 compounds, partial data including
the monitor sampling rate and working range are provided
in the compound guide.

The first examples here have been given for single
compounds. In the real world, mixtures are more often
encountered. To establish the performance of the monitor with
mixtures, gasoline was chosen. Complexity of the gasoline
mixture in terms of both number of compounds and their
similarity was the primary reason for its choice.

A fraction of the first 80% of distillate from gasoline
was used as being representative of the more volatile components.
Chromatograms of each indicate that the eluent from the monitor
more closely reflects the parent fraction in that the ratio
of high molecular weight to low molecular weight components is
less changed than the eluent from the charcoal tubes.

Comparative results with tubes were obtained for several components of the gasoline mixture. These results were quantified by preparing calibration standards for each component. In each case, results from the 3M Organic Vapor Monitor more closely match the test concentration and has a better standard deviation in all instances. This data can be found in Table I.

The capacity of the monitor for each individual compound is based on several factors, molecular structure, vapor pressure, environmental conditions, etc. The upper exposure limit is determined and specified in lieu of a back-up section.

The upper exposure limit for ethyl acetate was found experimentally. It was challenged with a tremendous exposure before significant deviation from the theoretical exposure was observed. This is especially significant when one considers the TLV for Ethyl Acetate is 400 ppm.

The Monitor is not a mechanical pump. Therefore, a minimum airflow is necessary across the Monitor face to prevent starvation and false negative readings. The airflow requirement is small. As long as flow is greater than 10 ft./min., accurate results will occur. (5).

The 3M Monitor was also compared to charcoal tubes in its ability to handle (integrate) high peak concentrations or excursions. Table II summarizes this data. Three concentration levels were chosen and achieved by injection:

1) 5 mL of 1,3-Butadiene (corresponding to 528 ppm) into the air stream of the calibration apparatus allowing it to be adsorbed by both charcoal tubes and OVM's, then analyzing the OVM and charcoal tubes.
2) The process was repeated using a challenge of three 5 mL injections (corresponding to 1585 ppm) and
3) The third was generated by a challenge of three 10 mL injections (corresponding to 3170), followed by analysis.

The analysis obtained by both methods again agree in their ability to handle such high excursions with reasonable accuracy and precision.

Field Test Data

Of more practical concern are some examples of field test data obtained with the 3M Organic Vapor Monitor. In addition to the examples given below, many other comparative studies, performed both in the laboratory and in the field, by companies and agencies other than 3M are available. Copies may be obtained from 3M upon request.

In laboratory tests conducted at the Technical Research Center of Finland (6) comparisons were made between the 3M Monitor and charcoal tubes (150 mg). The tests were monitored by both gas chromatography and Miran IA infrared instruments.

TABLE I

RESULTS OF GASOLINE EXPOSURE TEST

Components of Gasoline	Test Concentration (PPM)	3M No. 3500 Concentration (PPM ± 2σ)	Charcoal Tube Concentration (PPM ± 2σ)
Pentane	5.602	5.832 ± .408	5.053 ± .500
Heptane	0.573	0.570 ± .045	0.578 ± .082
Octane	0.294	0.307 ± .047	0.329 ± .056
Benzene	0.928	0.892 ± .160	0.873 ± .180
Toluene	1.768	1.710 ± .119	1.552 ± .220
Xylenes	1.153	1.230 ± .086	1.034 ± .160

TABLE II

INTEGRATION OF VARIABLE CONCENTRATION

1.3 Butadiene Generated Concentration (PPM)		3M No. 3500 Concentration (PPM±2σ)	Charcoal Tube Concentration (PPM±2σ)
528		527±44	546±78
1585		1690±74	1687±127
3170		3502±223	3489±179

Table III shows results obtained when 3M OVM and charcoal tubes
were exposed at the same time to known concentrations of toluene
in two separate tests. Comparability of the results are
corroborated by both instruments at both levels of concentration.

TABLE III

TOLUENE
SAMPLE TIME - 4 HOURS
RESULTS - PPM

	TEST A	TEST B
Miran IA	288	89
GC	289	86
3500 #1	297	86
#2	287	87
#3	–	87
#4	–	85
#5	–	88
#6	–	88
#7	–	87
#8	–	87
Tube #1	283	85
Tube #2	–	85

Table IV compares results of short term monitoring obtained
from the 3M OVM, and by both instrumental means. This test
demonstrates the short term sampling capability of the monitor.

TABLE IV

TOLUENE
RESULTS - PPM

SHORT TERM SAMPLING

TEST			
	A	B	C
SAMPLE TIME (MIN.)	20	15	10
Miran IA	560	560	560
GC	544	544	544
#3500 #1	550	600	503
#3500 #2	542	530	529

Table V reports results when a polar (isobutyl alcohol) and
non-polar compound is tested. These results again demonstrate
comparability between the four methods used.

TABLE V

MIXTURE TEST
MONITOR - 3M #3500
RESULTS - PPM
SAMPLE TIME - 240 MIN.

	ISOBUTYL ALCOHOL	TOLUENE
Miran IA	144	180
G.C.	143	175
Tube 1	150	164
Tube 2	138	168
Tube 3	138	169
Tube 4	130	169
#3500 1	138	166
2	148	178
3	144	176
4	142	165
5	146	180
6	144	170
Tube	139 $\pm$ 8.2	168 $\pm$ 2.4
#3500	144 $\pm$ 3.5	173 $\pm$ 6.4

Takada, et. al. (7), compared the 3M Organic Vapor Monitor
to charcoal tubes by studying the weight of contaminant
collected as a function of (1) exposure time at various
concentrations and (2) chamber concentration at various
exposure times. Their data is summarized in Figures 1,2,3,4.
Their study led to the following major conclusions:
(1) Regardless of the concentration variance between 8.7 and
190 ppm, adsorption on the monitor was proportional to
exposure time from fifteen minutes to eight hours, (2) similarly,
vapor adsorption remained proportional to concentration
(between 7.5 and 196.1 ppm) when the exposure time was varied
between one and eight hours, (3) With varying concentrations,
the monitors work well between fifteen minutes and eight hours,
(4) Linearity was maintained in comparison between average
adsorption in 4-hour-halving exposure and 4-hour continuous
exposure, as well as comparison of 4-hour tandem exposure and
8-hour continuous exposure.

Summary

The diffusional monitoring system is a much simpler, easier
method to present a sample of organic vapor to a charcoal
collector. It eliminates mechanical problems of pumps and in
the case of the 3M system, provides for elution
inside the monitor body, rather than forcing transfer of the
sorbent. It has a wide dynamic range, a quick response to

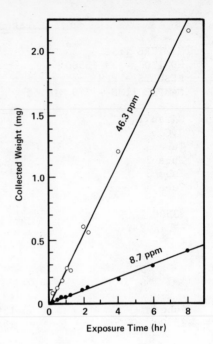

Figure 1. Relationship between weight
of toluene collected by the 3M Organic
Vapor Monitor and its exposure time

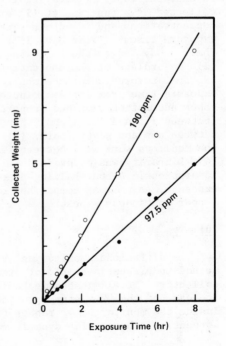

Figure 2. Relationship between weight
of toluene collected by the 3M Organic
Vapor Monitor and its exposure time

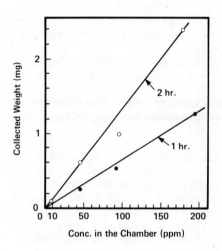

Figure 3. Relationship between weight of toluene collected by the 3M Organic Vapor Monitor and the chamber concentration

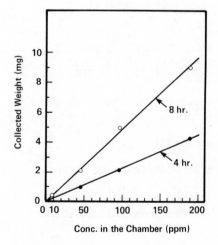

Figure 4. Relationship between weight of toluene collected by the 3M Organic Vapor Monitor and the chamber concentration

concentration changes, and is capable of integrating high
concentration excursions and complex mixtures with reasonable
accuracy and precision.

Abstract

The theory and practical use of a diffusional monitor
is described. The diffusional monitor can be used to
accurately measure personal exposures, as well as to determine
area concentrations of organic vapors.

The method consists of 1) attaching the monitor in the
breathing zone for a measured interval of up to eight hours
(STEL or ceiling level measurements can be made in fifteen
minute intervals), 2) terminating exposure and converting the
monitor to an analytical device by interchanging one part,
3) adding a measured volume of eluent, and 4) analyzing
by manual or automatic gas chromatography.

Environmental sampling is controlled by binary diffusion
such that the rate of sample collection is a function of
ambient concentration. The relationship between environmental
contaminant concentration and the weight of material collected
by the monitor is given by Fick's Law of Diffusion. Contact,
handling, and extraneous exposure of the analytical sample
is eliminated. Procedures qualifying the method for compounds
of industrial hygiene interest will be presented along with
results from laboratory and field evaluations comparing
the method to charcoal tubes and known concentrations.

Literature Cited

1. Palones, E.D.; Gunnison, A.F.; DiMattio, J.;
    Tomczyk, C. Am. Ind. Hyg. Assoc. J., 1976, 37, 570.

2. NIOSH Manual of Analytical Methods, Ed. 2, Part I
    DHEW (NIOSH) pub. no. 77-157-A, 1977, 1, 127.

3. Gilliland, E.R., Ind. Eng. Chem., 1934, 26, 681.

4. Nelson, G.O., "Controlled Test Atmospheres", Ann Arbor
    Sci. Publ., 1971.

5. McCammon, C.S. and Woodfin, J.W. Am. Ind. Hyg. Assoc. J.
    1977, 38, 378.

6. Technical Research Center of Finland, Lab. of Ind. Hyg.,
    Analysis Reports 91153, 01028, January, 1980.

7. Takada, T., Kadowaki, T., Takahashi, H., Suzuki, K.,
    Konishi, S., and Yosikawa, M., Work. Envir., 1980,
    (3) 35.

Appendix 1

To determine ppm, the following equation is used:

$$\text{ppm} = \frac{\text{mg}}{\text{m}^3} \times \frac{22.4 \text{ 1/mole}}{\text{MW g/mole}} \times \frac{T \text{ K}}{273 \text{ K}} \times \frac{760 \text{ mm Hg}}{P \text{ mm Hg}}$$

Based on 25°C and 760 mm Hg, the equation reduces to,

$$\text{ppm} = \frac{\text{mg}}{\text{m}^3} \times \frac{24.451}{\text{MW}}$$

RECEIVED November 3, 1980.

An Evaluation of Organic Vapor Passive Dosimeters Under Field Use Conditions

R. S. STRICOFF and C. SUMMERS

Arthur D. Little, Incorporated, Cambridge, MA 02140

The advent of diffusion-type organic vapor sampling devices has made available to the industrial hygienist a tool which appears to have several significant advantages over the commonly employed charcoal tube/personal sampling pump technique. Diffusion samplers are small and light weight (providing improved wearer acceptance), require no maintenance or calibration, and do not depend upon batteries which need regular recharging. The diffusion samplers potentially offer particular benefits to the industrial hygienist who travels to a number of remote locations, or who wants relatively untrained persons at a remote location to assume responsibility for sample collection.

Despite their potential advantages, diffusion samplers have received a cautious reception in the industrial hygiene community. The sampling technique that diffusion devices would supersede (charcoal tube/personal sampling pump) is widely accepted as a norm and, while it is well known that the tube/pump method does not offer ideal precision or accuracy, the sources and magnitudes of potential error in tube/pump sampling are generally perceived as being both understood and acceptable. Laboratory comparison of diffusion samplers to tube/pump sampling has been the subject of several previously published studies (1, 2, 3, 4, 5); however, little has appeared in the literature documenting diffusion sampler performance under field conditions (6).

Comparing the performance of diffusion samplers to tube/pump sampling under field conditions is a difficult task. In the field the true value of the exposure level is unknown, and neither the tube/pump nor the diffusion device can be assumed to provide a true value. However, field performance is important since factors such as wearer bending and stretching (which can "flip" a diffusion sampler, hiding its face from the workplace); sampling device handling, transportation, and climatic exposure; and uneven patterns of exposure to the challenge contaminant are not generally tested in laboratory studies.

0097-6156/81/0149-0209$05.00/0

The evaluation of diffusion samplers reported in this paper was initiated as an internal project at Arthur D. Little, Inc., to develop for ourselves a base of data which would help us to judge the utility of diffusion samplers in our work. Diffusion samplers were purchased from the two firms active in the market- place when the field study began (early 1979), Abcor and 3M. A program was devised wherein, during each of approximately 50 sur- veys being undertaken in plastic fabrication plants, four to six diffusion samplers would be exposed, each adjacent to a charcoal tube personal sample. At the conclusion of the field sampling, a data base of over 100 pairs of samples (each pair including a diffusion sample and a charcoal tube sample) had been compiled for each of the two diffusion samplers.

Materials and Methods

Abcor "Gasbadge" organic vapor samplers and 3M 3500 Organic Vapor Monitors were purchased for use in this study. Abcor bad- ges were loaded prior to each field trip in an office environment. Standard SKC charcoal tubes (100 mg collection section with 50 mg backup) were used, with DuPont P-200 pumps calibrated prior to each survey.

Within the plants surveyed, styrene was the principal air- borne contaminant of concern. Since these surveys were made in operating plants, there were other substances present; however, styrene was the dominant vapor and the only one potentially of concern relative to accepted exposure limits (TLV's).

Desorption efficiencies were determined for each sampling device in our laboratory. Analysis of passive dosimeter samples and calculation of concentrations were done as recommended by each device's manufacturer (7, 8). Charcoal tube samples were analyzed as recommended by NIOSH Sampling and Analysis Method S-30 (9).

Our previous experience in the use of small numbers of dif- fusion samplers paired with tube/pump samples had suggested that a "learning curve" phenomenon occurred wherein the first few sample pairs exhibited considerably more scatter than later pairs. It was speculated that this "learning curve" could be at- tributed to the acquisition of dexterity in handling diffusion samplers by both the field teams and laboratory personnel. With- out attempting to prove the actual occurrence of this phenomenon, it was decided in advance to exclude the data from sample pairs exposed in the first two plants in which each type of diffusion sampler was used.

Sample durations tended to be short, since the field survey team did not have advance knowledge of exposure levels to be en- countered and wished to avoid overloading any of the sampling media. In most cases, the exposures measured proved to be relatively low and, as a result of the short sampling times, the loading of each sample was low. Figures 1 and 2 illustrate the

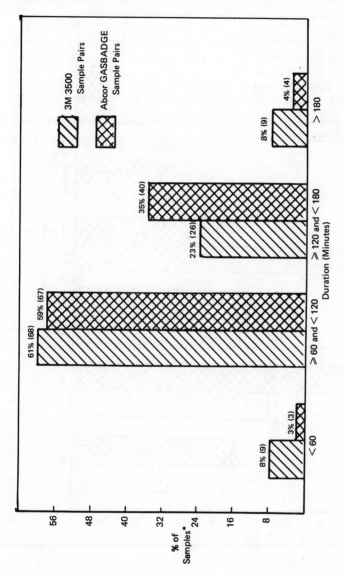

Figure 1. Distribution of samples by duration

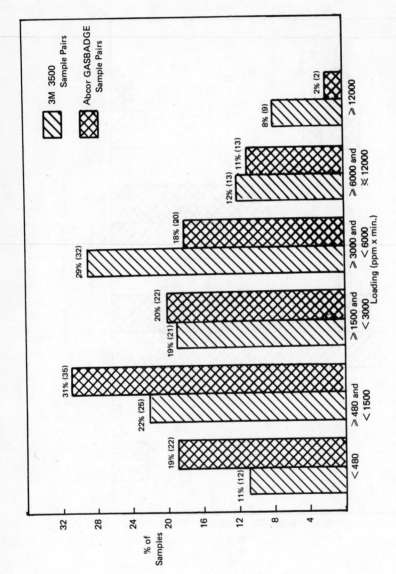

Figure 2.   Distribution of samples by loading on tube; percentages based upon number of pairs for each diffusion dosimeter

range of sampling durations and sample loading that characterized
the samples collected during this study. Samples from each individual field survey were blank corrected.

Results and Discussion

Figures 3 and 4 display the paired results obtained with Abcor and 3M diffusion-type samplers, respectively. These results
were analyzed through use of the "t" test for paired samples and
the calculation of correlation coefficients and regression equations, with the results of these analyses shown in Table I. A
statistically significant correlation is seen between the data
set for each type of diffusion sampler and the corresponding tube/
pump sample data set, and the "t" test fails to refute the null
hypothesis that there is no significant systematic difference between each of the diffusion sampler data set and the corresponding tube/pump data set.

Subsets of each data set were examined to determine whether
duration or loading played an identifiable role in determining
the comparability of the data sets. Results of analyses of the
data sets displayed in Figures 5, 6, and 7 are given in Table II.
These data indicate that the 3M diffusion sampler results for
samples of 60 minutes or longer duration correlate well with the
tube/pump data, and that higher loadings result in better correlation. 3M sampler loadings above 6,000 ppm-min resulted in
regression coefficients less than one and negative "t" values,
while loadings below 6,000 ppm-min resulted in positive "t"
values and regression coefficients greater than one. Abcor sampler results also indicate improved correlation at longer sample
duration; however, use of the "t" test on the Abcor sample results
from each of the "loading" and "result" subsets indicates that
the null hypothesis (i.e., that there is no significant difference
between diffusion and tube/pump data sets) is refuted.

Figures 8-11 display the difference between each diffusion
sampler and its corresponding tube/pump sample, plotted against
the tube/pump sample result. These figures show the difference
as a function of the magnitude of the nominally "true" (or, at
least, accepted) value. Lines indicating the bounds of ±10% and
±25% of the tube/pump sample are shown.

Conclusion

Statistical evalaluation of the data indicates that there is
general agreement between each passive sampler data set and the
corresponding tube/pump data set. However, examination of subsets of the data indicated that the 3M sampler results are more
likely at low loadings than at high loadings to differ from tube/
pump samples. The subsets suggest that the Abcor sampler is even
more likely to differ at low loadings and shorter durations from
the tube/pump samples.

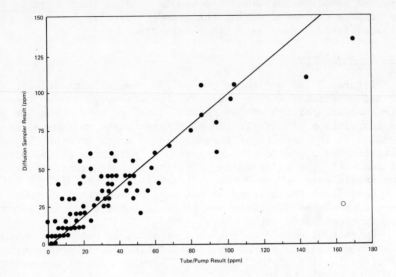

Figure 3.   Abcor gasbadge data vs. corresponding tube/pump data

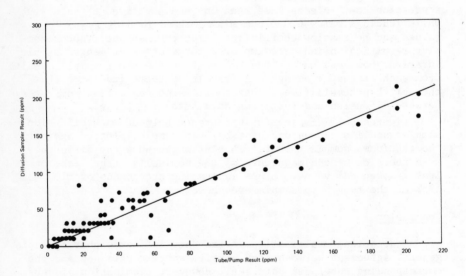

Figure 4.   3M Organic vapor monitor data vs. corresponding tube/pump data

Table I.   Statistical Evaluation of Data Sets

	Diffusion Sampler	
	3M	Abcor
Number of sample pairs	112	114
t - value	1.51[*]	-0.08[*]
Regression equation	D = 0.96C + 5.4	D = 0.69C + 8.15
Correlation coefficient	0.90	0.84

[*]t - value for p = 0.5 is approximately 1.66.

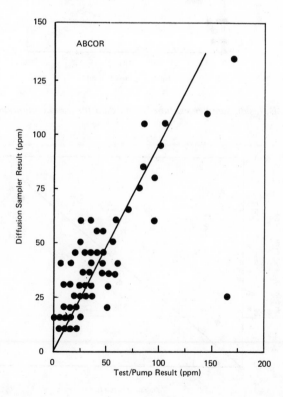

Figure 5.   Diffusion sampler vs. tube/pump data for sample duration $\geq$ 60 min

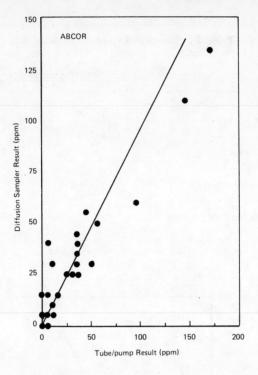

Figure 6.   Diffusion sampler vs. tube/pump data for sample duration ≥ 120 min

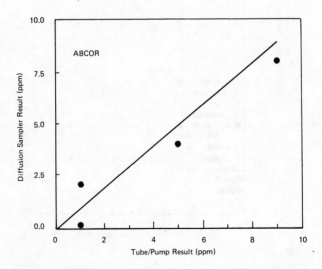

Figure 7.   Diffusion sampler vs. tube/pump data for sample duration ≥ 180 min

Table II. Statistical Evaluation of Data Subsets

Sampler	Data Set	# Pairs	t-Value	Regression Equation	Correlation Coefficient
3M	duration ≥ 60 min	103	0.855*	D = 0.91C + 4.97	0.97
	duration ≥ 120 min	35	1.05*	D = 0.88C + 5.25	0.95
	duration ≥ 180 min	9	1.36*	D = 1.5C - 0.86	0.87
	result ≥ 50 ppm	35	0.08*	D = 0.94C + 7.33	0.76
	result ≥ 30 ppm	56	0.72*	D = 0.92C + 9.62	0.83
	result < 30 ppm	56	2.92	D = 1.23C + 1.11	0.74
	loading ≥ 12,000 ppm-min	9	-1.40*	D = 0.86C + 12.3	0.85
	loading ≥ 6,000 ppm-min	22	-1.68*	D = 0.93C + 1.83	0.93
	loading < 6,000 ppm-min	90	2.27	D = 1.24C - 0.58	0.81
	loading < 3,000 ppm-min	58	2.24	D = 2.22C - 10.7	0.90
	loading < 1,500 ppm-min	37	2.00	D = 1.57C - 0.77	0.67
Abcor	duration ≥ 60 min	111	-0.40*	D = 0.69C + 7.54	0.85
	duration ≥ 120 min	44	-0.44*	D = 0.75C + 4.64	0.96
	duration ≥ 180 min	4	-0.87*	D = 0.86C + 0.05	0.97
	result ≥ 50 ppm	16	-1.99	D = 0.45C + 30.2	0.51
	result ≥ 30 ppm	39	-1.78	D = 0.53C + 21.0	0.70
	result < 30 ppm	75	3.21	D = 1.25C + 1.4	0.70
	loading ≥ 6,000 ppm-min	15	-1.76	D = 0.4 C + 37.2	0.48
	loading < 6,000 ppm-min	99	2.51	D = 0.87C + 4.83	0.80
	loading < 3,000 ppm-min	79	3.66	D = 1.27C + 1.39	0.79
	loading < 1,500 ppm-min	57	3,26	D = 1.77C - 0.54	0.73

*$p > .05$

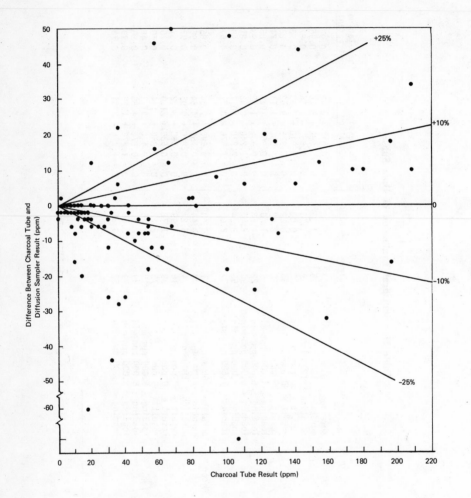

Figure 8.   Pair difference vs. charcoal tube result for 3M samples

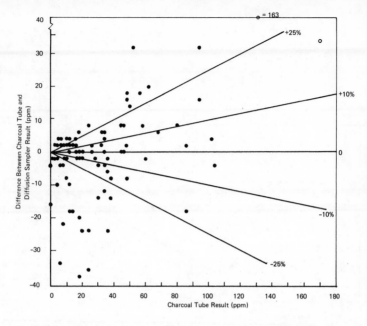

Figure 9.   Pair difference vs. charcoal tube result for Abcor samples

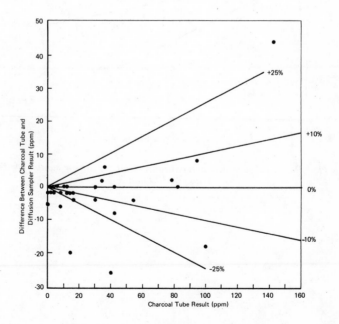

*Figure 10.   Pair difference vs. charcoal tube result for 3M samples with duration
≥ 120 min*

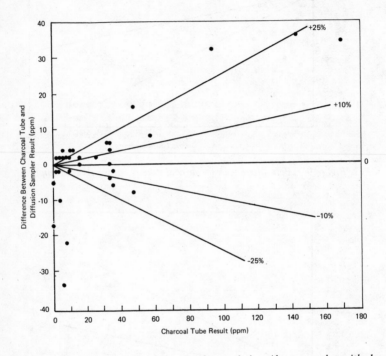

*Figure 11.   Pair difference vs. charcoal tube result for Abcor samples with dura-
tion ≥ 120 min*

These data suggest that diffusion samplers can be success-fully utilized under properly chosen field conditions. However, this study involved samples which were lightly loaded and were exposed in atmospheres containing only one significant hydrocarbon vapor. Other types of use conditions should be studied to gain more complete insight into the capabilities of diffusion samplers.

Literature Cited

1. Tompkins, F.C., Jr., and Goldsmith, R.L. Am. Ind. Hyg. Assoc. J. 1977, 38, 371-377.

2. Silverstein, L.G., "Validation of Abcor 'GASBADGE' for Acrylonitrile and Improved Desorption Efficiency " Am. Ind. Hyg. Assoc. J. 1977, 38, 412-413.

3. Evans, M., et al, The Practical Application of the Praton Diffusion Sampler for the Measurement of Time-Weighted Average Exposure to Volatile Organic Substances in Air, Annals Occup. Hyg. 1977, 20, 357-363.

4. Bailey, A., and Hollingdale-Smith, P.A., A Personal Diffusion Sampler for Evaluating Time-Weighted Exposure to Organic Gases and Vapors Annals Occup. Hyg. 1977, 20, 345-356.

5. Bamberger, R.L., Esposito, G.G., Jacobs, B.W., and Mazur, J.F. A New Personal Sampler for Organic Vapors Am. Ind. Hyg. Assoc. J. 1978, 39, 701-708.

6. Boeniger, M.F., Zaebst, D.D., Ludwig, H.R., Crandall, M.S., and Vongrongseman, P., A Field Comparison of Two Passive Organic Vapor Monitors with Charcoal Tubes Under Singular and Multiple Exposure Conditions, Unpublished paper, presented at 1980 American Industrial Hygiene Conference.

7. "GASBADGE Organic Vapor Dosimeter Use and Analysis Instructions," Abcor, Inc., Wilmington, MA, 1977.

8. "3M Brand Organic Vapor Monitor Product Information and Usage Guide," 3M Company, St. Paul, MN.

9. "NIOSH Manual of Analytical Methods, Second Edition," U.S. Department of Health, Education, and Welfare, 1977.

RECEIVED October 3, 1980.

The Role of Biological Monitoring in Medical and Environmental Surveillance

CARL B. MONROE

Corporate Health and Safety, Rohm and Haas Company, Box 584, Bristol, PA 19007

Biological Monitoring is used by environmental engineers
to assess the control of workplace exposures to toxicants and
by medical personnel to detect medical conditions associated
with exposure to these substances. Linch (1) has made an
extensive review of techniques that are available for use in
biological monitoring. This paper will explore the application
of biological monitoring in industrial hygiene and occupational
medical surveillance programs, relate biological monitoring to
other occupational medical surveillance techniques, and discuss
the importance of informing workers of the purpose for the
tests and the practical meaning of test results. The approach
to occupational health surveillance through biological
monitoring is diagramed in Figure 1.

Definition of Occupational Biological Monitoring

Biological monitoring in occupational health surveillance
consists of bioassay or other tests that are administered to
workers with known or presumed exposures to potential toxicants
in the workplace for the purpose of measuring body uptake of
the toxicant, an index of exposure, or detecting pathophysiologic effects of the toxicant, an index of toxicity. Since
exposure and toxicity are the determinants of the hazard
potential of a toxic substance, the occurrence of either one of
these phenomena is reason for preventive health measures.
Examples of biological monitoring tests that are applied in
environmental and medical surveillance programs are listed in
Table I.

Application of Biological Monitoring in Environmental and Medical Surveillance

The multidisciplinary approach to the evaluation and
solution of occupational health problems may involve the use of

0097-6156/81/0149-0223$05.00/0
© 1981 American Chemical Society

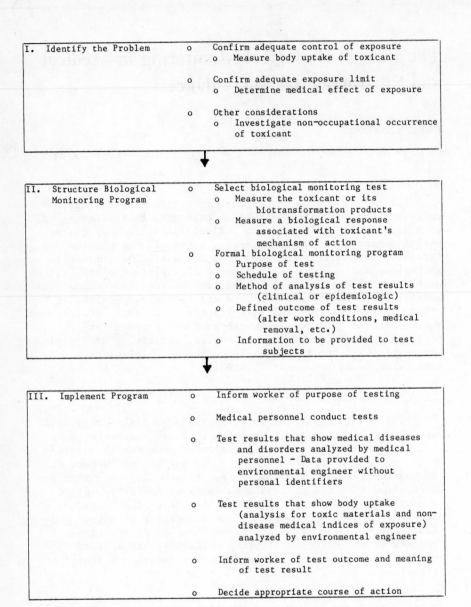

I. Identify the Problem
- o Confirm adequate control of exposure
- o Measure body uptake of toxicant

- o Confirm adequate exposure limit
- o Determine medical effect of exposure

- o Other considerations
- o Investigate non-occupational occurrence of toxicant

II. Structure Biological Monitoring Program
- o Select biological monitoring test
- o Measure the toxicant or its biotransformation products
- o Measure a biological response associated with toxicant's mechanism of action
- o Formal biological monitoring program
- o Purpose of test
- o Schedule of testing
- o Method of analysis of test results (clinical or epidemiologic)
- o Defined outcome of test results (alter work conditions, medical removal, etc.)
- o Information to be provided to test subjects

III. Implement Program
- o Inform worker of purpose of testing

- o Medical personnel conduct tests

- o Test results that show medical diseases and disorders analyzed by medical personnel – Data provided to environmental engineer without personal identifiers

- o Test results that show body uptake (analysis for toxic materials and non-disease medical indices of exposure) analyzed by environmental engineer

- o Inform worker of test outcome and meaning of test result

- o Decide appropriate course of action

Figure 1. Flow diagram illustrating the use of occupational biological monitoring

TABLE I

EXAMPLES OF BIOLOGICAL MONITORING TESTS USED IN

ENVIRONMENTAL AND MEDICAL SURVEILLANCE

Occupational Exposure	Biological Monitoring Test for Body Uptake	Biological Monitoring Test for Medical Effect of Exposure
Pb	Blood Pb	Blood Zinc Protoporphyrin
Cholinesterase-Inhibiting Pesticide	Blood Level of Pesticide or Biotransformation Product	Blood Cholinesterase Activity
Cd	Blood and Urinary Cd	Urinary Beta-2 Microglobulin
Diesel Exhaust	Expired Air-CO	Carboxyhemoglobin measured in blood or extrapolated from expired air-CO

a single tool or concept in dissimilar ways within the several specialty areas (industrial hygiene, medicine, etc.) represented on the occupational health team. Biological monitoring of a toxicant or its biotransformation products to assess the integrity of work practices, engineering controls, and personal hygiene can be useful in environmental surveillance. Medical surveillance programs often incorporate biological monitoring tests to screen for hypersusceptible workers and to detect medical conditions in the earliest and most curable stages. The particularity of biological monitoring is evident in the OSHA promulgation on employee access to exposure and medical records (2) which categorizes biological monitoring test results as an exposure record when the absorption of a substance is measured and as a medical record when pathophysiologic effects are determined.

The role of biological monitoring in environmental and medical surveillance can be differentiated on the basis of the purpose and outcome of testing in the two situations. In environmental surveillance, the goal of biological testing is to evaluate the exposure quantity of the hazard potential posed by a toxic substance. An environmental surveillance program might include the routine performance of biological tests to detect body uptake of workplace contaminants to confirm the effectiveness of industrial hygiene practices that appear to be sufficient by other measurements, such as area or personal sampling. The routine assessment of urinary cadmium in workers manufacturing plastic sheet containing cadmium pigments when environmental measurements have shown airborne cadmium levels to be well within acceptable limits illustrates this type of environmental monitoring. Alternatively, environmental surveillance that is targeted to work situations where there is a significant potential for overexposure to a toxic material, e.g., uranium mining operations with high airborne concentrations of uranium, can be enhanced considerably by biological monitoring because biological absorption is probably the best parameter to show the effectiveness of good personal hygiene, housekeeping, and the proper use of respirators or protective clothing in this type of exposure situation. The outcome of biological monitoring in environmental surveillance is the definition of the extent to which a toxic material has been absorbed by the body. When non-occupational sources of contamination have been excluded, this information allows the environmental engineer to judge whether environmental control measures are adequate or need refinement.

In medical surveillance, biological monitoring is a preventive medical tool that is used to discern the human toxicity of potential toxicants at exposure levels present in the workplace. Evidence of human toxicity can result from individual hypersusceptibility, inadequate exposure limits which might have been derived from animal data, or unsuspected

excursions in workplace contamination. Medical monitoring might
be done when industrial hygiene measurements show that a
potential toxicant is present in the workplace at levels that are
considered to be safe. The routine measurement of blood
cholinesterase levels among workers in well-contained
organophosphate pesticide manufacturing operations is an example.
When the workplace is contaminated with a potential toxicant to
such an extent that a significant potential exists for
overexposure, biological monitoring is one medical surveillance
tool that can be used to assure that environmental protective
measures have prevented the occurrence of health effects.
Biological monitoring of blood cholinesterase levels among field
applicators of cholinesterase-inhibiting pesticides examples this
type of monitoring situation. The outcome of medical monitoring
is the determination of the presence or absence of a medical
marker of a toxicologic response to a specific workplace
contaminant. The presence of a medical effect of overexposure
leads to medical actions aimed to interrupt the toxicant's impact
upon the worker's health. Such medical actions include
appropriate medical therapy, medical removal protection, and
medical advisements on the appropriateness of exposure limits.

A properly designed biological monitoring program should be
developed conjointly by the environmental engineer and medical
staff utilizing environmental data about the nature of the
workplace exposure and medical knowledge about the prevalence of
medical conditions that might alter the susceptibility of
individuals to the environmental contaminant. The evaluation of
the body's uptake of toxicants which are done for environmental
surveillance purposes can be used in conjunction with medical
findings to provide information on the validity of current
assumptions about the human toxicity of the substance at exposure
levels experienced in the workplace. In the same manner, medical
monitoring yields information which the environmental engineer
can use along with environmental measurements to judge the
adequacy of industrial hygiene practices.

It is essential to formulate biological monitoring programs
on the basis of prior environmental information about known or
potential workplace exposures to toxicants since few materials
are sufficiently unique to the workplace that biological
monitoring can be undertaken without prior documentation of the
potential for exposure. Assessments of occupational exposure can
be confounded by other environmental sources of toxic agents.
For a metal such as arsenic that can be absorbed from common
seafood items in the diet, potential non-occupational sources
should always be considered in the development of biological
monitoring programs (3).

The consultative approach to the design of biological
monitoring programs can aid in the determination of the type of
surveillance that is appropriate. For instance, environmental
data showing a chronic exposure to Pb might lead to

environmental monitoring with blood Pb levels and medical
monitoring for elevated blood zinc protoporphyrin levels since
the latter is a sensitive marker of an excessive body burden of
Pb (4), whereas environmental findings consistent with low
intermittent exposures might result in environmental
surveillance with blood Pb levels alone to monitor for
acute overexposures (5).

As shown in Figure 1, the use of biological monitoring in
environmental or medical surveillance programs should begin with
the selection of a test with an outcome that is appropriate to
the type of surveillance being conducted. For a designated
workplace contaminant, the validity of a biological monitoring
procedure is dependent upon (1) the sensitivity of the test, the
probability that the test result will be positive when the
factor to be measured is present; and, (2) the specificity of
the test, the probability that the test result will be negative
when the factor to be measured is absent; and, (3) the presence
of the potential toxicant in the workplace where monitoring is
conducted; and, (4) in the instance of medical monitoring, the
absence of other materials in the workplace that might cause the
same medical effect that the toxicant is known to cause. The
usual biological monitoring test is an assay of biological
specimens (semen (6), urine (3), blood (7), and expired air
(8)). In environmental surveillance, a bioassay directed to the
detection of a substance or its biotransformation product is the
usual test method. Skin sensitization tests (9), physiological
changes in pulmonary function (10), and characteristic effects
of dusts on chest x-rays (11) supplement the use of the bioassay
in monitoring for early medical manifestations of exposure to
hazardous substances.

The judicious use of biological monitoring entails the
scheduled administration of the test within the context of the
exposure history. The usefulness of urinary concentrations of
heavy metals, such as uranium, can be highly dependent upon the
time between the collection of the specimen and the last
exposure (12). The assessment of blood cholinesterase levels
is similarly dependent upon the temporal relationship between
the administration of the test and the exposure (13). The
designation of an exposure or medical parameter as normal or
abnormal can be made in relationship to the individual's
baseline status (3,13) or in reference to results from a
statistically generated reference group (14,15).

Test results that show bioabsorption of toxicants (e.g.,
blood Pb levels) or reveal early medical effects of
overexposure that do not constitute a disease entity (e.g.,
blood zinc protoporphyrin levels) can be given to the
environmental engineer in personally identifiable form.
Biological monitoring test results that identify a disease
state (e.g., oligospermia) constitute confidential information
about the health status of the individual which is not to be

transmitted outside the medical department without the
permission of the test subject. The physician should share non-
personally identifiable compilations of such data as well as
conclusionary statements about the test results with the
environmental engineer so that proper correlations can be made
between workplace exposures and any medical effects that are
discerned.

Relationship of Biological Monitoring to Other Medical Surveillance Activities

Biological monitoring tests are among the most valuable
medical surveillance tools. Although general tests of body
system function, such as renal function tests (e.g., blood urea
nitrogen), are done on biological specimens and have a role in
medical surveillance programs for occupational exposure to
systemic toxins (16), these procedures differ qualitatively from
biological monitoring tests which satisfy the four criteria
described earlier for valid biological monitoring procedures. A
biological monitoring test should be a reliable and valid
measure of a pathophysiologic change incurred by the action of a
toxic substance upon a specific metabolic or physiologic
pathway. The biological response should appear in most all
individuals exposed to unsafe levels of the harmful material.
For example, when cadmium is the only heavy metal to which
workers are exposed, the elevated urinary excretion of beta-2-
microglobulin in cadmium intoxication reflects a selective
action of cadmium upon renal handling of this low molecular
weight protein (17). In the absence of exposure to other heavy
metals which might confound the relationship between cadmium and
this biological effect, the use of beta-2-microglobulin
determinations for biological monitoring purposes is based on
the relative specificity of elevated urinary excretion of this
protein to cadmium nephrotoxicity as well as an extremely high
prevalence of this abnormality among persons overexposed to this
metal.
    Linch included pulmonary function tests and other non-
bioassay procedures in his compendium of biological monitoring
tests (18). When a strong identity exists between a toxicant
and a medical effect, laboratory procedures other than a
bioassay (pulmonary function tests, chest radiographs, etc.)
can be used to monitor the interaction between the toxic agent
and a particular physiologic or metabolic system. The
demonstration of an acute fall in ventilatory capacity after
TDI exposure (10) and calcified pleural plaques after asbestos
exposure (11) example the use of valid non-bioassay procedures
to monitor biological responses to workplace exposures. Since
certain respiratory abnormalities that are associated with
exposures to dusts or other respiratory toxins may be easily
confounded by other factors, such as cigarette smoking (19, 20,

21), the chest x-ray and pulmonary function tests should be
used as a biological monitoring technique when an exposure to a
respiratory toxin results in a pathophysiologic response that
can be readily ascribed to that material. The fulfillment of
similar criteria should be met before techniques other than
chest radiography or pulmonary physiologic testing are adopted
for use in biological monitoring programs.

Eckhardt (22) has remarked that the evolution of
biological tests to screen for the hypersusceptible worker
represents one of the important preventive health vistas in
occupational health. The development of sensitive and specific
biological markers of early medical sequelae of harmful
exposures holds promise to allow the identification of
inadequate exposure containment and unsafe exposure limits
before irreversible impairment of health has occurred.

Medical-Legal Aspects of Biological Monitoring in Medical Surveillance

As with other medical surveillance procedures, there is a
medical-legal ramification to the use of biological monitoring
in medical surveillance. Medical surveillance is a preventive
medical activity that screens for hypersusceptible individuals
and monitors the health experience of persons presumably
exposed to safe levels of substances known to have the
potential to cause disease. When such medical surveillance is
undertaken, there is a presumption of a relationship between
exposure to the potential toxicant for which medical
surveillance is being done and medical conditions that are
found to fit the toxicity profile of the agent. For instance,
the discovery of oligospermia during medical surveillance for
occupational exposure to dibromochloropropane (DBCP)
constitutes reasonable medical evidence that exposure to DBCP
is the cause of this medical finding (6). Outside the context
of medical surveillance for occupational exposure to DBCP and
in the absence of an occupational history of exposure to this
agent, the most common cause of oligospermia might be a
varicocele or endocrine dysfunction.

The Informed Worker

The necessity for the worker to cooperate in the conduct
of biological monitoring procedures is evident need for
communication to occur between the worker and the occupational
health professional conducting the program. The extent of
participation within a worker population and the quality of
compliance among individual workers might be enhanced by
informing participants of the purpose of the program and the
anticipated use of the test results. Monitoring that is done
for environmental and medical surveillance differs considerably

in the meaning of the test result (acceptable exposure vs. unacceptable exposure; normal medical finding vs. abnormal medical finding). Therefore, it is of utmost importance to provide the worker with sufficient information about the program so that the value of participation can be understood and reasonable expectations can be developed about the outcome to be achieved from the test results. In particular, the purpose of biological monitoring procedures that are done to detect potential medical effects of workplace exposures should be discussed with the worker by the physician within the context of the exposure and present medical knowledge about the individual's health. Information about the relationship of the possible outcomes of the test and the meaning for the health of the individual, i.e., reversible or irreversible pathophysiology, should also be disclosed in detail before the biological monitoring activity is begun.

Summary

Biological monitoring is useful in environmental and medical surveillance activities. Its use in environmental surveillance is to provide the environmental engineer with information about the control of potentially harmful exposures. The function of biological monitoring in medical surveillance is to identify medical effects consequent to such exposures. Biological monitoring entails the bioassay or other measurement of toxicants or biotransformation products and medical markers that are altered by the presence of the toxicant (or its biotransformation products). By virtue of a predictive effect of the toxicant upon the pathophysiologic response measured, biological monitoring tests are among the most desirable of medical surveillance techniques. Because environmental monitoring relates to the matter of exposure and medical monitoring to the occurrence of medical abnormalities consequent to exposure, it is advisable to inform the worker of the purpose of the test that is to be done as well as to indicate beforehand the meaning of potential test outcomes.

Acknowledgement

Mrs. Lorraine Meyers and Miss Patricia Abbe are thanked for their help in preparing the manuscript.

Literature Cited

1. Linch, A. L., "Biological Monitoring for Industrial Chemical Exposure Control", CRC Press, Inc.: West Palm Beach, FL, 1974.

2.   Final Rule on Access to Employee Exposure and Medical
     Records, 29 CFR part 1910, Federal Register, 1980, 45,
     35212-35303.

3.   Schrenk, H. H.; Schreibes, L., Urinary Arsenic Levels as an
     Index of Industrial Exposure, Am. Ind. Hyg. Assoc. J.,
     1958, 19, 225-228.

4.   Eisinger, J.; Fischbein, A.; Blumberg, W. E.; Lilis, R.;
     Selikoff, I. J., Zinc Protoporphyrin in Blood as a
     Biological Indicator of Chronic Lead Toxication, J.
     Environ. Pathol. Toxicol., 1978, 1, 897-910.

5.   Cavalleri, A.; Minoia C.; Pozzoli, L.; Baruffini, A.,
     Determination of Plasma Lead Levels in Normal Subjects
     and in Lead-Exposed Workers, Br. J. Ind. Med., 1978,
     35, 21-26.

6.   Whorton, D.; Krauss, R. M.; Marshall, S.; Milby, T. H.,
     Infertility in Male Pesticide Workers, Lancet, 1977, 2,
     1259-1261.

7.   Wills, J. H., The Measurement and Significance of Changes
     in the Cholinesterase Activities of Erythrocytes and
     Plasma in Man and Animals, CRC Crit. Rev. Toxicol.,
     1972, 1, 153-202.

8.   Breysse, P. A.; Bovee H. H., Use of Expired Air-Carbon
     Monoxide for Carboxyhemoglobin Determination in
     Evaluating Carbon Monoxide Exposures Resulting from the
     Operation of Gasoline Fork Lift Trucks in Holds of
     Ships, Am. Ind. Hyg. Assoc. J., 1979, 30, 477-483.

9.   Rostenberg, A.; Bairstow B.; Luther T. W., A Study of
     Eczematous Sensitivity to Formaldehyde, J. Invest.
     Dermatol., 1952, 19, 459-462.

10. Wegman, D. H.; Pagnotto, L. D.; Fine, L. J.; Peters J. M.,
     A Dose-Response Relationship in TDI Workers, J. Occup.
     Med., 1978, 16, 258-260.

11. Edge, J. R., Asbestos Related Disease in Barrow-in-
     Furness, Environ. Res., 1976, 11, 244-247.

12. Lippmann, M.; Ong, L. D. Y.; Harris, D. B., The
     Significance of Urine Uranium Excretion Data, Am. Ind.
     Hyg. Assoc. J., 1964, 25, 43-54.

13. NIOSH Criteria Document for Occupational Exposure to
     Carbaryl, U.S. Department of Health, Education, and
     Welfare, September 1976; p.85.

14. Kowal, N. E.; Johnson D. E.; Kraemer,, D. F.; Pahren, H.
     R., Normal Levels of Cadmium in Diet, Urine, Blood, and
     Tissues of Inhabitants of the United States, J.
     Toxicol. Environ. Health, 1979, 5, 995-1014.

15. Rider, J. A.; Hodges, J. L.; Swader J.; Wiggins, A. D.,
     Plasma and Red Cell Cholinesterase in 800 "Healthy"
     Blood Donors, J. Lab. Clin. Med., 1957, 50, 376-383.

16. NIOSH Criteria Document for Occupational Exposure to
     Inorganic Mercury, U. S. Department of Health,
     Education, and Welfare, 1973; p. 2.

17. Piscator, M., Proteinuria in chronic cadmium poisoning – 2. The applicability of quantitative and qualitative methods of protein determination for the demonstration of cadmium proteinuria, Arch. Environ. Health, 1962,5, 325-332.

18. Linch, A. L., "Biological Monitoring for Industrial Chemical Exposure Control", CRC Press, Inc.: West Palm Beach, FL, 1974; p. 4.

19. Krumholz, R. A.; Chevalier R. B.; Ross, J. C., Changes in Cardiopulmonary Functions Related to Abstinence from Smoking – Studies in Young Cigarette Smokers at Rest and Exercise at 3 and 6 Weeks of Abstinence, Ann. Intern. Med., 1965, 62, 197-207.

20. Weiss, W., Cigarette Smoking and Diffuse Pulmonary Fibrosis, Arch. Environ. Health, 1967, 14, 564-568.

21. Theriault, G. P.; Peters, J. M.; Johnson, W. M., Pulmonary Function and Roentgenographic Changes in Granite Dust Exposure, Arch. Environ. Health, 1974, 28, 23-27.

22. Eckardt, R. E., Evaluation of the Worker – Tools and Techniques for the Future, Am. Ind. Hyg. Assoc. J., 1964, 25, 126-132.

RECEIVED September 19, 1980.

Permeation of Protective Garment Materials by Liquid Halogenated Ethanes and a Polychlorinated Biphenyl

R. W. WEEKS, JR., and M. J. McLEOD

Industrial Hygiene Group, Los Alamos Scientific Laboratory, Los Alamos, NM 87545

Chlorinated ethanes are chemical compounds of the general structure $C_2H_xCl_{6-x}$ ($X \leq 5$) and are widely used in the industrial world as solvents, chemical intermediates, cleaning fluids, fumigants, and for numerous other purposes including uses in petroleum refining. Likewise, the chemicals in the class of compounds known as polychlorinated biphenyls $C_{12}H_{10-y}Cl_y$ ($2 \leq y \leq 10$) (PCBs) have found widespread use as fluids for heat transfer systems, hydraulic systems, and fire retardants; but by far their widest use has been as dielectric fluids in capacitors and transformers. Both the chlorinated ethanes and the polychlorinated biphenyls have shown evidence of chemical carcinogenicity and other toxic effects, (1-9) and as such must be handled and utilized with extreme care. Despite the hazards involved in the use of these compounds, the halogenated ethanes find continued widespread use and the PCBs are widely encountered, although their original roles are being satisfied through replacement by other substances and they are no longer being manufactured nor imported into the United States.

Because of the wide scope of usage of these materials and their associated risks, the present study was performed to determine those protective garment materials which would afford the highest degree of protection against the halogenated ethanes and the particular PCB (Aroclor-1254) chosen for this study.

Experimental

The work described here was performed in a laboratory which had been specially modified and equipped to handle chemical carcinogens (10). Strict care was taken to avoid worker exposure to hazardous chemicals during the course of these studies. Reagent grade chemicals used for this work were: 1,2-dichloroethane (Alfa Products, Danvers, MA); 1,1,1-trichloroethane (Matheson, Coleman & Bell, Norwood, OH); 1,1,2-trichloroethane (J.T. Baker Chemical Co., Phillipsburg, NJ); n-heptane and n-hexane (Burdick & Jackson Laboratories, Inc., Muskegon, MI). Also, used was a polychlorinated biphenyl whose chemical composition as determined by GC/MS were consistent with of Aroclor 1254 (Monsanto Chemical Co., Inc., St. Louis, MO). to which trichlorobenzene had been added as a 40% diluent.   CAP/ASTM Type I water was used through- out these experiments. The protective garment materials (butyl rubber, surgical rubber latex, neoprene rubber latex, nitrile rubber latex, milled nitrile rubber, polyethylene (medium density), poly(vinyl alcohol)[unsupported], Teflon, Viton, and the composite/ bonded materials butyl-coated nylon, polyethylene-coated Tyvek, polyurethane-coated nylon, and poly(vinyl chloride)-coated nylon) used in these studies were obtained from: Clean Room Products (Bay Shore, NY); David's Gloves (Springfield, OH); Edmont-Wilson (Coshocton, OH); Interex Corporation (Natick, MA); Norton Safety Products (Charleston, SC); Pioneer Rubber Co. (St. Louis, MO); and Surety Rubber Co. (Carrolton, OH). As such, these materials were broadly representative of those materials which were readily avail- able from commercial sources, but their choice does not represent an endorsement by the Los Alamos Scientific Laboratory. Each of the halogenated ethane determinations was performed in duplicate to ensure statistical validity. For those studies in which polyethylene-coated Tyvek was employed, a single determination was run with Tyvek toward the water-containing side of the permeation cell and the second determination was effected with the polyethy- lene toward the water.

A.   Protective Garment Material Weight Change Experiments
     Following Immersion in Solvent

          Samples (2-20 g) of protective garment materials
     were immersed in the particular halogenated ethane or PCB
     at room temperature (23 ± 1°C) for total time periods of
     seven days. During this period, they were removed and
     weighed at 24 and 168 h post immersion to determine the
     materials percent weight increase or decrease. For post
     immersion weighings, the materials were patted dry immed-
     iately after being removed from the respective liquids in
     order to remove excess solvent, and weighings were taken
     within 5 min. The net gain (loss) of weight was noted
     and reported as percent weight change relative to the
     initial sample weight.

B.   Volume Change of Protective Garment Material Upon
     Remaining in Solvent

          To determine the volume change of a given material
     following immersion, 1" by 1" pieces of the subject
     materials were individually placed in beakers which
     contained the halogenated ethanes and the particular PCB
     of this study. At 1, 4, and 24h following immersion, the
     material was removed from its solvent and dimensional
     measurements obtained within 1 min. following removal.

The reason for the rapid measurement was that certain
of the test materials would begin to shrink very soon
after removal from the solvent due to solvent evaporation
from within. Thickness measurements were to the nearest
mil whereas length and width measurements were to the
nearest 1/64th inch. The change in volume of the glove
following immersion was then calculated relative to its
original volume.

C.   Solvent Permeation Through Protective Garment Material

Experiments were performed to determine the rate at
which the halogenated ethanes and the PCB used in this
study permeated various protective garment materials.
Discrete, rather than continuous, sampling was employed
for these studies and because of the hazardous nature of
the compounds employed, the experiments were performed in
a chemical fume hood having a face velocity >125 linear
ft./min.

The permeation cells employed in these studies were
functionally identical to those described in the Arthur
D. Little, Inc., report for Contract 210-76-0130 to the
National Institute for Occupational Safety and Health(11)
and are shown in Fig. 1. The cell was constructed of
Pyrex 1720 glass with stainless steel membrane holders
and stainless steel top covers. Gaskets for the cell were
constructed from Teflon (trademark DuPont) with the test
material placed between the gaskets and with the bolts
tightened to a torque of 5 ± 2.5 pounds. This torque was
chosen because values less than 2.5 in pounds did not
seal the protective material firmly,and torques greater
than 7.5 in pounds would put such pressure on the material
that it would tear or degrade rapidly·at the points of
contact.

A constant temperature was maintained during each
permeation test by immersing the cells.in the water bath
of an Exacal 300 constant temperature bath (Neslab
Instruments, Inc., Portsmouth, NH) maintained at 26 ± 1°C
as shown in Figure 2. Stirring of the aqueous side of
the cell was effected by a Teflon stirring bar driven by
a Model MS-7 Micro-Submersible Magnetic Stirrer (Tri R
Instruments, Rockville Centre, NY, or Ace Glass, Vineland,
NJ). Breakthrough determinations were run with either one
or two permeation cell units in the constant temperature
bath.

A zero time base for the halogenated ethane permeation studies was established as follows: The aqueous
side of the cell was filled with 100 mL of water and
allowed to reach thermal equilibrium with the water bath.
Likewise, the particular organic compound being tested

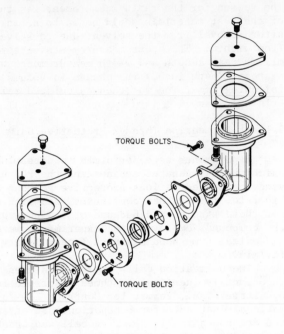

Figure 1. Permeation cell used to measure protective garment material break-through times

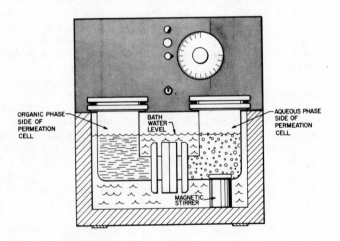

Figure 2. Cross-sectional drawing of water bath, magnetic stirrer, and permeation cell

was allowed to reach test temperature equilibrium in the
bottle in which it came from the manufacturer. When it
was desired to begin a study, 100 mL of the thermally
equilibrated organic was poured rapidly into the permeation cell and this was defined as the zero point in time
for a given study. Samples (3 mL for 1,1,2- and 1,1,1-
trichloroethane and 5 mL for 1,2-dichloroethane) were
taken via pipet from the aqueous phase of the permeation
cell at predetermined times. Each sample was placed in a
5-mL Mini-Vial (Applied Science Laboratories, State
College, PA) and capped with a Teflon faced silicone
rubber septum screw cap (Teflon toward solution).

The halogenated ethane concentration in the aqueous
phase of the permeation cell was obtained using a gas
chromatograph (GC) [Perkin-Elmer Model 900 (Norwalk, CN
equipped with a Tracor (Austin , TX) linearized Ni-63
electron capture detector] following solvent extraction.
Particularly, the water aliquot from 1,2-dichloroethane
was extraced with 2 mL of n-hexane and the trichloro-
ethanes were extracted with 3 mL of n-hexane to collect
the halogenated ethanes. The hexane was then analyzed by
GC in a manner similar to that reported by Reding,
et al. (12), for halogenated methanes. Each water aliquot
removed was replaced with an equal amount of water.
Calibration curves were prepared daily and calibration
standards were prepared at appropriate intervals. Numerical integration of peak areas was performed using a
Spectra-Physics (Santa Clara, CA) Minigrator electronic
integrator. The conditions for these determinations are
given in Table I.

TABLE I

GAS CHROMATOGRAPHIC ANALYSIS[a]
CONDITION FOR HALOGENATED ETHANES

	Analyte		
	1,2-Dichloro-ethane	1,1,1-Trichloro-ethane	1,1,2-Trichloro-ethane
Injector temperature, $^\circ$C	300	200	200
Column temperature, $^\circ$C	150	150	180
Detector temperature, $^\circ$C	300	300	300
Lower limit of quantitation (ppm)[b]	50	10	10
Lower limit of detection (ppm)[b]	8	2	1

[a]Perkin-Elmer Model 900 gas chromatograph equipped with a Tracor
linearized Ni-63 electron capture detector and using a 2-ft. by
1/8-in. o.d. stainless steel Chromosorb 107 column.
[b]Per injection of 0.2 μL.

The determination of the breakthrough time and the rate at
which the particular PCB of this study permeated the various
garment materials was performed by placing 110 mL of PCB (T=25°C)
in the organic phase side of the permeation cell with the test
material in place and with 120 mL of water and 40 mL of n-heptane
atop the water in the cells' aqueous phase side. The aqueous
phase was stirred vigorously to allow the heptane to continuously
extract the PCB from the water. This procedure was necessary
because of the limited solubility of PCB constituents in water.
The 5-mL heptane aliquots were used for each determination and
after each sample withdrawal n-heptane was added to replace that
which had been removed and to thus maintain constant heptane
volume.

Determinations of PCB concentrations in the n-heptane were
performed using either a 1 or a 10-cm far UV quartz cell in conjunction with a DK-2A UV-visible-near IR range scanning spectrophotometer (Beckmann Instruments Co., Fullerton, CA). Quantitation
was effected using the 288 nm peak of the mixture.

For both the halogenated ethanes and the PCBs, in those cases
where a positive organic concentration value was obtained, i.e.,
after breakthrough, it was necessary to mathematically correct the
observed concentration to compensate for dilution effects. Because
of sequential dilutions during sampling, the true organic concentration for each discrete sample was calculated from the formula.([13]):

$$C_n = c_n + (V_s)/(V_p) \sum_{i=1}^{n-1} c_i,$$

where
i   = an indexing number for each discrete sample,
c_i = the organic concentration in discrete sample number i, (ppm),
n   = the number of the most recent discrete sample taken,
c_n = the organic concentration in sample n, (ppm),
C_n = the organic concentration in the aqueous phase at the time of
      sample number n, (ppm),
V_s = volume of sample, (mL), and
V_p = volume of fluid from which sample was removed (mL).

Results and Discussions

Because of their widespread use in the American workplace,
butyl rubber, nitrile latex, neoprene latex, poly(vinyl alcohol),
surgical rubber latex, and Viton elastomer were chosen for the
present studies. The composite/bonded substances of this study
were not in all cases presently available as commercial material
for protective garments, but rather were chosen to determine their
potential for resistance to solvent permeation. Likewise, Teflon
gloves were included in these studies simply because they are
commercially available.

In studies involving protective garments constructed of
neoprene, nitrile, natural or surgical rubber, one must note that
giving a particular composition is not sufficient, by itself, to
describe the materials. In addition to its composition per se,
one must state whether the material was prepared as a latex, i.e.,
colloidal material deposited from an aqueous suspension or was a
so-called solvent/cement dipped or milled material(13) which denotes
production by a molecular layering from an organic solvent. This
industry jargon is further complicated by the fact that a "milled"
material may mean either that the material was prepared from an
organic solvent or that at some stage in processing the raw
material was actually subjected to physical milling by a crushing/
abrading type operation. An exaggerated illustration of these two
physical forms is given in Fig. 3.

In general, the conditions of the present study did not show
that the latex materials were more porous than the "milled" or
"cement dipped" materials for a given chemical composition, but as
shown in Table II, the percent weight change of milled nitrile
immersed in 1,2-dichloroethane was much less than that of nitrile
latex immersed in this same solvent and such would be expected
from physical forms such as those of Fig. 3 wherein there is less
surface area for sorption to occur in the cement dipped material
than with the colloidal particles of the latex. As will be discussed later in this section, there was a correlation between
permeation and weight or volume gain. This leads to the possibility that permeation occurred more rapidly through the latex,
but as mentioned above, the experimental conditions of the present
work coupled with the relatively rapid 1,2-dichloroethane breakthrough would not allow definite differentiation of the breakthrough

TABLE II

WEIGHT CHANGE OF PROTECTIVE GARMENT MATERIAL FOLLOWING IMMERSION IN VARIOUS LIQUIDS

	Per Cent Weight Change[a]							
	1,2-Dichloroethane		1,1,1-Trichloroethane		1,1,2-Trichloroethane		Polychlorinated Biphenyl (Aroclor 1254)	
	24 h	168 h	24 h	168 h	24 h	168 h	24 h	168 h
A. Homogeneous, Nonbonded Materials								
Butyl rubber	34	34	319	328	80	80	48	176
Neoprene rubber latex	182	190	230	241	290	291	170	287
Nitrile rubber latex	655[b]	[b]	237	263	[b]	[b]	112	208
Nitrile rubber milled	340	440[b]						
Polyethylene (medium density)	0.2	16	8.2	21	5	16	46	66
Poly(vinyl alcohol), unsupported	0.3	0.4	0.2	0.2	0.8	0.9	2	2
Surgical rubber latex	211	226	550	564	464	473	519	513
Teflon	0.2	1	0.0	0.1	0.3	0.4	-0.3	-0.2
Viton elastomer	6	6	12.8	13.5	4	5	0.6	0.8
B. Coated/Bonded Materials								
Butyl-coated nylon	24	27	121	143	47	49	77	90
Polyethylene-coated Tyvek	74	100	47	75	131	147	171	190
Polyurethane-coated nylon	26	86	49	55	58	79	92	89
Poly(vinyl chloride)-coated nylon	251[b]	265[b]	11	15	227	273	18	16

[a]Arithmetic mean of two samples. Duplicate samples evidenced 168-h weight change which did not differ by more than 5% relative in any of the samples whose weight gain was more than 2%.
[b]Sample disintegrated.

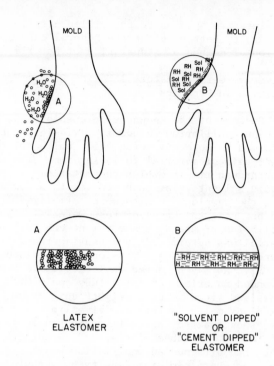

Figure 3.   *Illustration depicting the physical form of "latex" protective garment material and "cement or solvent dipped" protective garment material and showing how a latex is deposited as colloidal particles whereas the "cement dipped" materials are deposited as molecular layers*

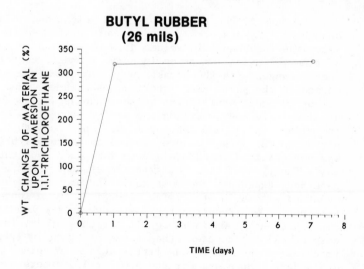

Figure 4.   *Weight change of butyl rubber as a function of time following immersion in 1,1,1-trichloroethane*

times for the latex as opposed to the milled nitrile. Further
work is necessary to better define the general case of permeation
being a function of method of preparation of material.

A. Protective Garment Material Weight Change Following
    Immersion in Organics

         Because of the known relationship between permea-
    tion and equilibrium solubility, (14-16) studies were
    performed to note any correlation which existed between
    weight gain of material following immersion and the
    breakthrough time at which the material was permeated by
    organic solvent. For such studies to allow comparisons
    among the samples, results were normalized to breakthrough
    time per unit thickness of test material. The so-called
    equilibrium or steady-state weight change for each
    material was determined by measuring the weight change at
    24 and 168h following immersion and is discussed in Sec.
    III.C.3. As may be noted in Table II, of the homogeneous,
    nonbonded materials of this study, only PVA, surgical
    latex, and Viton reached weight equilibrium before 24h.
    Typical examples of this are given in Figs. 4,5, and 6.
    These show that following immersion in 1,1,1-trichloro-
    ethane, neoprene rubber latex, and cement dipped butyl
    rubber had attained an equilibrium weight gain after 24h
    immersion; whereas, medium density polyethylene experienced
    a continuing weight increase through day seven.

B. Protective Garment Material Volume Change Following
    Immersion in Organics

         Because of the relation between weight and volume
    changes of given systems, studies were performed to
    determine the change in volume of a material as a function
    of time following immersion in solvent and to correlate
    these changes with the breakthrough time when the solvent
    permeated the particular material. As was the case for
    weight changes, equilibrium volume changes were a function
    of time following immersion and are presented in Table III
    for times of 1,4, and 24h of immersion. As with the
    situation for weight change, volume change equilibrium
    was not obtained immediately upon immersion, but rather
    was a function of time following immersion. As might
    have been expected, the most rapid increase in volume
    occurred within 1h following immersion, but a longer
    time period was required for a steady-state volume to be
    attained. As shown in Table III for this study, in the
    case of all halogenated ethanes except Viton in 1,1,1-
    trichloroethane, there was little change in volume between
    4 and 24h . However, for the Aroclor 1254 there was a

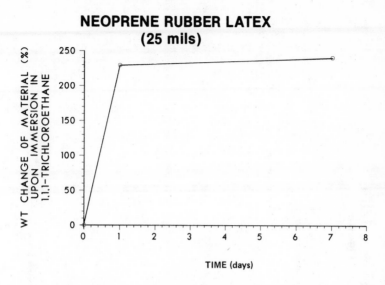

Figure 5.   *Weight change of neoprene rubber latex as a function of time following immersion in 1,1,1-trichloroethane*

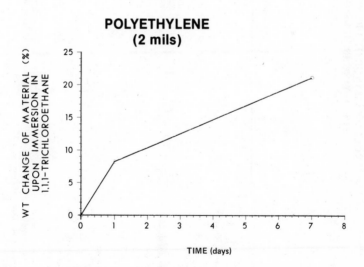

Figure 6.   *Weight change of medium density polyethylene as a function of time following immersion in 1,1,1-trichloroethane*

TABLE III

VOLUME CHANGE OF PROTECTIVE GARMENT MATERIAL FOLLOWING IMMERSION IN VARIOUS LIQUIDS

Per Cent Volume Changes[a]

	1,2-Dichloroethane			1,1,1-Trichloroethane			1,1,2-Trichloroethane			Polychlorinated Biphenyl (Aroclor 1254)		
	1 h	4 h	24 h	1 h	4 h	24 h	1 h	4 h	24 h	1 h	4 h	24 h
A. Homogeneous, Nonbonded Materials												
Butyl rubber	19	19	19	249	260	263	10	7	-1	-2.6	-2.6	22.5
Neoprene rubber latex	123	141	142	213	239	246	158	140	158	32	69	192
Nitrile rubber latex	259	275	286	182	208	214	277	339	355	31	41	138
Nitrile rubber milled	252	252	254	---	---	---	---	---	---	---	---	---
Polyethylene (medium density)	8	20	20	9.6	9.6	9.6	-23	-16	-20	9.6	-4.1	2.7
Poly(vinyl alcohol), unsupported	0.3	1.5	1	0	3	4	0	5	5	0	4.4	5.3
Surgical rubber latex	118	118	124	334	429	425	146	154	154	203	276	297
Teflon	0	0	0	0	0	0	0	0	0	0	0	0
Viton elastomer	0	9	11	0	8.6	20.5	16	16	19	0	-12	-21
B. Coated/Bonded Materials												
Butyl-coated nylon	25	25	25	11	153	181	42	44	44	18	26	71
Polyethylene-coated Tyvek	4.3	9	9	-5	13.3	12.3	14	4	-1	27	14	16
Polyurethane-coated nylon	-3	1.5	0	14.3	15.7	13.7	-5	-5	-5	-2	0.6	1
Poly(vinyl chloride)-coated nylon	b	b	b	0	0	22	b	b	b	-5.5	-5.5	-11.8

[a]Arithmetic mean of two samples. Duplicate samples evidenced 24-h volume changes which did not differ by more than 15% relative in any of the samples whose volume change was more than 5%.
[b]Sample disintegrated.

noticeable change in volume between 4 and 24h immersion, and this points out that care must be taken in defining an equilibrium volume change. Examples of volume change following immersion are given in Figs. 7 and 8. Figure 7 illustrates the volume change of Viton elastomer following immersion in 1,1,2-trichloroethane. The relatively small 24h volume change (<20%) for Viton is in agreement with its relatively long breakthrough time as evidenced against this solvent, i.e., even at 24h post challenge there was no 1,1,2-trichloroethane found in the aqueous phase of the permeation cell. At the other extreme in the relationship between volume change and permeability, Fig. 8 shows that for surgical rubber latex there was a volume change of ~150% following 24h immersion in 1,1,2-trichloroethane, and this correlates well with the fact that 1,1,2-trichloroethane was observed in the aqueous phase of the permeation cell within minutes following challenge, i.e., the breakthrough time was inversely related to the equilibrium volume change for the homogeneous, nonbonded materials.

C.   Permeation of Organics Through Protective Garment Material

Most of the homogeneous, nonbonded protective garment materials chosen for these studies were obtained from commercial sources and were thus representative of materials available to workers in the U.S. The composite materials were not generally used as glove materials, but rather were chosen as being representative of material available for use in coats or aprons or were experimental or prototype materials. In the present work, the results of these permeation studies are presented in three ways:

1.   The time required for the first detectable quantity of organic to permeate the given material into the aqueous phase, i.e., the so-called breakthrough time (T_B).

2.   A normalized breakthrough time in which the breakthrough time per unit thickness of the subject material is reported (T_B/X).

3.   The diffusivity, D, i.e., the diffusion coefficient for each material is presented.

Each of the above is of value for determining the relative efficacy of a given material in protecting the worker against exposure to the particular chemical being studied. The breakthrough time (T_B) gives the "bottom line" for the particular materials included in this study. It represents the combined effects of material imperviousness and material thickness and, thus, the relative degree of protection a worker is afforded by those materials of this study. The diffusivity is defined (11,17-19) as the thickness squared, divided by 6 times the (breakthrough time):

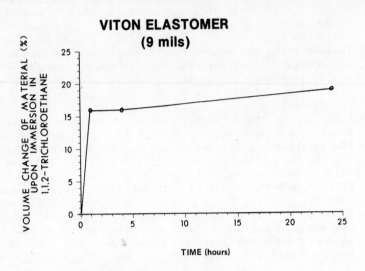

Figure 7.   Volume change of Viton elastomer as a function of time following immersion in 1,1,2-trichloroethane

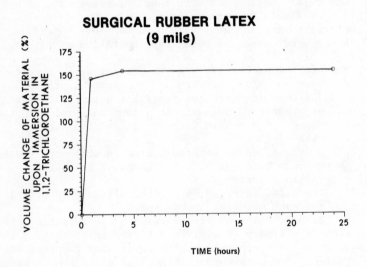

Figure 8.   Volume change of surgical rubber latex as a function of time following immersion in 1,1,2-trichloroethane

$$D = (X^2)/(6\ T_B),$$

D = diffusion coefficient or diffusivity,

X = thickness of membrane, cm, and

T_B = breakthrough time, minutes.

Both the normalized breakthrough time and the diffusivity values provide information which eliminates the thickness constraint and which concern the analytes' permeation rate as a function of the material per se.

In those cases in which the first measured organic concentration (following zero concentration values) was considerably higher than the lower limit of detection values, the breakthrough time was estimated by an extrapolation to zero concentration of a plot of aqueous phase permeation cell concentrations vs time following challenge. This provided the short-term diffusion coefficient (11) whose values would be slightly larger than the steady-state diffusion coefficient obtained through the time lag method. Because of the nature of these experiments and the fact that T_B was determined in terms of minutes at best, the short-term diffusion coefficient is reported in terms of cm^2min^{-1} rather than cm^2s^{-1}, its standard dimensions. The resulting value, D, will in terms of diffusion theory, allow a calculation of permeant flux when the concentration of the permeant is known (18,19).

As may be seen in Tables IV, V, VI, and VII the normalized breakthrough times for the materials and permeants studied varied from a low value of 0.04 min./mil of thickness for the permeation of Aroclor 1254 through surgical rubber latex to a high value of >720 min/mil for the permeation of 1,1,2-trichloroethane through uncrumpled Teflon. These values illustrate the rate of permeation per unit thickness and show how with common protective garment materials it was possible to observe permeation rates and hence degrees of worker protection, which varied by several orders of magnitude. From these values and as mentioned above, for the systems of this study, one may see that the amount of a given analyte permeating the various membranes in a given time period will vary by several orders of magnitude.

1. Protective Garment Materials Permeation by Halogenated Ethanes (1,2-Dichloroethane; 1,1,1-Tricloroethane; and 1,1,2-Trichloroethane). Tables IV, V, and VI present the protective garment breakthrough times for 1,2-dichloroethane; 1,1,1-trichloroethane; and 1,1,2-trichloroethane for both the homogeneous, non-bonded materials and for the coated/bonded composite materials. For these compounds the breakthrough time was defined as that time at which the aqueous phase organic concentration exceeded the Lower limit of detection (LLD) values listed in Table I or was extrapolated to zero concentrations as mentioned above for those

TABLE IV

1,2-DICHLOROETHANE PERMEATION RATE
PROPERTIES OF PROTECTIVE GARMENT MATERIALS

	Thickness (mils mm)	Breakthrough[a] Time (min)	Normalized Breakthrough Time (min mil^{-1})	Diffusivity (cm^2/min^{-1})
A. Homogeneous, Nonbonded Materials				
Butyl rubber	(22 0.56)	140	6.36	3.74×10^{-6}
Neoprene rubber latex	(23 0.58)	20	0.87	28×10^{-6}
Nitrile rubber latex	(8 0.20)	2.5	0.31	27×10^{-6}
Nitrile rubber, milled	(12 0.30)	2.5	0.21	60×10^{-6}
Polyethylene (medium density)	(2 0.05)	2.3	1.15	$2. \times 10^{-6}$
Poly(vinyl alcohol) unsupported	(15 0.38)	22	1.47	11×10^{-6}
Surgical rubber latex	(8 0.20	1.5	0.19	45×10^{-6}
Teflon	(2 0.05)	---	---	---
Crumpled	---	90	45	0.05×10^{-6}
Unwrinkled	---	>1440	>720	$<3 \times 10^{-9}$
Viton	(10 0.254)	820	82	$<0.005 \times 10^{-6}$
B. Coated/Bonded Materials				
Butyl-coated nylon (Mil C-12189)	(15 0.38)	70	4.7	3.5×10^{-6}

[a]Breakthrough time defined as that time following challenge when the permeation cells aqueous phase concentration reached 8 ppm.

TABLE V

1,1,1-TRICHLOROETHANE PERMEATION RATE
PROPERTIES FOR PROTECTIVE GARMENT MATERIALS

	Thickness (mils mm)	Breakthrough[a] Time (min)	Normalized Breakthrough Time (min mil^{-1})	Diffusivity (cm^2 min^{-1})
A. Homogeneous, Nonbonded Materials				
Butyl rubber	(22 0.56)	60	2.72	8.71×10^{-6}
Neoprene rubber latex	23 0.58)	45	1.96	12.5×10^{-6}
Nitrile rubber latex	(8 0.20)	30	3.75	2.2×10^{-6}
Polyethylene (medium density)	(2.0 0.05)	3	1.5	1.4×10^{-6}
Surgical rubber latex	(8 0.20)	4	0.50	16.7×10^{-6}
Viton	(10 0.25)	>1440	>144	$<0.07 \times 10^{-6}$
B. Coated/Bonded Materials				
Butyl-coated nylon (Mil C-12189)	15 0.38)	25	1.67	10×10^{-6}
Polyethylene-coated Tyvek	(5 0.13)	---	---	---
Tyvek toward water	---	10	2.0	2.8×10^{-6}
Polyethylene toward water	---	12	2.4	2.3×10^{-6}
Polyurethane-coated nylon	(4 0.10)	2	0.5	8.3×10^{-6}
Poly(vinyl chloride)-coated nylon	(10 0.26)	3	0.3	38×10^{-6}

[a]That time at which the aqueous phase 1,1,1-trichloroethane concentration exceeded 2 ppm.

TABLE VI

1,1,2-TRICHLOROETHANE PERMEATION RATE
PROPERTIES FOR PROTECTIVE GARMENT MATERIALS

	Thickness (mils mm)	Breakthrough[a] Time (min)	Normalized Breakthrough Time[a] (min mil^{-1})	Diffusivity (cm^2 min^{-1})
A. Homogeneous, Nonbonded Material				
Butyl rubber	(22 0.56)	50	2.27	10×10^{-6}
Neoprene rubber latex	(23 0.58)	7	0.30	80×10^{-6}
Nitrile rubber latex	(8 0.20)	2	0.25	33×10^{-6}
Polyethylene (medium density)	(2 0.05)	3.5	1.8	1.2×10^{-6}
Poly(vinyl alcohol), unsupported	(15 0.38)	15	1.0	16×10^{-6}
Surgical rubber latex	(9 0.23)	1	0.11	88×10^{-6}
Teflon	(2 0.05)	---	---	---
Crumpled	---	175	87	0.024×10^{-6}
Unwrinkled	---	>1440	>720	$< 0.0029 \times 10^{-6}$
Viton	(10 0.25)	>1440	>144	$< 0.072 \times 10^{-6}$
B. Coated/Bonded Materials				
Butyl-coated nylon (Mil C-12189)	(15 0.38)	45	3	5×10^{-6}
Polyurethane-coated nylon	(4 0.10)	<1	< 0.25	$> 17 \times 10^{-6}$

[a]That time following test initiation when 1,1,2-trichloroethane was observed at a concentration of 1 ppm in the aqueous phase of the permeation cell.

TABLE VII

POLYCHLORINATED BIPHENYL (PCB)[a] PERMEATION RATE
PROPERTIES FOR PROTECTIVE GARMENT MATERIALS

	Thickness (mils.mm)	Breakthrough[b] Time (min)	Normalized Breakthrough Time (min mil^{-1})	Diffusivity (cm^2 min^{-1})
A. Homogeneous, Nonbonded Materials				
Butyl rubber	(22 0.56)	2.5	0.114	209×10^{-6}
Neoprene rubber latex	(23 0.58)	0.5	0.022	1121×10^{-6}
Nitrile rubber latex	(8 0.20)	1	0.125	67×10^{-6}
Polyethylene (medium density)	(2 0.05)	0.8	0.40	5×10^{-6}
Poly(vinyl alcohol), unsupported	(11 0.28)	0.6	0.05	218×10^{-6}
Surgical rubber latex	(8 0.20)	0.3	0.0375	222×10^{-6}
Viton	(10 0.254)	60	6.0	2×10^{-6}
B. Coated/Bonded Materials				
Butyl-coated nylon (Mil C-12189)	(15 0.38)	3	0.20	80×10^{-6}
Polyurethane-coated nylon	(4 0.10)	0.5	0.125	33×10^{-6}
Poly(vinyl chloride)-coated nylon	(10 0.25)	0.5	0.05	208×10^{-6}

[a]Aroclor 1254.

[b]That time following test initiation when a PCB concentration of 1 ppm was observed present in the aqueous phase of the permeation cell.

cases in which the first measured organic concentration was considerably higher than the LLD. For 1,2-dichloroethane the breakthrough time was defined as that time at which its aqueous phase concentration exceeded 8 ppm. Likewise, the breakthrough times for 1,1,1-trichloroethane and 1,1,2-trichloroethane were defined as those times wherein their aqueous phase concentrations exceeded 2 and 1 ppm, respectively.

As may be seen in Table IV and Figs. 9-18, the apparent breakthrough times for 1,2-dichloroethane varied from less than 2 min for surgical rubber latex to more than 24h (1440 min) for non-crumpled Teflon. However, for this same Teflon which had been crumpled, breakthrough occurred at 90 min. Because in an actual work situation a crumpling of the Teflon would likely occur to an extent greater than that of these tests, the Teflon should be used with caution for most work situations.

Of the remaining materials in Table IV, only Viton, neoprene rubber latex,poly(vinyl alcohol), butyl rubber, and butyl-coated nylon exhibited at least a 20-min breakthrough time for1,2-dichloro-ethane permeation   to   occur. The nitrile rubber latex, cement dipped nitrile rubber, polyethylene (medium density), and surgical rubber latex were all penetrated by 1,2-dichloroethane in less than 3 min and would be of little use in situations requiring the garment to be in constant contact with 1,2-dichloroethane. From the above, butyl rubber or Viton appear to be the best materials to protect the worker against 1,2-dichloroethane, but because of apparent lot-to-lot variations(20) in butyl properties, Viton appears to be the best suited material of these studies to protect the worker from this chemical.

The results for the permeation of 1,1,1-trichloroethane through the subject protective garment materials are presented in Table V. Of the homogeneous, nonbonded materials of this study, only medium density polyethylene and surgical rubber latex gave breakthrough times less than 10 min. Similarly, the composite/bonded materials polyurethane-coated nylon, poly(vinyl chloride)-coated nylon, and polyethylene-coated Tyvek all gave breakthrough times of 12 min or less. Butyl-coated nylon (Mil C-12189) gave the longest 1,1,1-trichloroethane breakthrough time (25 min) of the composite/bonded materials of this study. Somewhat intermediate in breakthrough time for the homogeneous, nonbonded materials were butyl rubber (60 min), neoprene rubber latex (45 min), and nitrile rubber latex (30 min). The longest breakthrough time of the materials studied was Viton elastomer whose 1,1,1-trichloroethane break-through time was in excess of 24h.

The breakthrough times of Table VI present information concern-ing the breakthrough of 1,1,2-trichloroethane. There was a greater range in breakthrough times for this halogenated ethane than there was for 1,1,1-trichloroethane. Although unsupported poly(vinyl alcohol) showed a breakthrough time of 15 min, neoprene rubber latex, nitrile rubber latex, medium density polyethylene and surgi-cal rubber latex all had breakthrough times of 7 min or less as did

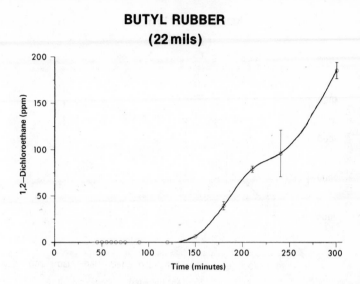

BUTYL RUBBER
(22 mils)

Figure 9.   Aqueous phase concentration (± one σ) of 1,2-dichloroethane as a function of time following challenge to butyl rubber

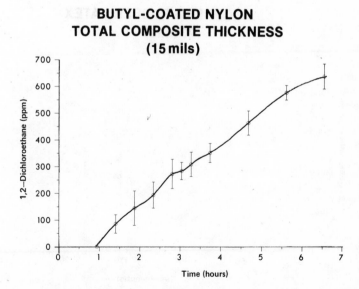

BUTYL-COATED NYLON
TOTAL COMPOSITE THICKNESS
(15 mils)

Figure 10.   Aqueous phase concentration (± one σ) of 1,2-dichloroethane as a function of time following challenge to butyl-coated nylon composite material

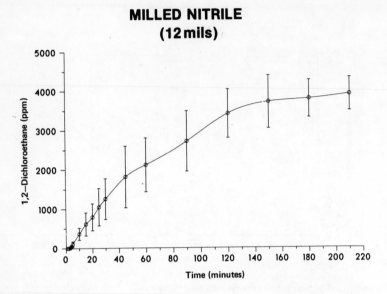

Figure 11. *Aqueous phase concentration (± one σ) of 1,2-dichloroethane as a function of time following challenge to milled nitrile material*

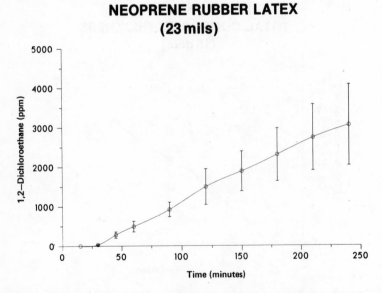

Figure 12. *Aqueous phase concentration (± one σ) of 1,2-dichloroethane as a function of time following challenge to neoprene latex*

NITRILE RUBBER LATEX
(8 mils)

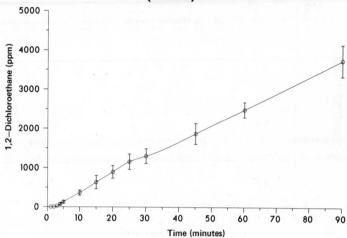

Figure 13.   Concentration (± one σ) of 1,2-dichloroethane in aqueous phase of cell as a function of time following challenge to medium density polyethylene

POLYETHYLENE
(2 mils)

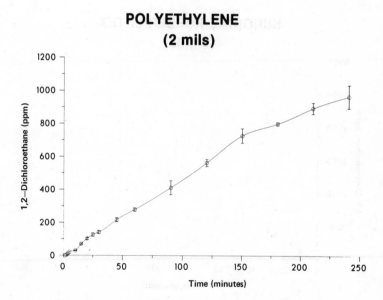

Figure 14.   Concentration (± one σ) of 1,2-dichloroethane in aqueous phase of cell as a function of time following challenge to nitrile rubber latex

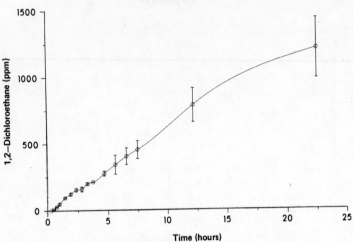

Figure 15.   Concentration ($\pm$ one σ) of 1,2-dichloroethane in aqueous phase of cell as a function of time following challenge to poly(vinyl alcohol)

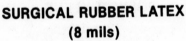

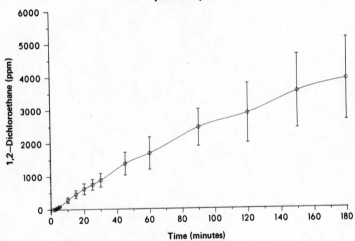

Figure 16.   Concentration ($\pm$ one σ) of 1,2-dichloroethane in aqueous phase of cell as a function of time following challenge to surgical rubber latex

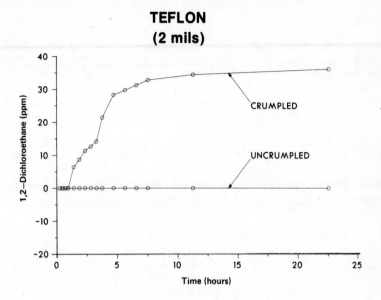

Figure 17.   *Concentration of 1,2-dichloroethane in aqueous phase of permeation cell following challenge to Teflon*

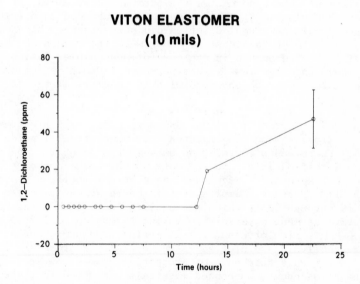

Figure 18.   *Concentration of 1,2-dichloroethane in aqueous phase of permeation cell following challenge to Viton*

polyurethane-coated nylon. Both the crumpled (T_B = 175 min) and
the uncrumpled (T_B >1440 min) Teflon, the butyl rubber (T_B = 50
min), and the butyl-coated nylon (T_B = 45 min) had relatively long
breakthrough times, but with the exception of uncrumpled Teflon,
Viton again had the longest breakthrough time of the materials
tested.

  2. Protective Garment Material Permeation by Aroclor 1254.
        Because the PCB sample of this study was actually a
complex mixture of chlorinated aromatics, that constituent which
permeated the garment first could not be designated by the particular analytical procedure used here. For most of the materials
of this study the PCB breakthrough was so quick that it is unlikely
a molecular permeation differential occurred to any appreciable
extent for the different constituents of this PCB mixture. As
presented in Table VII, the breakthrough times were quite rapid
for all materials except Viton. Butyl rubber, neoprene rubber
latex, nitrile rubber latex, polyethylene, poly(vinyl alcohol),
butyl-coated nylon, polyurethane-coated nylon, and poly(vinyl
chloride)-coated nylon all had Aroclor 1254 breakthrough times of
3 min or less. Viton's 60-min breakthrough was the only material
tested that would afford an appreciable degree of worker protection
against Aroclor 1254.
        As has been noted in other studies, (21,22) a double or
multi-S shaped curve resulted in many cases when the concentration
of analyte which had permeated the garment material was plotted as
a function of time. Of the homogeneous, nonbonded materials of
this study only noncrumpled Teflon did not evidence this non-
Fickian behavior to at least some degree. The reason for not
observing the phenomenon in the case of noncrumpled Teflon was
likely that the organics did not permeate it to a degree such that
this behavior could be observed with the analytical techniques
employed here.

  3. Correlation of Weight Change With Material Breakthrough
        Time. As may be expected from theoretical considerations
(14-16,) there was a correlation between the solubility of a permeant in a test material and the rate at which the permeant permeated the test material. Such cases have been identified previously
(14,15,21). Further examples of this correlation are presented in
Figs. 19 through 22 in which the equilibrium weight change of
protective garment materials following immersion in 1,2-dichloro-
ethane; 1,1,1-trichloroethane; 1,1,2-trichloroethane; and Aroclor
1254 respectively are plotted vs the normalized breakthrough time
(minutes per unit thickness of protective garment material).
For the homogeneous, nonbonded materials evaluated in these studies
(butyl rubber, neoprene latex, nitrile latex, polyethylene, poly-
(vinyl alcohol), surgical rubber latex, and Viton elastomer), only
the poly(vinyl alcohol) did not correlate well in this relationship. As has been shown in Fig. 6, most likely the polyethylene

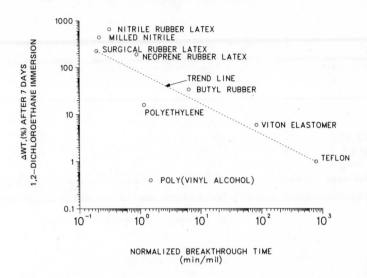

Figure 19.   Log–log plot of equilibrium weight change of homogeneous nonbonded protective garment material immersion in 1,2-dichloroethane

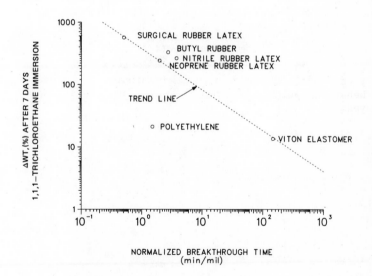

Figure 20.   Log–log plot of equilibrium weight change of homogeneous nonbonded protective garment material following immersion in 1,1,1-trichloroethane

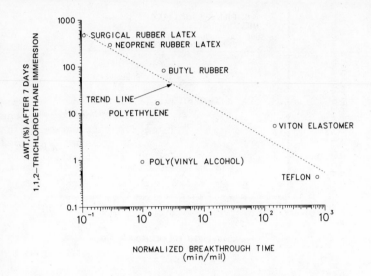

*Figure 21.   Log–log plot of equilibrium weight change of homogeneous nonbonded
protective garment material following immersion in 1,1,2-trichloroethane*

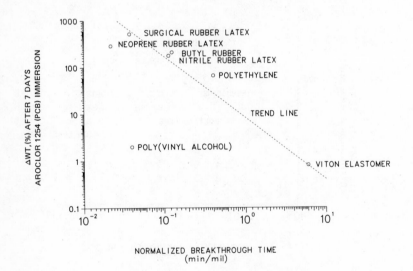

*Figure 22.   Log–log plot of equilibrium weight change of homogeneous nonbonded
protective garment material following immersion in a polychlorinated biphenyl
(Aroclor 1254)*

(PE) did not reach equilibrium weight following 7-day immersion in 1,1,1-trichloroethane. If this were also the situation for the 1,2-dichloroethane and the 1,1,2-trichloroethane, then upon attaining weight equilibrium, the PE percent weight increase would be shifted to a higher value and thus fall closer to the trend lines for the other halogenated ethanes. Because the permeation test method used in these studies required an aqueous media in contact with the test material, the PVA would not have been expected to respond as did the other test materials. Particularly, since PVA is known to chemically react with water (whereas none of the other test materials did) the organic permeation rate would not have been expected to respond in the same fashion as was observed for those garment materials showing no reaction with water.

    4.   Correlation of Volume Change with Material Breakthrough Time. Intuitively one would expect a direct relationship to exist between weight changes and volume changes for elastomeric materials such as the homogeneous, nonbonded materials of the present study. Such is the general case in these investigations as may be seen by comparing trends in weight (Table II) and volume (Table III) changes for given materials with specific solvents. In view of this and from theoretical considerations (14-16), the same type of inverse relationship might be expected to exist between volume change and normalized breakthrough time as existed with weight change and normalized breakthrough time. That this was the case is shown in Figs. 23-26 in which the logarithm of the 24h volume change of the materials is plotted against the log of their normalized breakthrough times. As with the equilibrium weight changes with the exception of PVA, good correlation exists in these plots for the homogeneous, nonbonded materials in these studies.

    As shown above, the normalized breakthrough time vs weight or volume change correlations do allow insight into how the homogeneous materials behaved with respect to permeation. However, because the normalized breakthrough values have eliminated thickness as a variable, these correlations cannot be applied as a universal rule concerning the individual garments because garments constructed from given materials are often found to be of different thicknesses.

Conclusions and Recommendations

    The present work has shown that for a homogeneous, nonbonded material, an inverse relationship exists between the weight change the materials will experience upon immersion in a given solvent and the rate at which this solvent will break through a material. Such was not the case for the bonded/composite materials and from theoretical considerations (14-16) would not necessarily have been expected. From these observations one can conclude that a noniron-clad rule-of-thumb for the relative rate of permeation of given solvents through given materials is, "The more the material swells

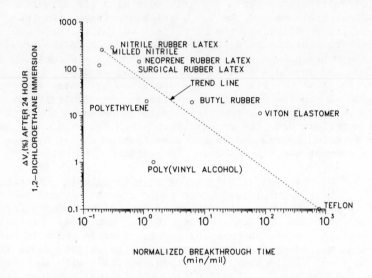

Figure 23. Log–log plot of equilibrium volume change of homogeneous non-bonded protective garment material immersion in 1,2-dichloroethane

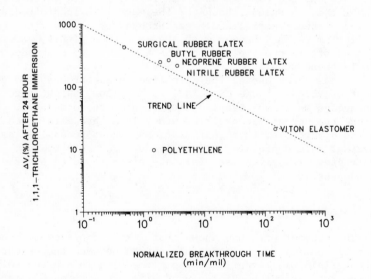

Figure 24. Log–log plot of equilibrium volume change of homogeneous non-bonded protective garment material following immersion in 1,1,1-trichloroethane

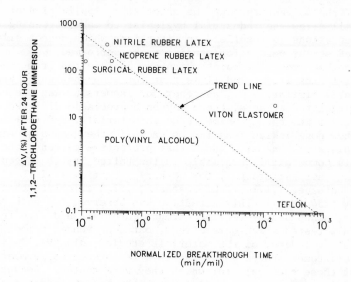

Figure 25. Log–log plot of equilibrium volume change of homogeneous non-bonded protective garment material following immersion in 1,1,2-trichloroethane

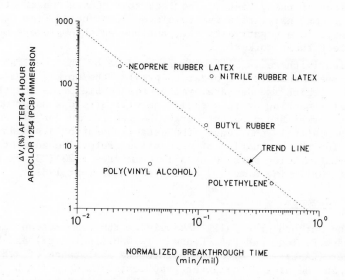

Figure 26. Log–log plot of equilibrium volume change of homogeneous non-bonded protective garment material following immersion in a polychlorinated bi-phenyl (Aroclor 1254)

(gains weight) in a given solvent, the greater the rate of permeation of the solvent through the material". Application of this rule will allow the industrial hygienist to make screening test recommendations in the field in situations which do not allow a detailed set of permeation studies to be performed.

Another important point which may be inferred from the limited data of Tables IV, V, and VI is that for compounds which are chemically similar, e.e., congeners or isomers as was the case for the halogenated ethanes, there may be appreciable differences in the rate at which they permeate given materials, and in many cases will show non-Fickian behavior as shown in the graphs of Fig. 9-18.

This study has also shown that protective garment material which is commercially available in the United States is, generally speaking, not satisfactory for worker protection against 1,2-dichloroethane; 1,1,1-trichloroethane; 1,1,2-trichloroethane; or Aroclor 1254 (PCB). This situation was observed for both homogeneous, nonbonded protective garment materials and also for those materials tested which were either coated or bonded composite materials. Indeed, of the materials studied only PVA, Teflon, and Viton elastomer showed an appreciable resistance to permeation against those chemicals for which they were tested. Because PVA reacts with moisture and Teflon (in a thickness conducive to manual dexterity) mechanically degrades rather readily, of the materials studied in this work only Viton can be recommended as being capable of affording adequate worker protection under the conditions of these tests. The fact that gloves constructed from Viton are expensive relative to gloves constructed from other elastomers is a negative factor, but one which employers must face to protect their workers.

Clearly, there is much work to be done by the American protective garment industry in order to provide the worker protective equipment which is adequate, yet of reasonable price and the successful meeting of this challenge will afford a safer workplace for the worker and a better worker health record for his employer. To help meet this challenge, the establishment of a standard method for the evaluation of permeation protective garment material is currently being pursued by ASTM Committee F-23 (23) with the goal of publishing a method during calendar year 1981.

Acknowledgments

The authors gratefully acknowledge the comments and assistance of W.F. Todd and E.R. Kennedy (NIOSH/Cincinnati) and H.J. Ettinger (LASL). We also appreciate the assistance of W.B. Nelson (LASL) in the area of computer graphics, G.O. Wood (LASL) in the area of gas chromatography, and H.H. Kutac and S.E. Medina in the typing and assembling of this document for the printer.

Abstract

The halogenated ethanes 1,2-dichloroethane; 1,1,1-trichloro=
ethane;and 1,1,2-trichloroethane are used as chemical intermediates
and in metal working operations, and polychlorinated biphenyls
(PCBs) have in the past been used by the tens of millions of
pounds in various roles in American industry. Because of the
widespread use and hazardous or potentially carcinogenic nature of
these compounds, a study was performed to determine the degree of
protection which was afforded against these compounds by certain
protective garment materials. The materials evaluated in these
studies have included: butyl rubber, milled nitrile rubber,
neoprene rubber latex, nitrile rubber latex, polyethylene, poly-
(vinyl alcohol), surgical rubber latex, Teflon, and Viton as well
as the following composite or multilayered materials; butyl-coated
nylon, polyethylene-coated Tyvek, polyurethane-coated nylon, and
poly(vinyl chloride)-coated nylon. The breakthrough time at
which each liquid phase compound permeated these materials was
studied by the time lag method. For the noncomposite materials,
the results of these breakthrough studies were correlated with
their equilibrium weight changes following immersion in the test
liquids. Results of these studies have shown that most materials
currently used in the construction of protective garment material
in the United States are of a generally unsatisfactory nature with
respect to protecting the worker against the halogenated ethanes
and the PCB used in this study.

Literature Cited

1. Revised Recommended Standard...Occupational Exposure to
Ethylene Dichloride (1,2-Dichloroethane) DHEW (NIOSH) Publication
No. 78-211, September 1978.

2. Criteria for a Recommended Standard...Occupational Exposure
to Polychlorinated Biphenyls (PCBs), DHEW (NIOSH) Publication
No. 77-225, September 1977.

3. Criteria for a Recommended Standard...Occupational Exposure
to 1,1,1-Trichloroethane (Methyl Chloroform), DHEW (NIOSH)
Publication No. 76-184, February 1979, (IARC).

4. IARC Monographs on the Evaluation of the Carcinogenic Risk
of Chemicals to Humans: Polychlorinated Biphenyls and Polybrominated
Biphenyls, Vol. 18, ISBN 92 832 1218 5, October 1978.

5. Nisbet, I.C.T., "Criteria Document for PCBs", U.S. Environmental Protection Agency report No. EPA-440/9-76-021, Washington,
D.C., 1976.

6. Current Intelligence Bulletin, No. 7, National Institute for

Occupational Safety and Health (November 3, 1975), reprinted from J. Occupational Medicine, 1976, 18(2), 109-113.

7. Drury, J.S.; Hammons. A.S., "Investigations of Selected Environmental Pollutants: 1,2-Dichloroethane", U.S. Environmental Protection Agency report No. EPA-560/2-78-006, Washington, D.C., 1979.

8. Durfee, R.L.; Cantos, G.; Whitmore, F.C.; Hackman, III, E.E. Westin, R.A., "PCBs in the United States: Industrial Use and Environmental Distributions, U.S. Environmental Protection Agency report No. EPA 560/6-76-005, Washington, D.C., 1976.

9. Hutzinger, O.; Safe, S.; Zitko, V., "The Chemistry of PCBs", CRC Press, Inc., Boca Raton, 1974, pp. 1, 252.

10. Rappaport, S.M.; Campbell, E.E., "The Interpretation and Application of OSHA Carcinogen Standards for Laboratory Operations", Am. Ind. Hyg. Assoc. J., 1976, 37, 690.

11. Coletta, G.C.; Schwope, A.D.; Arons, I.J.; King, J.W.; Swak,A., "Development of Performance Criteria for Protective Clothing Used Against Carcinogenic Liquids", Contract No. 210-76-0130, National Institute for Occupational Safety and Health, DHEW (NIOSH) Publication No. 79-106, A.D. Little, Inc., Cambridge, MA, 1977.

12. Reding, R.; Kollman, W.B.; Weisner, M.J.; Brass, H.J., "Trihalomethanes in Drinking Water: Analysis by Liquid-Liquid Extraction and a Comparison to Purge and Trap", Measurement of Organic Pollutants in Water and Wastewater, ASTM STP 686, C.E. Van Hall, Ed., American Society for Testing and Materials, Philadelphia, PA, 1979, pp. 36-41.

13. Townsend, H.E., Norton Company, Safety Products Division, Charleston, SC, personal communication, July, 1979.

14. Crank, J.,"The Mathematics of Diffusion", Second Edition, Oxford University Press, London, 1975, pp. 44-68.

15. Fuller, T.P.; Easterly, C.E., "Tritium Protective Clothing", Oak Ridge National Laboratory report ORNL/TM-6671, June, 1979.

16. Stannett, V.; Yasuda, H., "Liquid vs Vapor Permeation Through Polymer Films", Polymer Letters, 1963, 1, 289.

17. Hildebrand, J.H.,"Viscosity and Diffusivity", John Wiley and Sons, Inc., NY, 1977, p. 67.

18. Curry, J.G.,"Transient Permeation of Organic Vapors Through Elastomeric Membranes", Ph.D. Dissertation, University of Alabama, Tuscaloosa, 1972.

19. Moelwyn-Hughes, E.A.,"Physical Chemistry", 2nd Edition, Pergamon Press, NY, 1961, p. 1207.

20. Weeks, Jr., R.W.; McLeod, M.J., Los Alamos Scientific Laboratory unpublished data, 1979.

21. Weeks, Jr., R.W.; McLeod, M.J., "Permeation of Liquid Benzene Through Protective Garment Materials", Los Alamos Scientific Laboratory report LA-8164-MS, December, 1979.

22. Weeks, Jr., R.W.; Dean, B.J., "Permeation of Methanolic Aromatic Amine Solutions Through Commercially Available Glove Materials", Am. Ind. Hyg. Assoc. J., 1977, 38, 121.

23. Kohn, P.M., "Standards Are In Store For Protective Clothing", Chem. Engineering, April 21, 1980, p. 76.

RECEIVED October 30, 1980.

The Use of a Fiberoptics Skin Contamination Monitor in the Workplace

T. VO-DINH and R. B. GAMMAGE

Health and Safety Research Division, Oak Ridge National Laboratory, Oak Ridge, TN 37830

Utilization of coal and oil shale to produce liquid and gaseous synfuels results in the generation of many hazardous subtances. Workers in these synfuel plants are likely to be exposed to potentially carcinogenic materials present in coal tars and oils. Among the various pathways of exposure, skin contamination by direct contact transfer or by adsorption of vapors and particulates into the skin presents a serious occupational health hazard. The skin irritant and potential carcinogenic properties of raw syncrudes and their distillate fractions have been reported (1, 2, 3).

This paper reports on research involved the design, construction, and evaluation of a portable instrument, a "luminoscope", for detecting skin contamination by coal tars via induced fluorescence. The instrument has been used in the laboratory to measure the fluorescence of various coal tars and recycle solvents from liquefaction processes spotted on filter paper on rat and on hamster skin. The practical use of the devices in field test measurements to monitor skin contamination of workers at coal gasifier is discussed. The paper also discusses the practicality and usefulness of the luminescence method for detecting skin contamination.

Experimental

Description of the instrument. Figure 1 shows the schematic diagram of the luminoscope that is based upon a bifurcated fiber-optic waveguide. Fiberoptic waveguides have been widely used for electro-optic communication but not to as great an extent for analytical luminescence measurements. In this instrument, the dual purpose of the bifurcated lightpipe (Ealing Corp.) is to transmit the ultraviolet (UV) excitation radiation onto the surface area being monitored and to convey the fluorescence emission back onto the detector. The lightpipe bundle is composed of individual 60 μ-diameter fibers that are thinly clad with a low refractive index glass material to provide total internal reflection within the fiber. The use of fiberoptics is most suitable for remote sensing with a central data recording unit connected to one or several waveguide outlets located in convenient areas within a plant. This possibility for remote sensing is presently under investigation.

The light from a small 125-watt mercury lamp (PBL Electro-Optics, Inc., Model HG-125) is focused onto the excitation entrance of the waveguide. A broadband filter (Ealing Corp., Model OX1) that transmits light at 360 nm with a bandwidth of 50 nm was used in conjunction with the light source. A set of broadband interference filters (Rolyn Optics Co.) transmitting light from 400 nm to 700 nm were used for selecting the spectral regions at which the fluorescence emission is to be monitored. The tip of the common leg of the bifurcated waveguide is mounted into a stethoscopic cap. Several types of collimating optics can be used for focusing the excitation and emission radiations. The standard method of coupling light into a fiber in electro-optic communication generally involves centering a flat fiber end over the light-emitting spot. The use of a microsphere lens to reduce the power loss in such contact coupling (butt coupling) has been recently described (4). In this instrument, an optical lens mounted in the stethoscopic cap serves to focus the excitation light from the lightpipe tip onto the area being monitored and to transmit the luminescence from the same surface back into the fiberoptics tip. During the measurement, the open end of the stethoscopic cap is pressed against the targeted area of the skin.

The single-photon counting (SPC) technique is used for detection. This method uses high speed electronic circuitry to detect individual photon pulses. The SPC technique has proved to have several advantages over the conventional analog method in low level light detection (5, 6, 7, 8). The digital data can be processed directly in a discrete manner suitable for further computational treatment. The gathering of information by digital circuitry is also less susceptible to long-term drifts that usually limit analog systems. The SPC technique also provides the ability to optimize the signal-to-noise ratio by discriminating against photomultiplier dark currents. The luminoscope utilizes a photon counting detector commercially available from Research Support Instruments Inc. (Model 2G150). The photomultiplier tube used in the first prototype instrument has a bi-alkali photocathode that has a spectral sensitivity between 300 nm to 650 nm. This spectral range can be extended with the use of a multi-alkali photomultiplier (Hamamatsu, Model R761) that has a spectral sensitivity ranging from 180 nm to 800 nm. The photon pulse width is 70 ns and the maximum pulse rate is 5 MHz. The photon pulses produce Transistor-Transistor-Logic (TTL) signals that are fed into a digital counting circuit with a visual display. The signals can also be displayed on an analog current meter. The detailed description of the signal conversion and recording electronic circuits are given elsewhere (9).

The excitation lamp is enclosed within a compartment which is designed to shield the detector and its associated electronic circuits from the heat generated by the lamp. A small fan flushes cool air into the upper compartment that houses the detection devices. The trigger mounted at the instrument's handle is used to operate a dual shutter that opens and closes the excitation and emission apertures simultaneously. The hand-held instrument is low cost ($\sim$\$2,000), simple to operate, and weighs only a little over 1 kg without the power supply ($\sim$5 kg). A photograph of the prototype instrument is shown in Figure 2.

Materials. The luminescence emission from a large variety of samples was monitored by the luminoscope. All materials received were diluted in ethanol or in methylene chloride (reagent grade). Commercially available Perylene (Aldrich Chemical Co., 99% purity grade) was used without further purification. Coal tar materials of various origins were obtained through the Oak Ridge National Laboratory--Environmental Protection Agency (ORNL-EPA) Repository: University of Minnesota in Duluth (UMD) coal tar, Solvent Refined Coal (SRC-II) fuel oil blend, the Zinc Chloride ($ZnCl_2$) product distillate, centrifuged shale oil, H-coal distillate, Wilmington petroleum crude oil, and SRC light organic liquid. These materials varied in color (from dark brown to almost black) and viscosity. Their solubility in ethanol and methylene chloride also varied considerably.

Measurements and Discussions

The luminescence method. While photoacoustic and infrared spectroscopies can also be used for surface measurements, luminescence is most suitable for detecting traces of coal tars and oils deposited on surfaces. This is due to the fact that these materials contain appreciable amounts of polycyclic aromatic hydrocarbons (PAH) produced by incomplete combustion of organic substances. Since most of the PAH compounds are strongly luminescing species, they are most susceptible to detection by the luminescence technique. The method of monitoring skin contamination using induced fluorescence has, therefore, been used in the past. However, until now it has only been performed in a coarse manner by simple visual examination in industrial hygiene practices. By this method, the body parts of the workers are illuminated with a hand-held lamp ("black light" lamp) and the induced fluorescence is examined in a dark room with the naked eye. The contaminants on the skin are detected when they exhibit a fluorescence signal brighter than that of the skin. This method has several drawbacks. First, the method has to rely upon the sensitivity and discriminating ability of the operator's eyes. Second, the intensity of the hand-held black light lamp is high enough so as to be the source of some concern about the carcinogenic activity of long-wavelenth radiation acting synergistically with phototoxic agents.

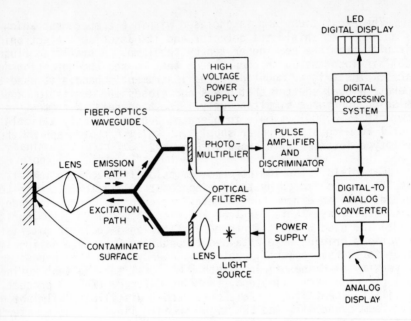

Figure 1.   Schematic of the luminoscope

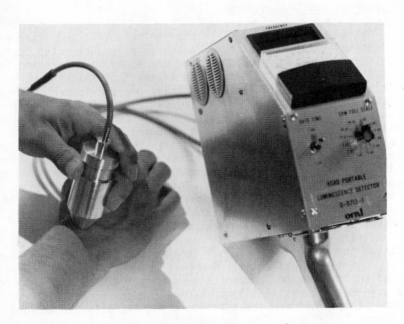

Figure 2.   Photograph of the prototype instrument

The use of ultraviolet radiation for skin illumination. The
illumination of the human body parts by UV radiation requires careful consideration. First, direct and excessive exposure to UV
radiation can be damaging to the eyes. The drastically dimini-
shing response of the eye in the UV spectral range (<350 nm) increases the risk of accidental exposure since the presence of the
UV light is poorly detected. This problem has been addressed
through the use of an optical system based upon the bifurcated
lightpipe. With such a system, the UV radiation is contained within the waveguide. This feature prevents inadvertent illumination
of the person being monitored. The risk of accidental exposure of
the eyes to UV radiation is particularly significant when areas on
the forehead are being monitored. Secondly, the use of flexible
lightpipes greatly facilitates the measurements in hard-to-reach
locations (areas under the neck and the chin; underarms).

Another health hazard associated with exposure to UV radiation is the potential cocarcinogenic activity of UV light with the
contaminant on the skin. Past studies have found that exposure to
UV radiation results in a significant enhancement of the effects
of chemical carcinogens such as 7,12-dimethylbenzanthracene (13)
and benzo[a]pyrene (14). Even normally innocuous compounds such
as anthracene, n-decane and n-tetradecane can develop tumorigenic
activity in mice under irradiation with long-wavelength (>350 nm)
UV light that is generally considered to be noncarcinogenic of
itself (15,16). Considering these potential health hazards, the
intensity of the UV excitation light has to be selected with care.
The American Conference of Government Industrial Hygienists has
set the threshold limit value (TLV) for occupational exposure to
radiation in the near UV spectral region (320 nm - 400 nm) at
10^7 ergs/cm^2 for an exposure time less than 1000 seconds (17).
This value should not apply to individuals concomitantly exposed
to photosensitizing chemicals (17). Since synfuel products contain photosensitizing substances, a lower radiance level must be
used. One factor which provides the guidance for a safe exposure
is the intensity of sunlight. Our intent was that the illumi-
nating intensity of the luminoscope at 365 nm should be less than
that of sunlight in the spectral region between 350 - 400 nm. The
intensity of the excitation light of the luminoscope at 350 nm
with a 50 nm bandwidth was set at 10^2 ergs/cm^2/sec. This value
is only 1/100th of the radiant flux of sunlight (350 - 400 nm)
at sea level (18). For a normal 10-second survey period, the
radiance is only 10^3 ergs/cm^2, or 10^{-4} the TLV dose.

Detection of coals and tars spotted on paper. The performance of the luminoscope was evaluated by measuring the luminescence of various oils and tars spotted on paper. The spectral
characteristics of the luminescence from these materials were first
investigated by recording their fluorescence spectra using a commercial spectrofluorimeter (Perkin Elmer Corp., Model 43A). These
measurements were used to determine the optimal conditions for

excitation and emission to be utilized with the luminoscope.
The results from spectral measurements of a Wilmington petroleum
crude oil and an SRC-II fuel oil blend are depicted in Figure 3.
These spectra illustrate the usually diffuse and broadand structure of most spectra from oils and tars. Since these materials are
extremely complex substances that often contain hundreds of organic species, spectral overlap of the luminescence from individual components results in an overall diffuse structure. A
technique based on synchronous scanning of both the excitation
and emission wavelengths can improve the spectral resolution of
the luminescence measurements (19, 20, 21). The prototype luminoscope, however, employs fixed-wavelength excitations. The use
of optical broadband filters is adequate in most cases to monitor
the gross fluorescence from oil and tar materials.

Measurements of oils and tars from a variety of processes
have been made with the lightpipe luminoscope using an emission at
425 nm and an excitation at 360nm. The sensitivity and ability
to quantify a particular type of oil or tar is depicted in Figure
4. This figure shows the variation of the fluorescence intensity
of the solvent-refined coal (SRC-I) recycle solvent with variation
in concentrations. The sample was dissolved in ethanol and
spotted on a paper having low background fluorescence. The
linearity of response at low amounts of analyte/cm^2 offers the
potential for quantifying the amount of the contaminant at low
levels (<300ng/cm^2). At higher concentrations, saturation effects
observed in Figure 4 are probably due to self-absorption and
quenching. Table I gives the results of detection of various oils
and coal products.

TABLE I

Limits of Optical Detection (LOD)
associated with the Lightpipe Luminoscope (LPL)
for several oil and tar products

Substance	LOD
Centrifuged Shale oil	0.23 nl/cm^2
UMD Coal Tar Sample A	0.062 μg/cm^2
UMD Coal Tar Sample B	0.11 μg/cm^2
H-Coal Distillate	0.22 nl/cm^2
ZNCl$_2$ Product Distillate	0.063 nl/cm^2

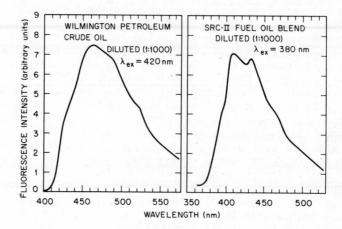

Figure 3. Measurements of the fluorescence spectra of Wilmington petroleum crude oil and SRC-II fuel oil blend

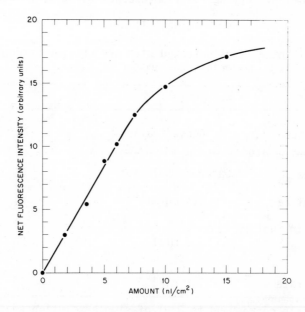

Figure 4. Measurements of the luminescence of SRC-I recycle solvent by the luminoscope

Experiments with rat and hamster skin. Measurements of various oils and tars were conducted with the luminoscope on rat and hamster skin in order to study their fluorescence characteristics on animal skin. (Figure 5). The two animals used in our experiments, a twelve-week-old Fisher 344 strain (white hair) male rat (ORNL Biology Division, Oak Ridge, TN) and a twelve-week-old "LVG" (random bred) brown hair male hamster (Charles River Lakeview, Newfield, NJ), were injected with sodium pentabarbital (35 mg per kg of mouse weight for the mouse and 40 mg per kg for the hamster). Hair on the back was removed with electric shavers and with razor blades. Samples of solvated coal tar (3 μl) were applied to the animal's back using a micropipette. The residual white hair left on the rat exhibited a strong fluorescence that interfered significantly with the measurement of spotted materials. The skin background fluorescence of the white hair rat was so intense that the animal was not used in further studies. Typical results from the measurements with the hamster for various oil and tar materials, such as the UMD tar, SRC-II fuel oil and $ZnCl_2$ product distillate, are given in Table II. The experiments with the hamster indicate that oils and tars can be detected on natural animal skin at trace levels (ng/cm^2). The color and complexion of the skin affect the measurements by varying the fluorescence background intensity.

TABLE II

Signal intensities detected with the LPL for various
materials on hamster skin[a]

Substance	Concentration	Relative Signal Normalized
Perylene	Saturate Solution	100
UMD Coal Tar	1.4 mg/ml [b]	64
SRC-II in Fuel Oil Blend	2.5 ml/l	96
$ZnCl_2$ Distillate	1.0 ml/l	118

[a]Signals correspond to 3 μl of sample solution spotted on mouse skin.

[b]Concentration unit given in mg/ml because of the extremely viscous nature of the tar.

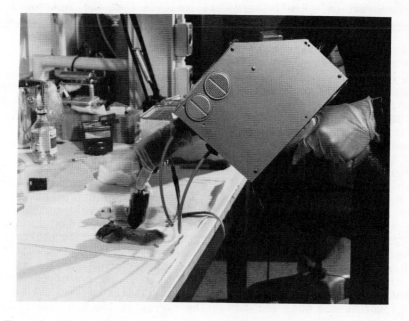

Figure 5. Photograph showing the use of the luminoscope to monitor tar and oil contamination on rat and hamster skin

Monitoring skin contamination at a coal gasifier. Recently
the luminoscope was field tested at a coal gasifier facility. The
purpose was to evaluate the performance of the prototype apparatus
in a real-life workplace environment and to test the applicability
of the instrumental concept in actual measurements. The skin contamination survey was carried out on six workers during two work-
shifts. All measurements were carried out before and after washing.
The portions of the workers bodies most likely to be directly exposed to coal and tar are those not protected by clothing, e.g.,
hands, arms, and faces. The measurements performed during this
field trip were restricted only to arms and hands.

The field study at the coal gasifier produced several pieces
of information. Preliminary data indicate that the background
fluorescence of clean skin is not constant. It varies by as much
as 10 to 35% between two individuals. Background correction is,
therefore, necessary to determine the net fluorescence from the
contaminant. For each individual, the fluorescence background also
varies by 10 to 15%. In addition to this normal background variation, the luminoscope has detected several specific areas in the
hands that are highly fluorescent (e.g., $\geq$ 50% above the average
signal level). This high fluorescence signal appears to originate
from areas where the skin has a different complexion, e.g., cal-
luses, dried skin, and finger tips. It is, therefore, important
to keep data files for each individual indicating the highly
fluorescent specific areas. Unlike contaminated areas, which also
exhibit high fluorescence, the luminescence signal due to skin
complexion did not decrease after washing. The instrument recorded two cases of a doubling in fluorescence level produced by
residual contamination. Unlike the fluorescing residual contamination, thick deposits of coal tar and coal dust usually decreased
the luminescence background of the skin by absorbing the excitation light. In practice, this type of heavy contamination was
visible to the naked eye and did not require the use of the instrument. Table III summarizes the characteristics of various types
of fluorescence observed during the survey at the gasifier (22).

TABLE III

Luminescence characteristics of various types of skin
monitored on workers at a low-Btu coal gasifier.

Nature of skin areas being monitored	Luminescence Characteristics
Background fluorescence of of normal clean skin	The signal intensity varied by 10-15% over the hand and arm area of each individual. Randomly distributed. The signal level varied by 10-35% from one individual to another.
Fluorescence from patches of clean dried skin (calluses)	Intensity >150% of the background level. <1 cm^2 area. Located in specific and defined areas. Signal persisted with same level after washing. Detected in five workers.
Fluorescence of contaminated skin	Intensity >150% of the background level. 1-3 cm^2 area. Signal decreased to normal background level after washing. Detected in two workers.

Conclusion

The problem of skin contamination is one of the principal occupational concerns of the synthetic fuels industry because it is associated not only iwth production but also with transportation and utilization of synthetic fossil fuels. Considering the importance of the risk associated with skin contamination and the lack of adequate monitoring instrumentation, there is an urgent need for an industrial hygiene device like the luminoscope. Such a device will be extremely useful in verifying the efficacy of process controls and the adherance by workers to good hygiene practices. Furthermore, the use of such instrument will provide data to determine the extent of personnel exposure to hazardous pollutants. Further investigation will be performed in order to standardize the technique for routine use in real-world situations.

Acknowledgements

This research is sponsored by the Office of Health and Environmental Research, U. S. Department of Energy under contract W-7405-eng-26 with the Union Carbide Corporation.

Abstract

This paper describes the luminoscope, a simple laboratory-constructed, portable luminescence detector designed specifically for monitoring occupational skin contamination. The instrument design is based upon a fiberoptics waveguide. The instrument is suitable for detecting trace amounts of various coal tars and has recently been field tested at a coal conversion facility.

Literature Cited

1. Scott, A. "Eighth Scientific Report of the Imperial Cancer Fund", London, 1923, 85.
2. Cook, J. W.; Hewett, C. L.; Hieger, I. J. Chem. Soc., 1933, 395.
3. Sexton, R. J. Arch. Environ. Health, 1960, 14, 42.
4. Ackenhusen, J. G. Appl. Opt., 1979, 18, 21.
5. Savager, C. M.; Maker, P. Appl. Opt., 1971, 10, 955.
6. Malmstadt, H. V.; Franklin, M. L.; Horlick, G. R. Anal. Chem., 1972, 44, 63A.
7. Vo-Dinh, T.; Wild, U. P. Appl. Opt., 1973, 12, 1286.
8. Vo-Dinh, T.; Wild, U. P. Appl. Opt., 1974, 13, 2879.
9. Vo-Dinh, T.: Design and Evaluation of a Fiberoptics Based Luminescence Detector (to be published).
10. Epstein, J. H.; Epstein, W. L. J. Invest. Dermotol., 1965, 39, 455.

11. Epstein, J. H. J. Natl. Cancer Inst., 1956, 34, 741.
12. Heller, W. Strahlentherapie, 1950, 81, 529.
13. Baird, W. M. Int. J. Cancer, 1978, 22, 292.
14. Cavalieri, E.; Calvin, M. Photochem, and Photobiol., 1971, 14, 641.
15. Blackburn, G. M.; Taussig, P. E. Biochem. J., 1975, 149, 289.
16. Bingham, E.; Nord, P. J. Natl. Cancer Inst., 1977, 58, 1099.
17. "Threshold Limit Values for Chemical Substances and Physical Agents in the Workroom Environment", Amer. Conf. Government Ind. Hyg., 1978.
18. Gates, D. M. Amer. Scientist, 1963, 51, 327.
19. Vo-Dinh, T. Anal. Chem., 1978, 50, 396.
20. Vo-Dinh, T., in "Modern Fluorescence Spectroscopy", Vol. 4, Wehry, E. L., Ed., Plenum Press: New York (in press).
21. Vo-Dinh, T.; Gammage, R. B.; Hawthorne, A. R.; Thorngate, J. H. Environ. Sci. Technol., 1978, 12, 1297.
22. Vo-Dinh, T.; Gammage, R. B.: The Lightpipe Luminoscope for Monitoring Occupational Skin Contamination (to be published).

RECEIVED November 17, 1980.

A Health Hazard Evaluation of Nitrosamines in a Tire Manufacturing Plant

J. D. McGLOTHLIN, T. C. WILCOX, and J. M. FAJEN

National Institute for Occupational Safety and Health, 4676 Columbia Parkway, Cincinnati, OH 45226

G. S. EDWARDS

New England Institute for Life Sciences, Waltham, MA 02154

In a preliminary study we have recently reported (1) that the air in three rubber and tire industry plants was contaminated with several N-nitroso compounds. The compounds included N-nitrosomorpholine (NMOR), N-nitrosodimethylaminne (NDMA), and N-nitrosodiphenylamine (NDPhA). The latter compound is used as a vulcanization retarder, so its presence near processes employing it was not surprising. Bismorpholine-carbamylsulfonamide, a cross-linking accelerator used in rubber tires, was found in tire factories and may be contaminated with NMOR. The source of the NDMA was not identified, but it could arise from nitrosation of amines which may be decomposition products of diamine based accelerators, as pointed out by Yeager, et al (2).

Our present work, reported here, covers the results of four separate survey visits made to a single tire plant. It strongly suggests that NMOR may be generated by transnitrosation of morpholine by NDPhA when these two chemicals are used together. Although such transnitrosation has been shown to occur experimentally (3,4,5), this is the first instance we are aware of where this occurrence may result in human exposure to NMOR, a known animal carcinogen (6). The chemical structures of four nitrosamine compounds found in this tire plant and three typical vulcanization accelerators and stabilizers used in the tire industry are in Figure 1.

Efforts to improve the worker environment through engineering controls and chemical substitution, and the results of a brief survey of biological samples (blood, urine, and feces) obtained from the workers during two of the NIOSH visits are also reported.

Background. The tire plant in Maryland produces bias-ply passenger, truck, and off-road tires 24 hours per day, 7 days per week. On the average, the tire company mixes approximately

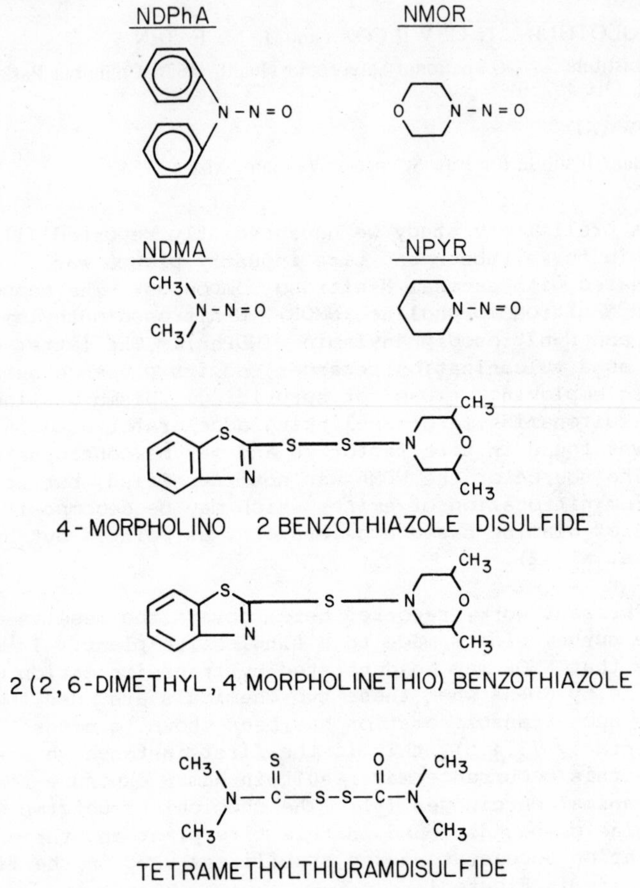

Figure 1. *Volatilized N-nitrosamines found in this tire plant and typical vulcanization accelerators and stabilizers used in the tire industry*

6000 batches of rubber per month for passenger and off-roadtires, and another 2000-2,500 batches per month for truck tires. Batch weights range between 400-500 pounds each. During the manufacture of truck tire tread and bias-ply rubber approximately two pounds of NDPhA, a retarding agent which controls the time of rubber cure, are added to each rubber batch. In August 1979, NIOSH took short term workplace air samples (approximately 2 to 3 hours) of truck tire rubber batches only. On subsequent NIOSH visits workplace air sampling was longer (approximately 5 to 7 hours), and included passenger tire rubber batches (65 to 75 percent of all batches sampled) which did not contain NDPhA. Nitrosamines found during NIOSH surveys were primarily in "hot process" areas where rubber is heated by friction and compression from milling, extruding, and curing operations. The milling and calendering temperatures range from 200-230°F, while extruding and curing operations range from 300-350°F. The term "process sampling" used in this report, refers to workplace air samples collected approximately one foot away from a tire manufacturing process. The word "calendering" refers to the sandwiching of rubber onto nylon fabric to make the plies for bias-ply tires.

Materials and Methods

Air Samples. Airborne nitrosamines were collected with a Thermo-Sorb/N* air sampler (7) connected to a battery-operated pump (DuPont, model P-4000)* which had been calibrated using a 500 ml bubble burett. The pumps were operated between 1.5 and 3.0 L/min. Air sampling ranged from 1 to 8 hours. The air collectors were tightly capped and returned to the laboratory for analysis of nitrosamines. They were eluted with 2 mL of methanol-dichloromethane (1:3, v/v)--and directly injected into a gas chromatograph (GC) and/or high performance liquid chromatograph (HPLC), each equipped with a TEA Thermal Energy Analyzer; (Thermo Electron*, Waltham, MA) detector.

The GC-TEA conditions used for the detection of volatile nitrosamines have been described by Fine and Rounbehler (8). A 14' x 1/8" stainless steel column packed with 5% Carbowax 20M containing 2% NaOH on Chromosorb W HP (80-100 mesh) was operated at 175°C with argon gas as the carrier at a flow rate of 15 mL/min. A TEA was used as the detector with dry ice/ethanol as the cold trap. The HPLC-TEA was constructed by sequentially connecting a high pressure pump (Altex, model 110), an injector (Waters, model U6K), a μPorasil column (Waters), and a TEA. The operation of HPLC-TEA has been described by Fine, et al.(9).

The samples were screened using two different solvent systems: 4% acetone and 96% isooctane for NMOR quantitation, and 0.5% acetone in isooctane for the determination of NDPhA.

Biological Samples. There were three types of biological samples obtained from workers at the plant: urine, whole blood, and feces. All urine and blood samples were internally "spiked" at the factory with 1 μg/mL of a nitrosopiperidine (NPiP) standard. NPiP was used for spiking because it has a similar stability and recovery characteristic to nitrosomorpholine, and to provide a means of gauging the accuracy of the analytical methods. Due to the inability to perform homogeneous mixing on-site, the feces samples were not spiked until they were thawed upon return to the laboratory. Ethyl acetate extracts of urine samples were examined for the presence of N-nitrosodiethanolamine (NDEIA), a metabolite of NMOR, by HPLC-TEA. All samples were immediately frozen at the plant $(-80^{\circ}C)$ and kept at this temperature until analysis.

Urine Samples.
Analysis for NMOR: Ten mL of thawed urine were placed on a Preptube cartridge (Thermo Electron Corp.) and eluted with 60 mL of dichloromethane (DCM). The Preptube was pre-wet with DCM before receiving the sample. The resulting solution was concentrated to a volume of 1 mL at $55^{\circ}C$ using a Kuderna-Danish apparatus. The concentrate was analyzed for NMOR by GC-TEA. Recoveries of the internal standard (NPiP) were typically 80-100%.

Analysis for NDElA: Ten mL of the thawed urine were placed on another Preptube, pre-wet with ethyl acetate. The sample was washed with 60 additional milliliters of ethyl acetate and the effluent dried (rotary evaporator) to 1 milliliter. It was then analyzed by HPLC-TEA using a μNH_2 column eluted with isooctane:dichloromethane:methanol (60:30:7). Recoveries for NDElA using this method were approximately 70%. The percent recovery was judged by an internal spike of N-nitrosodipropanolamine.

Blood Samples. Ten mL blood samples were analyzed for NMOR, using the method described for urine samples. Recoveries of the NPiP standard were more variable, ranging from 32 to 87%.

Feces Samples. Twenty to 45 grams of samples were weighed out and ground to a fine powder in a blender containing liquid nitrogen. The resulting homogenate was placed in a 500 mL distillation flask with 50 milliliters of mineral oil (containing 1 mg/mL of α-tocopherol [to prevent nitrosation during distillation]) and allowed to thaw. The contents of the

flask were mixed with 500 ng of NPiP to determine the recovery
of the method. The feces samples were distilled under a vacuum
of 2.2 torr at up to 130°C for 1 hour. Recoveries of NPiP
were low, approximately 20%.

Results
Air Samples: NMOR, NDMA, and NPYR were found during
the first NIOSH visit in air samples collected at a tire
manufacturing plant in Maryland. One process sample, collected
at a feedmill, contained 250 $\mu g/M^3$ of NMOR, a level several
times higher than has been reported for any airborne nitrosamine
at any industrial site (1). Maximum concentrations of NDMA and
NPYR found in the hot process areas were 4.4 $\mu g/M^3$ and 3.4
$\mu g/M^3$, respectively. Over the following 7 months, ventilation
improvements and changes in chemical formulation of the rubber
resulted in a 200-fold reduction in NMOR levels and elimination
or reduction of other nitrosamines at most sites. Results are
shown in Figure 2, and Table I.

Personal (breathing zone) air samples obtained in October
1979, showed feed mill and calendering operators to be most
heavily exposed to nitrosamines; one worker had a time-weighted
average NMOR exposure of 25 $\mu g/M^3$. Workers in other hot
process such as warm-up mills, extruding machines, and curing
processes were determined to have substantial personal exposure
to airborne nitrosamines (Table II). Personal exposures to
airborne NMOR and NDMA were also detected in the truck tire
building and tire shipping area. Although these nitrosamine
levels are not very high (1.9 $\mu g/M^3$ NMOR, 0.1 $\mu g/M^3$ NDMA),
they demonstrate the residual effect of nitrosamines still
volatilizing off from storage of freshly cured tires. When
compared to passenger tire builders, truck tire builders on the
average had 3 times the NMOR exposure even though 4 to 5
passenger tires could be built to every truck tire.

By November, 1979, there were strong indications that the
source of the high levels of airborne NMOR was the thermal
decomposition of the retarding agent NDPhA, and the subsequent
reaction of its nitroso group with other rubber additives
(preformed morpholino compounds). The results were most
striking when two short term air samples were collected from the
feedmill and calendering area - one rubber batch contained
NDPhA, and the other did not. NMOR levels from the NDPhA batch
were 14 times higher (120.3 $\mu g/M^3$) than the rubber batch
without NDPhA (NIOSH Interim Report No. 2, HE 79-109). In
December 1979, ventilation improvements (3-sided canopy
enclosures and new fan motors) to the feed mills which process
rubber for the calendering of bias-plies, and installation of
local exhaust on the top and bottom of the tire tread extrusion

Table I.　Area and Process Samples in $\mu g/M^3$[a]

Location	Nitrosamines	Aug. '79 Highest	Avg.	Oct. '79 Highest	Avg.	Dec. '79 Highest	Avg.	Feb. '80 Highest	Avg.
Banbury	NMOR	2.1	1.8	N.S.[2]	--	N.S.	--	0.3	--
	NDMA	0.1	--	N.S.	--	N.S.	--	0.1	--
	NPYR	N.D.	--	N.S.	--	N.S.	--	N.D.	--
	NDPhA	N.D.	--	N.S.	--	N.S.	--	N.S.	--
Feed Mill & Calender	NMOR	250	160	120	64	63	25	1.3	1.0
	NDMA	1.9	1.5	2.9	1.6	1.1	0.4	0.8	0.3
	NPYR	3.4	2.3	3.9	2.0	1.0	0.6	N.D.	--
	NDPhA	N.D.	--	12	--	N.S.	--	N.D.	--
Warm-up Mills	NMOR	5.2	3.5	N.S.	--	N.S.	--	14	4.6
	NDMA	0.7	0.4	N.S.	--	N.S.	--	5.5	2.7
	NPYR	N.D.	--	N.S.	--	N.S.	--	N.D.	--
	NDPhA	N.D.	--	N.S.	--	N.S.	--	N.D.	--
Extruders	NMOR	32	18	N.S.	--	N.S.	--	N.S.	--
	NDMA	4.4	--	N.S.	--	N.S.	--	N.S.	--
	NPYR	N.D.	--	N.S.	--	N.S.	--	N.S.	--
	NDPhA	N.D.	--	N.S.	--	N.S.	--	N.S.	--
Curing Room	NMOR	6.4	5.3	2.0	--	N.S.	--	N.S.	--
	NDMA	0.2	0.2	0.1	.07	N.S.	--	N.S.	--
	NPYR	N.D.	--	N.D.	--	N.S.	--	N.S.	--
	NDPhA	N.D.	--	N.D.	--	N.S.	--	N.S.	--

Hot Processes

Other Proceses									
Tire Storage	NMOR	—	N.S.	—	N.S.	—	0.6	N.S.	—
	NDMA	—	N.S.	—	N.S.	—	0.1	N.S.	—
	NPYR	—	N.S.	—	N.S.	—	N.D.	N.S.	—
	NDPhA	—	N.S.	—	N.S.	—	N.D.	N.S.	—
Lunch Room	NMOR	—	N.S.	—	N.S.	—	Trace	N.S.	—
	NDMA	—	N.S.	—	N.S.	—	.02	N.S.	—
	NPYR	—	N.S.	—	N.S.	—	N.D.	N.S.	—
	NDPhA	—	N.S.	—	N.S.	—	N.S.	N.S.	—
Outside Plant	NMOR	—	N.S.	—	N.S.	—	N.D.	N.S.	—
	NDMA	—	N.S.	—	N.S.	—	N.D.	N.S.	—
	NPYR	—	N.S.	—	N.S.	—	N.D.	N.S.	—
	NDPhA	—	N.S.	—	N.S.	—	N.D.	N.S.	—

NMOR = N-nitrosomorpholine, NDMA = N-nitrosodimethylamine, NPYR = N-nitrosopyrrolidine,
NDPhA = N-nitrosodiphenylamine

[1] N.D. = Not Detected
[2] N. = Not Sampled

Trace: <.002 $\mu g/M^3$

Detection Limit: 1 part per billion

[a] $\mu g/M^3$ = micrograms per meter cubed.

Table II. Personal Samples in $\mu g/M^{3a}$

Location	Nitrosamines	Oct. '79 Highest	Avg.	Dec. '79 Highest	Avg.	Feb. '80 Highest	Avg.
Banbury	NMOR	N.D.[1]	--	N.S.[2]	--	N.S.	--
	NDMA	N.D.	--	N.S.	--	N.S.	--
	NPYR	N.D.	--	N.S.	--	N.S.	--
	NDPhA	N.D.	--	N.S.	--	N.S.	--
Feed Mill & Calender	NMOR	25	17	18	17	1.0	0.1
	NDMA	0.4	0.3	0.1	0.1	0.2	0.1
	NPYR	0.8	0.5	0.2	0.2	N.D.	N.D.
	NDPhA	N.D.	N.D.	13	9.9	N.S.	N.S.
Warm-up Mills	NMOR	0.8	0.8	N.S.	--	1.3	1.0
	NDMA	0.2	0.1	N.S.	--	0.5	0.4
	NPYR	N.D.	--	N.S.	--	N.D.	--
	NDPhA	N.S.	--	N.S.	--	N.D.	--
Extruders	NMOR	1.0	--	N.S.	--	N.S.	--
	NDMA	0.1	--	N.S.	--	N.S.	--
	NPYR	N.D.	--	N.S.	--	N.S.	--
	NDPhA	N.D.	--	N.S.	--	N.S.	--
Curing Room	NMOR	1.8	--	0.4	--	N.S.	--
	NDMA	0.1	--	0.3	--	N.S.	--
	NPYR	N.D.	--	N.D.	--	N.S.	--
	NDPhA	N.S.	--	N.S.	--	N.S.	--

———— Hot Processes ————

Tire Building	NMOR	1.9	1.6	N.S.	--	N.S.	--
	NDMA	0.1	0.1	N.S.	--	N.S.	--
	NPYR	N.D.	--	N.S.	--	N.S.	--
	NDPhA	N.S.	--	N.S.	--	N.S.	--
Shipping	NMOR	0.6	--	1.7	1.3	N.S.	--
	NDMA	.04	--	0.1	0.1	N.S.	--
	NPYR	N.D.	--	N.D.	--	N.S.	--
	NDPhA	N.S.	--	N.S.	--	N.S.	--
Receiving	NMOR	N.D.	--	0.3	--	N.S.	--
	NDMA	N.D.	--	N.D.	--	N.S.	--
	NPYR	N.D.	--	N.D.	--	N.S.	--
	NDPhA	N.S.	--	N.S.	--	N.S.	--
Tire Storage	NMOR	N.S.	--	0.7	--	N.S.	--
	NDMA	N.S.	--	0.1	--	N.S.	--
	NPYR	N.S.	--	N.D.	--	N.S.	--
	NDPhA	N.S.	--	N.S.	--	N.S.	--

Other Processes

NMOR = N-nitrosomorpholine, NDMA = N-nitrosodimethylamine, NPYR = N-nitrosopyrrolidine, NDPhA = N-nitrosodiphenylamine

[1] N.D. = Not Detected
[2] N.R. = Not Sampled

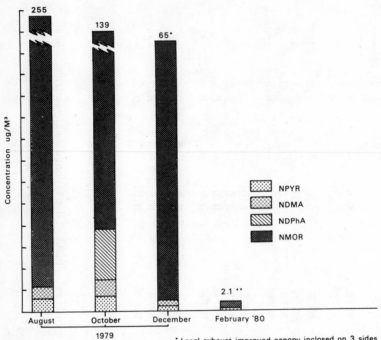

Figure 2.   Highest reported airborne N-nitrosamine levels in feedmill and calendering area

machines, significantly reduced airborne nitrosamine exposure
(Tables I & II). By February 1980, a phthalimide derivative was
substituted for NDPhA. In the feed mill and calendering area,
this resulted in a 99.5% NMOR reduction in process sample
concentrations when compared to the August 1979, results, and a
96% NMOR reduction in the highest personal exposure when
compared to the October results. In the same area, NPYR, which
had measured 3.4 $\mu g/M^3$ in August and 3.9 $\mu g/M^3$ in October,
was not detectable after removal of NDPhA. In general, the NPYR
levels usually were detected when high levels of NMOR were found
and were not detectable when NMOR levels diminished. The
appearance and disappearance of NPYR seemed to be related to
NDPhA, but the source of NPYR in these samples is not yet
known. NDMA levels did not decrease substantially between
August 1979 and February 1980. In fact, NDMA and NMOR levels
increased slightly in the tire tread warm-up mill area. We
suspect that the reason for the increase is due to a higher
percentage of rubber stock containing NDMA additives on the day
we sampled, and also due to the continued use of NDPhA in truck
tire tread.

NDPhA was detected in one process sample on the second
survey and in two personal samples on the third survey. The
fact that NDPhA was not detected more frequently was probably
due to its high molecular weight and relatively low volatility.

On the fourth NIOSH survey, process and personal nitrosamine
samples were collected from rubber batches made at a Banbury
(rubber batch mixer) where only starting ingredients such as
natural and synthetic rubber, and waxes, were mixed. No
retarding, accelerating, or stabilizing additives were added
during this phase of mixing. The purpose of this environmental
sampling was to determine if nitrosamines were being generated
without rubber additives. Four samples were collected; three
process, one personal. NMOR and NDMA were detected in two of
the four samples, but at very low levels. The highest NMOR
process sample was 0.27 $\mu g/M^3$, and 0.09 for NDMA. The
personal sample was 0.24 $\mu g/M^3$ and 0.12 $\mu g/M^3$ for NMOR and
NDMA, respectively. Thus, it seems that the majority, if not
all of the nitrosamines are derived from rubber additives,
either as a raw chemical contaminant, and/or from
transnitrosation of various compounds during the final mixing
and tire manufacturing stage.

In summary, nitrosamines were detected in every area of the
tire plant where NIOSH sampled. Only outside the plant, next to
the guard house, nearly 400 yards away, were nitrosamines not
detected. Generally, highest nitrosamine levels were in the hot
process areas, in particular, the feedmill and calendering
area. With the exception of the guard house, the lowest

nitrosamine levels found were in the receiving area (0.3 µg/M^3
NMOR) and lunch room (trace <.002 µg/M^3 NMOR). Results from
the nitrosamine levels found throughout the plant, and over the
past four NIOSH surveys are in Figure 3. The tire building and
tire storage areas were the only locations where appreciable
NMOR levels were still off-gassing, after the tires had been
cured. Personal NMOR exposures of truck tire builders were 3
times higher than personal NMOR exposures of passenger tire
builders. NDMA, however, did not decrease in the tire tread
warm-up mill area because there were no significant ventilation
improvements and NDPhA continued to be used for truck tire
tread. Nitrosamine results from rubber batches without
additives seems to indicate that natural and synthetic rubber,
and other basic ingredients are not sources of significant
levels of nitrosamines. Finally, over a 7 month period, process
sample NMOR levels decreased 200 fold and personal sample NMOR
exposure in the feedmill and calendering area decreased by 96%.
NPYR was reduced to non-detectable levels. The reduction was
caused by improved ventilation and substitution of NDPhA with a
phthalimide derivative.

   Biological Samples. In December 1979, urine, and either
blood, or stool specimens were collected from 15 non-smoking
workers at the end of their work shift. Based upon
environmental results from the previous two surveys, workers
were selected according to high, medium and low nitrosamine
exposure. The high exposure area was the feedmill and
calendering area. The medium exposure areas were the truck tire
tread warm-up mills, truck tire tread extrusion area, truck tire
curing presses, and truck tire building area. The low exposure
areas were receiving, shipping, passenger tire building, and
Banbury area. Workers donating specimens for nitrosamine
analysis were selected from at least one of these varied
exposure areas. Results of analyses for nitrosamines were
negative for all samples (detection limit = 1 ppb).
Breathing-zone nitrosamine measurements were taken
simultaneously for all workers volunteering biological
specimens. The time-weighted average (TWA) exposure for these
workers ranged from 18.2 µg/M^3 to 0.78 µg/M^3 for NMOR; from
0.180 to non-detectable for NDMA; and from 0.228 to
non-detectable for NPYR.

   In February 1980, urine samples were obtained from nine
workers for mutagenicity testing by the Ames Salmonella test
(10). Four hundred milliliters from each specimen were put
through an XAD-2 column and the adsorbed material was eluted.
Methylene chloride extracts of the post-column urine eluate were
made and tested for mutagenicity with the Salmonella/
Mammalian Microsome Mutagenicity test. The TWA breathing-zone
concentrations for nitrosamines taken from six workers donating

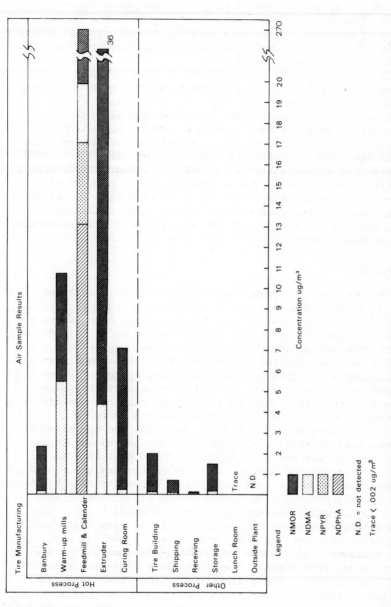

Figure 3.   Highest reported N-nitrosamine air samples during various stages of tire manufacturing

urine specimens ranged from 1.3 $\mu g/M^3$ to 0.64 $\mu g/M^3$ for
NMOR; and from 0.49 to 0.16 $\mu g/M^3$ for NDMA. NPYR was not
detected in any environmental samples. The remaining three
urine specimens were taken from truck tire builders for the
purpose of monitoring mutagenic activity resulting from possible
skin absorption (hands and forearms) of Nitrosamines. None of
the samples tested were mutagenic.

The negative findings for nitrosamines in these biological
samples may reflect the fact that they were collected only
during the last two visits, at which time the airborne
nitrosamine levels had been greatly reduced. For example,
assuming a respiration rate of 10 liters per minute, even the
most exposed workers during the third survey (18 $\mu g/M^3$) would
have breathed approximately 1 μg NMOR per kilogram body weight
per shift. NMOR is relatively lipophilic, but even if all of
this nitrosamine had been absorbed and evenly distributed in all
body components, and none metabolized or excreted during the
shift, its concentration in blood (a relatively aqueous medium)
would be below our detection limit (1 ppb).

Discussion

These results are generally in agreement with previous
reports (1) that NMOR and NDMA are present in tire factories.
NDPhA and NPYR are reported here for the first time as being
present in tire factory air. What is unique about this factory
is that NMOR was found at 250 $\mu g/M^3$--a level higher than any
nitrosamine ever reported in any industrial site--and that its
formation depended upon the thermal decomposition of NDPhA and
the subsequent reaction of its nitroso group with preformed
morpholino compounds. Company management response was effective
in reducing nitrosamine levels first through improved
ventilation, then by reducing or eliminating the use of NDPhA.
However, NDMA is still a problem since it has remained above 1
$\mu g/M^3$ despite process changes that reduced NMOR levels.
Laboratory testing of commercial grade amines has shown NDMA to
be present as a contaminant (11). Most recently, NDMA has been
shown to be generated from heated rubber stock containing a
rubber accelerator tetra-methyl thiuram disulfide (2). The
precursor chemicals and bulk process samples from this factory
environment have not, as yet, been tested to confirm that these
are the sources of the NDMA found. Transnitrosation of the
nitroso group from NDPhA is known to occur under a variety of
conditions (3,4,5). The nitrosation of the accelerators and
stabilizers used in vulcanization (which are designed to break
apart at elevated temperatures yielding the amines) was expected.

A second source of nitrosamine formation could be
nitrosation of amines by NO_x, which has been clearly

demonstrated in laboratory experiments (12). In industrial processes, NO_x was shown to be responsible for nitrosation of amines in foodstuffs dried with gas-fired burners (13), and also for the appearance of nitrosamines in diesel engine crankcase emissions (14). It is possible the NO_x could be nitrosating the amine rubber additives. It has been speculated that the source of NO_x is from the combustion of gasoline-powered forklift trucks used to transport rubber from one work station to another. Unfortunately, NO_x levels were not measured during any of the surveys to test this hypothesis.

A number of epidemiological studies conducted in the tire industry have shown workers to be at excess risk for cancer (15,16,17). In particular, excess cancer of the stomach and lung has been found in "hot process" areas in the tire plants (17). Although the majority of N-nitroso compounds have been shown to be animal carcinogens, including all nitrosamines found in this plant, they have not been directly associated with human cancer because no definable exposed population groups have been identified. Assessing the cancer risk for nitrosamine exposure among tire workers is confounded by past exposure to potential carcinogenic agents such as asbestos, benzene, beta-naphthalamine, and polynuclear aromatic hydrocarbons from carbon black. Although some of these agents have been reduced or removed from the tire industry, their confounding effect may obscure the health effects of nitrosamine exposure in future epidemiological studies.

Despite the epidemiological pitfalls, the nitrosamine levels found in this study are highly significant. Initially, NIOSH found the worst case exposures were typically among the feed mill and calendering operators; one worker had a time-weighted average NMOR exposure of 25 $\mu g/M^3$, an NDMA exposure of 0.37 $\mu g/M^3$, and an NPYR exposure of 0.78 $\mu g/M^3$. By February 1980, however, exposure to a feed mill and calender operator was reduced to 1.00 $\mu g/M^3$ for NMOR, and 0.2 $\mu g/M^3$ for NDMA. Compared to the worst case exposure from the fourth NIOSH survey, a worker in this area now would inhale approximately 6.0 μg of nitrosamines per shift, an exposure equivalent to eating a few strips of bacon and drinking a liter of beer.

Literature Cited

1. Fajen, J.M.; G.A. Carson; D.P. Rounbehler; T.Y. Fan; R. Vita; U.E. Goff; M.H. Wolf; G.S. Edwards; D.H. Fine; V. Reinhold; K. Bieman. *Science*, 1979, *205*, 1262.

2. Yeager, F.W.; N.N. van Gulick; B.A. Lasoski. *Am*. *Ind*. *Hyg*. *Assn*. *J*., 1980, *41*, 148.

3. Welzel, P. *Chem*. *Ber*., 1971, *104*, 808.

4. Buglass, A.J.; B.C. Challis; M.R. Osborne. N-Nitroso Compounds in the Environment (P. Bogovski and E. Walker, eds.), International Agency for Research on Cancer, Lyon, France, 1974, 94.

5. Singer, S.S.; W. Lijinsky; G.M. Singer. Environmental Aspects of N-Nitroso Compounds (E.A. Walker, M. Castegnaro, L. Griciute, R.E. Lyle, and W. Davis, eds.), International Agency for Research on Cancer, Lyon, France, 1978, *19*, 175.

6. Magee, P.N. (In Searle, C.E., ed.) Chemical Carcinogens, Washington, D.C. American Chemical Society, 1976, (monograph) 173.

7. Rounbehler, D.P.; J.W. Reisch; J.R. Coombs; D.H. Fine. *Anal*. *Chem*., 1980, *52*, 273.

8. Fine, D.H.; D.P. Rounbehler. *J*. *Chromatogr*., 1975, *109*, 271.

9. Fine, D.H.; D.P. Rounbehler; A.P. Silvergleid. in Proceedings of the 2nd Symposium on Nitrite in Meat Products (B.J. Tinbergen; B. Krol, eds.), Pudoc, Wageningen, Netherlands, 1977, 191.

10. Durston, W.E.; B.N. Ames. *Proc*. *Nat'l*. *Acad*. *Sci*., 1974, *71*, *3*, 737–41.

11. Speigelhalder, B.; G. Eisenbrand; R. Preussman. *Angew*, *Chem*. *Int*. *Ed*., 1978, *17*, 367.

12. Challis, B.C.; A. Edwards; R.R. Hunma; S.A. Kyrtopoulos; J.R. Outram. Environmental Aspects of N-Nitroso Compounds, International Agency for Research on Cancer, Lyon, France, 1978, *19*, 127.

13. Kann, J.; O. Tauts; R. Kalve; P. Bogovski. Formation and Occurrence of N-Nitroso Compounds in the Environment, International Agency for Research on Cancer, Lyon, France (in press)

14. Goff, E.; J.R. Coombs; D.H. Fine. Anal. Chem. (in press)

15. Peters, P.M.; R.R. Monson; W.A. Burgess; L.J. Fine. J. Env. Health Pers., 1976, 17, 31-4.

16. McMichael, A.J.; R. Spirtas; J.R. Gamble; P.M. Tousey. J. Occup. Med., 1976, 18, 185.

17. Monson, R.R.; L.J. Fine. J. Nat'l. Cancer Inst., 1978, 61, 1047.

RECEIVED November 7, 1980.

Sampling Methods for Airborne Pesticides

ELLEN C. GUNDERSON

SRI International, Menlo Park, CA 94025

Airborne pesticides have been collected by a variety of techniques using filters, impingers, bubblers, solid sorbents, polyurethane foams, and combinations thereof. Sampling for pesticides in air is complicated because specific pesticides may be present as particulate or as a vapor or both, depending on the concentration of the pesticide in the atmosphere, the equilibrium vapor concentration, and the temperature (1-6).

Personal sampling is often used to determine pesticide exposure in the workplace environment. Thus, the convenience of the sampling method becomes a major factor in selection of the best sampling medium or media. Impingers and bubblers containing a liquid absorbing solution may perform well, but they are very cumbersome to the worker and the industrial hygienist. Also, shipping regulations prohibit the mailing of most organic solvents. Filters have been shown to efficiently collect some pesticide vapors, but this has not been consistently demonstrated. In general, personal sampling for pesticides is best done by collecting pesticide aerosols on filters, vapors on solid sorbents, and aerosol/vapor mixtures with filter/sorbent sampling trains.

A universal sampler applicable to the majority of pesticides would be an ideal sampling device. In this study, personal sampling and analytical methods were developed and validated in the laboratory for determining workplace exposure to several pesticides. The major objectives of the study were to standardize on specific sampling media and to develop and validate methods using filter/sorbent sampling trains.

Sampling Device Criteria

The following criteria were used to define a good sampling device:

(1) The sampling medium or media must be compatible with the analytical method. For example, the sampling media should not dissolve in the most appropriate solvent for analysis and the media should not interfere in obtaining optimum detection of the pesticide.

(2)   Greater than 90% of the pesticide should be recovered from the
      sampling device.

(3)   The sampler should collect the sample efficiently and have
      adequate capacity.

(4)   Collected samples should be stable on the device for at least
      seven days before analysis.

(5)   The sampling media should not interfere with good precision
      and accuracy of the overall method.

      To meet the criteria for a good sampling device as well as a
good overall sampling analytical method, each method was tested
extensively with the pesticide of interest. The testing procedure
involved the following:

(1)   An analytical method was developed. Optimum conditions were
      established for maximum sensitivity and selectivity.

(2)   The potential sampling medium (or media) was selected based on
      the particular chemical properties of the pesticide, its ex-
      pected physical state, and the analytical method to be used,
      keeping in mind that we wanted to standardize on the materials
      used.

(3)   Recovery of the pesticide from the sampling media was deter-
      mined with spiked samples; greater than 90% recovery was
      desired. For filter/sorbent methods we have found that the
      sample losses due to volatilization are minimized if the fil-
      ter and sorbent are combined after sampling. The combined
      sample is then analyzed. This requires that the recoveries
      from filter and sorbent be statistically the same.

(4)   Test atmospheres of known concentration of the pesticide in
      air were dynamically generated at levels of twice, one-half,
      and at the OSHA standard for the specific pesticide.

(5)   The collection efficiency of a filter sampler was demonstrated
      by sampling test atmospheres with a backup collector at the
      proposed sampling rate and time, and analyzing the collected
      samples. For sorbents or filter/sorbent sampling trains, the
      breakthrough volume was determined (to demonstrate capacity)
      at 80% relative humidity.

(6)   Storage stability was demonstrated by collecting a set of sam-
      ples of a test atmosphere at a known concentration and analyzing

half of the samples immediately and the other half after seven
days storage at room temperature. A difference of no more
than 10% between the results was acceptable.

(7) The precision and accuracy of the overall method was assessed
     by collecting and analyzing three sets of samples from test
     atmospheres of known concentration. An overall coefficient of
     variation of 10% for all analytical data and accuracy of ±10%
     was required for method validation.

Sampling Methods

The methods developed and validated in our studies can be
classified for discussion into three categories--filter, sorbent,
and filter/sorbent methods. These methods should not be considered
absolute since they were validated over the concentration range of
one-half to two times the OSHA standard for each pesticide and at
or near room temperature. However, many may be applicable for a
wide concentration and temperature range with some additional testing or knowledge of vapor pressure data. For example, a filter
method for an aerosol may be inadequate if a high temperature and/
or low concentration of the material results in a significant fraction of the material being present as vapor. The method would then
require a backup collector for vapor.

The problem of sampling for coexisting particulate and vaporous forms of a toxic substance, as discussed by Taylor et al. (7),
becomes important when the OSHA environmental limit for vapor-pro-
ducing particulates is low compared to the substance's vapor pressure. To determine if a mixture may be present in workplace air, a
comparison of the equilibrium vapor concentration (EVC) with the
OSHA standard is helpful. For a specific compound the EVC is calculated as follows:

$$EVC = \frac{(VP)(MW)10^6}{(760)(24.47)} \quad mg/m^3 \text{ at } 25°C$$

where VP = vapor pressure in mm Hg at 25°C.
      MW = molecular weight.

If the ratio EVC/std is in the range 0.05 to 100-300, then a mixture of particulate and vapor may be present (7). A ratio above
this range indicates the presence of vapor alone and below the
range, particulate. The reliability of this determination depends
on the accuracy of vapor pressure data. In method development and
validation studies, it is often necessary to perform special tests
with generated test atmospheres at different temperatures and concentrations to demonstrate the physical form of the substance.

   Filter Methods. The filters most commonly used were mixed-
cellulose ester (MCE), glass fiber (GF), and polytetrafluoroethylene (PTFE). Glass fiber and PTFE filters were used almost

exclusively in the aerosol methods because high-performance liquid
chromatography (HPLC) was chosen as the best analytical method.
Clearly, the sampling device must be compatible with the analytical
method. The most commonly used mobile phases in HPLC analyses were
methanol-water and acetonitrile-water. To avoid adding interfering
substances to the sample, we limited the extraction solvents for
the filter samples to those used in the HPLC mobile phase or we
used the mobile phase itself. Since MCE filters dissolve in many
of these organic solvents, MCE filters were not acceptable for use
with the exception of ethylene glycol.

Table I summarizes the compounds for which methods were developed and validated using filters as a collection medium. Each
method was tested over the concentration range listed. All methods
demonstrated good recovery (>95%) of the analyte from the filters,
excellent collection efficiency, and good storage ability. Precision of the method is indicated by the total coefficient of variation (CV_T) for the total (sampling and analytical) method. The
detailed sampling and analytical methods for the specific compounds
are published in the "NIOSH Manual of Analytical Methods" (8,9)
under the compound's method number (as listed in Table I). Supporting experimental data obtained in the validation effort for
each method are included in the method's "Backup Data Report" (10).
In general, air samples are collected from the worker's breathing
zone with 37-mm diameter filters contained in cassette filter holders. A calibrated personal sampling pump draws air through the
filter at a flow rate of 1 to 2 liters/min to obtain a prescribed
sample volume. After sampling is completed, the filter holder is
sealed and prepared for storage or shipping to the analytical laboratory. In the laboratory, the compound of interest is extracted
from the filter with the appropriate solvent, and the resulting
solution is analyzed by the prescribed analytical method.

The use of glass fiber and PTFE filters is not interchangeable.
Besides the possibility of introducing fibers into the HPLC system
(which cannot be tolerated) and thus having to filter samples carefully, a significant recovery problem with glass fiber filters
could occur. This was demonstrated for ANTU (alpha-napthyl-thiorea)
and thiram. Reduced recoveries were noted when samples were stored
for very short periods of time. We felt this may have been due to
decomposition of the sample on the filter surface--possibly because
it is slightly alkaline (pH 8.5-9). In addition, a background of
interfering peaks was noted from the glass fiber filters in the
rotenone method. In these cases, the PTFE filters were satisfac-
tory--their inert surface and fiber-free property justified their
use.

The method developed for sodium fluoroacetate is based on collection of the sample with cellulose ester filters especially low
in extractables (Toyo-cel cellulosic ester membrane filters from
Nuclepore were used). The samples were extracted with water and
analyzed by ion chromatography. Other MCE filters contained high
levels of extractables (from wetting agents, surfactants) that

TABLE I. FILTER SAMPLING METHODS

Method No.	Substance	OSHA Std. (mg/m³)	Filter Collection Medium	Extraction Solvent	Range (mg/m³)	Total Coefficient of Variation (CV_T)
Analytical Method: HPLC/UV detection						
S276	ANTU	0.3	PTFE	Methanol	0.128–0.76	0.054
S294	Paraquat	0.5	PTFE	Water	0.256–1.03	0.088
S297	Pentachloro-phenol	0.5	MCE	Ethylene glycol	0.265–1.13	0.072
S298	Pyrethrum	5	GF	Acetonitrile	1.41–8.5	0.070
S300	Rotenone	5	PTFE	Acetonitrile	1.16–11.1	0.079
S302	Strychine	0.15	GF	Mobile phase (ion-pairing reagent in acetonitrile/water)	0.073–0.34	0.059
S279	2,4-D	10	GF	Methanol	5.1–20.3	0.051
S303	2,4,5-T	10	GF	Methanol	4.9–21.4	0.053
S256	Thiram	5	PTFE	Acetonitrile	3.0–12.2	0.055
P&CAM 313	Warfarin	0.1	PTFE	Methanol	0.054–0.24	0.056
Analytical Method: Ion chromatography/electrolytic conductivity detection						
S301	Sodium floro-acetate	0.05	MCE (Toyo-cel)	Water	0.020–0.137	0.060

produced interferences. PTFE filters were not acceptable because
they are not wettable with water.

Another factor regarding recovery of the sample is that the
complete sampling device should be tested, not just the filter itself. Generation of test atmospheres may be necessary to perform
these tests. This became most apparent in the work on thiram where
the filter cassette top collected thiram to some degree. The
amount varied from 1% to 12% of the total sample. Fortunately, the
thiram on the cassette was stable so that cassettes containing filter samples could be stoppered and shipped for analysis. In laboratory analysis, a separate "cassette-rinse" was analyzed based on
a 5-ml wash of the cassette top with acetonitrile.

Sorbent Methods. The sorbent methods developed in this program are listed in Table II. These methods are strictly for vapors
of the listed pesticides. Complete methods are published in the
"NISOH Manual of Analytical Methods" (9,11).

Samples are collected with especially prepared sorbent tubes.
A glass tube (6-mm I.D. by 8-mm O.D. by 3-cm long) is packed with
approximately 100 mg of sorbent in a front section and 50 mg in a
backup section, each separated by glass wool plugs. Sorbents of
coarse mesh size ($\sim$20/40) are used to minimize the pressure drop
across the tube. A calibrated personal sampling pump draws air
through the sorbent tube at a flow rate of up to 1 liter/min.
Capped sorbent tubes containing the sample are shipped to the analytical laboratory, where the pesticide is desorbed from the sorbent with toluene and the solution is analyzed by gas chromatography.

The selection of a solid sorbent for personal sampling of pesticides was based on the factors of good recovery of the sample,
adequate capacity and storage stability, and contribution to overall precision and accuracy of the method as discussed earlier. The
sorbent should also be inert and free of background interferences.
Prewashing the sorbents before use by Soxhlet extraction with a
methanol/acetone solution and drying was done to remove residual
monomers, impurities, and any solvents.

The sorbents, Chromosorb 102 and XAD-2, which are styrene-
divinyl benzene cross-linked porous polymers, proved to be most useful in our studies. Capacity of the sorbent sampling tubes was not
a problem with the pesticides we studied since most were not extremely volatile. Sampling humid atmospheres of the pesticides
also did not affect the sorbent capacity since these porous polymers are hydrophobic.

We attempted to standardize on Chromosorb 102 for collection
and on desorption of the sample with toluene for all pesticides
amenable to analysis by gas chromatography. In a few cases, Chromosorb 102 was not acceptable, usually because of reduced recoveries;
in these cases, XAD-2 was used. The reason for reduced recoveries
was not apparent. The difference in behavior between Chromosorb
102 (Johns-Manville Corp.) and XAD-2 (Rohm and Haas) remains to be

TABLE II. SORBENT METHODS

Method No.	Substance	OSHA Std. (mg/m^3)	Sorbent	Desorption Solvent	Analytical Method	Range (mg/m^3)	Total Coefficient of Variation (CV_T)
S287	Heptachlor	0.5	Chrom 102	Toluene	GC/EC	0.23–1.0	0.066
S296	Phosdrin	0.1	Chrom 102	Toluene	GC/FPD-P	0.027–0.145	0.069
P&CAM 315	TEPP	0.05	Chrom 102	Toluene	GC/FPD-P	0.025–0.124	0.086
P&CAM 295	Dichlorvos (DDVP)	1	XAD–2	Toluene	GC/FPD-P	0.38–1.71	0.054

GC/EC   = gas chromatography/electron capture detector

GC/FPD-P = gas chromatography/flame photometric detector-phosphorus mode

determined. One would expect that both sorbents should exhibit similar or identical properties since Chromosorb 102 is prepared commercially from XAD-2 (12).

In studying heptachlor it was initially thought, based on vapor pressure data, that a filter/sorbent sampling train would be necessary to collect both vapor and aerosol components. A breakthrough test was performed by sampling a test atmosphere at 80% relative humidity of heptachlor at twice the OSHA standard. Several sampling devices, each consisting of an MCE filter followed by a Chromosorb 102 sorbent tube, were connected to the sampling chamber. The test air was sampled at the proposed flow rate, 1 liter/min, and samplers were removed at intervals throughout the test; the individual parts of the samplers were analyzed separately.

The heptachlor breakthrough test, illustrated in Figure 1, shows an interesting result. The amounts of heptachlor found on the filter and sorbent and the total are plotted versus the time of sampling. The amount of material collected on the filter appears to rise and level off to a constant amount after extended sampling. It was apparent that the filters were absorbing heptachlor vapor up to a saturation point and then allowing the remainder to pass through to the sorbent. Based on these results, it was decided that only heptachlor vapor was present and that a sorbent tube alone would be the most appropriate sampler. A second test, performed with sorbent tubes only, demonstrated that the sorbent had sufficient capacity.

Filter/Sorbent Methods. The methods developed using filter/ sorbent sampling trains are listed in Table III. The sampling train consists of a 37-mm diameter filter contained in a cassette filter holder followed by a sorbent tube as described above. Samples are collected at 1 liter/min to obtain the prescribed sample volume. After sampling is completed, the filter is removed from the filter holder and placed in a glass vial with the front sorbent section and capped. The combined sample is extracted with toluene and the resulting solution is analyzed by gas chromatography.

Again, we tried to standardize on sampling media and sample treatment, using MCE filters followed by Chromosorb 102 sorbent tubes and extraction of the sample from the collection media with toluene. In the case of demeton, poor recoveries were noted from Chromosorb 102; this was also found by other investigators (13). XAD-2 was demonstrated to be satisfactory.

The breakthrough test results for these compounds give the best illustration of the partitioning of material on the sampling train. These tests were performed with test atmospheres at twice the OSHA standard for each pesticide, 80% relative humidity, and at or near room temperature.

Figures 2 through 4 are graphical representations of the tests performed on the filter/sorbent sampling trains to test for sorbent capacity or breakthrough, and to best determine aerosol/vapor partitioning.

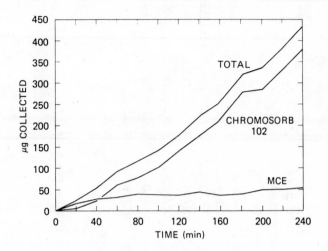

Figure 1.   *Heptachlor breakthrough test*

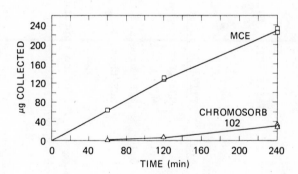

Figure 2.   *Chlordane breakthrough test*

TABLE III.   FILTER/SORBENT METHODS

Method No.	Substance	OSHA Std. (mg/m^3)	Filter/Sorbent	Solvent	Analytical Method	Range (mg/m^3)	Total Coefficient of Variation (CV_T)
S278	Chlordane	0.5	MCE/Chrom 102	Toluene	GC/EC	0.156-0.17	0.070
S280	Demeton	0.1	MCE/XAD-2	Toluene	GC/FPD-P	0.06-0.33	0.080
S284	Endrin	0.1	MCE/Chrom 102	Toluene	GC/EC	0.06-0.31	0.071
S299	Ronnel	10	MCE/Chrom 102	Toluene	GC/FPD-P	2.82-17.1	0.080

GC/EC   = gas chromatography/electron capture detector

GC/FPD-P = gas chromatography/flame photometric detector-phosphorus mode

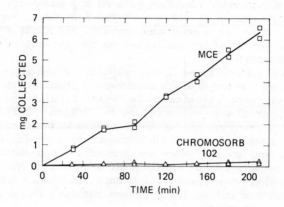

Figure 3.   Ronnel breakthrough test

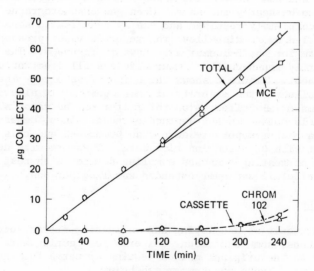

Figure 4.   Endrin breakthrough test

The first of the set is chlordane (Figure 2). In this case
chlordane was collected primarily on the mixed-cellulose ester filter with about 10% collected on the sorbent. If the sampling device was only tested over the proposed sampling period, one hour in
this case, vapor collected on the sorbent may not have been detected. Hence, extended sampling periods are necessary to realistically test collection efficiency of the individual sampling media
in the train. Also, at higher temperatures and lower chlordane
concentration, the fraction collected on the sorbent may be substantially increased.

The ronnel test (Figure 3) was similar to the chlordane test,
with about 5% of the total collected sample found on the sorbent.

The endrin test, illustrated in Figure 4, also showed that the
majority of material was collected on the filter. This was another
case in which a fraction of the total sample was collected on the
filter holder cassette parts. This amount of material was equivalent to the amount collected on the sorbent. In a short sampling
period these amounts may be undetected; however, at the longer sampling period, they were about 7% for each fraction. Again, at
higher temperatures and/or lower concentrations, the sorbent fraction may be much greater.

Demeton, or Systox, was one of the most interesting and challenging pesticides studied. Demeton consists of two isomers,
Demeton-S and Demeton-O. The vapor pressures of both isomers are
reported to be nearly the same. When gas chromatographed, the isomers are completely resolved and easily quantitated separately.
Our tests led us to disbelieve the reported vapor pressure data.
The results for the S-isomer are shown in Figure 5. The S-isomer
was primarily found on the filter with a small fraction on the
XAD-2 sorbent. Figure 6 shows the O-isomer results. The XAD-2
sorbent collected the majority of this isomer. It also appeared
that the material collected on the filter may be vapor since the
curve levels out as in the heptachlor test. Thus, Demeton-O probably has a higher vapor pressure than Demeton-S and this result is
consistent with GC retention time data. These results demonstrate
the value of testing proposed sampling devices with test atmospheres in detail and over extended sampling periods.

Conclusions

As a result of developing and validating filter/sorbent sampling methods, some additional criteria for testing were formulated
and added to the original method testing criteria for these sampling trains. These are summarized below:

(1) Recovery of the sample must be greater than 90% from all indi-
    vidual parts of the sampling train. Because the filter and
    sorbent should be combined for analysis, the recoveries from
    each must be quantitative. If >90% recovery is obtained, no
    recovery correction should be required.

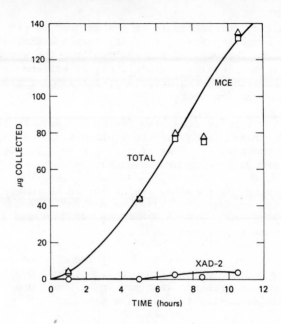

Figure 5. Breakthrough test—Demeton-S; MCEF/XAD-2: (△) total Dem-S, (□) MCE, (○) XAD-2

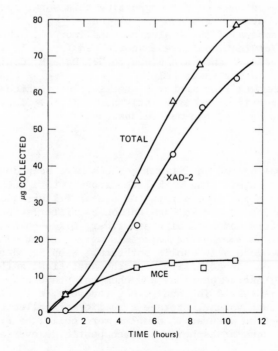

Figure 6. Breakthrough test—Demeton-O; MCEF/XAD-2: (△) total Dem-S, (□) MCE, (○) XAD-2

(2) The sampling device should be tested over an extended concen-
    tration range. That is, if little vapor contribution is ex-
    pected, the recovery of the analyte from the sorbent should be
    tested at low levels. In addition, if extreme temperatures
    are anticipated in the workplace, the sampler should be tested
    in such an environment.

(3) After sample collection, the filter and sorbent should be com-
    bined by transferring them to a glass vial. This ensures
    against the possible loss of a sample that may volatilize from
    the filter when being stored or shipped.

(4) All parts of the sampling device should be checked for sample
    adsorption, especially filter holder cassettes. If material
    collects on the cassette, complete recovery and storage sta-
    bility must be demonstrated.

In summary, it should be said that these methods may be applicable to other air sampling situations, not just workplaces, and
the principles and problems involved in aerosol/vapor sampling are
not in any way exclusive to pesticide sampling.

Acknowledgments

The author wishes to thank the National Institute for Occupational Safety and Health for supporting this work, and especially
Dr. Laurence Doemeny, the project officer.
Staff members of the Analytical and Inorganic Chemistry Department at SRI International are acknowledged for their work on this
project. Special thanks are given to Dr. Dale M. Coulson and C.
Clarine Anderson for their help.
Also, the staff at Arthur D. Little, Inc., which was a sub-
contractor on this work, is greatly appreciated. They performed
work on about half of the pesticides.

Abstract

Air sampling and analytical methods have been developed and
validated for determining workplace exposure to a number of pesticides. The methods incorporate the use of filters and solid sorbents independently and in combination as filter/sorbent sampling
trains. Filters composed of glass fiber, mixed-cellulose ester,
and polytetrafluoroethylene materials were studied, in addition to
XAD-2 and Chromosorb 102 solid sorbents. Sampling devices were
chosen based on compatibility with the analytical methods and the
specific substances' physical and chemical properties. The methods
developed were tested for analytical recovery, collection efficiency, breakthrough volume, storage stability of collected samples,
and precision and accuracy. Samples were collected from dynamically generated test atmospheres of each pesticide over a known

concentration range and analyzed to test the overall methods. Criteria were developed for testing pesticides that have significant vapor and particulate contribution at workplace concentrations.

Literature Cited

1. Van Dyk, L. P.; Visweswariak, K. Residue Reviews, 1975, 55, 91.

2. Miles, J. W.; Fetzer, L. E.; Pearce, G. W. Environ. Sci. Technol., 1970, 4, 420.

3. Thomas, T. C.; Seiber, J. N. Bull. Environ. Contam. Toxicol., 1974, 12, 17.

4. Farwell, S. O.; Bowes, F. W.; Adams, D. F. J. Environ. Sci. Health, 1977, B12, 71.

5. Turner, B. C.; Glotfelty, D. E. Anal. Chem., 1977, 49, 7.

6. Melcher, R. G.; Garner, W. L.; Severs, L. W.; Vaccaro, J. R. Anal. Chem., 1978, 50, 251.

7. Taylor D. G.; Doemeny, L. J.; Heitbrink. W. A. "Developing and Validating Methods for the Sampling and Analysis of Vapor-Particulate Mixtures," presented at the 18th Annual American Industrial Hygiene Conference, Los Angeles, CA, 1978.

8. Taylor, D. G. (Manual Coordinator) "NIOSH Manual of Analytical Methods," 2nd ed., Vol. 4, DHEW(NIOSH) Publication No. 78-175, Cincinnati, Ohio, 1978.

9. Taylor, D. G. (Manual Coordinator) "NIOSH Manual of Analytical Methods," 2nd ed., Vol. 5, DHEW(NIOSH) Publication No. 79-141, Cincinnati, Ohio, 1979.

10. Available through National Technical Information Service by Method No.

11. Taylor, D. G. (Manual Coordinator) "NIOSH Manual of Analtyical Methods," 2nd ed., Vol. 6, DHEW(NISOH) Publication (to be published).

12. Private communication with S. Dave (Johns-Manville Corp.).

13. Hill, R. H.; Arnold, J. E. "A Personal Air Sampler for Pesticides" (to be published).

RECEIVED October 27, 1980.

SPECIAL TOXICANTS

Occupational Exposure to Polychlorinated Dioxins and Dibenzofurans

CHRISTOFFER RAPPE

Department of Organic Chemistry, University of Umeå, S-901 87 Umeå, Sweden

HANS RUDOLF BUSER

Swiss Federal Research Station, CH-8820 Wädenswil, Switzerland

Polychlorinated dibenzo-p-dioxins (PCDDs) and dibenzofurans (PCDFs) are two series of tricyclic aromatic compounds which exhibit similar physical and chemical properties. Some of these compounds have extraordinary toxic properties and were the subject of much concern. They have been involved in accidents like the Yusho accident in Japan 1968 (1), the intoxication at horse arenas in Missouri, USA in 1971 (2) and the accident near Seveso, Italy in 1976 (3). The chemical structures and the numbering of these hazardous compounds are given below.

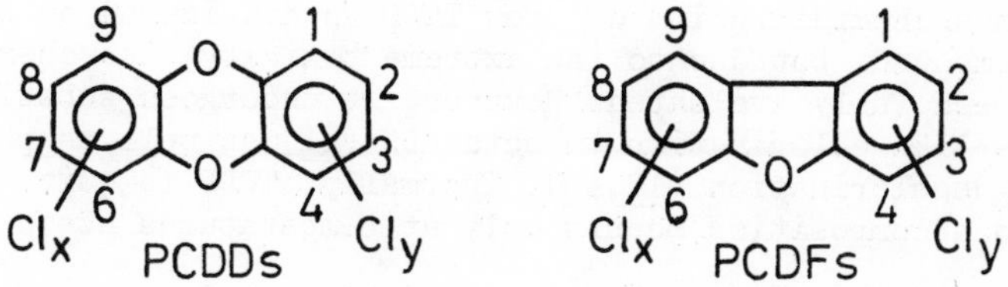

The number of chlorine atoms in these compounds can vary between one and eight. In all, there are 75 PCDD and 135 PCDF isomers as shown in Table I, ranging from the mono- to the octachloro compounds.

A large number of individual PCDD and PCDF isomers have been synthesized by various methods and characterized mainly by gas chromatography – mass spectrometry (4-7). As a general trend in both series, solubility in most solvents and volatility decrease with increasing number of chlorine atoms.

The first synthesis of TCDD was reported by Sandermann et al (8) using catalytic chlorination of the unchlorinated dioxin. TCDD has also been prepared in good yields by the dimerization of 2,4,5-trichlorophenol salts (9).

PCDDs with symmetrical chlorine substitution (one, two, three or four chlorines in each carbon ring) were prepared by the pyrolyses of different chlorophenates. Unsymmetrically substituted

Table I. Possible number of positional isomers of PCDDs and PCDFs.

Chlorine substitution	Number of isomers	
	PCDDs	PCDFs
mono-	2	4
di-	10	16
tri-	14	28
tetra-	22	38
penta-	14	28
hexa-	10	16
hepta-	2	4
octa-	1	1
	75	135

PCDDs were prepared by a mixed pyrolysis. In this case, the additionally expected symmetrically substituted PCDDs were also formed, see Figure 1 (4,5).

The most toxic and most extensively studied PCDD and PCDF isomer is 2,3,7,8-tetrachlorodibenzo-p-dioxin (2,3,7,8-tetra-CDD or TCDD). The melting point is 305-306° (10). No boiling point has been given for TCDD, and the volatility must be quite low, but it can be analyzed by gas chromatography. Although TCDD is lipophilic, it is only slightly soluble in most organic solvents and very slightly soluble in water.

From a chemical point of view TCDD is considered to be a stable compound, but due to its extreme toxicity, its chemistry has not been fully evaluated. However, it undergoes substitution reactions like chlorination to octa-CDD (11) as well as photochemical dechlorination (12,13). Thermally, TCDD is quite stable, and rapid decomposition occurs only at temperatures above 750°C (14).

Toxicity and Metabolism of PCDDs and PCDFs

There is a pronounced difference in biological and toxicological effects between different PCDD and PCDF isomers which is contradictory to the chemical and physical properties of these compounds discussed above. The isomers with the highest acute toxicity are 2,3,7,8-tetra-CDD, 1,2,3,7,8-penta-CDD, 1,2,3,4,7,8-, 1,2,3,6,7,8- and 1,2,3,7,8,9-hexa-CDD, 2,3,7,8-tetra-CDF, 1,2,3,7,8- and 2,3,4,7,8-penta-CDF and 2,3,4,6,7,8-hexa-CDF, see Figure 2. All these isomers have their four lateral positions substituted for chlorine, and they all have LD_{50} values in the range 1-100 μg/kg for the most sensitive animal species (15-17). The same isomers have been reported to have the highest biological potency (18).

Isomers of PCDDs and PCDFs vary highly in their acute toxicity and biological activity (15-19). A factor of 1 000-10 000

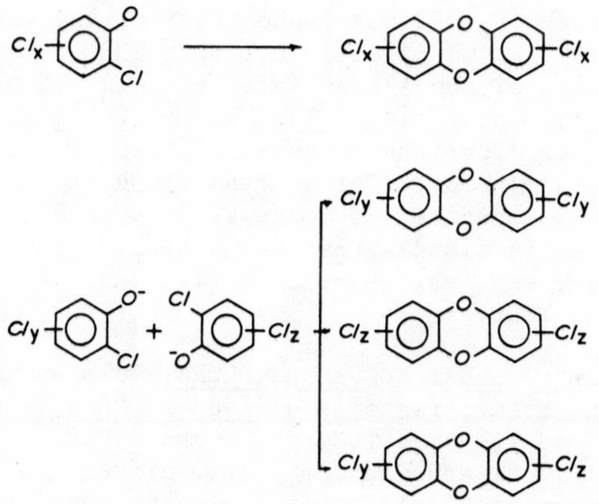

Figure 1.   Formation of PCDDs by pyrolytic dimerization of chlorophenate

can be found for so closely related isomers as 2,3,7,8- and
1,2,3,8-tetra-CDD.

Metabolism of TCDD. No metabolites of TCDD have been identified so far. It has recently been reported by Guenthner et al.(20)
that TCDD can be metabolized by the mouse liver cytochrome P-450
system to reactive intermediates, which easily bind covalently
to cullular proteins. It is suggested that this extreme reactivity inhibits the formation of normal metabolites like phenols,
dihydrodiols, or conjugated products.
Following a single oral dose of ^{14}C-TCDD in rats, Rose et
al. (21) were able to detect ^{14}C activity only in feces and not
in urine. The half-life of ^{14}C activity in the body was about 31
days and the major part of the TCDD was stored in liver and fat.
After repeated oral doses the major route of excretion was again
found to be feces, but the urin contained 3-18% of the total [14]C
activity. The half-life of ^{14}C activity in these rats was about
24 days, and most of TCDD was found in liver and fat. The experiments indicated that materials other than TCDD constituted a significant fraction of the ^{14}C activity excreted in the feces, but
no metabolite was identified (21). Van Miller et al. (22) reported on the tissue distribution and excretion of ^{3}H TCDD in monkeys and rats. A marked difference was found in the tissue distribution in the two species. In monkeys, a large percentage of the
dose was located in tissues that had a high lipid content, i.e.
in skin, muscle, and fat; whereas in rats these tissues had much
lower levels of TCDD.

Metabolism of other PCDDs and PCDFs. Tulp and Hutzinger have
studied the rat metabolism of a series of PCDDs (23). 1- And
2-mono-, 2,3- and 2,7-di-, 1,2,4-tri, and 1,2,3,4-tetra-CDD are
metabolized to mono- and dihydroxy derivatives, whilst in the
case if the two monochloro isomers, also sulphur containing meta-
bilites are excreted. It has also been shown that the primary
hydroxylation exclusively takes place in the lateral positions
(2-, 3-, 7- and/or 8-positions) in the molecule. In none of the
experiments metabolites resulting from a fission of the C-O-C
bonds were detected. No metabolites were found from octa-CDD.
The results are rationalized in terms that the metabolism
of the PCDDs occurs mainly via 2,3-epoxides. In the octa-CDD as
in 2,3,7,8-tetra-CDD these positions are blocked, consequently
the reaction is less likely to take place or takes place at a
highly reduced rate (23).
A similar relationship between PCDF isomers retained and
apparently excreted has been observed for patients with the
Yusho disease, an intoxication by a rice oil contaminated with
PCBs and PCDFs. The contaminated rice oil and liver samples from
two of the patients were analyzed and all the major PCDFs were
identified. A comparison revealed that none of the isomers
retained had two vicinal hydrogenated C-atoms in any of the two

C-rings of the benzofuran system. Most of these isomers had all their <u>lateral</u> positions chlorinated. Contrary, all the PCDF isomers apparently excreted had two <u>vicinal</u> hydrogenated C-atoms in at least one of the two rings, and these unblocked positions are involved in the metabolism by forming epoxides, see Figure 3 (<u>23</u>). The data discussed here show a striking similarity between the most toxic PCDD and PCDF isomers and the isomers most efficiently retained.

Kuroki and Masuda have estimated that 0.37% of 2,3,6,8--tetra-, 0.006-0.03% of 2,3,7,8-tetra- and 0.9% of the 2,3,4,7,8--penta-CDF ingested were retained in the liver of one of the Yusho patients when he died 44 months after the use of the rice oil had been discontinued (<u>24</u>).

Analytical Methods

Due to the extreme toxicity of some of the PCDDs and PCDFs, very sensitive and highly specific analytical techniques are required. Detection levels in biological and environmental samples should be orders of magnitude below the usual detection limits obtained in pesticide analysis. Any analysis at such low levels is complicated by the presence of a multitude of other, possibly interfering compounds. The best available separation techniques followed by highly specific detection means have to be used for an accurate determination of these hazardous compounds. The different isomers of PCDD or PCDF may vary sinificantly in their biological and toxicological properties and therefore their separation and identification becomes important.

In recent years, many analytical methods were developed for the analysis of PCDDs, PCDFs and especially 2,3,7,8-tetra-CDD in environmental and industrial samples, the most specific methods making use of mass spectrometry (<u>25</u>). Prerequisites for best analyses are efficient extraction and sample purification followed by good separation, ultrasensitive detection and - very desirably - confirmation. A technique for analyzing individual PCDDs and PCDFs has been described and discussed in detail (<u>26</u>, <u>27</u>, <u>28</u>). It involves one or two steps by column chromatographic clean-up followed by high-resolution gas-chromatography separation using glass capillary columns and detection and quantitation using mass spectrometry (mass chromatography and/or mass fragmentography). Artifacts usually disturb the analytical work at these extreme low concentration levels, but the risk can be minimized by a careful inspection of the complete mass spectrum. For a correct structure assignment of the PCDDs, the low mass ions have shown to be useful (<u>4</u>). Additional information can be obtained by a comparison of the retention times using high-resolution GC with authentic standards, but the number of synthetic standards is still limited. This analytical technique has also been used for the separation of the 22 tetra-CDD isomers (<u>5</u>).

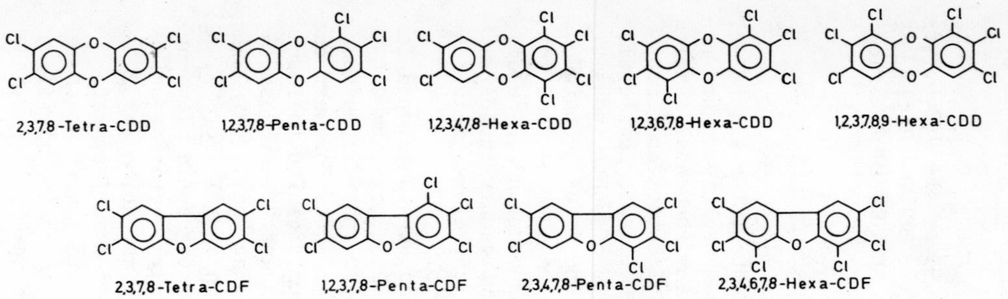

Figure 2. The most toxic PCDD and PCDF isomers

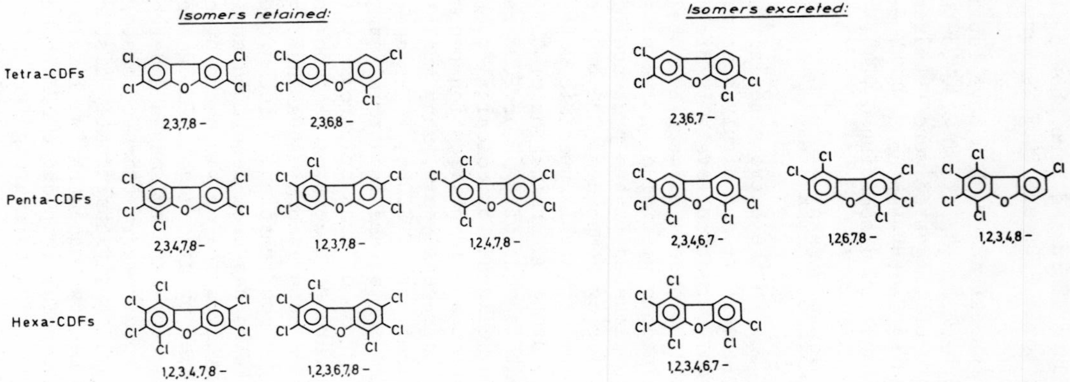

Figure 3. PCDF isomers retained and excreted from the liver of Yusho patients

Occurrence of PCDDs and PCDFs in Idustrial Chemicals

A. Phenoxy Acids (2,4,5-T). In 1971 Courtney and Moore (29) observed teratogenic effects for the phenoxy acid 2,4,5-T. They related these effects to the presence of 30 µg/g of 2,3,7,8- -tetra-CDD found in the particular sample used in that study. TCDD is formed during the industrial preparation of 2,4,5-trichlorophenol from 1,2,4,5-tetrachlorobenzene. This reaction takes place at about 180° and, when the solvent is methanol, the presure rises to about 7 KPa. The formation of TCDD is an unwanted side reaction, and a 2,4,5-T sample, possibly prepared by this method, was found to be contaminated by 6.0 µg/g of TCDD (30). The amount of TCDD formed increases at higher temperatures (31). The reaction to 2,4,5-trichlorophenol is exothermic, higher temperatures may result in uncontrolled conditions leading to explosions. At these occasions the amounts of TCDD formed are much higher than normally.

In order to diminish the levels of TCDD, the experimental conditions were changed in some German factories. Still using methanol as solvent, the temperature was kept at 157°C (32).

In some factories, ethylene glycol is used as a solvent in order to avoid the high pressure. As already pointed out by Milnes (31), however, use of this solvent requires special precautions because of the occurrence of a base-promoted polymerization of ethylene glycol and decomposition reactions that produce ethylene oxide. These reactions are also exothermic; they may start spontaneously at above 180°C and proceed rapidly and uncontrollably to result in the formation of TCDD. It has been suggested that this reaction sequence caused the accident at Bolsover, UK (31). It has also been suggested that in the accident at Seveso, Italy, this series of reactions began when part of the ethylene glycol had been distilled off from the alkaline solution at 170°C, i.e. at stage during which there was considerable risk of the occurrence of exothermic reactions (33).

After most of the solvent has been distilled off, the reaction mixture is acidified; the trichlorophenol can be free from TCDD by one or two distillations, with the result that the TCDD is concentrated in the residues. An episode involving accidental poisoning in horse arenas in Missouri, USA, in 1971, clearly shows the hazards of such residues (2).

The levels of 2,3,7,8-tetra-CDD in drums of Herbicide Orange placed in storage in the USA and in the Pacific before 1970 have been found to vary between 0.1 and 47 µg/g (34). Since Herbicide Orange was formulated as a 1:1 mixture of the butyl esters of 2,4-D and 2,4,5-T, the levels of 2,3,7,8-tetra-CDD in individual 2,4,5-T preparations used in the 1960's could be as high as 100 µg/g. As a result of governmental regulations, efforts were made during the 1970's to control and to minimize the formation of 2,3,7,8-tetra-CDD, and now all producers claim that their products contain less than 0.1 µg/g of 2,3,7,8-tetra-CDD.

In analyses using high-resolution GC/MS and MS confirmation, Rappe et al. (35) and Norström et al. (30) have reported that in other samples of Herbicide Orange, as well as in European and US 2,4,5-T formulations from the 1950's and 1960's, 2,3,7,8--tetra-CDD was the dominating compound of this group. Only minor amounts of other PCDDs and PCDFs could be found, primarily lower chlorinated PCDDs in samples of Herbicide Orange, see ref 35. The European samples are possibly prepared by a low-temperature process, while the US sample is prepared by a high-temperature process, see Table II.

Table II. Levels of 2,3,7,8-tetra-CDD in 2,4,5-T acid and 2,4,5-T ester formulations (30, 35).

Sample	Origin	2,3,7,8-tetra-CDD $\mu g/g$
2,4,5-T acid	1952, Sweden	1.10
2,4,5-T ester	unknown, Sweden	0.50
2,4,5-T ester	unknown, Sweden	< 0.05
2,4,5-T ester	1960, Sweden	0.40
2,4,5-T ester	1962, Finland	0.95
2,4,5-T ester	1966, Finland	0.10
2,4,5-T ester	1967, Finland	< 0.05
2,4,5-T ester	1967, Finland	0.22
2,4,5-T ester	1967, Finland	0.18
2,4,5-T acid	1964, USA	4.8
2,4,5-T acid	1969, USA	6.0
Herbicide Orange	unknown, USA	0.12
Herbicide Orange	unknown, USA	1.1
Herbicide Orange	unknown, USA	5.1

B. Hexachlorophene. The bactericide hexachlorophene is prepared from 2,4,5-trichlorophenol, the key intermediate in the production of 2,4,5-T. Due to additional purification, the level of 2,3,7,8-tetra-CDD in this product is usually < 0.03 mg/kg (36). However, hexachlorophene also contains about 100 mg/kg of a hexachloroxanthene, the 1,2,4,6,8,9-substituted isomer (37).

C. Chlorophenols. Chlorophenols have been extensively used since the 1930's as insecticides, fungicides, mould inhibitors, antiseptics and disinfectants. The annual production volume is estimated to be in the order of 150'000 tons. In the US pentachlorophenol is the second heaviest in use of all pesticides (38). The most important use of 2,4,6-tri, 2,3,4,6-tetra- and pentachlorophenol (or their salts) is for wood protection. Pentachlorophenol is also used as a fungicide for slime control in the manufacture of paper pulp and for a variety of other purposes such as in the tanning process of leather and an additive in cutting oils and fluids, paint, glues and out-door textiles.

2,4-Di- and 2,4,5-trichlorophenol are used for the production of 2,4-D and 2,4,5-T herbicides (phenoxy acids), and hexachlorophene.

Chlorophenols may contain a variety of by-products and contaminants such as other chlorophenols, polychlorinated phenoxyphenols and neutral compounds like polychlorinated benzenes and diphenyl ethers, PCDDs and PCDFs (39). Some of these contaminants may also occur in chlorophenol derivatives like phenoxy acids, other pesticides and hexachlorophene. The possible presence of PCDDs and PCDFs is of special significance because of their extraordinary toxicological properties.

Buser and Bosshardt reported on the results of a survey on the PCDD and PCDF contents of pentachlorophenol (PCP) and PCP-Na from commercial sources in Switzerland (40). From the results, the grouping of the samples into two series can be observed: a first series with generally low levels (hexa-CDD < 1 ppm) and a second series with much higher levels (hexa-CDD > 1 ppm) of PCDDs and PCDFs. Samples of high PCDD contents had also high PCDF contents. For most samples, the contents of these contaminants were in the order tetra- ~ penta < hexa < hepta ≲ octa--CDD/CDF. The ranges of the combined levels of PCDDs and PCDFs were 2-16 and 1-26 ppm, respectively, for the first series of samples, and 120-500 and 85-570 ppm, respectively, for the second series of samples. The levels of octa-CDD and octa-CDF were as high as 370 and 300 ppm, respectively (40).

Some PCP-Na samples analyzed (40) showed the unexpected presence of a tetra-CDD (0.05-0.25 ppm), which was later identified as the unusual 1,2,3,4-substituted isomer (4). PCP and PCP-Na samples of high PCDD content (hexa-CDD > 1 ppm) were reanalyzed on a high-resolution GC column for the presence of individual PCDD isomers (26). As reported earlier (40), all samples showed an almost identical pattern of hexa- and hepta-CDD isomers. The major hexa-CDD isomers were identified as 1,2,3,6,7,8-hexa--CDD one of the most toxic isomers, see Figure 4. In addition 1,2,4,6,7,9- and 1,2,3,6,8,9-hexa-CDD or their Smiles-rearranged products (1,2,4,6,8,9- and 1,2,3,6,7,9-hexa-CDD, respectively), were found. These three isomers were always present in an almost constant isomeric ratio of 50:40:10. Both of the hepta-CDD isomers were present in these samples in a ratio of 15:85 with the biologically most active (17) 1,2,3,4,6,7,8-hepta-CDD as the major constituent. All hexa-CDD isomers found in these samples were dimerization products of 2,3,4,6-tetrachlorophenol, the expected precursor of PCP in the chlorination starting from phenol (26).

Rappe et al. have reported on the analysis of two commercial chlorophenate formulations from Scandinavian sources using a 50 m OV-17 column (7). The tetrachlorophenol was known to contain approximately 5% 2,4,6-tri-, 50% 2,3,4,6-tetra- and 10% pentachlorophenol as their sodium salt. The combined levels of PCDDs and PCDFs (tetra- to octa-) were 10 and 160 µg/g, respectively. Whereas on the earlier analyzed PCP samples (40) the

levels of PCDDs and PCDFs were comparable, we found here significantly higher levels of PCDFs. Differences were also seen in the
distribution of the individual PCDD isomers in these two types
of samples. The main hexa-CDD isomers in the Scandinavian samples
were 1,2,4,6,7,9- and 1,2,3,6,8,9-hexa-CDD (or their Smiles-
-rearranged products), and 1,2,3,4,6,8-hexa-CDD (26). The latter
hexa-CDD was completely absent in the earlier analyzed PCP samples. 1,2,3,6,7,8-Hexa-CDD, which was the major hexa-CDD isomer
in the PCP samples, was only a minor isomer in the Scandinavian
samples. A similar difference was seen in case of the hepta-
-CDDs. The major isomer in the Scandinavian samples was
1,2,3,4,6,7,9-hepta-CDD, whereas in all of the PCP samples it was
the 1,2,3,4,6,7,8-substituted isomer.

Using the same analytical technique the major PCDFs in the
Scandinavian 2,4,6-trichloro- and 2,3,4,6-tetrachlorophenol
samples have been identified and quantified. In addition a US
PCP-Na formulation was also analyzed, and the quantitative results are collected in Table III (7). In this table we have also
included the results of a few other investigations (17, 41).

The chlorophenols in the formulations analyzed differed in
their degree of chlorination and were likely synthesized in different ways. Nevertheless, the same penta-, hexa- and hepta-CDF
isomers were found as the main PCDF components in all three samples although in somewhat different proportions (7), see Figure
5.

D. PCBS. Vos et al. in 1970 were able to identify PCDFs
(tetra- and penta-CDFs) in two samples of European PCBs but not
in a sample of Aroclor 1260 (42). The toxic effects of the PCBs
were found to parallel the levels of PCDFs present. Bowes et al.
examined a series of Aroclors as well as the samples of Aroclor
1260, Phenoclor DP-6 and Clophen A-60, that had previously been
analyzed (42). They used packed column GC and mass spectrometry,
and found that the most abundant PCDFs had the same retention
time as 2,3,7,8-tetra- and 2,3,4,7,8-penta-CDF; their results
are collected in Table IV (43). Using high-resolution GC and MS,
Rappe and Buser (unpublished) have analyzed a number of commercial PCBs, and the results are also collected in Table IV. In
general the PCBs contained quite a complex mixture of PCDFs, up
to 40 different isomers. The highest level of PCDFs was found in
a Japanese PCB used for two years in a heat exchange system,
which was found to have about 10 µg/g. The dominating isomer was
identified as 2,3,7,8-tetra-CDF at a level of 1.25 µg/g (44).

Using the synthetic standards now available the major PCDFs
have been identified, see Figure 6. (Rappe and Buser, unpublished).

E. Fly ash. Olie et al. reported in 1977 on the occurrence
of PCDDs and PCDFs in fly ash and flue gas samples from municipal
incinerators in the Netherlands (45). No quantitative data were
given in this report, but Buser and Bosshardt made a

Table III. Levels of PCDFs in commercial chlorinated phenols (µg/g).

	PCDFs					Σ PCDFs	Σ PCDDs
	tetra-	penta-	hexa-	hepta-	octa-		
2,4,6-Trichlorophenol, Sweden (7)	1.5	17.5	36	4.8	-	60	< 3
2,4,6-Trichlorophenol, USA a)	1.4	2.3	0.7	< 0.02	-	4.6	0.3
2,3,4,6-Tetrachlorophenol, Finland (7)	0.5	10	70	70	10	160	12
Pentachlorophenol, USA (7)	0.9	4	32	120	130	280	1000
Pentachlorophenol, USA (17)	-	-	30	80	80	190	2625
Pentachlorophenol, USA (41)	≤ 0.4	40	90	400	260	790	1900
Pentachlorophenol, Germany a)	-	-	0.03	0.8	1.3	2.1	6.8

a) Rappe, Buser, Nygren, unpublished work.

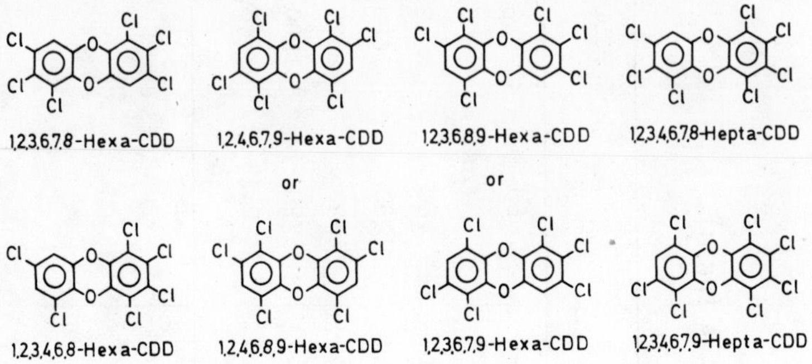

Figure 4.   PCDD isomers identified in commercial chlorinated phenols

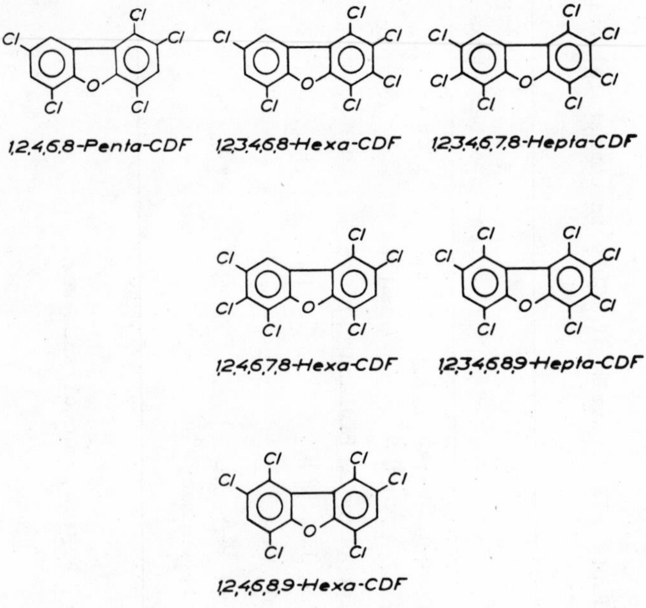

Figure 5.   PCDF isomers identified in commercial chlorinated phenols

Table IV. Levels of PCDFs in commercial PCBs.

Sample	3-Cl μg/g	4-Cl μg/g	5-Cl μg/g	6-Cl μg/g	7-Cl μg/g	Total μg/g
Aroclor 1248,1969 ([43])	–	0.5	1.2	0.3	–	2.0
Aroclor 1254,1969 ([43])	–	0.1	0.2	1.4	–	1.7
Aroclor 1254,1970 ([43])	–	0.2	0.4	0.9	–	1.5
Aroclor 1254 [a]	0.10	0.25	0.70	0.81	–	1.9
Aroclor 1254 (lot KK 602)[a]	–	0.05	0.10	0.02	–	0.2
Aroclor 1260,1969 ([43])	–	0.1	0.4	0.5	–	1.0
Aroclor 1260 (lot AK 3)([43])	–	0.2	0.3	0.3	–	0.8
Aroclor 1260 [a]	0.06	0.30	1.0	1.10	1.35	3.8
Aroclor 1016,1972 ([43])	–	< 0.001	<0.001	<0.001	–	–
Clophen A 60 ([43])	–	1.4	5.0	2.2	–	8.4
Clophen T 64 [a]	0.10	0.30	1.73	2.45	0.82	5.4
Phenoclor DP-6 ([43])	–	0.7	10.0	2.9	–	13.6
Prodelec 3010 [a]	0.41	1.08[b]	0.35	0.07	–	2.0
Mitsubishi (used) [a]	2.13	4.00	3.30	0.53	–	10.0

a) Rappe and Buser, unpublished

b) Major isomer 2,3,7,8-tetra-CDF

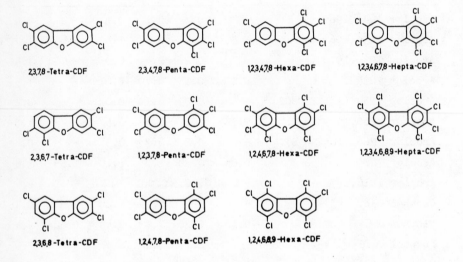

Figure 6.   PCDF isomers identified in commercial PCBs

quantification that the total amount of PCDDs and PCDFs in fly ash from a municipal incinerator in Switzerland was 0.2 µg/g and 0.1 µg/g. respectively, and in the fly ash from an industrial heating facility, also in Switzerland, it was 0.6 µg/g and 0.3 µg/g, respectively (46). In additional studies Buser et al. have shown that the number of individual isomers was quite large with up to 30 PCDD and over 60 PCDF isomers. The highly toxic PCDDs (2,3,7,8-tetra-, 1,2,3,7,8-penta-, and 1,2,3,6,7,8- and 1,2,3,7,8,9-hexa-CDD) were only minor constituents whereas the known toxic PCDFs (2,3,7,8-tetra- and 2,3,4,7,8-penta-CDF) where major constituents, see Figure 2 (5, 47, 48).

Human Exposure and Risk

Human exposure to PCDDs and PCDFs may be due to either specific exposure, mainly of occupational origin, or due to a general exposure of the public.

Occupational exposure may occur:
in chemical plants producing chlorinated phenols or PCBs, in factories utilizing these chemicals for the production of other substances, in the process of using these chemicals under various occupational conditions such as spraying of phenoxy herbicides, using chlorinated phenols for a variety of applications especially as wood perservative or in the tanning process, in the use of hexachlorophene in sanitation applications, in factories manufacturing or repairing transformers or capacitors containing PCBs, in factories having heat exchange systems containing PCBs, in factories utilizing casting waxes containing PCBs, in offices utilizing carbonless copy paper containing PCBs.

Occupational exposure to 2,3,7,8-tetra-CDD can occur during the production of 2,4,5-trichlorophenol and the subsequent production and use of 2,4,5-T acid and esters. The commercial production of 2,4,5-T started in the US in 1944, and the use of this herbicide increased in the 1940's and 1950's. However, the dioxin problem was not recognized until 1957 (8, 49, 50).

The most heavy exposure is most likely during purification as the residues are by far more contaminated than the purified produts (2). However, only limited information is available on the levels of PCDD contamination of products prepared prior to the 1970's and no information at all is available on the levels of TCDD in the residues. Consequently it is quite difficult to estimate the levels of exposure. Moreover, the TCDD content seems to be dependent on the experimental conditions during the preparation of 2,4,5-trichlorophenol, but these conditions are very seldom given in the literature.

In 1977 the Swedish Parliament totally banned the phenoxy

acid 2,4,5-T. The dioxin problems were the main reason for this
action. This year (1980) the Swedish Parliament passed a temporary ban for all aereal spraying in the forestry (including all
phenoxy acids). The chlorinated phenols are also banned in Sweden, although the import of chlorophenol treated timber, planks
and fabrics is still allowed.

Occupational Exposure to Chlorinated Phenols

The chlorinated phenols are widely used in a variety of working operations. They may be highly contaminated with PCDDs and
PCDFs; the levels of these impurities for most products on the
market is in the range of 100 - 3000 µg/g, see Table III. Consequently special interest should be given the occupational exposure to chlorinated phenols.

The wood industry is the major consumer of technical chlorophenols. In the United States and in Canada it has been assumed
that more than 80% of pentachlorophenol (PCP) is used for wood
perservation and wood protection (38,51). PCP dissolved in various solvents (mineral spirits, fuel oil, kerosene and methylene
chloride) is the major compound used for wood perservation. This
procedure involves the use of pressure and vacuum cycles to obtain deep and optimum retention of the perservative. This process
is used to produce a product which will have a long period of
service such as railway ties, pilings and hydropoles.

However, a substantial proportion of the processed wood does
not require long term perservation. Fresh-cut lumber is protected against attack by fungi and molds by passing the lumber
through a spray tunnel, or by dipping. The chemicals used for
this purpose are the sodium salts of PCP (US, Canada) or 2,3,4,6-
-tetra- and 2,4,6-trichlorophenol (Scandinavia).

The use of chlorinated phenols in the saw mill industry is
known to cause occupational health problems. In an investigation
carried out in Sweden it was noted that the workers in the trim-
ming-grading plant, where sawn timber is handled after chlorophenol treatment, often complain of cutanious irritations, respiratory difficulties and headache (52, 53).

In a French ivestigation (54) it was reported on occupational intoxication, which afflicted men dipping timber in a solution of chlorinated phenates. Some workers showed only cuta-
neous symptoms; others showed symptoms of anesthesia, loss of
appetite and respiratory difficulties.

In a series of epidemiological investigations (case-control
studies) Hardell and collaborators have shown that saw mill workers exposed to chlorophenols and spraymen exposed to phenoxy
herbicides have a higher risk for special malignant tumors than
the matched controls (55, 56, 57). The values for the risk ratio
are given in Table V. Confounding factors like smoking, DDT,
diesel oil and exhaust gases are excluded.

Table V. RR-Values From Case-Control Studies (55, 56, 57).

| Occupation/Exposure | Risk Ratio (RR) | | | |
| | Malignant Mesenchymal Sarcomas | Malignant lymphomas | | |
		High doses	Low doses	Total
Saw mill/Cl-phenols	6.6	9.3	2.5	4.6
Spraymen/2,4,5-T	5.3			
Spraymen/2,4-D; MCPA	5.5	7.0	4.3	4.8

The level of occupational exposure to chlorophenols and contaminants has been estimated by analyses of dust and air samples and the urine and blood from exposed workers.

Wyllie et al. (58) have studied PCP levels in the air of a small US wood treatment plant (pressure treatment). PCP could be found in all samples, the highest values (0.2-15 $\mu g/M^3$) were found in samples taken inside the pressure treatment building. Analyses of the PCP-levels in the urine and blood serum from the exposed workers were also included in this study. The urine values ranged between 0.04-0.76 µg/ml and the blood serum values were between 0.35-3.55 µg/ml. A good parallellity was observed between the urine and the blood values, and the highest values were found for a pressure treater and a welder. The dioxin impurities were not discussed in this study.

Levin et al. (52) have studied the levels of chlorophenols and dibenzofurans in wood dust samples and sludge from two Swedish sawmills. Both plants were using the Finnish 2,3,4,6-tetrachlorophenate, see Table III for the levels of impurities. The results from this study are collected in Table VI.

Table VI. Chlorinated contaminants in sawdust (52).

Position	Application	Cl_4-Phenol (µg/g)	PCDFs (µg/g)
Trimming	Spraying	300	6
Grading	-"-	100	3
Packaging	-"-	70	1
Trimming	Dipping	450	< 0.5
Grading	-"-	50	< 0.5
Packaging	-"-	125	< 0.5

In the sawmill where the spraying application was used, a hundredfold enrichment of PCDFs was found in the sawdust in comparison to the chlorophenol content when compared to the composition of the fungicide formulation used (1:100 vs 1:10 000).

The highest levels of contaminants in the sawdust were found in the beginning of the feeder-line (52). The major PCDFs were identified as 1,2,3,4,6,8- and 1,2,4,6,8,9-hexa- and 1,2,3,4,6,7,8- and 1,2,3,4,6,8,9-hepta-CDF (7).

In the sawmill where the dipping application was used, no enrichment of PCDFs in the sawdust could be observed, see Table VI. A sludge found on the bottom of the dipping tank in this factory was also analyzed, the level of PCDFs in this sludge was 700 µg/g. The ratio of PCDFs to chlorophenols in the sludge was 300 times higher than in the chlorophenol formulation used (52).

In a sawmill where PCP was used, levels of up to 100 µg/g of octa-CDD can be found in the sawdust (59).

In Table VII the levels of chlorophenols found in the urine from a variety of occupationally exposed groups have been collected.

Table VII.   Levels of Chlorophenols in Urine of Exposed Workers.

Branch	Range µg/ml	Mean µg/ml	Number of analyses	Source
Saw mill a)	0.12-10.3	3.2	12	60
Saw mill b)	0.03-0.50	0.20	22	61
Tannery c)	0.10-10.5	2.7	20	"
Textile c)	0.01-0.80	0.30	15	"
Seamstresses d)	0.01-0.35	0.20	20	"
Seamstresses e)	0.01-0.05	0.02	38	"

a)   loading newly treated timber
b)   exposed to dust from imported chlorophenol treated timber
c)   impregnation of fabrics
d)   sewing impregnated fabrics
e)   unexposed

In an ongoing study we have used the analytical technique now available for the determination of the levels of PCDDs and PCDFs in the blood of workers exposed to chlorophenols (C. Rappe, M. Nygren, H.R. Buser and T. Kauppinen, unpublished results).

One of the branches under study is the saw-mill industry. In Finland 2,3,4,6-tetrachlorophenol is used for bluestain control of the timber during the summer season (May - October). It is known that PCDFs are the major contaminants in this particular formulation, see Table III. The sampling was performed twice, the first set of samples was taken after 6 months of non-exposure. The other sampling was performed after one month of chlorophenol exposure. In Table VIII we have collected the levels of chlorophenols found in the urine samples as well as the levels of PCDDs and PCDFs found in the blood samples.

Table VIII. Levels of PCDDs and PCDFs in blood samples from workers in saw mill industry after exposure to 2,3,4,6-tetrachlorophenate.

Person	Profession	Chlorophenols in urine μg/ml	PCDDs (pg/g of blood)				PCDFs (pg/g of blood)			
			octa	hepta[a]	hexa	penta	octa	hepta[b]	hexa	penta
Blank A		—	7	<2	<3	—	<3	2	<3	—
-"- B		—	<2	<1	<1	<1	<2	<1	<1	<1
1	Loader [c]	0.04	5	<2	<3	—	<3	40	<3	—
-"-	[d]	5.2	7	<1	<1	<1	<2	22	<1	<1
2	Cleaner [c]	<0.02	5	2	<3	—	<3	30	<3	—
-"-	[d]	0.23	22	8	<1	<1	<2	17	<1	5
3	Loader [c]	0.03	18	10	3	—	<3	18	<3	—
-"-	[d]	0.83	3	<1	<1	<1	<2	17	<1	1
5	Packade [c]	<0.05	<3	<2	<3	—	<3	7	<3	—
-"-	[d]	0.11	4	<1	<1	2	<2	12	<1	<1
7	Control	<0.01	3	<1	<1	<1	<2	3	<1	<1

a. Major isomer 1,2,3,4,6,7,8-hepta-CDD
b. Major isomer 1,2,3,4,6,7,8-hepta-CDF
c. Sampling 6 months after latest exposure
d. Sampling after one month of exposure

Table IX. Levels of PCDDs and PCDFs in blood samples from workers in textile and leather industry after exposure to PCP or PCP derivatives.

Person	Profession	Chlorophenols in urine µg/ml	PCDDs (pg/g of blood)				PCDFs (pg/g of blood)			
			octa	hepta[a]	hexa	penta	octa	hepta[b]	hexa	penta
3	Textile	3.12	304	59	<1	<1	10	33	<1	<1
4	_"_	<0.01	3	<1	<1	<1	2	<1	<1	<1
5	_"_	<0.01	10	1	<1	<1	2	<1	<1	10
6	_"_	0.42	105	15	<1	<1	2	<1	<1	<1
7	_"_	0.16	30	6	<1	<1	2	<1	<1	1
10	Tannery[c]	0.55[d]	20	7	<3	-	3	7	<3	-
15	_"_[c]	0.04[d]	80	30	3	-	7	18	3	-
16	_"_[c]	0.03[d]	12	4	<3	-	3	3	<3	-
17	_"_[c]	-	7	2	<2	-	3	3	<3	-

a Major isomer 1,2,3,4,6,7,8-hepta-CDD

b Major isomer 1,2,3,4,6,7,8-hepta-CDF

c Blood sampling 8 months after last exposure

d Urine sampling 6 months after last exposure

A similar study was performed in the textile industry where the workers were exposed during the fabrics impregnation, and in the leather industry where the workers were exposed during the tanning process. In these two branches pentachlorophenol or pentachlorophenol laurate were used, products highly contaminated by PCDDs, the level of PCDFs being much lower, see Table III. The values of the blood and urine analyses are given in Table IX.

The following conclusions could be drawn although this investigation is not yet finished.

1. PCDDs and PCDFs can be detected in blood of exposed workers, the total levels can be as high as 400 pg/g of blood, plasma.

2. PCDDs and PCDFs were also found at similar levels in blood following a 6-8 months' long period of non-exposure as compared to samples taken after one month of exposure, although the level of chlorophenols in the urine increased up to 100 times after the period of exposure. Consequently no good correlation was found between the blood and urine analyses.

3. A difference is observed in the pattern of PCDDs and PCDFs in the blood between the workers exposed to 2,3,4,6-tetrachlorophenol or pentachlorophenol. This difference parallels the difference in contaminants in these two products.

4. The same PCDD and PCDF isomers can be found in the blood samples as in the products used.

5. Studying occupational exposure to highly contaminated chlorophenols, the analysis of PCDDs and PCDFs in blood samples seems to be a better parameter to follow than the level of chlorophenol in the urine.

Literature Cited

1   Higuchi, K. (Ed.) "PCB Poisoning and Pollution" Kodansha-
     -Academic Press: Tokyo-New York-San Francisco-London,1976;
     pp. 1-123.
2   Carter, C.D.; Kimbrough, R.D.; Liddle, J.A.; Cline, R.F.;
     Zack, Jr., M.M.; Barthed, W.F.; Koehler, R.E. and Phillips,
     P.E. Science, 1975, 188, 738.
3   Pocchiari, F. Ecol. Bull. (Stockholm), 1978, 27, 67.
4   Buser, H.R. and Rappe, C. Chemosphere, 1978, 7, 199.
5   Buser, H.R. and Rappe, C. Anal. Chem, In press.
6   Buser, H.R. and Rappe, C. Chemosphere, 1979, 9, 157.
7   Rappe, C.; Garå, A. and Buser, H.R. Chemosphere, 1978, 7,
     981.
8   Sandermann, W.; Stockmann, H. and Carsten, R. Chem. Ber.,
     1957, 90, 690.
9   Buu-Hoi, N.P.; Saint-Ruf, G.; Bigot, P. and Mangane, M.
     Comp. rend. Ser. D.,1971, 273, 708.
10   Pohland, A.E. and Yang, G.C. J. Agric. Food Chem., 1972,
     20, 1093.
11   Baughman, R.W. "Tetrachlorodibenzo-p-dioxins in the Envi-
     ronment. High Resolution Mass Spectrometry at the Picogram.
     Level" Thesis, Harvard University, Cambridge, MA, 1974, pp.
     13-23.
12   Crosby, D.G. and Moilanen, K.W. Bull. Environ. Contam.
     Toxicol., 1973, 10, 372.
13   Crosby, D.G. and Wong, A.S. Science, 1977, 195, 1337.
14   Stehl, R.H.; Papenfuss, R.R.; Bredesweg. R.A. and Roberts,
     R.W. "Chlorodioxins - Origin and Fate." Blair, E.H. (Ed.)
     Adv. Chem. Ser., 1973, 120, 119.
15   McConnel, E.E.; Moore, J.A.; Haseman, J. and Harris, M.
     Toxicol. Appl. Pharmacol., 1978, 44, 335.
16   Moore, J.A.; McConnell, E.E.; Dalgard, D.W. and Harris, M.W.
     Ann. N.Y. Acad. Sci., 1979, 320, 151.
17   US EPA "Report of the ad hoc Study Group on Pentachloro-
     phenol Contaminants" EPA/SAB/78/001.
18   Poland, A.; Greenlee, W.F. and Kende, A.S. Ann. N.Y. Acad.
     Sci., 1979, 320, 214.
19   Bradlaw, J., FDA Washington. Personal communication.
20   Guenther, T.M.; Fysh, J.M. and Nebert, D.W. Pharmacology,
     In press.
21   Rose, J.Q.; Ramsey, J.C.; Wentzler, T.H.; Hummel, R.A. and
     Gehring, P.J. Toxicol. Appl. Pharmacol., 1976, 36, 209.
22   Van Miller, J-P.; Marlar, R.J. and Allen, J.R. Food Cosmet.
     Toxicol., 1976, 14, 31.
23   Rappe, C.; Buser, H.R.; Kuroki, H. and Masuda, Y. Chemos-
     phere, 1979, 8, 259,
24   Kuroki, H. and Masuda, Y. Chemosphere, 1978, 7, 771.
25   McKinney, J.D. Ecol. Bull. (Stockholm), 1978, 27, 55.

26 Buser, H.R. "Polychlorinated Dibenzo-p-dioxins and Diben-
    zofurans: Formation, Occurrence and Analysis of Environ-
    mentally Hazardous Compounds." Thesis, University of Umeå,
    Sweden, 1978.

27 Rappe, C. and Buser, H.R. "Chemical Properties and Analy-
    tical Methods" in Kimbrough, R.D. (Ed.) "Halogenated Bi-
    phenyls, Terphenyls, Naphthalenes, Dibenzodioxins and Re-
    lated Products" Elsevier/North Holland Biomedical Press:
    Amsterdam, 1980; pp. 41-75.

28 Rappe, C.; Buser, H.R.; Stalling, D.L.; Smith, K.M. and
    Dougherty, R.D. Nature, Submitted.

29 Courtney, K.D. and Moore, J.D. Toxicol. Appl. Pharmacol.,
    1971, 20, 396.

30 Norström, Å.; Rappe, C.; Lindahl, R. and Buser, H.R. Scand.
    J. Work Environ. Health, 1979, 5, 375.

31 Milnes, M.H. Science, 1971, 232, 395.

32 Holmstedt, B. Arch. Toxicol., 1980, 44, 211.

33 Homberger, E., Reggiani, G., Sambeth, J. and Wipf, H.K.
    Ann. Occup. Hyg., 1979, 22, 327.

34 Firestone, D. Ecol. Bull. (Stockholm), 1978, 27, 39.

35 Rappe, C.; Buser, H.R. and Bosshardt, H.-P. Chemosphere,
    1978, 7, 431.

36 Baughman, R. and Newton, L. "Analysis for Tetrachlorodi-
    benzo-p-dioxin in a French Talcum Powder - Hexachlorophene
    Formulation Implicated in the Death of a Number of Infants"
    Department of Chemistry, Harvard University, Cambridge, MA,
    1972.

37 Göthe, R. and Wachtmeister, C.A. Acta Chem. Scand., 1972
    26, 2523.

38 Cirelli, D.P. in Ranga Rao, K. (Ed.) "Pentachlorophenol:
    Chemistry, Pharmacology and Environmental Toxicology"
    Plenum, New York, 1978, pp. 13-18.

39 Nilsson, C.-A.; Norström, Å; Andersson, K. and Rappe, C.
    in Ranga Rao, K. (Ed.) "Pentachlorophenol: Chemistry,
    Pharmacology and Environmental Toxicology" Plenum, New York,
    1978, pp. 313-323.

40 Buser, H.R. and Bosshardt, H.-P. J. Assoc. Offic. Anal.
    Chem., 1976, 59, 562.

41 Goldstein, J.A.; Friesen, M.; Linder, R.E.; Hickman, P.;
    Hass, J.R. and Bergman, H. Biochem. Pharmacol., 1977
    26, 1549.

42 Vos, J.G.; Koeman, J.H.; van der Maas, H.L.; ten Noever de
    Brauw, M.C. and de Vos, R.H. Food Cosmet. Toxicol., 1970,
    8, 625.

43 Bowes, G.W.; Mulvihill, M.J.; Simoneit, B.R.T.; Burlingame,
    A.L. and Risebrough, R.W. Nature, 1975, 256, 305.

44 Buser, H.R.; Rappe, C. and Garå, A. Chemosphere, 1978, 7,
    439.

45 Olie, K.; Vermeulen, P.L. and Hutzinger, O. Chemosphere, 1977, 6, 455.
46 Buser, H.R. and Bosshardt, H.-P. Mitt. Geb. Lebensmittel-unters. u. Hyg., 1978, 69, 191.
47 Buser, H.R.; Bosshardt, H.-P. and Rappe, C. Chemosphere, 1978, 7, 165.
48 Buser, H.R.; Bosshardt, H.-P.; Rappe, C. and Lindahl, R Chemosphere, 1978, 7, 419.
49 Kimmig, J. and Schulz, K.H. Dermatologica, 1957, 115, 540.
50 Kimmig, J. and Schulz, K.H. Naturwiss., 1957, 44, 337.
51 Hoos, R.A.W. in Ranga Rao, K. (Ed.) "Pentachlorophenol: Chemistry, Pharmacology and Environmental Toxicology" Plenum, New York, 1978, pp. 3-11.
52 Levin, J.-O.; Rappe, C. and Nilsson, C.-A. Scand. J. Work Environ. Health, 1976, 2, 71.
53 Levin, J.-O. "Kemiska Riskmoment i Sågverksindustrin" Report University of Umeå ASF 74/102, 1975. In Swedish.
54 Truhaut, R.; L'Epée, P. and Boussemart, E. Arch. Mal. Prof. Med. Trav. Secur. Soc., 1952, 13, 567.
55 Hardell, L. and Sandström, A. Brit. J. Cancer, 1979, 39, 711.
56 Eriksson, M.; Hardell, L.; Berg, N.O.; Möller, T, and Axelson, O. Läkartidningen, 1979, 76, 3874.
57 Hardell, L.; Eriksson, M. and Lenner, P. Läkartidningen, 1980, 77, 209.
58 Wyllie, J.A.; Grabicae, J.; Benson, W.W. and Yoder, J. Pest Monitoring, J., 1975, 9, 150.
59 Levin, J.-O. and Nilsson, C.-A., Chemosphere, 1977, 6, 443.
60 Kauppinen, T., Lappeenranta Institute of Occupational Health, Lappeenranta, Finland. Unpublished data.
61 Levin, J.O. National Board of Occupational Safety and Health, Umeå, Sweden. Unpublished data.

RECEIVED October 30, 1980.

Occurrence of Nitrosamines in Industrial Atmospheres

D. P. ROUNBEHLER, J. W. REISCH, J. R. COOMBS, and D. H. FINE

New England Institute for Life Sciences, 125 Second Avenue, Waltham, MA 02154

J. FAJEN

National Institute of Occupational Safety and Health, Division of Surveillance, Hazard Evaluation and Field Studies, Cincinnati, OH 45226

Until as recently as 1975, the primary interest in the environmental occurrence of the carcinogenic N-nitroso compounds (nitrosamines) centered around nitrite preserved meats, foodstuffs, cheese products, fish, fish meal and biological samples. This emphasis began to change when N-nitrosodimethylamine (NDMA), one of the more potent of these carcinogens (1), was found in the atmosphere near a Baltimore, Maryland facility producing 1,1-dimethylhydrazine (a rocket fuel) (2,3) and in the atmospheres near West Virginia (4) and West Germany (5) facilities producing dimethylamine.

It has been demonstrated that NDMA is produced from the mixing of gas phase dimethylamine (DMA) and oxides of nitrogen, even when the concentration of the initial substances are at the ppm level (6,7,8). These reactions have also been shown to occur in either organic solvents or basic aqueous solutions when gas containing oxides of nitrogen are passed through the solvent containing secondary amines (9). In a test of air sampling sorbents, it was shown that surface bound secondary amines form nitrosamines when gas containing mixtures of NO and NO_2 are passed over them (10). It is apparent from these studies that nitrosamines can be formed in environments that have the needed precursors either in the air, in solution, or bound on surfaces.

The reactions for the formation of nitrosamines from secondary amines is as follows:

$$R_2 N-H + NOX \rightarrow R_2 N-NO + HX \quad (X = Cl, I, NO_3, NO_2, etc.)$$

Further discoveries of consumer and industrial products contaminated with N-nitroso compounds (11) led to the speculation that industrial workers that either use or manufacture these products may be exposed to significant amounts of these carcinogenic agents. These discoveries of environmentally occurring N-nitroso compounds plus increased understanding of the mechanism by which they can be formed from their widely available precursor amines (12) and oxides of nitrogen (13), has resulted in this study of worker exposure to N-nitroso compounds sponsored by the National Institute for Occupational Safety and Health (NIOSH).

During Part I of this study a total of 40 site visits were made to 28 separate manufacturing facilities representing five categories of industrial activity. The industries included in this study were:

- Leather manufacturing
- Rubber manufacturing
- Use and manufacture of synthetic metal working fluids
- Azo dye manufacture
- Fish processing plant

The basis for selecting the industries included in this study were: the known or suspected use of N-nitroso compounds, the use of products or manufacture of products shown to be contaminated with N-nitroso compounds, the use of chemical or manufacturing processes which could give rise to these compounds, the consideration of epidemiological data which suggested worker exposure to a chemical carcinogen and the results of this study as it proceeded. For example, the leather tanning industry, was included in this study because dimethylamine sulfate (a precursor to NDMA) is used in some leather tanneries as a depilatory agent in the unhairing process (12) and some leather workers have been reported to have an increased risk of cancer (14).

N-nitroso compounds were found in 21 of 28 plants surveyed. The highest levels of airborne nitrosamine was the finding of 47 $\mu g/m^3$ of NDMA in a leather tannery (15) and 27 $\mu g/m^3$ of N-nitrosomorpholine (NMOR) in a rubber tire plant (16). The results of the overall study have also been presented (17). In this report we discuss our findings of N-nitroso compounds in five leather manufacturing facilities (Table I). While the findings of the first tannery have been reported, the results from the other four tanneries surveyed in this study have not been previously discussed. In order to present a more complete report of our survey for N-nitroso compounds in the leather industry and to examine why these compounds are present as contaminants, we include the findings from all five tanneries in this paper. Because NDMA had been previously found in more than one tannery we selected tanneries with limited or specific operations (Fig. 1) in an effort to determine the source of this discovered NDMA or to determine the mechanism by which it was produced. If the mechanisms which produced NDMA in the atmosphere of leather tanneries are found not to be unique to this industry, then other plants and industries which use nitrosatable amines may also be contaminated with N-nitroso compounds. With the exception of the use of the artifact-free ThermoSorb /N air cartridges (10), the experimental method and apparatus used to conduct these surveys and to analyze the samples are essentially the same as previously reported (15).

Table I. Summary of *N*-Nitrosodimethylamine in Each Tannery

| Tannery | Airborne ($\mu g/m^3$) | | Bulk Samples (ppb) | | | |
	Maximum	Average	Waste Water	Process Water	Chemicals	Other[1]
A[2]	33	13	6	–	–	–
A[3]	47	18	–	1.5	500[4]	N.D.[6]
A[5]	3.4	2.4	–	N.D.	–	–
B	8	1.5	N.D.	–	–	N.D.
C	2.0	1.6	N.D.	N.D.		N.D.
D	trace	–	–	–		–
E	N.D.	–	–	–		–

1. Other samples include raw hide, tanned leather, condensed steam, etc.

2. Tannery A was the only tannery visited that was using Dimethylamine sulfate in the hide dehairing process.

3. Re-visit to Tannery A two days after the first

4. 500 ng/ml of NDMA was found in a 36% solution of Dimethylamine sulfate

5. Third visit to Tannery A 50 days after the first site visit. At this time they had ceased using Dimethylamine sulfate.

6. N.D. - None detected, detection limit of 0.05 $\mu g/m^3$ for NDMA in air and 0.5 ng/g for NDMA in bulk samples.

Tannery Description and Results

Tannery A. The first tannery surveyed was located in New England. Approximately 300 workers were processing about 2000 hides per day into finished leather. This tannery used methods of tanning that are characteristic of the industry (New England Tanners Club, 1977). In addition, they used dimethylamine sulfate in the unhairing process. All of the 300 or so tanning operations were performed in a one-story building divided into wet and dry operations areas (Figure 2). Three visits were made to the factory and the entire leather tanning and finishing processes were examined for the presence of N-nitroso compounds. During the first visit (April 11, 1978) six area air samples representing a cross section of the air in the tannery were collected and during the following visit (April 13, 1978) 20 air samples were taken along with several bulk samples. The third visit was made on June 1, 1978 when 10 air samples and numerous bulk samples of process water were collected. A total of 27 bulk samples were examined for N-nitroso compounds. These samples consisted of 11 chemicals or chemical mixes, two from the hide and leather, two of the waste water, and 12 of the process water from the wet operation of the tannery.

Results. During the first visit to this facility, NDMA was found in all six air samples at levels ranging from 6 $\mu g/m^3$ in the spray finishing area to 33 $\mu g/m^3$ near the chrome tanning operation. All of the inside air samples taken on the second visit, two days following the first, contained NDMA with levels ranging from 0.1 $\mu g/m^3$ in the lunch room, 1.4 $\mu g/m^3$ in the shipping room to 47 $\mu g/m^3$ in the re-tanning area. The average amount of airborne NDMA found on these two visits was 19 $\mu g/m^3$. This figure is based on the area air volumes. On the second visit, in addition to NDMA, N-nitrosomorpholine was found at a level of 2.0 $\mu g/m^3$ in three air samples taken in the leather finishing area. During the third visit, 50 days after the first, NDMA was again found in the atmosphere at all the sampled sites. However, these levels were considerably reduced, ranging from 1.1 to 3.4 $\mu g/m^3$ in the re-tanning area. The highest levels of airborne NDMA and their locations with respect to the tanning process are shown in Figure 3.

Only four of the 27 bulk samples contained NDMA (Table 2). The highest level of NDMA (0.5 $\mu g/ml$) was found in a sample from a 36.5% aqueous solution of dimethylamine sulfate (DMAS). According to plant personnel, 117 liters of DMAS are used each day in the hide unhairing process. This would amount to less than 60 mg NDMA/day potential exposure from the contaminated DMAS. NDMA (0.0015 $\mu g/ml$) was also found in the re-lime pit (unhairing vat) process water. This process water contained 5.7 liters of DMAS in 8000 L of lime-saturated water. The other samples which contained NDMA (0.004 $\mu g/ml$ and 0.006 $\mu g/ml$) were waste-water from the tannery outlet pipe collected at the local municipal waste-

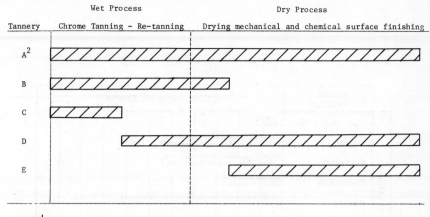

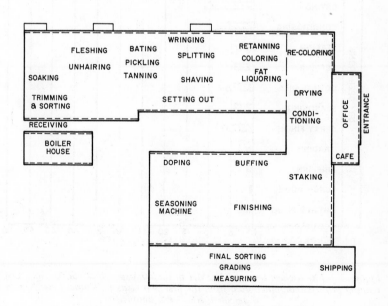

Figure 1. Leather tanning operations[1] of the five tanneries surveyed for N-nitroso compounds

Figure 2. Process flow diagram for Plant A

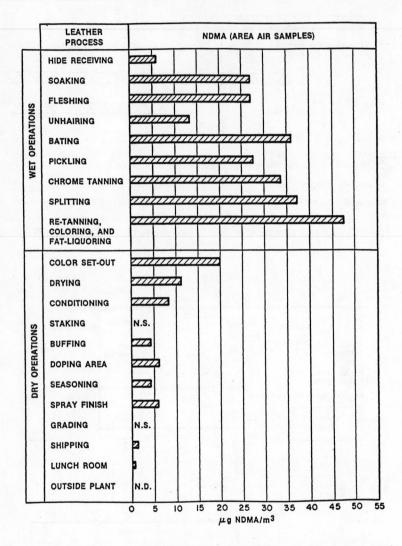

Figure 3.   The highest levels of NDMA in the atmosphere at various stages of the tanning process (combined data from the first two visits to Tannery A): ND = not detected; NS = not sampled

Table II. NDMA Levels in Liquid and Solid Bulk Samples

Sample Description	NDMA (µg/g)
Chemicals	
Azo Rubine Dye	ND
Penetrator L-219	ND
Nigrosine Blue L	ND
Polar Sol 5	ND
Betz Formula NA-6	ND
Boiler rush inhibitor mix	ND
Ammonia paste wash	ND
ρ-Nitrophenol	ND
KITO-40 (fungicide)	ND
Fresh brine	ND
Aqueous dimethylamine sulphate (36.5%)	0-5
Leather Samples	
Chrome-tanned leather	
Raw-salted cowhide	ND
Waste Water	
Beam-house waste water	0-004
Tanning-house waste water	0-006
Process Water	
Re-lime pit	0-0014
Bating solution (two samples)	ND
Pickling solution (two samples)	ND
Chrome-tanning solution (two samples)	ND
Final rinse from chrome tanning (two samples)	ND
Wash out of colouring	ND
First rinse from fat-liquoring	ND
Final rinse from fat-liquoring	ND

ND = Not detected

water treatment plant. It waw reported that the waste–water from
the tannery amounted to about 4×10^6 liters/day. Approximately
20 g of NDMA would be required to produce 0.006 µg/ml NDMA in this
volume of water. A sample of the salted cowhide was also examined
for NDMA, but none was detected (detection limit 0.05 µg/g). In
order to test the salted cowhide for any material which could possibly nitrosate dimethylamine, a 1 g piece of cowhide was placed
in 5 ml sulfuric acid pH (3) containing 50 mg of dimethylamine
hydrochloride for 4 hrs at 25°C. At a detection limit of 0.05
µg/g neither NDMA, nor any other TEA-responsive compounds were
found. Finally, process water was sampled from all phases of
the wet tanning operations and, except for the re-lime pit, none
of these contained NDMA at a detection limit of 0.005 µg/ml.

During the first and second visit to the tannery, a strong
odor (ammonia-like) was noted and the air appeared saturated with
moisture. In addition, several propane fork-lift trucks were
exhausting their gases to the atmosphere. Also during these
visits DMAS was being used in the re-lime pits. By the third
visit, the outside weather was warmer and the tannery was being
better ventilated. During this visit there were no strong odors
in the air and the tannery had discontinued using DMAS. However,
the propane driven fork-lift trucks were still operating within
the hide receiving and wet process area.

Using the plant dimensions supplied by the plant personnel,
the calculated air volume within the plant is about $1 \times 10^5 m^3$.
In order to produce the airborne concentration of NDMA observed
during the first two visits there would have to be about 2 grams
of NDMA in the air at any given time. If one assumes an air
change each 30 min, then on a daily basis there is about 100
grams of NDMA in the air. The waste water content of 20 grams/
day brings the total amount of NDMA needed to explain our findings
to 120 g/day. The NDMA impurity found in the DMAS would amount
to only 60 mg/day which is insufficient to account for the observed NDMA. We have failed to find a source for the NDMA in this
plant except for the possibility for airborne or surface formation from the dimethylamine moiety of the DMAS and airborne NO_x
from exhaust of the propane driven fork-lift trucks. We did not
measure the dimethylamine or nitrogen oxide levels in the air
during this survey. A crude measurement of the nitrosation capacity of the air in the tannery was made at a later date by drawing
air through cartridges containing magnesium silicate coated with
morpholine (10). We did find that N-nitrosomorpholine was indeed
formed on these cartridges, thus indicating an airborne nitrosating agent was present in the tannery. The greatest amount of this
airborne nitrosating agent was found in the wet process area where
the previous high levels of NDMA were found. It should be noted
that this is the same area where the propane driven fork-lift
truck operated and where DMAS is used. As previously mentioned,
the formation of NDMA in the gas phase and on surfaces from
dimethylamine and nitrogen oxides has been demonstrated.

Tannery B. This tannery, also located in New England, employs 80 workers who process about 700 hides per day producing fully tanned and colored leather which are shipped to a separate facility, Tannery E, for surface and mechanical finishes. During the site visit on October 12, 1978, 19 air samples and 4 bulk samples were collected. The air samples were collected at all stages of the tanning operation. The bulk samples consisted of two waste water specimens, one steam system condensate and one floor scraping from the dye room.

Results. NDMA was found in the atmosphere of this plant at all stages of production except the new beam house. The highest level of 8 $\mu g/m^3$ was found in an unused loft above the unhairing process. In the production area, 3 $\mu g/m^3$ was found in the dye storage room with the remaining samples ranging from 1.2 $\mu g/m^3$ near the the unhairing process to 0.03 $\mu g/m^3$ in the hide drying area. The sample locations and NDMA concentrations are shown in Figure 4. The farther away from the unhairing process that the air samples were taken the lower the levels of NDMA. The finding of nearly 3 $\mu g/m^3$ of NDMA in the small dye room near the re-tanning area with levels of 0.1 - 0.2 $\mu g/m^3$ just outside (a 15 to 30-fold difference) suggests that there may be a source (contamination) for the NDMA within the dye room. The highest level of 8 $\mu g/m^3$ in the loft above the unhairing process is hard to explain unless the levels of NDMA in the plant were higher in the past. None of the bulk samples examined contained any N-nitroso compounds.

This plant does not presently use dimethylamine sulfate, but it has used this compound in the recent past. Another possible source for the dimethylamine precursors may be the hides or it may be formed during the unhairing step. The only potential nitrosating agent identified in any of the processes (except dyeing) was the antifungal agent Paranitro-phenol. Some of the dyes used by this plant contained C-nitro groups which could transnitrosate dimethylamine to form NDMA. However, an air sample taken inside one of the coloring drums was no higher in NDMA levels than the air outside of it. We did observe propane operated fork-lift trucks being used within the tannery and these will contribute nitrogen oxides to the plant atmosphere.

Tannery C. This tannery, located in the Midwestern United States, employs 135 workers who process about 3000 hides per day into chrome tanned hides. These "blue hides" are then shipped to Tannery D for re-tanning, coloring and final finishing. This tannery has used dimethylamine on an experimental basis but was not using it during this survey. During the site visit to this facility on January 16, 1979, 10 area air samples and 5 process air samples were collected along with 8 bulk samples. The bulk samples consisted of 6 process water specimens, one plant wastewater specimen and one steam system condensate. The process air samples were collected inside the hide processing drum. These hide processing drums resembled large cement mixers and were ven-

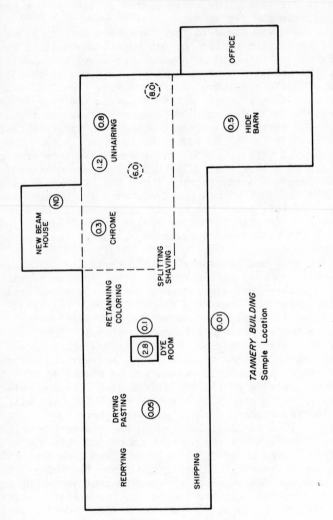

Figure 4.   Air sample locations within Plant B. The circled numbers refer to $\mu g/m^3$ of N-nitroso-dimethylamine (NDI-1A).

tilated to the outside air. All of the operations of this tannery, except the chemical mixing, were performed in a large 40,000 sq. ft unpartitioned building.

Results. Again NDMA was found in all of the air samples taken in this tannery with levels ranging from 0.2 μg/m^3 in the chemical processing room (sulfide stripping) to over 3 μg/m^3 in the center of the plant on a deck above the water treatment area. All other NDMA levels ranged from 1 to 2 μg/m^3 with a mean of 1.5 μg/m^3. These results indicate a fairly uniform distribution of the NDMA in the plant atmosphere. Air samples collected within the hide process drums had NDMA levels that were at or below the mean level within the plant. None of the bulk samples contained NDMA at a detection limit of 0.05 μg/ml.

A source for the NDMA was not found in this tannery. It was reported that they did use dimethylamine sulfate on an experimental basis and nitrosation of three of the bulk process water samples did result in the formation of 0.0015 to 0.0025 μg/ml of NDMA. We also collected three air samples using acid pH traps and after nitrosating the content, found twice the level of NDMA, thus indicating the presence of the NDMA precursor amine in the air. In a further experiment we sampled air over magnesium silicate coated with morpholine and found that from 3-20% of the 30 μg of morpholine had been converted to N-nitroso morpholine thus indicating an airborne nitrosating agent.

This plant also uses propane driven fork-lift trucks and at the time of the survey direct gas fired heaters were being used to warm the air in the plants. Both of these combustion sources would contribute nitrogen oxides to the air.

Tannery D. This facility, located in New England, consists of a two-story building with 150,000 sq. ft of floor space employing 560 workers who process about 8-10,000 hides per day. This facility is a re-tanning and finishing operation that receives chrome tanned hides from other plants which it re-tans, re-colors and surface finishes. This facility was visited on February 5, 1979 when 21 air samples were collected at all stages of the operation. There was no reported use of dimethylamine sulfate or any other amines in this facility.

Results. Two air samples were found to contain N-nitrosomorpholine (NMOR) at levels of 0.1 and 0.25 μg/m^3 and in addition 0.05 μg/m^3 NDMA was found in the sample containing 0.1 μg/m^3 of NMOR. These levels of N-nitroso compounds are considerably lower than what has been found in other tanning operations.

Tannery E. This tannery, located in New England on the third floor of a three-story brick building, occupies 40,000 sq. ft and employs 60 workers who apply surface finishes to leather that has been fully tanned and colored at Tannery B. On the day this plant was visited (October 13, 1978) 17 air samples were taken at all stages of production. While there were many chemicals and dyes being used at this plant there were no known sources for amines or nitrosating agents.

   Results. No N-nitroso compounds were found in any of the
air samples collected at the facility.

Conclusion

   The tannery industry was selected to be surveyed for N-
nitroso compounds in their environment because of its reported
use of dimethylamine sulfate in the hide unhairing process. Upon
examination of a tannery using this compound we did find two
N-nitroso compounds in its air, N-nitrosodimethylamine (NDMA) and
N-nitrosomorpholine (NMOR). Other tanning operations which
either did not use DMAS or had ceased its use had greatly reduced
levels of these compounds in their environment. We were unable
to identify any specific tanning process within the industry
which was responsible for the observed NDMA. However, the data
strongly suggests that it is being formed in the air or on surfaces. If this hypothesis is proven to be correct then any
environment where nitrosatable amines and nitrogen oxides are
found in a confined atmosphere (not just tanneries) are likely
to be contaminated with N-nitroso compounds.

Acknowledgments

   The authors thank Nancie Bornstein for her editorial assistance. We also thank all of our co-workers at the New England
Institute for Life Sciences for their help and support. The
work described here was supported by the National Institute for
Occupational Safety and Health (NIOSH) under Contract 210-77-0100.
Any opinions, findings, conclusions and recommendations expressed
are those of the authors and do not necessarily reflect the views
of the National Institute of Occupational Safety and Health.

Literature Cited

1. Magee, P.N.; Montesano, R. and Preussmann, R., "N-nitroso Compounds and Related Carcinogens" IN: ACS Monograph 17 - Chemical Carcinogens (C.E. Searle, Ed.), American Chemical Society, Washington, 1976 491-625.

2. Fine, D.H.; Rounbehler, D.P.; Pellizzari, E.D. Bunch; J.E.; Berkeley, R.W.; McCrae, J.; Bursey, J.T.; Sawicki, E; Krost, K. and DeMarrais, G.A.,"N-nitrosodimethylamine in Air" Bull. Environ. Contam. Toxicol., 1976, 15, 739-746.

3. Pellizzari, E.D.; Bunch, J.E.; Bursey, J.T; Berkeley, R.E.; Sawicki, E. and Krost, K.,"Estimation of N-nitroso-dimethylamine Levels in Ambient Air by Capillary Gas-Liquid Chromatography/ Mass Spectrometry, Anal. Lett. 1976, 9(6) 579-594.

4. Bretschneider, K. and Matz, J.,"Nitrosamines in the Atmosphere Air and in the Air of the Places of Employment" Archiv. Geschurilstforsch., 1973, 43, 36-42.

5. Fine, D.H., "An Assessment of Human Exposure to N-nitroso Compounds" IN: Environmental Aspects of N-nitroso Compounds (E.A. Walker, M. Castegnaro, L. Griciute, R.E. Lyle and W. Davis, Eds.), IARC Scientific Publication No. 19, 1978, Lyon, France, p. 267-278.

6. Bretschneider, K. and Matz, J. "Occurrence and Analysis of Nitrosamines in Air" IN: Environmental N-nitroso Compounds Analysis and Formation (E.A. Walker, P. Bogovski, L. Griciute W. Davis, Eds.), IARC Scientific Publication No. 14, 1976. Lyon, France, p. 395-400.

7. Hanst, P.L.; Spence, J.W.; Miller, M.,"Atmospheric Chemistry of N-nitrosodimethylamine" Env. Sci. & Tech. 1977, 11, 403.

8. Glasson, W.A.,"An Experimental Evaluation of Atmospheric Nitrosamine Formation" Env. Sci. & Tech. 1979, 13, 1145.

9. Challis, B.C. and Kyrtopoulos, S.A.,"Nitrosation of Amines by Two-Phase Interaction of Amines in Solution with Gaseous Oxides of Nitrogen" J.C.S. Perkin I, 1979, 299.

10. Rounbehler, D.P.; Reisch, J.W.; Coombs, J.R. and Fine, D.H., "Nitrosamine Air Sampling Sorbents Compared for Quantitative Collection and Artifact Formation" Anal. Chem. 1980, 52, 273-276.

11. Fine, D.H.,"Exposure Assessment to Preformed Environmental
     N-nitroso Compounds from the Point of View of our Own
     Studies" Oncology, 1980, 37, 199-202.

12. Walker, P.; Gordon, J.; Thomas, L. and Ouellette, R.,
     "Environmental Assessment of Atmospheric Nitrosamines"
     MITRE Corporation Report MTR-7152 EPA Contract 68-02-2495
     1976.

13. Anonymous, "Air Quality Criteria for Nitrogen Oxides"
     Environmental Protection Agency, Air Pollution Control
     Office Publication No. AP-84, 1971.

14. Acheson, F.D. Prevent. Med., 1976, 5, 295.

15. Rounbehler, D.P.; Krull, I.S.; Goff, U.E.; Mills, K.M.;
     Morrison, J.; Edwards, G.S.; Fine, D.H.; Fajen, J.M.;
     Carson, G.A. and Reinhold, V.,"Exposure to N-nitroso-
     dimethylamine in a Leather Tannery" Fd. Cosmet Toxicol.
     1979, 17, 487-491.

16. Fajen, J.M.; Carson, G.A.; Rounbehler, D.P.; Fan, T.Y.;
     Vita, R.; Goff, U.E.; Wolf, M.H.; Edwards, G.S.; Fine, D.H.;
     Reinhold, V. and Biemann, K.,"N-nitrosamines in the Rubber
     and Tire Industry" Science 1979, 205, 1262-1264.

17. Fajen, J.M.; Fine, D.H. and Rounbehler, D.P.,"N-nitros-
     amines in the Factory Environment" Sixth International
     Meeting on Analysis and Formation of N-nitroso Compounds,
     Budapest, Hungary, October 16-21 1979.

RECEIVED October 27, 1980.

Gas Chromatography–Mass Spectrometric Characterization of Polynuclear Aromatic Hydrocarbons in Particulate Diesel Emissions

DILIP R. CHOUDHURY and BRIAN BUSH

Division of Laboratories and Research, New York State Department of Health, Albany, NY 12201

Polynuclear aromatic hydrocarbons (PAH), of widespread occurrence in the environment, result from incomplete combustion of carbon- and hydrogen-containing substances. Many PAHs are well-recognized carcinogens and mutagens. Several industrial processes, such as fossil fuel conversion and production of aluminium and ferroalloys, can produce PAHs and result in their occurence in the working environment.

There is considerable current interest in development of state-of-the-art methodology for characterization of PAHs in various matrices (1). Gas chromatography and gas chromatography-mass spectrometry are two techniques widely used for PAH analysis because the necessary equipment is available in most laboratories. The development of high resolution glass capillary columns has significantly advanced the compound-characterizing capability of these techniques.

The major problem in PAH analysis is separation and conclusive identification of individual isomeric compounds, since the biological properties of many PAHs are isomer-specific. Another problem is the unavailability of many reference standards, making optimization of GC-operating parameters and column preparation methods for isomer separation difficult. The best possible separation efficiency is crucial for identification and quantitation of PAHs in any environmental sample. In addition, PAHs must be separated from other classes of compounds mostly encountered in environmental samples.

Diesel emission particulates, mostly of respirable size, are assuming increased importance as a source of atmospheric particulate pollutants. They may also present significant hazard in certain working environments such as bus garages and mines.

In this paper application of glass capillary gas chromatography (GC) alone and in conjunction with mass spectrometry

(GC-MS) is described for characterization of particulate-bound PAHs in diesel emissions. The term 'PAH' will refer to the parent and alkyl-substituted PAHs. The analytical methodology may also be adaptable to other types of sample matrices.

Experimental

Preparation of the Particulate Extract and Isolation of the PAH Fraction. Diesel emission particulates were collected by dilution tunnel technique on Pallflex T60A20 Teflon-coated glass fiber filters (20" x 20") (2). The particulates were Soxhlet-extracted from the filters with dichloromethane for 24 h. After filtration through a 0.2-μm Fluoropore filter under partial vacuum, the solvent was evaporated by gentle heating under vacuum. Several extracts were pooled to produce sample large enough for in-depth chemical characterization. Much of the work described here was performed on the PAH fractions from (a) S1, the pooled particulate extract from a Volkswagen (VW) Rabbit driven in a highway fuel efficiency test (HFET) mode, and (b) S2, the pooled extract from a Mercedes 300-D (federal test procedure [FTP]).

PAHs were isolated from the crude extracts by a two-step procedure. The neutral fraction was separated by simple acid-base partitioning and then chromatographed on a silica gel column. The column was first eluted with 70 ml of hexane. Subsequent elution with 200 ml of hexane containing 5% dichloromethane gave the PAH fraction. The solvent was carefully evaporated to produce the dry extract. The PAH fraction of S1 and S2 were designated as S1-C2 and S2-C2, respectively.

GC and GC-MS Analyses. Glass capillary columns were prepared in our laboratories as described briefly elsewhere (3). Aliquots (1-2 ul) of the PAH fraction dissolved in a small volume of chloroform were injected without stream splitting into the Hewlett Packard 5840A gas chromatograph. Injection port temperature was held at 250°C, and the column oven temperature was started at 100°C. Two minutes after injection a multistep temperature program was initiated; final temperature was 290°C. Nitrogen was the carrier and make up gas.

Results and Discussion

Chromatographic Resolution. To optimize column-coating conditions and operating parameters glass capillary columns coated with various silicone-based stationary phases were tested with difficult-to-separate groups of PAH standards and Complex samples. The SE54-coated columns performed excellently with respect to separation efficiency, column bleed and long-term stability. Other observers have had similar results with this

stationary phase (4). Since many isomeric PAHs are expected to
occur in environmental samples, it is essential to obtain best
possible chromatographic separation to achieve conclusive
identification and quantitation.

A chromatogram of 21 synthetic PAHs is shown in Figure 1.
Peaks were sharp and separation was excellent between
benzo[a]anthracene and chrysene, benzo[b]fluoranthene and
benzo[k]fluoranthene, and benzo[e]pyrene and benzo[a]pyrene.
Separation of these three groups of PAHs is critically important,
since some are moderate to strong carcinogens, whereas others
are relatively innocuous. Calculated detection limits of low
molecular weight PAHs were in subnanogram to low nanogram range,
slightly higher detection limits were observed for high-molecular-
weight compounds. In our method coronene, a seven-ring compound,
eluted in 41 min, as opposed to 110 min reported elsewhere (4).

GC Analysis. The gas chromatographic profile of the PAH
fraction S1-C2 is shown in Figure 2A. A comparison of retention
times of the major constituents with those of parent PAHs showed
that 3 and 4 ring compounds predominated. Readily identifiable
peaks were phenanthrene (peak 1), anthracene (peak 2),
2-phenylnaphthalene (peak 7), fluoranthene (peak 11) and pyrene
(peak 13). Minor constituents could not be identified with any
significant confidence. Several abundant peaks between
phenanthrene and fluoranthene were not readily identifiable from
their retention times. The higher-molecular-weight PAHs were
present in rather low concentration. Since many of these are
toxicologically significant, they were preconcentrated for more
definitive identification.

A Zorbax-CN high performance liquid chromatography (HPLC)
column used with hexane as mobile phase produced desirable
results. Details of the HPLC procedure will be published
elsewhere (5). Three HPLC subfractions of S1-C2 were made:
S1-C2A, S1-C2B, and S1-C2C. Gas chromatographic examination of
each subfraction showed that S1-C2C contained the minor
constituents of interest. (see the GC profile, Figure 2B and
legend.). All of the important isomeric compounds were
separated.

The PAH isolate from sample S2 (S2-C2, Figure 3) showed a
GC profile similar to that for S1-C2 (Figure 2A). Their most
abundant components were similar, and high-molecular-weight PAHs
were apparently present in low concentration only. The recorder
attenuation was changed at 19 min to help detect the less
abundant PAHs. Examinations of several other samples indicated
similar patterns.

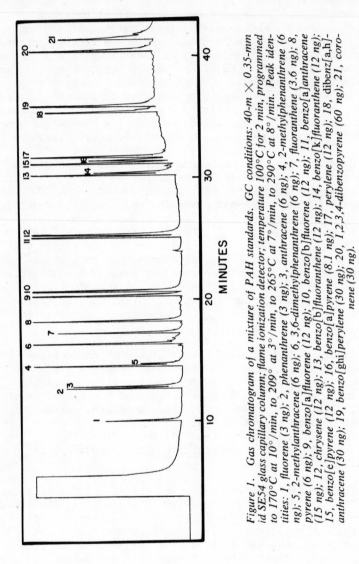

MINUTES

Figure 1. Gas chromatogram of a mixture of PAH standards. GC conditions: 40-m × 0.35-mm id SE54 glass capillary column; flame ionization detector; temperature 100°C for 2 min, programmed to 170°C at 10°/min, to 209° at 3°/min, to 265°C at 7°/min, to 290°C at 8°/min. Peak identities: 1, fluorene (3 ng); 2, phenanthrene (3 ng); 3, anthracene (6 ng); 4, 2-methylphenanthrene (6 ng); 5, 2-methylanthracene (6 ng); 6, 3,6-dimethylphenanthrene (6 ng); 7, fluoranthene (3.6 ng); 8, pyrene (6 ng); 9, benzo[a]fluorene (12 ng); 10, benzo[b]fluorene (12 ng); 11, benzo[a]anthracene (15 ng); 12, chrysene (12 ng); 13, benzo[b]fluoranthene (12 ng); 14, benzo[k]fluoranthene (12 ng); 15, benzo[e]pyrene (12 ng); 16, benzo[a]pyrene (8.1 ng); 17, perylene (12 ng); 18, dibenz[a,h]-anthracene (30 ng); 19, benzo[ghi]perylene (30 ng); 20, 1,2,3,4-dibenzopyrene (60 ng); 21, coronene (30 ng).

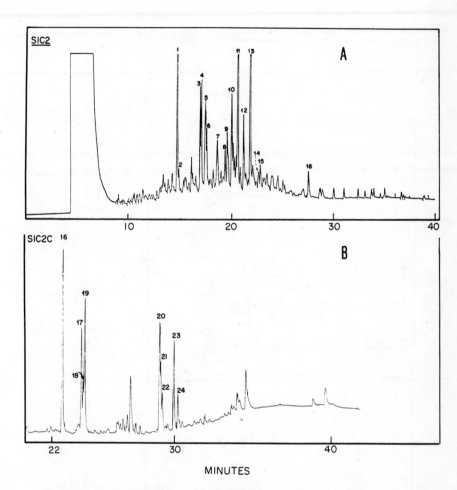

Figure 2. Gas chromatogram of A, PAH fraction of diesel particulate extract (S1-C2) and B, its HPLC subfraction C (S1-C2). GC conditions: 45-m × 0.35-mm id SE54 glass capillary column; flame ionization detector; temperature, 110°C for 2 min, programmed to 170°C at 10°/min, to 212°C at 3°/min, to 278°C at 8°/min. Peak identities: 1, phenanthrene; 2, anthracene; 3–6, methylanthracene/-phenanthrene; 7, 2-phenylnaphthalene; 8–10, dimethylanthracene/-phenanthrene; 11, fluoranthene; 12, aceanthrylene/acephenanthrylene; 13, pyrene; 14–15, trimethylanthracene/-phenanthrene; 16, benzo[ghi]fluoranthene; 17, benzo[a]anthracene; 18, triphenylene; 19, chrysene; 20, benzo[b]fluoranthene; 21, benzo[j]fluoranthene; 22, benzo[k]fluoranthene; 23, benzo[e]pyrene; 24, benzo[a]pyrene.

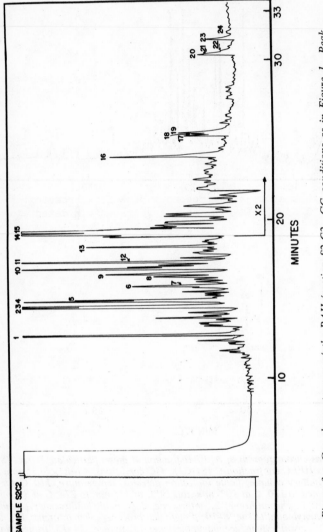

Figure 3. Gas chromatogram of the PAH fraction S2-C2. GC conditions as in Figure 1. Peak identities: 1, phenanthrene; 2–5, methylanthracene/phenanthrene; 6, dimethyldibenzothiophene; 7, 2-phenylnaphthalene; 8–11, dimethylanthracene/phenanthrene; 12, fluoranthene; 13, pyrene; 14–15, trimethylanthracene/phenanthrene; 16, benzo[ghi]fluoranthene; 17, benzo[a]anthracene; 18, triphenylene; 19, chrysene; 20, benzo[b]fluoranthene; 21, benzo[j]fluoranthene; 22, benzo[k]fluoranthene; 23, benzo[e]pyrene; 24, benzo[a]pyrene.

GC-MS Analysis.   PAHs produce stable molecular ions upon
electron impact; very little subsequent fragmentation takes
place.   This is advantageous in determining molecular weight and
hence the empirical formula and ring system of a PAH molecule.
However, lack of fragmentation makes it usually impossible to
differentiate between structural isomers.   Chromatographic
separation is therefore essential for identification isomeric
PAHs by GC-MS.

For definitive characterization of the GC-separated constituents, the total PAH isolate was examined by GC-MS using the
same capillary column.   Some loss of chromatographic resolution
occured in the GC-MS work, presumably because the injector of the
Finnigan instrument is significantly different from that of the
Hewlett Packard gas chromatograph.   The long transfer line could
be partly responsible too.   Operating parameters were separately
optimized for the GC-MS work.   The total ion chromatogram of the
PAH fraction S1-C2 is shown in Figure 4.

A thorough search of the reconstructed ion chromatogram
(RIC) was made to determine the mass spectra of all detectable
components.   The spectra were compared with those of reference
standards when available or with the spectra from the mass
spectral library.   Molecular weights of more abundant constituents
are shown in Figure 4, many additional minor components were also
characterized.   A complete list of compounds characterized in
this sample by GC-MS is given in Table I.

Phenanthrene, fluoranthene, and pyrene--the three abundant
parent PAHs identified by GC--were confirmed by MS.   The major GC
peaks between phenanthrene and fluoranthene were characterized as
methyl- and dimethylphenanthrene/anthracene.   Four compounds,
each having a nominal mass of 192 amu were detected.   Small
fragment ions at masses corresponding to $(M-1)^+$, $(M-27)^+$ and M^{2+},
were detected.   A general feature of these spectra was loss of a
methyl group from the parent ion.   The spectral features are
characteristic of dimethyl or ethylphenanthrene/anthracene.

Six compounds with parent ions at m/e 220 were detected.
All six spectra showed strong fragment ions at m/e 205 corresponding to loss of a methyl group.   The spectral features are in
general agreement with those of trimethyl or methylethylphen-
anthrene anthracene.   Exact isomeric structures could not be
determined because many isomers of the methyl-, C_2-alkyl-, and
C_3-alkylphenanthrene/anthracene are possible; but only a few
methylphenanthrene/anthracene standards were available to us.
The presence of chrysene, benzo[a]anthracene, and triphenylene
was confirmed.   These three appeared as one broad peak in the
RIC.   However, the presence of all three were established by

Table I

Compounds characterized by EI GC/MS of sample S1-C2

Scan No.	Compound
539	Methylfluorene
609	Phenanthrene
644	Dimethylfluorene
656	Dimethylfluorene
670	Dimethylfluorene
680	Methyldibenzothiophene
703	Methyldibenzothiophene
724	Methylphenanthrene/-anthracene
729	Methylphenanthrene/-anthracene
751	Methylphenanthrene/-anthracene
754	Methylphenanthrene/-anthracene
745	Cyclopenta[def]phenanthrene
789	C_2-Alkyldibenzothiophene
806	C_2-Alkyldibenzothiophene
814	2-Phenylnaphthalene
830	C_2-Alkylphenanthrene/-anthracene
841	C_2-Alkyldibenzothiophene
853	C_2-Alkylphenanthrene/-anthracene
860	C_2-Alkyldibenzothiophene
865	C_2-Alkylphenanthrene/-anthracene

Table I (cont.)

Scan	Compound
885	C_2-Alkylphenanthrene/-anthracene
911	C_2-Alkylphenanthrene/-anthracene
916	Fluoranthene
945	Acephenanthrylene/aceanthrylene
980	Pyrene
1032	C_3-Alkylphenanthrene/-anthracene
1041	C_3-Alkylphenanthrene/-anthracene
1057	C_3-Alkylphenanthrene/-anthracene
1064	C_3-Alkylphenanthrene/-anthracene
1077	Methylpyrene/-fluoranthene
1114	Benzo[a]fluorene
1143	Benzo[b]fluorene
1174	Methylpyrene/-fluoranthene
1184	Methylpyrene/-fluoranthene
1345	Benzo[ghi]fluoranthene
1353	Acepyrene
1434	Chrysene, benzo[a]anthracene, triphenylene
1724	Benzo[b,j,&k]fluoranthene
1775	Benzo[a]pyrene, benzo[e]pyrene

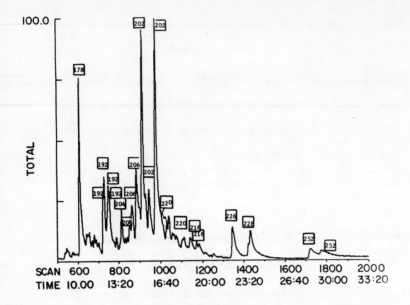

Figure 4.   Total ion chromatogram of the PAH fraction S1-C2 (3), peak identities are listed in Table 1

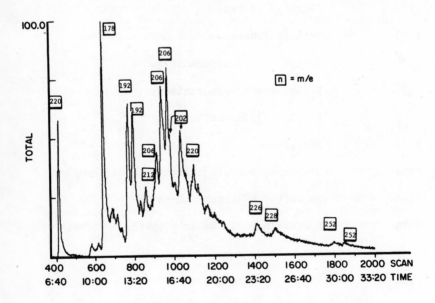

Figure 5.   Total ion chromatogram of the PAH fraction S2-C2 (3)

their GC retention times. The HPLC subfraction S1-C2C enriched with high-molecular-weight PAHs was not separately examined by GC-MS.

The peak with the parent ion of m/e 226 was assigned benzo[ghi]fluoranthene. The reference standard was not available, but its retention index was in agreement with that reported (389.6) by Lee et al. (6). The presence of benzo[b,j&k]fluoranthenes was also confirmed. They appeared as one broad peak, but the presence of all three was established by their GC retention time in the HPLC subfraction S1-C2C. Benzo[e]pyrene and benzo[a]pyrene were also characterized by MS. PAHs with higher ring systems were not detected in this sample.

Among minor constituents some O- and S-heterocycles and their alkyl derivatives were detected. No non-PAH-type compound or polar derivatives of PAHs (e.g., quinones) were detected in this fraction by mass spectrometry indicating the effectiveness of the isolation procedure.

GC-MS examination of the PAH fraction of sample S2 (S2-C2) gave very similar results; the total ion chromatogram is shown in Figure 5. Major constituents were phenanthrene, fluoranthene, pyrene, and methyl, dimethyl/ethylphenanthrene/anthracene. Relative abundance of some C_2-alkylphenanthrenes/anthracenes were higher in this sample than in S1-C2. Smaller quantities of benzo[ghi]fluoranthene, chrysene, benzo[a]anthracene, triphenylene, benzo[b,j,&k]fluoranthenes, and benzo[e&a]pyrenes and were characterized by MS. In addition, most compounds listed in Table 1 were also detected in this sample.

Thus glass capillary gas chromatography-mass spectrometry is a powerful technique for detecting PAHs and determining their ring system and empirical formula. Use of ultra-high-resolution gas chromatography significantly enhances compound identification capability. An efficient clean-up of the sample is also critical for obtaining interference-free mass spectra.

Acknowledgement

This work was partially supported by grant R805934010 awarded by USEPA. We thank Dr. R. Gibbs and his associates of the New York State Department of Environmental Conservation for collection and extraction of samples.

Literature cited

1. Jones, P. W. and Leber, P. E., Eds., "Polynuclear Aromatic Hydrocarbons," Ann Arbor Science: Ann Arbor, Mich., 1979.
2. Wotzak, G.; Gibbs, R.; and Hyde, J., "Proceedings of the International Symposium on the Health Effects of Diesel Engine Emissions," Cincinnati, December 1979, in press.
3. Choudhury, D. R. and Bush, B., "Proceedings of the International Symposium on the Health Effects of Diesel Engine Emissions," Cincinnati, December 1979, in press.
4. Bjorseth, A., in "Polynuclear Aromatic Hydrocarbons," Jones, P. E., Eds., Ann Arbor Science: Ann Arbor, Mich., 1979.
5. Choudhury, D. R., manuscript in preparation.
6. Lee, M. L.; Vassilaros, D. L.; White, C. M.; and Novotny, M., _Anal. Chem._, 1979, _51_, 768.

RECEIVED October 25, 1980.

Application of Glass Capillary Gas Chromatography for Determination of Potential Hazardous Compounds in Workplace Environments

G. BECHER, A. BJØRSETH, and B. OLUFSEN

Central Institute for Industrial Research, Forskningsvn 1, Blindern, Oslo 3, Norway

Presently the measurement and control of chemical hazards in the workplace environment is attracting increasing attention. Recent reports from the National Cancer Institute and the National Institute for Environmental Health Services in the USA state that at least 20% of all cancer in the United States may be job-related (1). This is a substantial increase from previous estimates that 1-5% of cancer results from occupational exposure to harmful substances. A trend in environmental chemistry is currently directed towards a better characterization of potential carcinogenic chemicals in the workplace environment. In this vein, considerable efforts have been invested by analytical chemists to develop methods for analyzing samples of airborne pollutants in the workplace environment. Furthermore, it is frequently necessary to supplement the air measurements data with determination of the pollutants or their metabolites in body fluids.

Recently the method of glass capillary gas chromatography (GC^2) has been developed as an extremely useful tool in characterizing multicomponent mixtures (2,3,4,5,6). GC^2 exhibits excellent reproducibility, high sensitivity and good resolution among individual compounds when compared to regular packed column gas chromatography. The purpose of this paper is to describe the application of GC^2 for determination of potential hazardous compounds in workplace environments. We have selected three classes of compounds to illustrate this; polycyclic aromatic hydrocarbons (PAH), chlorinated hydrocarbons and aromatic amines.

We also describe the use of different glass capillary columns and different detector systems.

Polycyclic Aromatic Hydrocarbons. PAH, their alkyl derivatives and their heteroatom analogues are among the largest single group of chemical carcinogens known (7). Consequently this class of pollutants has been in focus of interest for both analytical chemists and environmental toxicologists for a long time. Dibenz(a,h)anthra-

cene was the first carcinogen of defined chemical constitution to
be recognized (8). Shortly thereafter Hieger and coworkers (9)
undertook a large scale isolation of a fluorescent carcinogenic
constituent in tar by starting with two tons of gas work pitch.
This resulted in the identification of benzo(a)pyrene (BaP). By
1976 more than 30 parent PAH compounds and several hundred alkyl-
derivatives of PAH were reported to have carcinogenic effects (7).

PAH are formed by every high temperature reaction involving
organic materials (10). There are a number of industrial processes
where PAH can be identified in the workplace atmosphere. Well known
examples are coke plants, ferroalloy plants, aluminum plants,
secondary lead smelters and others (11,12). In some cases the
harmful effect of these compounds has been indicated by epidemiological studies (12).

The PAH pollution found in these cases consists of an extremely complex mixture of individual PAH compounds. In some cases
more than 100 compounds have been identified in the particulate
matter from workplace environments (13). This means that, for a
complete characterization of the potential hazardous compounds in
some of these workplace environments, analytical techniques with
very high resolving power, high sensitivity and good reproducibility for quantitative determinations are necessary. At present GC^2
seems to be the best method to meet these requirements (14).

Chlorinated Hydrocarbons. Chlorinated hydrocarbons are well
known environmental pollutants (15). They have been used as pesticides, high temperature stabilizing additives, plasticizers,
etc. These compounds exhibit high persistency to chemical and biological degradation and have a global distribution (15,16,17).
Occupational exposure to chlorinated hydrocarbons may represent
a serious problem for several reasons. These compounds are non-
polar and lipid soluble, they have a persistency to biodegrada-
tion, and hence, exhibit a high degree of bioaccumulation.
Furthermore, some chlorinated hydrocarbons are previously shown
to have mutagenic, carcinogenic and teratogenic effects (18,19).

Chlorinated aromatic hydrocarbons may be formed as unexpected
by-products of high temperature reactions involving carbon and
chlorine. The production of magnesium is an example (20). In this
process, MgO is first converted to $MgCl_2$ by carbon and Cl_2. Subsequently, magnesium is produced by anhydrous electrolysis of
magnesium chloride utilizing carbon electrodes. A large number
of chlorinated aromatic hydrocarbons are formed by this process.
Due to the high persistency of these compounds, body fluids may be
used as indicators to monitor the occupational exposure of the
chlorinated pollutants in the environment (21). However, to be
successful in this endeavour an analytical system exhibiting high
sensitivity is necessary. The combination of glass capillary gas
chromatography with an electron capture detector provides the
necessary sensitivity.

Aromatic Amines. Aromatic amines are compounds of considerable industrial and commercial importance. They are used as intermediates in the synthesis of numerous organic compounds including dyes, drugs, pesticides and plastics. Furthermore, they are constituents of several consumer goods such as most rubber products and hair dyes. The biological effects of aromatic amines vary from acute to chronic poisoning (22,23); some aromatic amines can give rise to cancer of the bladder or exert irritant effects on the urinary tract, and many of them induce allergies. The analytical problems related to determination of aromatic amines involve the development of a glass capillary column (24) suitable for analyzing the aromatic amines preferably without derivatization and the use of a detector with specific sensitivity for nitrogen compounds.

Experimental

Chemicals. All solvents were purified by distillation in glass apparatus. Standard compounds were supplied commercially and their purity checked by GC. If necessary standards were further purified by distillation in vacuo or by conventional recrystallization. Standard solutions were prepared by dissolving weighed amounts of PAH and chlorinated hydrocarbons in cyclohexane and of aromatic amines in 2-butanone. The standard mixtures were stored at 4 $^\circ$C in the dark.

Sampling. Samples of PAH were collected from an aluminum plant. Particulate matter was collected on a Gelman glass fiber filter. Vapors were collected in two impingers filled with ethanol and cooled with dry ice. The sampling device is reported elsewhere (25).

Blood samples for analysis of chlorinated hydrocarbons were obtained by glass and stainless steel syringes and transferred to prewashed centrifugal tubes. The samples were frozen immediately and stored at -25 $^\circ$C until analyzed.

Samples of aromatic amines were collected on commercially available silicagel tubes using personal pumps with a capacity of 0.2 1/min for 8 hrs.

Analytical Methods. The samples of PAH were extracted with cyclohexane, and the extract was subjected to liquid-liquid extractions with N,N-dimethylformamide as reported elsewhere (26). Following a concentration step, the extract was analyzed by GC[2] using a Carlo Erba Fractovap 2101 equipped with a flame ionization detector. The column was a 50 m x 0.32 mm i.d. persilanized glass capillary coated with OV-73 according to the Grob method (27).

The blood samples (about 25 g) analyzed for chlorinated hydrocarbons were added to 16 ml of hexane/isopropanol (1:1) in a centrifugal tube. The tube was kept in an ultrasonic bath for

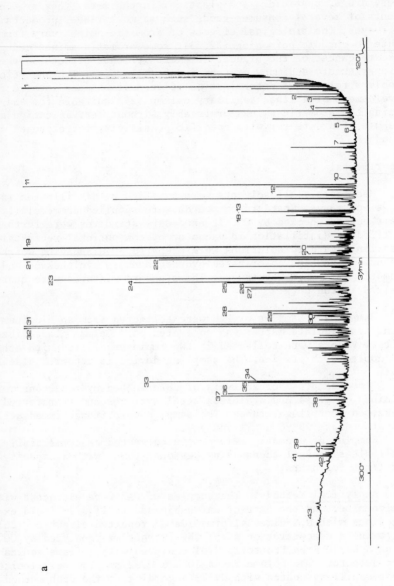

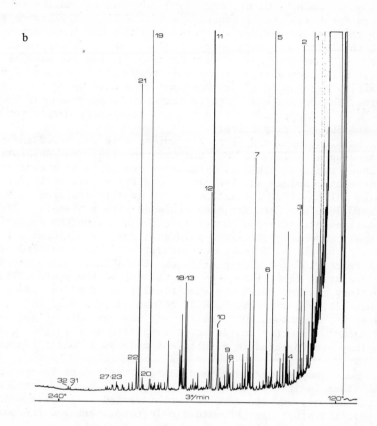

Figure 1. Gas chromatogram of PAH in a work atmosphere: a, particulate PAH; b, gaseous PAH. The peak identities are: 1, naphthalene; 2, 2-methylnaphthalene; 3, 1-methylnaphthalene; 4, biphenyl; 5, acenaphthene; 6, dibenzofuran; 7, fluorene; 8, 2-methylfluorene; 9, 1-methylfluorene; 10, dibenzothiophene; 11, phenanthrene; 12, anthracene; 13, methylphenanthrene/methylanthracene; 14, methylphenanthrene/methylanthracene; 15, 2-methylanthracene; 16, 4,5-methylenephenanthrene; 17, methylphenanthrene/methylanthracene; 18, 1-methylphenanthrene; 19, fluoranthene; 20, benzo(def)dibenzothiophene?; 21, pyrene; 22, ethylmethylenephenanthrene?; 23, benzo(a)fluorene; 24, benzo(b)fluorene; 25, 4-methylpyrene; 26, methylpyrene; 27, 1-methylpyrene; 28, benzothionaphthene?; 29, benzo(c)phenanthrene; 30, benzophenanthridine?; 31, benzo(a)anthracene; 32, chrysene/triphenylene; 33, benzo(b)fluoranthene; 34, benzo(j)fluoranthene; 35, benzo(k)fluoranthene; 36, benzo(e)pyrene; 37, benzo(a)pyrene; 38, perylene; 39, indeno(1,2,3-cd)pyrene; 40, dibenz(a,c/a,h)anthracenes; 41, benzo(ghi)perylene; 42, anthanthrene; 43, coronene/dibenzopyrenes.

20 min and subsequently shaken for 2 h. After addition of 20 ml
of distilled water, the mixture was shaken for several min and
centrifuged to separate phases. The hexane phase was transferred
to a small tube, 2 ml of conc. H_2SO_4 were added and the tube
shaken for 4 min. After centrifugation, 2 µl of the hexane solution
were injected into a Hewlett-Packard 5830 gas chromatograph equipped
with an electron capture detector. The glass capillary column was
10 m x 0.25 mm i.d. coated with SP-2100. Helium was used as carrier
gas at a flow rate of 3 ml/min and argon with 5 percent of methane
at a flow rate of 70 ml/min was used as a make-up gas for the detector. The column was linearly temperature programmed from 100 -
250 °C at 4 deg/min. The injector and detector temperatures were
275 and 300 °C, respectively.

Samples containing aromatic amines were extracted and
cleaned-up according to Wood and Anderson (28). For desorption 2-
butanone replaced ethanol. 2-butanone proved to be a good solvent
for aromatic amines and its minor polarity is desirable for a
longer column life. The extracts were usually analyzed on a Hewlett-
Packard 3830 gas chromatograph equipped with a flameless alkali-
sensitized nitrogen-phosphorus detector allowing the determination
of small amounts of aromatic amines in complex mixtures with other
organic compounds. Normally selectivity of this type of detectors
will exceed 10^4 g carbon/g nitrogen, the actual value depending on
the operating conditions used and the age of the source. The glass
capillary column, which was specifically designed for aromatic
amines (24,29), was a basic WCOT column (37 m x 0.32 mm i.d.)
coated with Carbowax 20 M. Helium was used as carrier gas at
1.2 kg/cm^2. Injector and detector temperatures were 250 °C. The
samples were injected in a splitless mode at a column temperature
of 100 °C.

Some samples were analyzed using a Carlo Erba Fractovap 2350
with flame ionization detector. Hydrogen was used as carrier gas
at 0.8 kg/cm^2, and the samples were injected in a split mode
(1:60). The other gas chromatographic conditions and the column
were as described above.

Results and Discussion

Polycyclic Aromatic Hydrocarbons. An application of GC^2 for
the analysis of PAH in the workplace environment of an aluminum
plant is given in Figure 1. Figure 1a shows the particulate PAH,
while Figure 1b shows the volatile PAH that pass the filter. The
chromatograms reveal the complexity of the PAH mixture. About
50 PAH compounds have been identified in the samples, based on
their retention time. More detailed identification based on glass
capillary gas chromatography/mass spectrometry has been reported
(13). Carcinogenic activities as revealed in animal experiments
(7) have been assigned to several of the compounds, i.e. benz(a)-
anthracene, benzo(c)phenanthrene, benzo(k)fluoranthene, benzo(a)-
pyrene, and dibenzopyrene. It is of particular importance that

GC^2 is capable of separating critical isomeric compounds such as benz(a)anthracene from chrysene/triphenylene, benzo(b)fluoranthene from benzo(j and k)fluoranthene and benzo(a)pyrene from benzo(e)-pyrene. In all these cases one isomer is carcinogenic while the other is not.

Previously, the main attention to PAH in workplace environments has been focused on the occurrence of benzo(a)pyrene. This compound has also been suggested as an indicator of the PAH level in work atmospheres containing tar and pitch volatiles (30). This study demonstrates the existence of other potentially hazardous materials as well.

The advantages using GC^2 are also illustrated in Figure 1. The high resolution power provides for a good profile analysis of the sample with a clear pattern and minor peak overlap compared to packed column GC. Furthermore, in view of the possible toxic, synergestic or antagonistic effects of the individual PAH, it is important to quantify each compound separately. The chromatograms also demonstrate the application of OV-73 to PAH analysis, a stationary phase similar to SE-52, but with improved temperature stability.

In our work with characterizing PAH in the work environment, we have felt that it is necessary to establish body doses through appropriate body fluid analysis for a better risk evaluation of occupational exposure. Analysis of metabolites and adducts between cellular macromolecules and PAH-metabolites is in progress.

Chlorinated Hydrocarbons. In the magnesium plant, samples from two different locations have been analyzed. Emissions from the chlorination plant were scrubbed with seawater and the particles collected as sewage. Figure 2 shows a chromatogram of chlorinated hydrocarbons in a sewage sample. As revealed by the chromatogram, the sewage contains a number of chlorinated benzenes and styrenes.

In order to study the possible occupational exposure to chlorinated aromatic hydrocarbons in the magnesium smelter, a number of blood samples from employees were analyzed by GC^2. A typical chromatogram is shown in Figure 3. As revealed by this chromatogram only the higher chlorinated compounds, penta- and hexachlorobenzenes, hepta- (HCS) and octachlorostyrenes (OCS), are observed. The presence of HCS and OCS is particularly noteworthy. To the best of our knowledge there is no commercial synthesis of these compounds. Nevertheless, they have been identified in biological samples from different countries in Europe and in USA (21,31,32).

The peaks corresponding to DDE and PCB are not due to the occupational exposure. Similar peaks are observed in samples from non-exposed persons and are due to the global distribution of these compounds.

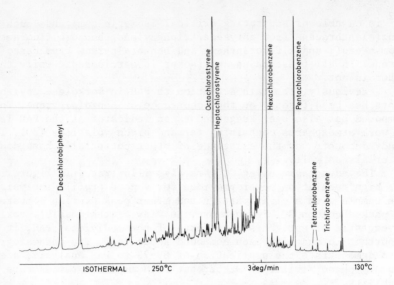

Figure 2. Glass capillary gas chromatogram of chlorinated aromatic compounds in a magnesium plant effluent

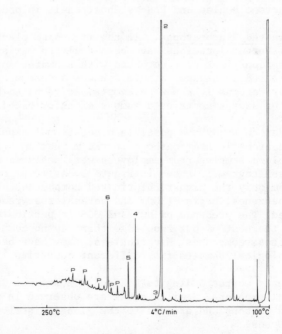

Figure 3. Capillary gas chromatogram of chlorinated hydrocarbons in a blood sample, magnesium plant worker using electron capture detector. Peak identities: 1, pentachlorobenzene; 2, hexachlorobenzene; 3, heptachlorostyrene; 4, octachlorostyrene; 5, unknown; 6, p,p'-DDE (p is probably PCB isomers).

Aromatic Amines. Analysis of aromatic amines represents a challenging problem. The basic, polar compounds may give rise to adsorption on the glass surface, leading to severe tailing of peaks or even loss of the component. As a part of this task it has therefore been necessary to develop a basic, polar glass capillary column suitable for aromatic amines (24). Furthermore, since aromatic amines frequently are found as minor components in a complex mixture of other organic contaminants, it was necessary to apply a nitrogen-selective detection using a flameless alkali-sensitized detector.

Figure 4 shows the chromatogram of a standard mixture of aromatic amines using a basic Carbowax 20 M column. Although the compounds were not derivatized, the peaks are eluting with negligible tailing from this column. Figure 5 shows a chromatogram of air-borne aromatic amines in a film processing laboratory, using a nitrogen-sensitive detector. A total of 5 compounds were detected, of which 4 were positively identified by gas chromatography/mass spectrometry.

Generally, the detector response will depend on the type of aromatic amine detected. Table I shows that response to N,N-dimethylaniline is about twice as high as the response to its primary isomer 2,6-dimethylaniline. Sensitivity for a selected number of aromatic amines was found to be increased by a factor 5-16 in comparison to a flame-ionization detector.

The use of a basic glass capillary column in the analysis of aromatic amines has several advantages:

- The high resolution power allows good separation of critical isomers, like 1- and 2-aminonaphthalene and yields accurate results in quantitative work (Figure 4).

- No derivatization is necessary, giving faster and simpler analysis and avoiding sample contamination and the necessity to determine synthesis yields.

It is plausible that analysis of urine samples may be used to monitor exposure to aromatic amines. In a model experiment, rats were treated with 2,4-diaminoanisole (2,4-DAA) and the urine analyzed for the amine and its most important metabolites (33). The results are shown in Table II. For the two doses used in this study, the major metabolite is 4-acetamido-2-aminoanisole, indicating that this compound may be used to monitor occupational exposure to 2,4-DAA.

Conclusion

It is shown that GC^2 may be successfully used in the characterization of potential harmful chemicals in the work atmosphere. The method is characterized by high resolution and good sensitivity and may be applied for analyzing different groups of pollutants with varying chemical properties. By modifying the inner

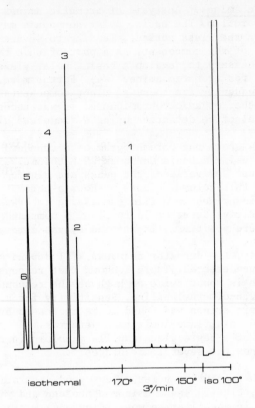

Figure 4.   Separation of selected aromatic amines on a basic WCOT glass capillary column with stationary phase Carbo wax 20 M; solvent, 2-butanone. Peak identities: 1, 1,2-diaminotoluene; 2, 2,4-diaminotoluene; 3, N,N-diphenylamine; 4, 1-aminonaphthalene; 5, 2-aminonaphthalene; 6, 2,4-diamino-1-methoxybenzene.

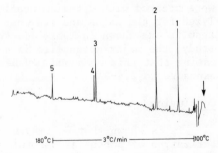

Figure 5.   Glass capillary gas chromatogram of airborne aromatic amines in a film processing laboratory; nitrogen selective detection. Peak identities: 1, N,N-diethylaniline; 2, 2,6-dimethylaniline (internal standard); 3, N,N-diethyl-1,4-diaminobenzene; 4, N,N'-diisopropyl-1,4-diaminobenzene; 5, not identified.

Table I

FID - and NPD - responses of selected aromatic amines

	relative responses NPD/FID	relative areas / weight	
		FID	NPD
N,N - dimethylaniline	13.6	0.78	1.86
aniline	10.4	0.75	1.36
2,6 - dimethylaniline	7.2	0.78	0.98
N,N - diethyl - 1,4 - diaminobenzene	15.3	1.00	1.57
2,4 - diaminotoluene	15.4	0.43	1.28
2,4 - diaminoanisol	16.6	0.23	0.74

Table II

Amount of metabolites of 2,4-diaminoanisol

	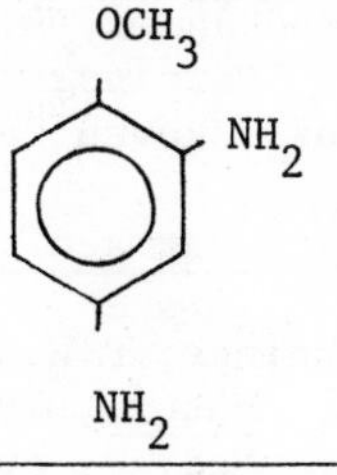 OCH_3 / NH_2 / NH_2	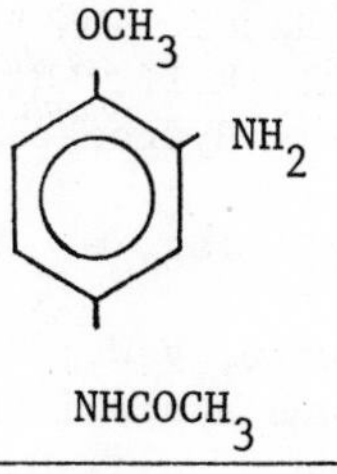 OCH_3 / NH_2 / $NHCOCH_3$	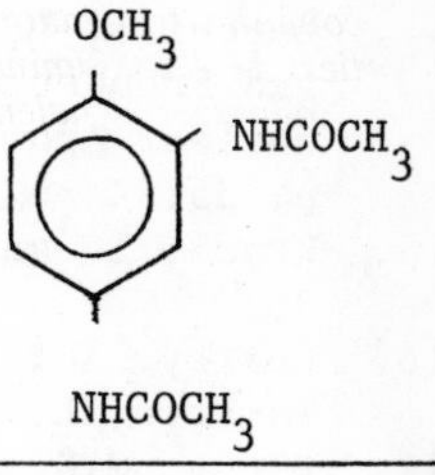 OCH_3 / $NHCOCH_3$ / $NHCOCH_3$
Yield at dose 100 mg/kg	23%	58%	19%
Yield at dose 10 mg/kg	not detected	87%	13%

glass surface, the column may be tailored to the chemical properties of the pollutants to be analyzed. The use of both general
and selective detectors provides additional information about the
sample composition.

The method may be applied to samples of air as well as body
fluids.

Literature Cited

1. Anonymous, Toxic Materials News, 1978, 5, 268.
2. Grob, K. and Grob, G. J. Chromatogr. Sci., 1969, 7, 584.
3. Grob, K. and Grob, G. J. Chromatogr., 1979, 62, 1.
4. Grob, K. and Jaeggi, J.J. Chromatographia, 1972, 5, 382.
5. Grob, K. and Grob Jr., K. J. Chromatogr., 1974, 94, 53.
6. Novotny, M.V. Anal. Chem., 1978, 50, 16A.
7. Dipple, A., in Searle, C.E., Ed. "Chemical Carcinogens",
    ACS Monograph 173, American Chemical Society, Washington, D.C.,
    1976; p. 245.
8. Kennaway, E.L. and Hieger, I. Br. Med. J., 1930, 1, 1044.
9. Cook, J., Hewett, C.L., and Hieger, I. J. Chem. Soc.,
    1933, 395.
10. Committee on Biological Effects, "Particulate Polycyclic
    Organic Matter", National Academy of Science, Washington, D.C.,
    1972.
11. Bjørseth, A. in Jones, P.W. and Leber, P., Eds., "Polynuclear
    Aromatic Hydrocarbons", Ann Arbor Sci., Ann Arbor, Mich., 1979;
    p. 371.
12. Criteria for a Recommended Standard - Occupational Exposure
    to Coal Tar Products, DHEW (NIOSH) Publication No 78-107,
    NIOSH, 1977.
13. Bjørseth, A. and Eklund, G. Anal. Chim. Acta, 1979, 105, 119.
14. Burchill, P., Herod, A.A. and James, R.G., in Jones, P.W.
    and Freudenthal, R.I., Eds., "Carcinogenesis, Vol. 3: Poly-
    nuclear Aromatic Hydrocarbons", Raven Press, New York, 1978;
    p. 35.
15. Kraul, I. and Karlog, O. Acta Parmacol. Toxicol., 1976, 38,
    38.
16. Curley. A.; Burse, V.W.; Jennings, R.W.; Villanueva, E.E.,
    Tomatis, L. and Akazaki, K. Nature, 1973, 242, 338.
17. Brady, M.N. and Siyali, D.S. Med. J. Austr., 1972, 1, 158.
18. de Bruin, A., "Biochemical Toxicology of Environmental
    Agents", Elsevier, New York, 1976.
19. Shepard, T.H., "Catalog of Teratogenic Agents", John Hopkins
    University Press, Baltimore, 1976.
20. Böckman, O.C.; Crowo, J.A.; Falck, S.T. and Johansen, J.G.,
    Proceedings, 12th Nordic Symposium on Water Research,
    Nordforsk, 1976; p. 349.
21. Lunde, G. and Bjørseth, A. Sci. Total. Environ., 1977, 8,
    241.

22. Scott, T.S., "Carcinogenic and Chronic Toxic Hazards of Aromatic Amines", Elsevier, New York, 1962.
23. "IARC Monographs on the Evaluation of Carcinogenic Risks of Chemicals to Man", Vol. 16, IARC, Lyon, 1978.
24. Olufsen, B. J. Chromatogr., 1979, 179, 97.
25. Bjørseth, A., Bjørseth, O., and Fjeldstad, P.E. Scand. J. Work Environ. & Health, 1978, 4, 212.
26. Bjørseth, A. Anal. Chim. Acta, 1977, 94, 21.
27. Grob, K, and Grob, G. J. High Resol. Chromatogr. & CC., 1979, 11, 677.
28. Wood, G.O. and Anderson, R.G. Amer. Ind. Hyg. Assoc. J., 1975, 36, 538.
29. Becher, G. and Olufsen, B., "10th Annual Symposium on the Analytical Chemistry of Pollutants", 28-30 May, 1980, Dortmund, FRG.
30. "Coke Oven Emissions", Occup. Safety Health Rep. Current Report 5., 1975; p. 283.
31. ten Noever de Brauw, M.C. and Koeman, J.H. Sci. Total Environ., 1972, 1, 427.
32. Kuehl, D.W.; Kopperman, H.L.; Veith, G.D. and Glass, G.E. Bull. Environ. Contamin. & Toxicol., 1976, 16, 127.
33. Becher, G. and Dybing, E., Unpublished results.

RECEIVED October 2, 1980.

Suitability of Various Filtering Media for the Collection and Determination of Organoarsenicals in Air

GREG RICCI, GEORGE COLOVOS, NORMAN HESTER,
and L. STANLEY SHEPARD

Rockwell International Environmental Monitoring and Services Center,
Newbury Park, CA 91320

JANET C. HAARTZ

National Institute for Occupational Safety and Health, Cincinnati, OH 45226

The use of arsenic and its organic derivatives as herbicides, pesticides, and wood preservatives has been increasing steadily each year. Large quantities of arsenical compounds are manufactured by the chemical industry and eventually find their way into the environment (1). About seventy percent of these chemicals are inorganic in form and the rest are organoarsenicals (2). Of the organoarsenicals, the most important species from the point of view of use and health effects are monomethylarsonic acid (MMA), dimethylarsenic acid (DMA), and p-aminophenylarsonic acid (p-APA).

Accurate estimates of worker exposure to organoarsenicals have not been possible because: 1) extensive studies of collection media for sampling organoarsenicals in air have not been performed; and 2) ruggedized analytical techniques capable of distinguishing between the different inorganic and organic species present in a sample have not been available.

A number of analytical techniques have recently been reported for analyzing mixtures of various arsenicals (2-9). Although the sensitivity of these techniques is excellent, most are unsuitable to use for routine application either because of poor precision and incomplete recoveries, or because they are too tedious and time consuming.

We have developed a highly sensitive, automated technique for separating and analyzing arsenic (III), arsenic (V), MMA, DMA, and p-APA in solution. The technique separates the components using ion exchange chromatography followed by continuous generation of the arsine derivatives and atomic absorption detection (10). The developed system provides an excellent means of analysis of environmental samples and has been applied directly to the determination of organoarsenicals in air.

In the present study, sampling of both particulate and vapor forms of MMA, DMA, and p-APA was investigated and it was found that these compounds exist in air mostly in the particulate form.

0097-6156/81/0149-0383$05.00/0

Collection of airborne particulates in workplace atmospheres can be achieved effectively by a variety of filtering media which are amenable to use with personal sampling devices. Extraction of the particulates from the filters with suitable solvents results in solutions which may then be analyzed by the developed analytical technique.

Three commercially available filtering media were evaluated for the collection of airborne particulates of the organoarsenical compounds. These were: cellulose ester, Fluoropore, and Mitex membrane filters. The evaluation was based on the following requirements: 1) the collection medium must quantitatively collect the particulates; 2) the collection medium must be capable of retaining the compounds without loss or chemical change before analysis; 3) the collected species must be extracted quantitatively from the collection medium for analysis; and 4) the method must be efficient at temperatures up to $50^{\circ}C$ and a relative humidity of 95%.

In the course of the studies, the cellulose ester filters were found to react with certain species and the Mitex filters were found not to collect the particulates effectively. Fluoropore membrane filters were found to be the most effective medium for collection of particulate organoarsenicals. The complete sampling and analysis procedure was tested for precision and accuracy using filter samples loaded in a dynamic aerosol generation and sampling system to simulate workplace atmospheres (11). The results have shown the developed method to be very precise and accurate for concentrations ranging from 5-20 μg As/m^3. In this paper, the studies conducted for the selection of an appropriate filtering medium and the evaluation of the combined sampling and analysis method are presented.

Experimental

The highly sensitive automated analytical method utilized for the determination of MMA, DMA, p-APA, As(III), and As(V) has been discussed in detail elsewhere (10). The optimized procedure for the determination of organoarsenicals in air is described below.

Sampling Procedure. Atmospheric particulate matter is collected quantitatively on 37mm membrane filters. This is done by sampling air at the rate of about 1.5 liters per minute through a three-piece polystyrene filter holder containing the appropriate filter and a cellulose support pad. This device is compatible with personal sampling pumps and can be attached to a worker's collar or lapel for the collection of a breathing zone sample. At the end of the sampling period, the air flow rate and collection time are recorded to calculate the total volume of air sampled. This sampling procedure was utilized in the present study for the collection of organoarsenical particulates generated in a dynamic aerosol generation and sampling system.

Analytical Procedure. The optimized analytical procedure for determination of organoarsenicals on the collected filter samples is as follows.

The organoarsenical particulates are extracted ultrasonically from the filters for 30 minutes in 25 mL of an aqueous carbonate/ bicarbonate/borate buffer (Eluent 1, Table I). After sonication the resulting extracts are ready for analysis and no further sample preparation is necessary.

A sample is injected onto a 3 x 500 mm anion-exchange separator column and the organoarsenicals are eluted with buffer at a flow rate of 2.6 mL/min. The organoarsenical species which are separated are reduced to their corresponding arsine derivatives as they exit the column in the eluent stream. This is achieved by mixing the eluent stream with a solution of 15% HCl acid saturated with potassium persulfate, and then reducing the arsenic species with 1% sodium borohydride ($NaBH_4$). The respective arsines of the organoarsenicals are carried to a heated quartz furnace (800^oC) aligned in the light path of the AAS where thermal decomposition of the arsine derivatives into free arsenic atoms occurs. The AAS instrumental parameters are adjusted to optimum following the manufacturer's recommendations. A detailed list of instrumental parameters for the total analytical system is given in Table I.

Two separate analyses of a sample, utilizing two eluent buffer systems of different ionic strength, are required for the complete characterization of a sample because As(III), if present in the sample, will interfere with the resolution and determination of DMA. The two eluent buffers which are used are: Eluent 1 - 0.0024 $\underline{M}$ $NaHCO_3$/0.0019 $\underline{M}$ Na_2CO_3/0.001 $\underline{M}$ $Na_2B_4O_7$; and Eluent 2 - 0.005 $\underline{M}$ $Na_2B_4O_7$.

Eluent 1 allows the separation and determination of MMA, p-APA, and As(V). If As(III) is not present in the sample, DMA is also effectively determined. The order of elution using Eluent 1 is: unresolved DMA/As(III), MMA, p-APA, and As(V). The entire chromatogram requires ∿10 minutes and the analysis can be performed at the rate of 5 samples per hour.

Eluent 2 is a lower ionic strength buffer and is used only to resolve DMA from interfering As(III). With this eluent, MMA, p-APA, and As(V) have very long retention times and will accumulate on the column typing up active resin sites. Therefore the column must be flushed with Eluent 1 after ∿10-15 samples have been analyzed and the column reequilibrated for 1 hour with Eluent 2 before further analysis. The analysis of DMA and As(III) can be performed at the rate of 10 samples per hour, and each chromatogram requires ∿3 minutes.

Results and Discussion

The contribution of vapor forms of organoarsenicals to the total atmospheric concentration depends on the vapor pressure of each compound and also on the temperature conditions. Data on the vapor pressures of the compounds studied were not available and therefore the following experiment was designed to estimate the concentration of vapor forms found in a workplace atmosphere.

Table I. Instrumental conditions for the analysis of
organoarsenicals in air particulate samples

Ion Chromatograph Parameters

Column	Dionex 3x500 mm Anion-Exchange
Mobile Phase Flow Rate	2.6 mL/min
Eluent 1	0.0024 $\underline{M}$ NaHCO$_3$/0.0019 $\underline{M}$ Na$_2$CO$_3$/ 0.001 $\underline{M}$ Na$_2$B$_4$O$_7$
Eluent 2	0.005 $\underline{M}$ Na$_2$B$_4$O$_7$
Injection Loop	Variable

Arsine Generation Parameters

15% HCl w/saturated potassium persulfate	0.8 mL/min
1% NaBH$_4$/0.2% KOH	2.0 mL/min
Argon Carrier Gas	300 cc/min

AAS Parameters

Arsenic EDL	8 watts
Wavelength	193.7 nm
Slit Width	0.7 nm
Quartz Furnace	800°C
D$_2$ Background Correction	None
Signal	Absorbance
Scale Expansion	3x
Recorder	10 mv full scale
Chart Speed	1 cm/min

An enclosed sampling train for generation and quantitative collection of vapors was assembled with the following components: (1) a glass impinger wrapped with heating tape to serve as a vapor generator and modified so that the air inlet tube was ~4 cm above the bottom; (2) a 47mm cellulose ester membrane filter in-line to collect particulates; (3) three bubblers in series to collect any vapors, each containing 15.0 mL of 0.1 $\underline{N}$ NaOH; and (4) a vacuum pump. MMA, DMA, and p-APA were placed in the heated impinger which was maintained at a temperature of 50°C (122°F) to simulate an extreme sampling temperature situation. Air, at the rate of 0.2 liters/min, was drawn over the heated salts and through the filter and the three bubblers. Two sampling experiments were performed: in the first, a total volume of 30 liters of air was sampled in 2.5 hours; in the second, a total volume of 100 liters of air was sampled in 8.0 hours. After sampling, each bubbler was acidified by adding 10.0 mL of 10% HNO_3 acid saturated with potassium persulfate. The bubbler samples were then analyzed for total arsenic using the automated hydride generation system of Vijan and Wood ($\underline{12}$). The results for both experiments showed that the extent to which any organoarsenical vapors may have been generated under these conditions was immeasurable and below the analytical detection limit (i.e. <0.4 $\mu g/m^3$ arsenic for a 2.5 hour sample, and <0.12 $\mu g/m^3$ arsenic for an 8-hour sample). Based on these results, it was concluded that the organoarsenical compounds tested do not exhibit enough vapor pressure to require special considerations for the sampling technique. Therefore, the only form of these organoarsenicals which need be considered is the particulate which can easily be collected by filtration.

Three filtering media commercially available from Millipore Corporation were evaluated for the collection of airborne particulates of the organoarsenical compounds. These were: (1) cellulose ester membranes; (2) Fluoropore; and (3) Mitex. The cellulose ester membrane filters (0.8 μm pore size) are the standard Millipore filters used for aerosol sampling and consist of a mixture of cellulose nitrate and cellulose acetate. The Fluoropore filters are made of Teflon (PTFE) and are bonded to a polyethylene net. Fluoropore filters with a pore size of 1.0 μm were selected for this study. The Mitex filters are also made of Teflon, but have no backing material. A 5 μm pore size Mitex filter was used because it is the smallest pore size available. Each of these is available in 37mm diameter disks and can be used in convenient, 3-piece polystyrene filter holders (Millipore Aerosol Monitors). Extracts of samples collected on these filters were tested for stability by conducting the following experiments.

In the first experiment, a total of 5 cellulose ester filters were spiked with the equivalent of 1.5 μg of arsenic from each of the 5 species, MMA, DMA, p-APA, As(III), and As(V). The filter spikes were allowed to dry and were then extracted ultrasonically in 25 mL of deionized water. The resulting extracts were stored at room temperature and analyzed three separate times over a period of one week. The first analysis was performed the same ·day that

the filters were spiked, the second after 5 days, and the third
after 7 days. Table II presents the results of this experiment,
which show that the filter extracts changed drastically.

Table II. Stability at room temperature of aqueous extracts
of spiked cellulose ester filters as a function of time.

| Species | Percent Recoveries ± Standard Deviation | | |
	1st Day	5th Day	7th Day
DMA/As(III) *	100.4 ± 1.5	124.7 ± 11.9	150.6 ± 5.6
MMA	106.7 ± 1.0	58.2 ± 4.1	23.2 ± 2.5
p-APA	95.5 ± 5.3	91.6 ± 6.6	85.1 ± 6.0
As(V)	143.6 ± 24.5	118.4 ± 16.1	119.8 ±18.3
	N = 2	N = 5	N = 4

*The percent recoveries shown are for unresolved DMA/As(III)

Over the 7-day period, the percent recovery for MMA decreased
to ∿23%, while the percent recovery for unresolved DMA and As(III)
increased to ∿150%. (The filter extracts were not analyzed a
second time to quantitate DMA and As(III) individually.) Since these
phenomena occurred only for extracts containing cellulose ester
filter material, it appeared that the filter material itself in
solution caused chemical changes. This same phenomenon might occur
also on the cellulose ester filter under ambient sampling conditions, with the rate of such a reaction dependent upon the temperature and relative humidity of the sampled air.

A second experiment was performed then to determine whether
the organoarsenical extracts would be more stable if stored in
acidic or basic solution, or under refrigeration. Fifteen cellulose
ester filters were spiked with the equivalent of 1.5 µg of arsenic
from each of the five species. The filters were allowed to dry
before extraction with 25 mL of one of the extraction solutions
shown in Table III. After sample preparation, extracts were either
stored at room temperature or refrigerated. Each experiment was
run in triplicate and the nonrefrigerated extracts were analyzed
four days and six days after extraction, whereas the refrigerated
extracts were analyzed six days after extraction. Table IV presents
the results of this experiment. The data show that chemical changes
are still occurring, but that acetate and borate buffers as well as
refrigeration improve the stability of MMA. No effort was made to
explain the high recoveries for p-APA and As(V) or the low recoveries for DMA and As(III).

Table III. Extraction solutions used in the stability studies

Extraction Solution	Description
A	H_2O
B	0.0021 $\underline{M}$ $NaHCO_3$/0.017 $\underline{M}$ Na_2CO_3/0.0015 $\underline{M}$ $Na_2B_4O_7$
C	0.005 $\underline{M}$ $Na_2B_4O_7$
D	0.001 $\underline{M}$ CH_3COONa

On the basis of the results of Table IV, it was decided to extract the filter samples with buffer B because: 1) this extraction medium yields greater stability of MMA at room temperature, and with refrigeration recoveries might be even more improved; and 2) this extraction medium is the eluent used for separation and analysis of MMA, p-APA, and As(V).

As shown above extracts of organoarsenicals obtained from cellulose ester filters and analyzed by the developed method suffer severely from lack of stability. To evaluate conclusively and also to find an appropriate filtering medium, experiments were performed to simultaneously test the stability of organoarsenicals on cellulose ester, Fluoropore, and Mitex filters. The experimental design is outlined in Figure 1.

This design was developed to test the variables associated with the sample preparation and storage for each of the three filtering media. Triplicate spiked filters were used for each parameter tested. A mixed standard solution containing MMA, DMA and p-APA was used for spiking. Each filter was spiked with the equivalent of 1.5 µg of arsenic from each species. The inorganic species As(III) and As(V) were not included on the filter spikes in anticipation that if any instability of the organoarsenical species was encountered, conversion to either of the inorganic species could be expected and interpretation of the analytical results would be simplified. A total of thirty filters were spiked in individual 50 mL beakers and extracted ultrasonically in 25 mL of buffer B for 30 minutes. The filter extracts were analyzed against calibration standards prepared in the same buffer matrix. The average percent recoveries and standard deviation of the triplicate filter spikes are presented in Table V for each species. From the results presented there, the following conclusions can be made.

1. The percent recoveries of DMA from each filter type are excellent in all cases (94.4-104.0%) and therefore no stability problem is seen for this species.

2. The percent recoveries of MMA and p-APA from cellulose ester filters which were stored three days before extraction and

Table IV. Stability of aqueous extracts of spiked cellulose ester filters as a function of extraction solution, storage temperature, and storage time.

Experimental Conditions	Percent Recovery ± Standard Deviation (N = 3) Species				
Extraction Solution Storage Temperature Storage Time	DMA	As(III)	MMA	p-APA	As(V)
A Refrigerate 6 Days	99.6 ± 0.75	79.0 ± 7.9	88.7 ± 10.6	111.0 ± 1.4	93.5 ± 10.2
A Room Temperature 6 Days	114.7 **	171.4 **	28.6 ± 8.1	112.6 ± 4.0	80.7 ± 45.2 106.8 ± 3.2*
B Room Temperature 4 Days	72.0 **	40.0 **	79.8 ± 16.1	106.1 ± 5.2	123.7 ± 7.4
C Room Temperature 4 Days	73.3 **	45.7 **	93.6 ± 9.6	117.5 ± 7.3	112.4 ± 5.7
D Room Temperature 4 Days	88.0 **	65.7 **	106.1 ± 5.0	110.0 ± 1.8	113.9 ± 4.6

* Two data points
** One data point

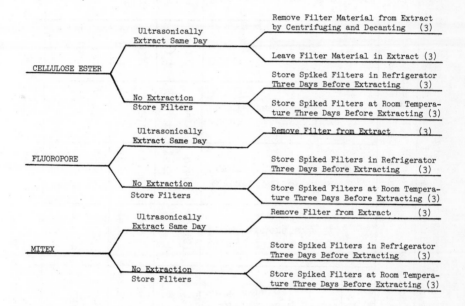

Figure 1. Experimental design used to test and compare the stability of organoarsenicals on cellulose ester, Fluoropore, and Mitex filters. Triplicate spiked filters
were used for each experiment.

Table V. Average percent recoveries of organoarsenicals from spiked cellulose ester, Fluoropore, and Mitex filters (N = 3).

Species	Millipore Cellulose Ester Membrane Filters				Fluoropore Polyethylene Backed Teflon Filters			Mitex Unbacked Teflon Filters		
	Extract Filters Same Day/ Remove Filter Material	Extract Filters Same Day/ Leave Filter Material	Store Filters 3 Days in Refrigerator	Store Filters 3 Days at Room Temperature	Extract Filters Same Day/ Remove Filter Material	Store Filters 3 Days in Refrigerator	Store Filters 3 Days at Room Temperature	Extract Filters Same Day/ Remove Filter Material	Store Filters 3 Days in Refrigerator	Store Filters 3 Days at Room Temperature
DMA	100.6 ± 6.4%	100.7 ± 8.4%	94.4 ± 2.1%	101.8 ± 4.4%	104.0 ± 7.4%	104.0 ± 6.4%	100.6 ± 6.4%	102.2 ± 8.2%	102.9 ± 5.2%	102.2 ± 5.0%
MMA	100.9 ± 0.8%	103.3 ± 6.5%	75.1 ± 9.6%	77.6 ± 6.4%	106.4 ± 4.3%	102.2 ± 2.0%	103.3 ± 4.4%	110.9 ± 3.1%	104.9 ± 3.0%	110.0 ± 5.3%
p-APA	83.3 ± 1.0%	76.7 ± 7.7%	61.7 ±11.5%	64.6 ± 7.3%	98.3* ± 7.2%	98.0* ± 4.8%	95.6 ± 2.1%	100.5 ± 8.2%	94.0 ± 6.8%	90.7 ±11.4%

*Calculated with the exclusion of one outlying data point

analysis are poor (62-78%), whereas the percent recoveries obtained for the same species from Fluoropore and Mitex filters stored for 3 days are excellent (better than 90%).

3. The percent recoveries of MMA and p-APA from cellulose ester filters are significantly better if the filter samples are extracted the same day as they are spiked rather than storing the spiked filters for 3 days.

4. Removal of the undissolved cellulose ester filter material from aqueous extracts did not appear to improve the stability of the organoarsenicals.

On the basis of these results, it is clear that a more inert filter material, Mitex or Fluoropore, should be used. As will be discussed below, the collection efficiency of the Fluoropore filters is better than that of the Mitex filters, and for this reason it was finally decided to use Fluoropore filters for the combined sampling and analytical procedure.

The collection efficiency of the Mitex and Fluoropore filters was tested by generating and collecting mixed aerosols containing MMA, DMA and p-APA under controlled conditions and analyzing the resulting filter extracts by the developed analytical technique. The aerosol generation/sampling system which was used for this was built and characterized in a previous study (11). The system was found to produce 90% of the particle mass in the size range of 0.1 to 10 µm.

To determine collection efficiency, each aerosol cassette monitor was assembled with a double stage filtering arrangement to test for "breakthrough". The aerosol cassettes were assembled in the following manner using either the 5 µm Mitex unbacked Teflon filters or the 1.0 µm Fluoropore polyethylene backed Teflon filters: Teflon filter #1 (upstream), supported by a cellulose pad and isolated from Teflon filter #2 (downstream), also supported by a cellulose pad. Both the Mitex and Fluoropore filters were used for simultaneous sampling of an atmosphere of 60 µg/m^3 as arsenic of each of the three organoarsenical species in the aerosol generation/ sampling system. After sample collection, the four filters from each cassette (both Teflon filters and both cellulose support pads) were extracted and analyzed. The results are presented in Table VI and are expressed in µg As/m^3 and an average and standard deviation were calculated for each species. The data show that no break-through is seen past the first Fluoropore filter, whereas significant breakthrough is seen on the cellulose support pad behind the first Mitex filter. The 5 µm pore size of the Mitex filter is simply too large to quantitatively collect the particulates. The collection efficiency of the Fluoropore 1.0 µm filters is better than 99%, whereas the collection efficiency of the Mitex 5.0 µm filters relative to the Fluoropore filters is only ∿65%. On the basis of these results, Fluoropore filters were selected for collection of organoarsenical particulates in air. Their collection

Table VI. Collection efficiency of Fluoropore and Mitex filters for sampling organoarsenicals in air (N = 5)

Organoarsenical Species	Filter Stage	Concentration Arsenic ($\mu g/m^3$) ± Std. Deviation	
		Fluoropore 1.0 μm	Mitex 5.0 μm
MMA	Membrane Filter 1	57.04 ± 5.32	37.11 ± 3.28
	Support Pad 1	< DL	3.26 ± 0.48
	Membrane Filter 2	< DL	< DL
	Support Pad 2	< DL	< DL
DMA	Membrane Filter 1	54.29 ± 2.67	36.67 ± 3.21
	Support Pad 1	< DL	3.42 ± 0.54
	Membrane Filter 2	< DL	< DL
	Support Pad 2	< DL	≤ DL
p-APA	Membrane Filter 1	50.77 ± 2.99	32.67 ± 3.62
	Support Pad 1	< DL	3.06 ± 0.76
	Membrane Filter 2	< DL	< DL
	Support Pad 2	< DL	< DL
Estimated Collection Efficiency		> 99%	~ 65%

DL = ~ 0.5 $\mu g/m^3$

efficiency is as good as the traditionally used cellulose ester filters and in addition are inert towards the various organoarsenical species.

The effectiveness of the proposed sampling procedure using 1.0 μm Fluoropore filters was further tested under conditions of high temperature (50°C) and high humidity (∼95% RH) as well as in the presence of high SO_2 concentrations. This was done by subjecting filters which previously had been loaded with organoarsenical aerosols to these conditions as follows. In one experiment, arsenic-free air of 50°C and 95% relative humidity was passed through pre-loaded filters for the same period of time as the initial sample collection time. In a second experiment, air containing 1 ppm SO_2 was passed through the preloaded filters. These filters were then analyzed along with control filters which had not been subjected to these conditions. The results of these experiments are presented in Tables VII and VIII, and show clearly that there is no significant effect of temperature, humidity or SO_2 on the samples.

The precision and accuracy of the combined sampling and analytical method was tested using filter samples loaded in the aerosol generation and sampling system. Replicate samples were analyzed using the developed analytical procedure, and also by Neutron Activation (NAA) and X-Ray Fluorescence (XRF) analyses performed by an independent laboratory. For this, aerosols containing MMA, DMA and p-APA were generated in concentrations of 5, 10 and 20 μg arsenic per cubic meter of air and sampled for 4 hours at 10 equivalent positions in the aerosol generation and sampling system. Three of these positions were used to collect samples for NAA and XRF analysis.

Data for the 7 replicates of each level, analyzed by the developed IC-AAS procedure, are presented in Table IX. The data are expressed in μg/m^3 and an average and standard deviation were calculated for each species. Depending on the concentration and the species, the relative standard deviation was found to range from ±14.4% at the lowest level to ±4.7% at the highest level.

Table VII. Effect of temperature (50°C) and humidity (95% RH) on particulate organoarsenicals collected on Fluoropore filters (N = 5)

| Species | Concentration Arsenic (μg/m^3) ± Standard Deviation | |
	Control Samples	Samples Exposed to 50°C/95% RH
MMA	24.36 ± 0.63	24.92 ± 0.70
DMA	23.97 ± 0.88	23.86 ± 0.27
p-APA	21.41 ± 1.79	19.24 ± 1.08

Table VIII. Effect of SO$_2$ on particulate organoarsenicals collected on Fluoropore filters (N = 5)

Organoarsenical Species	Filter Stage	Concentration Arsenic (μg/m^3) ± Std. Deviation	
		Control Samples	Samples Exposed to 1 ppm SO$_2$
MMA	Fluoropore Filter 1	55.76 ± 0.33	55.38 ± 3.84
	Support Pad 1	< DL	< DL
	Fluoropore Filter 2	< DL	< DL
	Support Pad 2	< DL	< DL
DMA	Fluoropore Filter 1	55.65 ± 0.48	55.43 ± 4.63
	Support Pad 1	< DL	< DL
	Fluoropore Filter 2	< DL	< DL
	Support Pad 2	< DL	< DL
p-APA	Fluoropore Filter 1	48.26 ± 0.72	49.53 ± 4.31
	Support Pad 1	< DL	< DL
	Fluoropore Filter 2	< DL	< DL
	Support Pad 2	< DL	< DL

DL = $\sim$ 0.5 μg/m^3

Table IX. Precision of combined sampling and analytical method using Fluoropore filters loaded in the aerosol generation/sampling system (N = 7)

Approximate Aerosol Concentration Level for Each Species ($\mu g/m^3$)	MMA Found ($\mu g/m^3$) ± Std. Dev.	DMA Found ($\mu g/m^3$) ± Std. Dev.	p-APA Found ($\mu g/m^3$) ± Std. Dev.	Total Arsenic ($\mu g/m^3$) ± Std. Dev.
5	4.93 ± 0.68	5.08 ± 0.65	4.16 ± 0.6	14.17 ± 0.64
10	10.27 ± 0.62	10.58 ± 0.32	8.07 ± 0.91	28.92 ± 1.25
20	22.03 ± 1.48	21.62 ± 1.01	19.49 ± 1.32	63.14 ± 1.66

The accuracy of the analytical method was established by independent analysis of the three additional filters from each of the 5, 10 and 20 $\mu g/m^3$ generation runs using both NAA and XRF analyses. Because NAA and XRF analysis techniques provide only a total arsenic measurement, the IC-AAS speciation results obtained for MMA, DMA and p-APA were used to estimate the total amount of arsenic. Table X presents the total arsenic obtained by the three techniques. The accuracy ranged from 90-120% of the values obtained by NAA and XRF.

Table X. Comparison of total arsenic concentrations on generated filters determined by IC-AAS, NAA, and XRF analysis techniques.

Approximate Total Arsenic Aerosol Concentration Level ($\mu g/m^3$)	Total Arsenic Concentration Found ($\mu g/m^3$)		
	IC-AAS (N = 7)	NAA (N = 3)	XRF (N = 3)
15	14.17 0.64	12.04 0.54	11.87 0.76
30	28.92 1.25	28.28 3.67	30.52 0.93
60	63.14 1.66	56.25 5.90	62.02 3.53

Literature Cited

1. "Pesticide Use Report", California Department of Agriculture. 1972.

2. Soderquist, C.J.; Crosby, D.G.; Bowers, J.B.; Anal. Chem., 1947, 46, No. 1, 155.

3. Braman, R.S.; Johnson, D.L.; Foreback, C.C.; Ammons, J.M.; Bricker, J.L.; Anal. Chem., 1977, 49, 621.

4. Talmi, Y.; Bostic, D.T.; Anal. Chem., 1975, 47, 2145.

5. Crecelius, E.A.; Anal. Chem., 1978, 50, 826.

6. Yamamoto, M.; Soil Sci. Soc. Am. Proc., 1975, 39, 859.

7. Elton, R.K.; Geiger, W.E., Jr.; Anal. Chem., 1978, 50, 712.

8. Henry, F.T.; Thorpe, T.M.; Anal. Chem., 1980, 52, 80.

9. Iverson, D.G.; Anderson, M.A.; Holm, T.R.; Starforth, R.R.; Environ. Sci. and Technol., 1979, 13, 1492.

10. "Determination of Inorganic and Organic Arsenic Species by Ion Chromatography Using an Atomic Absorption Detection System", Ricci, G.; Colovos, G.; Hester, N.; Shepard, L; <u>Anal. Chem.</u>, accepted for publication 1980.

11. Colovos, G.; Eaton, W.S.; Ricci, G.R.; Shepard, L.S., Wang, H.; "Collaborative Testing of NIOSH Atomic Absorption Method", DHEW (NIOSH) Publication No. 79-144, 1978.

12. Vijan, P.N.; Wood, G.R.; <u>At. Absorpt. Newsl.</u> 1974, <u>13</u>, 33.

RECEIVED October 27, 1980.

Determination of Aromatic Diamines and Other Compounds in Hair Dyes Using Liquid Chromatography

KENNETH JOHANSSON and WALTER LINDBERG

Department of Analytical Chemistry, University of Umeå, S-901 87 Umeå, Sweden

CHRISTOFFER RAPPE and MARTIN NYGREN

Department of Organic Chemistry, University of Umeå, S-901 87 Umeå, Sweden

The use of permanent hair dyes or the occupational exposure to these cosmetic products has recently been discussed because of a possible link to adverse long term effects like genetic effects or carcinogenic effects (1, 2, 3). Epidemiological studies seem to indicate an overrepresentation of lung and breast cancer among occupationally exposed people (3). Previously, 2,4-diaminotoluene (TDA), a very common product in hair dyes, has been proven to be an animal carcinogen after oral administration and subcutaneous injection (4).

The hair dyes consist of quite complex mixtures of a variety of ingredients, and are usually sold as two-component systems. One component is an oxidation agent, usually hydrogen peroxide. The other component is a mixture of vegetable and animal fats, detergents and aromatics holding oxidizable amino- and hydroxyl groups, which produce coloured pigments on oxidation. From a toxicological point of view much interest has been focused on a number of aromatic diamines used in most permanent hair dyes.

Several methods have been used for the analysis of aromatic diamines, including thin-layer chromatography (TLC) (5, 6, 7, 8), paper chromatography (9) and gas chromatography (GC) (7). Diaminotoluene isomers have been separated by GC (10, 11,12). Thin-layer chromatography with fluorometric detection has been reported as a sensitive method for 2,6- and 2,4-TDA in urethane foams (13). Unger and Friedman (14) were successful in separating the 2,6- and 2,4-TDA by adsorption chromatography on a silica column. The lengthy reequilibration and difficulties in controlling the amount of water in the eluent are wellknown with this mode of chromatography. Turchetto *et al.* (15) had some success with the separation of various diamines employing polar bonded phases and gradient elution. From the published chromatograms it is seen that one of their systems showed very low efficiency. The necessity of reequilibrating the column after each gradient run makes this approach time-consuming.

0097-6156/81/0149-0401$05.00/0

Recently Choudhary (16) has described a GC-method for the determination of 1,4-diaminobenzene, 2,5-diaminotoluene and 2,4-diaminoanisole in hair dyes after ethyl acetate extraction.

In this work we describe a method based on modern LC which avoids an extraction step. This technique is rapid and selective and gives, with multiwavelength detection, good qualitative and quantitative information.

The standard compounds considered to be of interest and/or likely to be found were 2,4-, 2,5- and 2,6-diaminotoluene, 2,4-diaminoanisole, resorcinol (1,3-dihydroxybenzene), hydroquinone (1,4-dihydroxybenzene) and α-naphthol.

Experimental

Reagents and Chemicals. Ethanol was of spectrograde or absolute quality, 99.5 %, Kemetyl AB, Sweden, hexanes and methylene chloride were HPLC grade, Fisons, Loughborough, England. Polygosil 10 μm CN modified silica, pore diameter 60 Å, pore volume 0.75 ml/g, Batch 9031 and Polygosil 10 μm NO₂ silica, Batch 8121 was obtained from Machery-Nagel, Düren, German Federal Republic. 2,4-Diaminoanisole, 98 %, 2,6-diaminotoluene 97 %, and 2,4-diamino-toluene, 98-99 %, were obtained from EGA-Chemie, German Federal Republic, and 2,5-diaminotoluenesulphate, pract., Fluka AG, Switzerland, α-naphthol, pro analysi (p.a.), J.T. Baker, resorcinol, p.a., Merck AG, hydroquinone, lab. grade, Baker, acetone, p.a., Merck, and chloroform, p.a., Merck, were also used. Samples of hair dyes were kindly supplied by the hairdressers school in Umeå, Sweden. These dyes were commercially available only to hair-dressers and not directly to the public.

Preparations of the Free Amine. Since 2,5-TDA was supplied as the sulphate salt it was necessary to prepare the free amine. This was done by dissolving the diamine salt in sodium chloride saturated distilled water acidified with a few drops of sulphuric acid. Methylene chloride was added to this solution in a separating funnel. Sodium hydroxide was added to adjust the pH to 10-11 to transfer the amine to the organic phase, the methylene chloride now containing the diamine was washed once with saturated sodium chloride and then evaporated in a rotating evaporator. Red crystals of 2,5-TDA resulted from the evaporation. The purity was checked chromatographically.

Equipment. A LDC (Laboratory Data Control) Constametric III pump was used together with a Rheodyne 7120 20 μl loop injection valve and two LDC Spectromonitor III, variable wavelength UV-detectors. A Stanstead constant pressure pump was used for packing the columns.

Preparation of the Columns. Columns were manufactured from 316 SS steel tubes 1/4" x 4 mm diam. Swagelok 1/8" x 1/4" 316 SS steel reducers were modified to zero dead volume and were used with 2 μm Alltech steel frits as column ends. One 250 x 4 mm column was prepared with Polygosil CN silica as follows. A slurry of 1.55 g silica in 75 ml acetone was ultrasonified for 3 min and rapidly poured into the slurry container. The column was subsequently packed by pressurizing to 300 atm. About 300 ml hexanes was pumped through.

One 195 x 4 mm column was packed with Polygosil NO_2 silica by the same method except that the slurry medium was 40/60 acetone/chloroform and the pressurizing solvent was acetone.

The efficiency of both columns was tested by eluting 1,2-dinitrobenzene with a k´ of at least 3.

The same columns were used throughout this investigation. The dead volume was determined by injecting pentane and measuring by the crossing of base line. Hexanes/ethanol mixtures, used as the mobile phase, were prepared by weighing. The columns were run with a flow rate of 2 ml min^{-1}.

The CN-column had 8100 theoretical plates and a peak skew of 1.4 (back/front calculated at 10 % peak height), the NO_2-column had 2700 plates and peak skew 2.1.

All measurements were made at ambient temperature but the temperature variations in the eluent bottle were measured continuously and were constant within 3-4 °C.

Procedure. A sample of 0.5 - 1 g of hair dye was added with 50/50 v/v hexanes/ethanol until it dissolved, a clear solution was obtained and no solids remained. It was then diluted to known volume. Filtration was made through a Fluoropore 0.2 μm filter with a glass syringe and a swinny filter adaptor. This was necessary to avoid clogging of the column inlet frit.

After preparation, the samples as well as the standards were transferred and kept in brown bottles in a refrigerator. Analyses were carried out on the same day as the preparation in order to avoid problems with diamine oxidation.

Results and Discussion

The Chromatographic Systems. Bonded phases caused the breakthrough of modern liquid chromatography as an analytical method due to their stability, fast reequilibration and ease of operation. In the initial part of the work presented in this paper two different polar bonded phases with CN and NO_2 groups respectively were investigated. The parent silica was in both cases Polygosil 10 μm irregular particles. Hexanes and ethanol mixtures were chosen as the mobile phase due to ready availability and low toxicity. Figures 1 and 2 show k´-values measured for the standard substances on the two columns. The NO_2-column gives more retention than the CN-column for the same mobile phase composition. It is also interesting to note the differences in selectivity; it can be seen that NO_2-column cannot differentiate between the 1,3- and 1,4-OH positions in resorcinol and hydroquinone. On both columns there is sufficient selectivity to separate the 2,4-TDA from the 2,4-DAA.

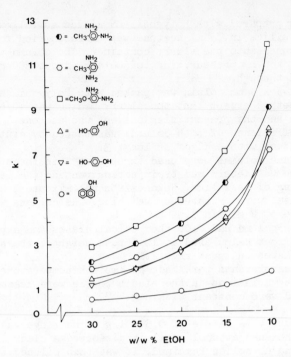

Figure 1.   Capacity ratio k' *as a function of the percentage w/w ethanol in hexanes
for the NO₂-column*

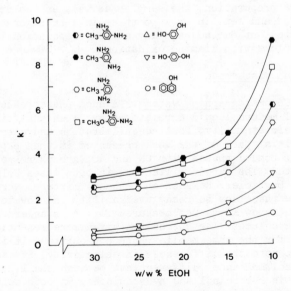

Figure 2.   Capacity ratio k' *as a function of the percentage w/w ethanol in hexanes
for the CN-column*

For the further work the CN-column was chosen with a mobile phase composition of 86.4/13.6 w/w % hexanes/ethanol, since this system could resolve all the standards without giving excessive retention times. Figure 3 shows the separation of the standard substances. More tailing is noted for the 2,5-TDA than with the column test procedure described earlier and it is likely that the chemical matching column vs eluent composition and sample can be improved. The properties of polar bonded phases have not yet been fully investigated and are beyond the aim of this study, but work is in progress in this laboratory which includes a more extensive investigation on the chromatography of diamines (17).

Detection and Linearity. Since the UV-detector measures $\partial c/\partial t$, quantitation by use of peak height measurements puts less stringent demands on the constancy of the flow rate. A fundamental requirement for adequate evaluation of data is that the chromatography is linear.

Investigation of the UV-spectra of the eluent and standards revealed that a wavelength of 235 nm should be the best compromise with respect to sensitivity and stability (noise level). Moreover, the sensitivity can be increased by using a wavelength in the lower UV-range. However, one should be aware of the fact, that when measuring on an edge of the UV-spectrum the broad spectral bandpass of these instruments might cause deviations from linearity already at low absorbance values.

Figure 4 shows calibration curves measured as the peak absorbance at 235 nm. This shows good linearity and high sensitivity for all substances. The detection limit is well below 10 ng for all substances and thus competes favourably with flame ionization detector - GC. The retention times were also measured at all concentrations since it was suspected, due to the apparent tailing, that the chromatography might be non-linear. This did not prove to be so.

Peak Identity Confirmation by Measurement of Absorbance Ratios. It is known that peak absorbance ratio measurements provide an alternative method of peak identity confirmation to running the sample on two different columns, fraction collection for MS and so on. If Beer's law is obeyed

$$A = c \cdot l \cdot \varepsilon_\lambda$$

one can obtain $\varepsilon_{\lambda_1}/\varepsilon_{\lambda_2}$ by measuring the peak height at two wavelengths and calculating the (peak height A.u.)$_{\lambda_1}$/(peak height A.u.)$_{\lambda_2}$ ratio. This should be constant for a certain molecule in a certain solvent. As variable wavelength detectors for LC have large spectral bandwidths one cannot determine true ε_λ-values. This means that the absorbance ratio measured will be specific for the bandwidth of the detector used.

When using one variable detector and performing successive

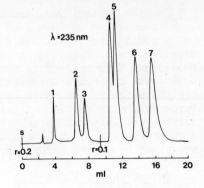

Figure 3.   Separation of 1, α-naphthol; 2, resorcinol; 3, hydroquinone; 4, 2,6-diaminotoluene; 5, 2,4-diaminotoluene; 6, 2,4-diaminoanisole; and 7, 2,5-diaminotoluene. Mobile phase, 86.4/13.6 w/w % hexanes/ethanol; linear flow velocity, 2.66 mm s⁻¹; pressure drop, 380 psi; λ = 235 nm; column, 250 × 4 mm Polygosil 10 μm CN.

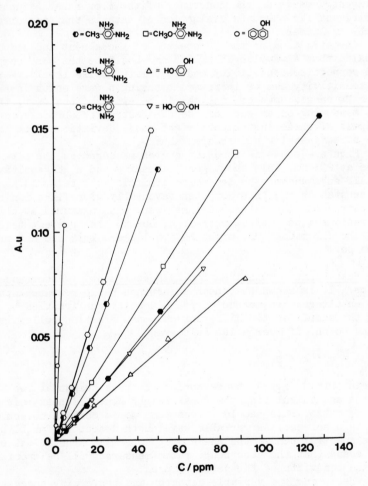

Figure 4.   Calibration curves as measured at 235 nm

runs at different wavelengths, the reproducibility of the wavelength setting will prove to be essential. To overcome this source of error we used two LDC Spectromonitor III detectors coupled in series to obtain two chromatograms at the same time on a two-channel recorder. According to the manufacturer, the special bandwidth of this detector is 8 nm and the flow cell of 10 μl volume. Negligible extra bandbroadening was caused by the extra detector due to the large volume half-width of the eluting peaks.

All the diamines have a UV-maximum in the 290-310 nm region and therefore 290 nm was the second detection wavelength chosen. The ratios $\varepsilon_{235}/\varepsilon_{290}$ were calculated for each standard substance and these values are found in Table I. The variation of the absorbance ratio with concentration is also included in the standard deviation shown in the table.

TABLE I. Peak Height Absorbance Ratios

Substance	$\varepsilon_{235}/\varepsilon_{290}$	S.d.	n
2,6-diaminotoluene	6.39	0.010	14
2,5-diaminotoluene	6.41	0.039	8
2,4-diaminotoluene	4.01	0.048	14
2,4-diaminoanisole	4.34	0.028	11
Resorcinol	2.37	0.037	14
Hydroquinone	0.88	0.022	14
α-Naphthol	7.3	0.31	14

When running the samples it was considered that if absorbance ratios and retention volumes coincide as statistically tested (t-test), the probability of positive identification is high. If retention volumes coincide but absorbance ratios do not, this indicates either that the peak is not pure or that it originates from a different substance. It is the opinion of the authors that a more extensive investigation of multiwavelength detection in LC is desirable. Manipulating the detection wavelength can provide a means of changing the resolution of the chromatogram when necessary instead of increasing k´ or trying to find proper selectivity.

Analytical Results. Figures 5 and 6 show the chromatograms obtained from samples (I) and (IV) as measured at 235 nm. The flow rate was 2 ml min^{-1} which means that one run takes about 10 minutes.

With the chromatographic system employed, nonpolar fats and detergents present in the dyes can be expected to have little or no retention and therefore should elute with the solvent front. Peak 1 in Figure 6 might partly consist of such substances. Most

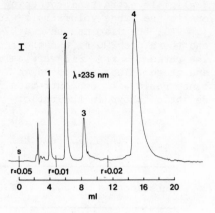

Figure 5.   Chromatogram of hair dye I: mobile phase, 86.4/13.6 w/w % hexanes/ethanol; flow rate, 2 mL min⁻¹ (2.66 mm s⁻¹); pressure drop, 380 psi; λ = 235 nm; column, 250 × 4 mm Polygosil 10 μm CN.

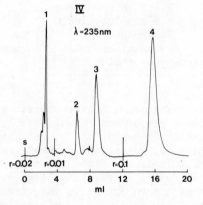

Figure 6.   Chromatogram of air dye IV: same chromatographic conditions as in Figure 5. Peaks: 1, unknowns eluting with solvent front; 2, resorcinol; 3, unknown; 4, 2,5-diaminotoluene.

TABLE II. Analytical Results

Hair dye analysed	Colour developing according to manufacturer	Substance found	A.u. peak height $\lambda = 235$	Std. dev.	n	Substance w/w % in hair dye
I	medium matt blonde	resorcinol	0.00727	0.000068	3	0.50
		2,5-TDA	0.0184	0.00042	3	1.17
II	dark ash blonde	resorcinol	0.0098	0.00048	3	1.66
		2,5-TDA	0.0330	0.0008	3	0.40
III	light brown	resorcinol	0.0195	0.00057	3	0.33
		2,5-TDA	0.0536		3	0.65
IV	mahogany	resorcinol	0.00258	0.000042	4	0.05
		2,5-TDA	0.070	0.0010	4	0.56
V	bordeaux red	α-naphthol (?)				
VI	black	recorcinol	0.0440	0.0016	5	3.52
		2,5-TDA	0.0155	0.0011	3	0.98
VII	medium brown	2,5-TDA	0.0095	0.00047	3	0.63

fats and detergents have little or no UV-absorbance at this wavelength and are thus not detected. In Figure 5 the chromatogram is rather clean in the solvent front and this might be explained as above. In both dyes (I) and (IV) resorcinol and 2,5- TDA could be identified, and these elute as peaks 2 and 4 in both chromatograms. Peak 3 in both Figure 5 and Figure 6 is the same substance since absorption ratios and retention volumes coincide. The unknown peak 1 in Figure 5 is not found in dye (IV) but it was found in several other dyes.

Table II shows the substances identified and amounts determined in the various dye samples. Resorcinol and 2,5-TDA were found in most samples. The identification of α-naphthol in sample (V) is somewhat uncertain since it was found that the absorbance ratios were not independent of the concentration for this substance. In general, few peaks could be seen on chromatographing all the samples and the same simple pattern showed up with several dyes.

As regards the quantitative determination it remains to be investigated to what extent inhomogeneity in the sample tubes influences the result. Batch to batch variations from the manufacture may also occur.

Conclusions

A simple and rapid method for the analysis of hair dyes has been developed. Owing to low wavelength UV-detection high sensitivity is achieved with detection limits well below 10 ng and at the same time fats and detergents in the dyes do not interfere with the detection. The simple dissolution procedure possible together with the chromatographic system used is an advantage in comparison with lengthy and uncertain extractions.

Identification is considerably improved by the use of multiwavelength detection and absorbance ratioing. It was possible to analyse some commercial hair dyes with the method described here. However, several substances were found that could not be identified with the present standard substances. Recent results in this laboratory make a possible improvement of the chromatography probable. Future work will consider this aspect and an increased number of standard substances (17) will be examined.

Acknowledgments

This work was supported by the Swedish Work Environment Fund.

Literature Cited

1. Ames, N.B.; Kammen, H.O.; Yamasaki, E. <u>Proc. Nat. Acad. Sci. U.S.</u>, 1975, <u>72</u>, 2423.
2. NIOSH – <u>Current Intelligence Bulletin</u> 19 US Department of Health Education and Welfare Public Health Service. Rockville MD, January, 1978.
3. <u>IARC Monographs on the Evaluation of the Carcinogenic Risk of Chemicals to Man</u>, Vol. 16, IARC, Lyon, 1977; p. 25.
4. <u>Ibid</u>., p. 83.
5. Kotteman, C. J. Assoc. Offic. Anal. Chem., 1966, <u>49</u>, 954.
6. Macke, G.F. <u>J. Chromatogr</u>., 1968, <u>36</u>, 537.
7. Glinsukon, T.; Benjamin, T.; Grantham, P.; Weisburger, E.; Roller, P. <u>Xenobiotica</u>, 1975, <u>5</u>, 475.
8. Glinsukon, T.; Benjamin, T.; Grantham, P.; Lewis, N.; Weisburger, E. <u>Biochem. Pharmacol</u>., 1976, <u>25</u>, 95.
9. Waring, R.H.; Pheasant, A.E. <u>Xenobiotica</u>, 1976, <u>6</u>, 257.
10. Boufford, C.E. <u>J. Gas Chromatogr</u>., 1968, <u>6</u>, 438.
11. Brydia, L.E.; Willeboordse, F. <u>Anal. Chem</u>., 1968, <u>40</u>, 110.
12. Willeboordse, F.; Quick, Q.; Bishop, E.T. <u>Anal. Chem</u>., 1968, <u>40</u>, 1455.
13. Guthrie, J.L.; McKinney, R.W. <u>Anal. Chem</u>., 1977, <u>49</u>, 1676.
14. Unger, P.D.; Friedman, N.A. <u>J. Chromatogr</u>., 1979, <u>174</u>, 379.
15. Turchetto, L.; Cuozzo, V.; Terracciano, M.; Papetti, P.; Percaccio, G.; Quercia, V. <u>Boll.Chim.Farm</u>., 1978, <u>117</u>, 475.
16. Choudhary, G. <u>J. Chromatogr</u>., 1980, <u>193</u>, 277.
17. Johansson, K.; Lindberg, W. to be published.

RECEIVED November 18, 1980.

High Performance Liquid Chromatographic Determination of Aromatic Amines in Body Fluids and Commercial Dyes

P. J. M. VANTULDER, C. C. HOWARD, and R. M. RIGGIN

Battelle–Columbus Laboratories, 505 King Avenue, Columbus, OH 43201

This paper focuses primarily on the application of High Performance Liquid Chromatography (HPLC) with electrochemical detection (EC) for the determination of aromatic amines. All of the studies described in this paper were conducted in the HPLC facility of Battelle Columbus Laboratories during the past two years. Much of the work is only cursory in nature (due to time and funding limitations) but as a whole the data presented are useful in projecting the range of applicability of HPLC/EC. The operational characteristics of HPLC/EC have been reviewed[1] and will only be described briefly in this paper. HPLC/EC, for the purposes of this paper, refer to the coupling of a HPLC with a thin-layer flow-through electrochemical cell, as shown in Figure 1. The working electrode (in most cases glassy carbon or carbon paste) is held at a fixed potential relative to the reference electrode. The counter or auxiliary electrode (in this case a platinum tube through which the column eluent flows) carries the required counter current. When an electrochemically active component elutes from the column it exchanges electrons with the working electrode surface, causing a current to flow. This current is converted to a voltage and recorded on a strip-chart recorder as a function of time, giving rise to a chromatogram. Since currents can be measured very sensitively using modern operational amplifiers the electrochemical detector can be extremely sensitive (e.g. 1-10 picograms injected) for electroactive compounds.

A typical electrochemical reaction for an aromatic amine (benzidine) is shown in Figure 2. Most of the HPLC/EC work to date has been conducted using carbon working electrodes and thus the oxidative mode of the detector has been exploited (e.g. for benzidine) to the greatest extent. However, platinum and mercury have been used successfully for electroreducible species such as metal ions[2] and parathion[3].

One of the most important features of HPLC/EC is the effect of electrode potential on detector response. Figure 3 illustrates this feature wherein the response of the detector as a

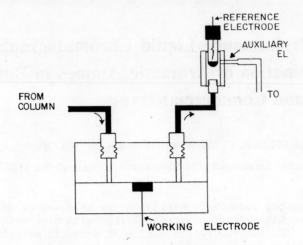

Figure 1.   Schematic of HPLC/EC cell

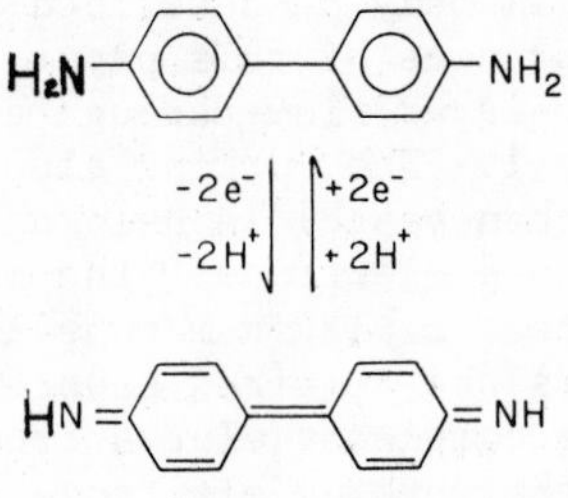

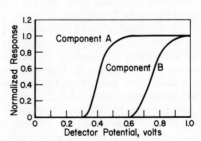

Figure 2.   Electrochemical reaction for
              benzidine

Figure 3.   Hypothetical HPLC/EC detector potential/response curve

function of working electrode potential for two components, A
and B, is plotted. Component A gives virtually no response
at 0.3 volts but its response increases as the potential is
increased and reaches a maximum at 0.6 volts. Component B,
however, still gives no response at 0.6 volts and requires a
potential of 0.9 volts to give maximum response. Therefore
one could analyze for A using a detector potential of 0.6 volts
with virtually no interference from B, even if the two components are not separated chromatographically.

Significance of Determining Aromatic Amines

Aromatic amines as a compound class account for many of the
known carcinogenic organic chemicals. Perhaps the most widely
publicized compounds are benzidine and 3,3'-dichlorobenzidine
(DCB)[4,5] which until recently were widely used in the manufacture of dyestuffs. Another widely publicized suspected carcinogenic aromatic amine is 4-methoxy-m-phenylenediamine
(MMPDA)[6] which is used as an ingredient in many permanent hair
dye formulations. 4,4'-methylenebis(2-chloroaniline) or
MBOCA[7] and 4,4-methylenedianiline (MDA) are widely used as
curing agents in polyurethane resins and are said to be carcinogenic.

Various halogenated and nitrated aniline derivatives are
used in the manufacture of dyestuffs and carbamate and urea
based pesticides. Many of these compounds are suspected to be
carcinogenic.

Since workers can be exposed to these compounds during their
manufacture and use, it is important to have reliable analytical
methods for determining the degree of exposure through body fluid
analysis. Additionally, since these compounds can be present
at significant levels in commercial products (derived from them)
it is desirable to monitor their level in such products (e.g.
dyestuffs) as well. Furthermore, many of the commercial
products can be metabolized to the original chemical (e.g.
benzidine based dyes can be metabolized to benzidine) making
it desirable to monitor the body fluids of workers exposed to
the commercial products.

Applicability Of HPLC/EC To The Determination Of Aromatic Amines

In order to examine the range of aromatic amines which
could possibly be determined by HPLC/EC, cyclic voltammograms
were run on a large number of aromatic amines. In this study
a Bioanalytical Systems DCV-3 potentiostat (designed for use
in cyclic voltammetry) and a Hewlett-Packard Model 7004B XY-
recorder were used. The working electrode was glassy carbon
(Model 9333 from Princeton Applied Research). A Ag/AgCl
reference electrode and platinum auxiliary electrode were used.
A scan range of 0-1.5 volts vs. Ag/AgCl at 250 mV/second was

used. The glassy carbon electrode was polished after each scan,
using fine grit alumina and emery cloth.

The compounds were prepared as 100 µg/mL solutions in 50%
acetonitrile/ 50% 0.2 M sodium acetate buffer, pH 4.7.

The data for the aromatic amines studied are given in
Table I. All data are reported as E_p (peak potentials) from
the cyclic voltammograms.

The lower the peak potential for oxidation of a compound,
the more specifically it can be determined (since a lower
detector voltage can be used). Therefore compounds such as
MMPDA, MDA, DCB, MBOCA, and benzidine (i.e. aromatic diamines)
can be more readily determined by HPLC/EC than are the aniline
derivatives. This does not imply that HPLC/EC is not useful
for the determination of anilines. In fact, a recent paper[8]
illustrates the usefulness of HPLC/EC for the determination
of halogenated anilines in urine.

The data in Table I shows that addition of a nitro group
to aniline increases the peak potential ∿200–250 mV whereas
addition of a halogen has very little affect. Therefore HPLC/EC
should be more useful for determining aniline and its halogen
substituted derivatives, than for nitro substituted anilines.

The data given in Table II support this conclusion.
Chromatographic parameters used to obtain the data in Table II
were as follows:

    Column – Lichrosorb RP-2, 5 µm particle diameter, 25 cm
             long x 0.46 cm ID
    Mobile Phase – 40% methanol/60% 0.2 M sodium acetate
                   buffer, pH 4.7
    Flow Rate – 1 mL/min
    Detectors
      ● UV–LDC UV III @ 254 nm, 0.016 AUFS
      ● EC–Bioanalytical Systems LC-2A, @ 1.0 volt vs.
            Ag/AgCl, equipped with a glassy carbon working
            electrode.

The detection limit for each compound is defined as the
quantity which gives a signal to noise ratio of 5.

These data illustrate that at a detector potential of 1 V
sensitivity for halogenated anilines is approximately 100–1000 X
greater than for their nitro-substituted counterparts. The
sensitivity for nitro-substituted anilines can be increased by
raising the detector potential, but only at the expense of
detector selectivity. It has been our experience that detector
voltages greater than 1.1–1.2 V are of little value due to a high
background signal and non-selectivity.

In summary, the data given in this section show that HPLC/EC
technique should be useful for the determination of aromatic
diamines and halogenated anilines but is of little value for
determining nitro-substituted anilines.

TABLE I. PEAK POTENTIALS FOR AROMATIC AMINES

Compounds	Peak Potential (mV)
4,4'-Methylenedianiline (MDA)	830
4,4'-Methylenebis(2-chloroaniline) (MBOCA)	870
Benzidine	550
3,3'-Dichlorobenzidine (DCB)	650
4-Methoxy-m-phenylenediamine (MMPDA)	370
Aniline	950
2-Chloroaniline	1000
3-Chloroaniline	1030
4-Chloroaniline	900
3,4-Dichloroaniline	980
2,4,5-Trichloroaniline	1000
4-Bromoaniline	900
2-Nitroaniline	1200
3-Nitroaniline	1080
4-Nitroaniline	1180
2,4-Dinitroaniline	1400
2-Chloro-4-nitroaniline	1150
4-Chloro-2-nitroaniline	1180
2,6-Dichloro-4-nitroaniline	1380
2-Chloro-4,6-dinitroaniline	1380
2,6-Dibromo-4-nitroaniline	1180
2-Bromo-4,6-dinitroaniline	1330
2-Bromo-6-chloro-4-nitroaniline	1150

TABLE II.   ESTIMATED HPLC DETECTION LIMITS FOR
            SELECTED ANILINES

Compound	Retention Time (min)	Est: Detection Limits (Nanograms Injected)	
		UV	EC
Aniline	5.2	2	0.1
4-Nitroaniline	7.0	2	10
3-Nitroaniline	7.0	0.5	5
2,4-Dinitroaniline	8.9	0.3	100
2-Chloroaniline	9.0	5	0.2
4-Chloroaniline	9.0	2	0.5
2-Nitroaniline	9.1	4	100
3-Chloroaniline	9.5	4	0.5
4-Bromoaniline	10.0	4	0.5
2-Chloro-4,6-dinitroaniline	10.8	5	100
2-Chloro-4-nitroaniline	11.5	5	100
2-Bromo-4,6-dinitroaniline	12.0	5	100
4-Chloro-2-nitroaniline	16.2	4	100
3,4-Dichloroaniline	16.5	4	2
2,6-Dichloro-4-nitroaniline	17.8	6	100
2,6-Dibromo-4-nitroaniline	23.0	8	100
2,4,5-Trichloroaniline	38.0	6	5

Determination Of Benzidine And DCB In Commercial Dyes

Two experiments were conducted to study the utility of HPLC in determining aromatic amines in commercial dyes. In the first experiment a benzidine based dye, Direct Blue 6, and a commercial hair dye formulation containing Direct Blue 6 were analyzed for benzidine. In the second experiment a DCB based pigment (diarylide yellow) was analyzed for residual DCB.

Determination of Benzidine in Hair Dyes. Experiments were conducted to determine whether or not benzidine could be detected in benzidine based dyes and/or hair dye products containing benzidine based dyes. The benzidine based dye Direct Blue 6 was obtained commercially as were two shades of a particular hair dye product. One shade, "Lucky Copper", contained Direct Blue 6 whereas another shade, "Silver Lining", did not contain any benzidine based dyes.

The hair dye formulations were extracted using the following procedure:

1. Adjust 100 µL of the hair dye formula to pH 7 with phosphate buffer and add 200 mL of distilled water.
2. Extract with 2 x 100 mL of chloroform. Centrifuge for 10 minutes at approximately 1000 x g to separate the chloroform.
3. Wash the chloroform layers with 40 mL of distilled water.
4. Add 40 mL of methanol to the chloroform and concentrate to 1 mL.
5. Add 10 mL of chloroform, wash with 3 mL of water and then back extract with 2 mL of 0.01 $\underline{M}$ HCl.
6. Assay extract for benzidine by HPLC with electrochemical detection.

The HPLC conditions used were as follows:

Column – Lichrosorb RP-2, 5 µm particle diameter, 25 cm long x 0.46 cm ID stainless steel
Mobile Phase – 50% acetonitrile/50% 0.2 $\underline{M}$ sodium acetate buffer, pH 4.7
Flow Rate – 0.8 mL/min
Injection Volume – 50 µL
Detector – Electrochemical (LC-2A Bioanalytical Systems) equipped with thin-layer glassy carbon electrode operated at 0.9 V.

The Direct Blue 6 was dissolved in pH 7 phosphate buffer (0.5 gram in 20 mL) and extracted in a similar manner.

Figure 4 shows the results for the Direct Blue 6 when it was extracted unspiked and after spiking with 1 ppm (500 ng per 0.5 gram of dye). A peak for benzidine corresponding to approximately 400 ppb was noted in the unspiked extracted (a process blank showed no response). Figure 5 shows the results for the hair dye formulation containing Direct Blue 6 (Lucky Copper) and Figure 6 shows the results for a second shade of the same product, not containing any benzidine based dyes. The Lucky Copper exhibits a response corresponding to approximately 2 ppb benzidine (Figure 5) whereas Silver Lining shows no such response (benzidine less than 0.2 ppb). Recoveries ranged from 50-80% for the samples spiked before processing.

The work illustrates the ability of HPLC/EC to detect trace quantities of benzidine in hair dye products, although positive results should be verified by GC-MS or other spectroscopic means to be certain of the component identity. Unfortunately, the sensitivities of such techniques are generally much poorer than HPLC/EC, thus making absolute confirmation impractical.

<u>Determination Of Residual DCB In Diarylide Yellow</u>. Diarylide yellow is a widely used pigment derived from DCB. The purpose of this study was to determine residual DCB in a lot of the commercial pigment being used in animal feeding experiments.

The chromatographic conditions used were the same as those given in the previous section. The extraction procedure was as follows:

Five grams of the dye was Soxhlet-extracted with 200 mL of methylene chloride ($MeCl_2$) for 48 hours. The $MeCl_2$ extract and 100 mL of $1\underline{M}$ H_2SO_4 were combined in a beaker and stirred for 10 minutes. The H_2SO_4 layer was decanted and the $MeCl_2$ fraction extracted once again with 100 mL of $1\underline{M}$ H_2SO_4. Acid fractions were combined and neutralized to pH 7 with 0.4 $\underline{M}$ Na_3PO_4 and $1\underline{M}$ NaOH in an ice bath. The neutralized fraction was extracted 2 times with 50 mL $CHCl_3$ (preserved with ethanol) and washed with 15 mL of H_2O. Fifteen milliliters of methanol were added to the extract and it was concentrated to 5 mL on a rotary evaporatory (room temperature), then to 0.2 mL with a vortex evaporator (40° C).

The extract was then diluted with 0.5 mL with 0.2 $\underline{M}$ sodium acetate buffer, pH 4.7 and analyzed by HPLC. Chromatographic conditions were the same as for the determination of benzidine in hair dye formulations. For the particular lot of diarylide yellow studied ~46 µg/kg of DCB was found. In an attempt to confirm the identity of the chromatographic peak, its response as well as the response for the authentic DCB standard was determined at several different electrode potentials. These data, shown in Figure 7, illustrate the ability of HPLC/EC to yield qualitative as well as quantitative information for unknown components.

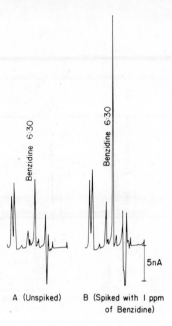

A (Unspiked)   B (Spiked with I ppm
                   of Benzidine)

Figure 4.   Chromatogram for Direct Blue 6 extract

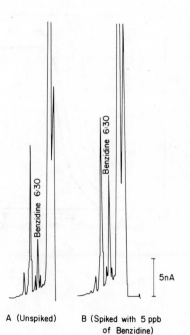

A (Unspiked)   B (Spiked with 5 ppb
                   of Benzidine)

Figure 5.   Chromatogram for extract of commercial hair dye formulation containing Direct Blue 6

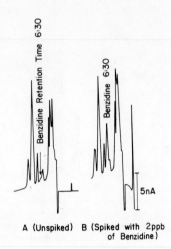

Figure 6. Chromatogram for extract of commercial hair dye formulation not containing benzidine-based dyes

A (Unspiked)   B (Spiked with 2 ppb of Benzidine)

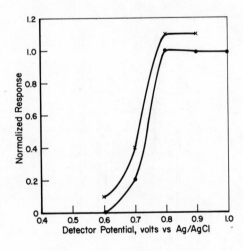

Figure 7. Detector potential/response curve for (●) DCB and (✕) component extracted from diarylide yellow (offset 1 cm for clarity)

Determination Of Aromatic Amines In Urine

In this section we will discuss preliminary work done at Battelle Columbus Laboratories wherein HPLC/EC was used to determine MMPDA, DCB, and MBOCA in urine. Other workers have demonstrated its usefulness for determining halogenated anilines[8] and benzidine[9] in urine.

Determination of DCB in Rat Urine. In this study a method for the determination of DCB in rat urine was developed. The sample extraction procedure employed was as follows:

1. Place 2 mL of urine in a 15 mL centrifuge tube.
2. Adjust the urine to pH 7 with 0.1 M HCl or 0.1 M NaOH.
3. Extract 3 times with 2 mL of benzene, centrifuging between each extraction.
4. Combine benzene fractions and wash with 1 mL of distilled water (discard aqueous layer).
5. Add 1 mL of methanol to the benzene extract.
6. Concentracte to 0.5 mL on vortex evaporator (40° C).
7. Add 1 mL methanol and concentrate to 0.2 mL.
8. Dilute to 0.5 mL with 0.2 M sodium acetate buffer, pH 4.7, mix well, and analyze by HPLC.

The HPLC conditions used were the same as described for the commercial hair dye experiments.

Figure 8 shows typical chromatograms for spiked and unspiked rat urine. Recovery was found to be 50, 55, and 35% for three replicate samples spiked at the 2 µg/L level, which we have found to be typical recovery values for trace levels of DCB in aqueous media (e.g. industrial wastewater).

Determination Of MMPDA In Human Urine. MMPDA is widely used in permanent hair dye formulations and has been shown to be carcinogenic in rodents [6]. A question exists as whether or not MMPDA can be absorbed through the scalp to enter the blood stream of persons using the dye. In order to answer this question a sensitive, reliable analytical procedure for MMPDA in urine is needed. For this reason we decided to investigate the usefulness of HPLC/EC in this problem.

The extraction procedure developed was as follows:

1. Adjust 2 mL of urine to pH 7 with 0.1 M HCl or 0.1 M NaOH.
2. Extract twice with 4 mL of MeCl$_2$ in a 15 mL screw capped centrifuge tube.
3. Combine the organic extracts and wash with 1 mL of 0.01 M ammonium citrate, pH 6.9. Discard aqueous layer.
4. Extract organic layer with 1 mL of 0.05 M HCl. Analyze

acid extract by HPLC.
The chromatographic conditions were as follows:

Column – Dupont Zorbax ODS, 5 μm particle diameter, 25 cm
         long x 0.46 cm ID
Mobile Phase – 30% methanol/70% 0.5 $\underline{M}$ ammonium citrate
                    buffer, pH 5
Flow Rate – 0.7 mL/min
Detector Voltage – 0.9 V.

Figure 9 shows a chromatogram for human urine spiked at 50 μg/L.
Recovery was found to be 58% at this spike level.

Determination Of MBOCA in Human Urine. MBOCA is commercially important as a curing agent for polyurethanes and
epoxy resin systems. Since MBOCA was found to be carcinogenic
in animals and is a suspected human carcinogen[7], it is
important to have a reliable method available for the determination of MBOCA in the urine of those workers who are
potentially exposed to this compound. In a NIOSH publication[7]
a GC method was described for the determination of MBOCA in
urine. However, since HPLC does not require derivatization and
a lower detection limit was expected, the GC method was modified
to be performed by HPLC/EC. In order to be able to compare
both methods, we used the same extraction procedure. The
extraction and sample preparation procedures are as follows:

1. Stabilize urine by adding 2 mL of an aqueous 30%
    citic acid solution to the bottle before collection.
2. Place 50 mL of stablized urine in a 125 mL separatory
    funnel and spike the urine with 1 mL of MBOCA standard
    solution in methanol. Swirl to mix and let stand
    5 minutes.
3. Add 10 mL of ethanol, shake to mix and let stand
    5 minutes.
4. Add 5 mL of an aqueous 10% sodium bicarbonate solution,
    shake to mix (vent CO_2 through stopcock) and test pH
    with testpaper. If not alkaline (pH 7.5–8) add
    stepwise additional 1 mL portions until alkaline.
5. Add 50 mL of diethyl ether and shake uninterruptedly
    for two minutes, venting periodically.
6. Let solutions stand for 5 minutes, drain off the lower
    urine layer and discard.
7. Add 5 mL of the 10% bicarbonate solution to the
    separatory funnel and shake for 10 seconds.
8. After 5 minutes drain off the bottom layer.
9. Place the ether extract in a 100 mL round bottom flask
    and evaporate to approximately 5 mL on a rotary
    evaporator at ambient temperature.

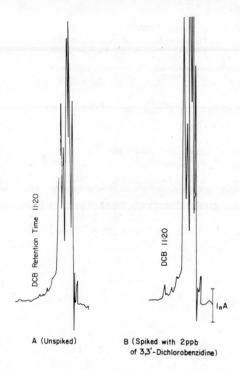

A (Unspiked)   B (Spiked with 2ppb of 3,3'-Dichlorobenzidine)

Figure 8.   Determination of DCB in rat urine

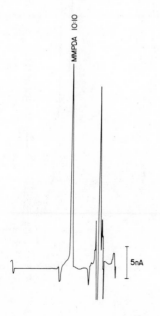

Figure 9.   Determination of MMPDA in spiked human urine

10. Transfer the 5 mL of extract quantitatively to a 15 mL conical screw-capped centrifuge tube. Rinse the flask with two portions of 2 mL of ether and combine with the original extract.
11. Evaporate the extract to 0.2 mL using a gentle stream of nitrogen.
12. Add mobile phase (see below) to 1 mL and analyze by HPLC.

The chromatographic conditions were the same as for the determination of benzidine in hair dye formulations, except for the flow rate, which was 1 mL/min. Figure 10 shows typical chromatograms for spiked and unspiked urine.

Preliminary results indicate that the detection limit for the HPLC/EC method will probably be in the order of 1-10 ppb. which is substantially lower than the 40 ppb detection limit[7] for the GC method.

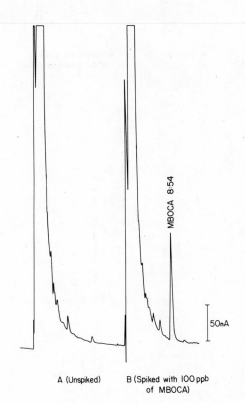

Figure 10.   Determination of MBOCA in
              spiked human urine

A (Unspiked)   B (Spiked with 100 ppb
                       of MBOCA)

LITERATURE CITED

1. Kissinger, P. T., "Amperometric and Coulometric Detectors for High-Performance Liquid Chromatography", Anal. Chem., 1977, 49, 447A-456A.

2. Johnson, D. C. and Larochelle, J., "Forced-Flow Liquid Chromatography with a Courlometric Detector", Talanta, 1973, 20, 959-971. Pergamon Press, Great Britian.

3. Stillman, R. and Ma, T. S., "Application of High-Speed Liquid Chromatography to Organic Microanalysis II. Separation and Polarographic Detection of Pesticides, Vitamins, and Analgesics", Mikrochimica Acta [Wien], 1974, 641-648.

4. Haley, T. J., "Benzidine Revisited: A Review of the Literature and Problems Associated with the Use of Benzidine and its Congeners", Clin. Toxicol., 1975, 8 (1), 13-42.

5. Price, J. M., "Benign and Malignant Tumors in the Urinary Bladder"; Medical Examination Publishing Co., Inc. Flushing, New York, 1971, p. 264.

6. "2,4-Diaminoanisole (4-Methoxy-m-Phenylenediamine) in Hair and Fur Dyes", DHEW (NIOSH) Publication No. 78-111, 1978.

7. "Special Hazard Review with Control Recommendations for 4,4'-Methylenebis(2-chloroaniline)", DHEW (NIOSH) Publication No. 70-188, 1978.

8. Lores, E. M., Bristol, D. W., and Moseman, R. F., "Determination of Halogenated Anilines and Related Compounds by HPLC with Electrochemical and UV Detection", J. Chromatogr. Sci., 1978, 16, 358-362.

9. Rice, R. L., and Kissinger, P. T., "Determination of Benzidine and its Acetylated Metabolites in Urine by Liquid Chromatography", J. Anal. Toxicol., 1979, 3, 64-66.

RECEIVED October 27, 1980.

QUALITY ASSURANCE

An Evaluation of Statistical Schemes for Air Sampling

S. M. RAPPAPORT, S. SELVIN, R. C. SPEAR, and C. KEIL

University of California, School of Public Health, Berkeley, CA 94720

The evaluation of hazards posed to human health by toxic airborne chemicals is one of the common tasks employed in industrial hygiene. This process requires the collection of air samples to estimate air concentrations of specific substances inhaled by workers which can then be compared with standards and guides of acceptable exposure. Thus air sampling directly influences the formulation of important decisions. If air samples underestimate exposures, the consequence may be death or occupational disease. Conversely, overestimating exposures may result in the institution of unnecessary controls. Since either form of error is undesirable, it is fundamentally important that air sampling accurately define the extent of hazard. This requires that air samples be collected according to scientific, unbiased schemes for estimating exposures to toxic airborne chemicals.

The body of knowledge surrounding this subject has grown tremendously during the last two decades. Published reports have identified two conceptual areas which are particularly relevant to the problem, namely, the variability of airborne exposures and the health hazards posed by exposures. These topics will be developed first as the framework for analysis of air-sampling schemes.

Then, the air-sampling scheme will be analyzed which was developed by governmental agencies in the United States to determine compliance with regulatory standards. It will be shown that this specialized scheme has limited scope and is poorly suited for use as a decision-making tool. Finally, air sampling schemes will be considered that offer an alternative for the evaluation of acute exposures.

Basic Concepts

Variability of Airborne Exposures. Air concentrations of chemicals released into occupational settings vary considerably in both time and space. Oldham (1) was apparently the first

0097-6156/81/0149-0431$06.25/0

to recognize that the distributions of levels of airborne dust
in coal mines were approximately lognormal, i.e., logarithms
of individual dust concentrations were more or less normally
distributed. Subsequently, other investigators (2-12) confirmed
Oldham's observation and showed that the lognormal distribution
was an excellent description of the distributions of sets of
air concentrations of airborne chemicals (gases, vapors and
aerosols) in the workplace. Esmen and Hammad (10) presented
theoretical proof that justifies the lognormal distribution
as a model for such data.

Lognormal distributions of air concentrations have been
observed for the following types of data: for workers' short-
term exposures during the day (1,2,5,7,9,11,12), for workers'
average daily exposures (4,8,10) and for air concentrations
at fixed locations in the work area (3,4,6,7,8,11). The major
sources of the variability are the intraday and interday
fluctuations in exposure resulting from perturbations in the
generation and dispersion of airborne contaminants as well as
from the range of tasks employed by a worker. By comparison,
the variability resulting from random errors in measurement
of air concentrations is quite small. These normally distributed
errors are usually within 10% of the magnitude of the measurement
whereas the lognormally distributed environmental fluctuations
often cover a range of between 10 and 30 times (ratio of the
the standard deviation to the mean).

Lognormal distributions are characterized by two parameters,
the geometric mean or median (μ_g) and the geometric standard
deviation (σ_g), which are defined by the following equations:

$$\ln \mu_g = \frac{1}{n} \sum_{i=1}^{n} \ln x_i$$

$$\ln \sigma_g = \left[\frac{1}{n-1} \sum_{i=1}^{n} (\ln x_i - \ln \mu_g)^2 \right]^{\frac{1}{2}}$$

Estimates of these parameters, based upon small samples
of data, are designated $\bar{x}_g$ and s_g respectively. The corresponding
parameters and their estimates of the normal distribution are
the arithmetic mean, μ and $\bar{x}$, and the standard deviation, σ
and s.

Consider a hypothetical situation where the air concentration
of a chemical, inhaled by a worker, was known at every instant
over an entire workday. Figure 1 depicts the exposure by
displaying air concentration versus time. The integrated 8-
hour time-weighted average (TWA) air concentration for this
exposure was 43.4 mg/m^3. The combined errors stemming from

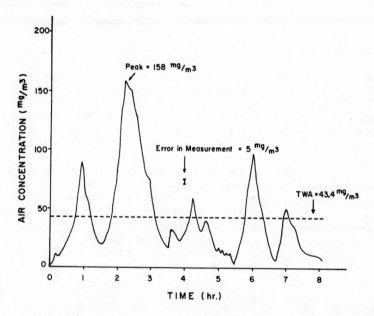

Figure 1. A record of one worker's hypothetical exposure to an airborne chemical on a given day

air sampling and analysis corresponded to an air concentration
of 5 mg/m^3. As shown in Figure 1, these errors in measurement
contributed relatively little to the overall variability in
air concentration on that day.

A common technique for displaying lognormal data
involves plotting the cumulative distribution
on log-probability paper. If the points fall along a straight
line, the data are assumed to follow, at least approximately,
a lognormal distribution. For instance, in Figure 2 the 48
instantaneous air concentrations from each 10 min interval in
Figure 1 were ranked from lowest (6 mg/m^3) to highest (158
mg/m^3) and each observation was plotted versus cumulative
probability [(rank − 0.4) ÷ 48] (13). The data hardly
deviated from a straight line, indicating that the air
concentrations could be described as approximately lognormally
distributed.

Consideration of the Hazard. A successful air-sampling
scheme must account for the nature of the hazard posed by a
chemical as well as the variability of exposures. Specifically,
the temporal relationship between the exposure and the potential
health effect must be assessed. Acute effects are characterized
by brief exposures to relatively large air concentrations of
a contaminant followed by the sudden onset of responses which
are usually of serious consequence. The chemical asphyxiant,
hydrogen cyanide, is a good example of an acute toxin since
the effect, blockage of the cytrochrome oxidase system, proceeds
within minutes of exposures. Other acute hazards which can
produce death from plausible short-term exposures include
hydrogen sulfide (respiratory paralysis), nitrogen dioxide and
phosgene (pulmonary edema) and arsine (intravascular hemolysis).
Respiratory irritants such as acid gases and ammonia are also
classified as acute hazards though the consequences of occasional
exposures above the threshold of irritation are less severe
than those resulting from exposures to lethal toxins.

When devising a scheme for evaluating acute hazards, it
is important that the intraday variability of exposures be
considered. Air sampling must be oriented towards determining
those brief periods of maximum exposure during the workday.
If the exposure illustrated in Figure 1 involved an acute hazard,
it would be critical to identify the peak air concentration
of 158 mg/m^3 occuring at 2-hr into the day.

Chronic effects arise from the cumulative dose of a chemical
which has resulted frcm integrated exposures over months or
years. The best examples of airborne chemicals which produce
only chronic effects are the fibrogenic dusts. However, most
systemic poisons also produce chronic effects although some
produce acute effects as well if inhaled in sufficient quantities
during short intervals. For example, many halogenated solvents
damage the kidney or the liver after long-term inhalation of
moderate air concentrations but produce anesthesia or narcosis

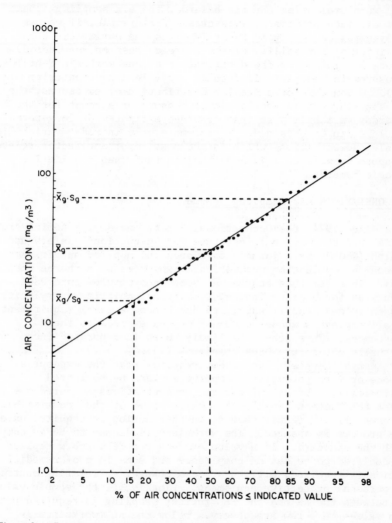

Figure 2. Cumulative distribution of instantaneous air concentrations from Figure 1 plotted on log–probability paper: $\overline{x}_g$ = geometric mean, s_g = geometric standard deviation

during brief periods of high exposure. In such cases the
exposure levels required to trigger acute effects are usually
orders of magnitude greater than those required for chronic
effects.

    When evaluating chronic hazards, the air sampling scheme
must estimate integrated exposures. For example, if we use
daily-exposure averages (8-hr TWAs) as the estimates, then
the interday variability of the averages must be considered.
Figure 3 illustrates the distribution of one worker's 8-hr TWA
exposures to inorganic lead in an alkyl lead manufacturing plant
(14). Given this or a similar distribution of exposures, the
air-sampling scheme should either measure or account for the
interday variability so that accurate estimates of long-term
exposure can be obtained. Furthermore, the intraday variability
of air concentrations which is the most important factor related
to acute hazards is relatively unimportant when considering
chronic hazards.

The OSHA/NIOSH Air-Sampling Scheme

    Since 1974, the Occupational Safety and Health Administration
(OSHA) and the National Institute for Occupational Safety and
Health (NIOSH) have jointly developed and applied an air sampling
scheme for evaluating occupational exposures to airborne chemicals
(15). Thus far this scheme has been incorporated into OSHA
standards for 5 contaminants, i.e., vinyl chloride, acrylonitrile,
benzene, lead, and arsenic, and has become a legal requirement
of all employers whose personnel may be exposed to these
substances. This scheme is likely to be incorporated into
virtually all future OSHA standards (15).

    Action Levels. The scheme requires that the exposures
of one or more potentially-exposed workers be monitored
periodically. If all of the air concentrations measured are
below the "action level" (AL), which is ∿½ of the "permissible
exposure limit" (PEL), then no further action is required unless
the process is changed. The workplace is deemed to be in compliance
with the standard. If a value exceeds the PEL, the workplace
is declared to be out of compliance and some form of remedial
action is required, e.g., a process change, engineering controls
or personal protective equipment. Finally, if the sample value
is between the AL and the PEL further sampling is required until
two values in a row are observed below the AL (workplace in
compliance) or one value is observed above the PEL (workplace
out of compliance).

    Table I summarizes the values of the AL and the PEL listed
by OSHA for the five chemicals previously mentioned. In each
standard, the scheme is essentially the same as described,
although the required frequency of monitoring varies. Significantly,
all of these chemicals are chronic-exposure hazards and decisions
are based upon PELs which are 8-hr TWA values. It is implied

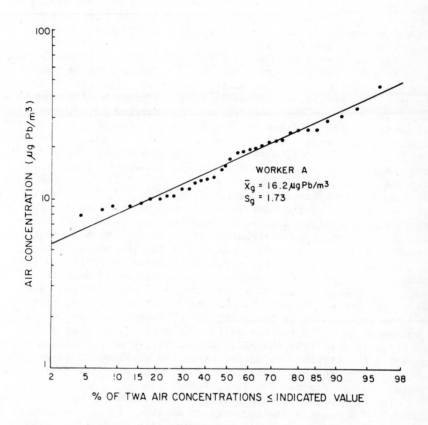

Figure 3. Cumulative distribution of one worker's 8-h exposures to inorganic lead in an alkyl lead manufacturing plant (14): TWA, 8-h time-weighted-average air concentration

by Leidel et al (15), though never explicitly stated, that this
scheme was developed solely for chronic hazards. Therefore,
it remains an open question as to whether the scheme would ever
be applied to acute hazards. Incongruously, three of the five
standards (vinyl chloride, acrylonitrile and benzene) also
contain ceiling limits which are air concentrations not to be
exceeded for 15 min intervals. However, these ceiling limits
are certainly unrelated to acute exposures, they play no part
in the air-sampling scheme and one can only speculate as to
the basis for their inclusion.

Performance of the OSHA/NIOSH Scheme. Rock (16) and Tuggle
(17) investigated the OSHA/NIOSH sampling scheme and reported
the probabilities of making various decisions regarding specific
lognormal distributions of sample values. Using a different
approach, shown in the appendix, we have derived equations which
allow these probabilities to be calculated. Assuming that for
a particular standard the PEL was 10 and the AL was   5 (e.g.
for arsenic), Table II shows the probability of declaring
the workplace as being in compliance for a series of lognormal
distributions (μ_g = 2,   5 and 10 with σ_g = 1.5, 2.0 and 3.5).
Two important observations emerge from Table II. First, when
the variability of the environment is minimal with σ_g = 1.5
the scheme is fairly accurate. That is, the probability of
declaring a complying environment to be noncomplying is small
(< .01 for μ_g = 2 and .14 for μ_g = 5) and the probability
of declaring a noncomplying environment to be complying is also
small (.04 for μ_g = 10).   However, when the environmental
variability increases, the probabilities of making both types
of errors also increase. For instance when μ_g = 5 and σ_g =
2.0 the probability of declaring the environment to be out of
compliance is 0.33; yet the arithmetic mean, μ, is only 6.4
or slightly over half the PEL. Conversely, when μ_g = 10 and
σ_g = 3.5 the probability of declaring the environment to be
in compliance is .31; yet μ is 21.9 or over twice the PEL.
(Note: $\mu = \exp[\ln(\mu_g) + \frac{1}{2}(\ln \sigma_g)^2]$)

The reason for the relatively poor performance of the
OSHA/NIOSH scheme in environments of moderate to high variability
is that it fails to consider the variability associated with
the exposures encountered. The scheme is biased towards declaring
the environment to be out of compliance as long as the initial
measurement is above the AL. In environments of slight variability
air concentrations are relatively uniform, i.e., the first
measurement will usually be less than the AL in complying
environments and will usually be greater than the AL in
noncomplying environments. Thus, the scheme works well in
these situations. However, in environments with moderate to
great variability, higher probabilities exist that the initial
sample will be above the AL even though the environment is
complying, or will be below the AL even though the environment
is noncomplying. Raising or lowering the AL would decrease
the likelihood of one type of error at the expense of the other.

Table I

OSHA Standards which Incorporate the OSHA/NIOSH
Air-Sampling Scheme

Chemical	Permissible Exposure Limit (8-hr TWA)	Action Level	Ceiling Limit	Standard*
Vinyl chloride	1.0 ppm	0.5 ppm	5.0 ppm	1910.1017
Acrylonitrile	2.0 ppm	1.0 ppm	10 ppm	1910.1045
Benzene	1.0 ppm	0.5 ppm	5.0 ppm	1910.1028
Lead	50 $\mu g/m^3$	30 $\mu g/m^3$	--	1910.1025
Arsenic	10 $\mu g/m^3$	5 $\mu g/m^3$	--	1910.1018

*Title 29 CFR Part 1910 Subpart Z.

Table II

Probabilities that the OSHA/NIOSH Scheme Will Declare
Various Lognormal Distributions of Sample Values to Be
in Compliance with the Standard
(PEL = 10, AL = 5)

σ_g	μ_g	μ*	P(Compliance)
1.5	2	2.2	>.99
	5	5.4	.86
	10	11	.04
2.0	2	2.5	.98
	5	6.4	.67
	10	13	.17
3.5	2	4.4	.87
	5	11	.58
	10	22	.31

$$*\mu = \exp[\ln(\mu_g) + \tfrac{1}{2}(\ln\sigma_g)^2]$$

Perhaps the ultimate failing of the OSHA/NIOSH scheme is that it bases important decisions on relatively small amounts of data. Intuitively, such a scheme would lead to incorrect conclusions in many cases. Table III gives the number of samples expected to be required for making decisions in various environments (calculated from the relationship derived in the appendix). As in the previous example the PEL is 10 and the AL is 5. In virtually all cases the number of samples is two or less. With such small sample sizes accurate prediction of the long-term rates of exposure is impossible without additional information or assumptions. Stated in slightly different terms, the interday variability of 8-hr TWA values cannot be measured or controlled for with information based strictly on such small sample sizes.

Application of the OSHA/NIOSH Scheme to Acute Hazards. Although the OSHA/NIOSH air sampling scheme has not been applied to acute hazards, it would be useful to illustrate the implications of doing so. Consider exposures to the acute toxin, hydrogen cyanide (HCN). Let us assume that the current exposure limit recommended by NIOSH of 5.0 mg/m^3 ([18]) is used as the PEL and that the AL is one half of this value (2.5 mg/m^3). The air concentration of HCN required to produce death in man in 10 min is $\sim$100 mg/m^3 ([18],[19]) or 20 times the PEL.

Table IV shows the probabilities of the OSHA/NIOSH scheme declaring lognormal distributions of 10-min exposures, in which the variability is great (σ_g = 3.5), to be in compliance. Table IV also lists the probabilities that exposures from the distributions would exceed 100 mg/m^3, that is, the probabilities that fatal exposures would occur, and the number of such exposures expected in a 21-day working month (1008 intervals). When μ_g = 5 mg/m^3 and 7.5 mg/m^3 the probabilities of declaring the environment "safe" are .31 and .20 respectively. Yet the environments are unquestionably hazardous since 8 or 20 fatal exposures could be expected in a working month. Even when μ_g is 2 mg/m^3, one fatal exposure per month would be expected. In this case the OSHA/NIOSH scheme would declare the environment "safe" 64% of the time!

The expected number of fatalities is hypothetical and is based on the assumption that exposure levels are independent and lognormally distributed over the working month, i.e., P (fatality) = 1 - (1-f)^{1008} where f = P (fatality in a single exposure). Nevertheless the above examples clearly demonstrate that the OSHA/NIOSH air-sampling scheme is prone to serious errors when environmental variability increases above minimal levels. This weakness renders it to be a relatively poor decision-making tool for the industrial hygienist.

Air Sampling Schemes for Acute Hazards

Little information is currently available to assist the indus-

Table III

Expected Sample Sizes* for the OSHA/NIOSH
Air Sampling Scheme
(PEL = 10, AL = 5)

μ_g	σ_g		
	1.5	2.5	3.5
2	1.1	1.6	1.7
5	4.5	2.3	2.0
10	2.2	1.9	1.8
15	1.3	1.5	1.6
20	1.1	1.4	1.5

*Calculated from the probability
distribution given in the
Appendix.

Table IV

Probabilities that Various Hazardous Distributions
of Exposures to HCN Would Be Declared In Compliance
by the OSHA/NIOSH Scheme and Expected Fatal
Excursions (σ_g = 3.5)

μ_g (mg/m^3)	P(Compliance)	P(X>100mg/m^3)[1]	Expected Excursions[2] >100mg/m^3/mo
2.0	0.64	9.0×10^{-4}	0.9
5.0	0.31	8.4×10^{-3}	8.5
7.5	0.20	1.9×10^{-2}	20

[1]Probability that a 10-min exposure would exceed 100mg/m^3.
[2]Number of 10-min exposures exceeding 100 mg/m^3 expected
per month (21 days = 1008 10-min intervals).

trial hygienist in evaluating acute exposures. One would logically look for guidance to published standards and guides established by OSHA, NIOSH (recommended standards), the American Conference of Governmental Industrial Hygienists (ACGIH) and the American National Standards Institute (ANSI). Yet invariably these documents merely suggest that "ceiling limits" defined as maximum short-term air concentrations of specific duration should not be exceeded. Table V lists a variety of ceiling limits proposed by NIOSH.

The only significant source of information on this subject was published by Leidel et al. (15) of NIOSH, who suggested air sampling schemes for both acute and chronic exposures. Regarding acute exposures the authors made an important distinction between environments in which the period of highest short-term exposure can be predicted and those where this is not possible. This ability or inability to predict the highest exposure dictates the type of air sampling scheme.

Predictable Peak Exposures. Ceiling limits, as defined, should be applied only to situations where the periods of maximum exposure can be prospectively identified. If the highest exposure can be anticipated, then its measurement is a physical problem, not a statistical one. All pertinent information related to the operation should be considered so that only air samples which are likely to represent peak air concentrations are collected. Consider, for instance, exposures of workers in a chemical plant to hydrogen sulfide (a reactant in a process). If analysis of the operation shows that the highest exposure should occur when a valve is opened to vent the reaction vessel, then only this period needs to be sampled to determine the peak exposure.

Leidel et al. (15) suggest that when several equivalent, high-exposure events occur during a day at least three should be randomly selected for sampling. If all of the measured exposures are below the ceiling limit, then they propose that the probability be estimated that an unsampled period could have exceeded the limit. The estimation procedure which they use is based upon the assumptions that the air concentrations during these events are independent and lognormally distributed. However, since the events are sampled in a way which ensures that air concentrations will be large, the distribution would almost certainly not be lognormal. In fact, since the statistical distribution which describes these air concentrations is truncated, rather complex estimation procedures could be required. Thus, it is recommended that analysis of this kind should not be made.

Unpredictable Peak Exposures. When periods of peak exposure cannot be predicted, the concept of an absolute ceiling limit becomes naive because for any lognormal distribution of sample values ($\sigma_g > 1$) there is always a probability (however remote) that the limit will be exceeded. It is suggested, therefore, that in these situations acute-exposure limits be interpreted as air concentrations which should be exceeded only rarely, perhaps 5% of the time. This does not imply that it would be acceptable to allow lethal or even irritating exposures to occur 5% of the time. As shown in Table V most ceiling limits include safety factors; i.e., the limit is set at a fraction of the air concentration which is expected to cause the toxic effect. Since the safety factors are relatively large for lethal toxins ($\sim$10-100) and small for irritants ($\sim$1-5), and assuming excursions above the exposure limit are only rarely encountered, then the likelihood of exceeding the corresponding toxic limit should be acceptable in virtually all cases. That is, the probability of allowing lethal exposures to occur would be extremely remote whereas short periods of irritation could be allowed to occur infrequently, perhaps once or twice a week. The use of 5% as the criterion of acceptability for exceeding a limit seems reasonable but is only illustrative. The important point is that this criterion and the safety factor can be chosen to provide the desired level of coverage.

The essence of the air sampling problem is to gain sufficient knowledge of the distribution of short-term exposures by random sampling to ensure that the acute-exposure limit is not exceeded more than the accepted fraction of time. If only one worker is exposed or if a highest-risk worker can be identified from a group of workers, then only one individual's exposure needs to be sampled. In all other cases workers should be divided into uniform-exposure zones (20) and should be randomly selected for monitoring according to the requirements of the sampling scheme. The sampling period is the 8 to 10 hr workday or that fraction of the day in which a worker is potentially exposed. As in the previous case, the sampling interval would be defined by an appropriate standard or guide. The number of samples to be collected is dictated by the sampling scheme.

The NIOSH Scheme - Tolerance Sets. Leidel et al. (15) of NIOSH suggest that the random selection of workers and of short-term sampling intervals during the workday be based upon tolerance sets. This approach leads to a situation in which it is possible to state that some given fraction of the distribution of air sample values lies above (below) the largest (smallest) of n sample values with given confidence. Regrettably, in most air-sampling applications, the number of short-term samples required to allow reasonable coverage of one worker's exposures is quite large. If workers must also be randomly selected by the same procedure, the total number of samples required could be impractically large. For example, 18 workers from a population of 50 would be selected for monitoring to insure that at the 90% confidence level at least one worker chosen would be in the highest 10% of exposures on that day. If it were desired to be equally certain that at least one sampled interval for each worker selected was in the highest 10% of all 10-min intervals for the day, 17 samples would be required per worker. This results in a total of 306 air samples to be collected in a single day!

If any of the sample values exceed the ceiling limit, Leidel et al. (15) propose that the environment be declared unsafe. If

Table V

Ceiling Limits and Approximate Safety Factors
Recommended by NIOSH for Several Airborne Chemicals

Chemical	Toxic Effect	Ceiling Limit	Duration (min)	Safety Factor[1]	Publication Number[2]
Hydrogen Cyanide	Chemical Asphyxiation	$5mgCN^-/m^3$	10	20	77-108
Hydrogen Sulfide	Respiratory Paralysis	10 ppm	10	10	77-158
Nitrogen Dioxide	Pulmonary Edema	1 ppm	15	100	76-149
Phosgene	Pulmonary Edema	0.2 ppm	15	75	76-137
Chlorine	Pulmonary Edema	0.5 ppm	15	100	76-170
Ammonia	Irritation	50 ppm	5	2	74-136
Hydrogen Fluoride	Irritation	$15mgF^-/m^3$	15	5	76-143
Sodium Hydroxide	Irritation	$2mg/m3$	15	1	76-105

[1]Approximate ratio of the PEL to the threshold of the toxic effect listed.
[2]USDHEW, NIOSH Publication Number.

all of the sample values are below the ceiling limit they suggest
that the probability be estimated that an unmeasured interval could
have exceeded the limit. This estimate is based upon the assumption that the sample values are lognormally distributed. If the
probability that an observation could exceed the limit during an
unsampled interval is less than 10%, the authors contend that the
worker's exposures are acceptable.

Another Scheme - Limiting Distributions. A fundamental requirement of a usable air sampling scheme is that the number of
samples needed for a decision to be made should not be excessive.
As shown in the previous example, the use of tolerance sets can
require large numbers of samples. This stems primarily from the
fact that with this approach, decisions are based almost entirely
on the highest sample collected and no assumptions are made about
the sampled distribution (non-parametric). Another approach would
be to predict high exposures on the basis of relatively few samples
by making the assumption that the sample values are lognormally
distributed (parametric). The assumption that the distribution
of sample values is lognormal effectively adds information and
necessarily leads to greater efficiency in terms of sample size.
When evidence suggests this assumption to be valid, the question
is raised of how best to take this approach.

We propose to approach this problem by addressing the question
of "safe" levels via a "limiting distribution" rather than a ceiling limit. The limiting distribution we define as that distribution of air sample values that is the upper limit of all safe distributions. That is, since we assume the workplace to be characterized by a statistical distribution of air sample values, there
are safe and unsafe distributions. The limiting distribution,
within the context of acute exposures, is that which allows the
highest short-term airborne concentration consistent with safety.
Figure 4 shows an example of what is meant by the upper limit using
cumulative distributions. The solid line represents a limiting
distribution whose parameters were selected to represent the highest distribution of acceptable short-term exposures for a particular exposure ($\mu_g^o = 3.2$, $\sigma_g^o = 2.0$). The dotted lines represent
distributions of sample values. Distribution No. 1 with a large
percentage of values above those of the limiting distribution
represents an "unsafe" condition whereas distribution No. 2 with a
large percentage of values below those of the limiting distribution represents a "safe" condition. As will be seen there is some
flexibility in defining the limiting distribution, but the general
idea should be clear.

The advantage of the limiting distribution is that it allows
one to approach the sampling problem in the context of hypothesis
testing. For example, the hypothesis can be tested that the sample
values come from the limiting distribution against the alternative
that the samples are from a distribution whose high values lie
mostly below those of the limiting distribution, in the context
of Figure 1.

The hypothesis to be tested requires an appropriate test statistic. Since acute toxins are being considered here, it is essential to choose a statistical measure that is likely to identify lognormal distributions that potentially produce large values, even if these values are improbable. A sensitive statistic tic must combine both the overall level (mean) and the intrinsic variability (variance). A test statistic with this property is the estimated 95th percentile $\hat{B}$ defined as

$$\hat{B} = \bar{x}_g \, s_g^{1.645}$$

The value $\hat{B}$ clearly varies from sample to sample and is a slightly biased (likely to underestimate the true value) estimate of the 95th percentile of the lognormal distribution, B, i.e. $\mu_g \sigma_g^{1.645}$.

The actual hypothesis to be tested must be carefully chosen. The basic issue is to ensure a safe working environment and we interpret this in the sense that safety must be statistically demonstrated or the environment is considered unsafe. Thus a hypothesis is constructed that the 95th percentile of the distribution of air concentrations B is equal to the 95th percentile of the limiting distribution B_0 (null hypothesis) and an environment is not considered safe until statistical evidence demonstrates that $B < B_0$ (alternate hypothesis) in a lower one-sided test. (Note $B_0 = \mu_g^o \sigma_g^o{}^{(1.645)}$ where μ_g^o and σ_g^o are parameters of the limiting distribution.)

If a large number of sets of samples of given size is drawn from the limiting distribution, a distribution of 95th percentiles is generated. Figure 5 shows distributions of 95th percentiles derived from the limiting distribution shown in Figure 4 (B_0 = 10; μ_g^o = 3.2; σ_g^o = 2.0) for sets of 5, 10 and 20 samples. If samples are then collected from the environment and $\hat{B}$ is calculated, the hypothesis can be tested that the data arose from the limiting distribution ($B = B_0$) against the alternative that statistical evidence suggests that the data came from a distribution "below" the limiting distribution ($B < B_0$). If the value of $\hat{B}$ lies in the shaded area of Figure 5 we infer that the sample comes from a distribution whose 95th percentile value is less than the corresponding value of the limiting distribution with 95% confidence. The procedure produces a 5% chance of falsely declaring an unsafe environment to be safe (type I error) when in fact the data arose from the limiting distribution. The probability of a type I error will be less if the data arose from a distribution with a $\sigma_g < \sigma_g^o$.

Returning to the conceptual issues surrounding the limiting distribution, it was earlier pointed out that there is some flexibility in its specification. In the case of acute exposures we are concerned with the upper tail of the actual distribution of air sample values. The flexibility arises in the specific choice of μ_g^o and σ_g^o to obtain appropriate protection yet to maximize the probability of declaring the condition safe if it is so. If the

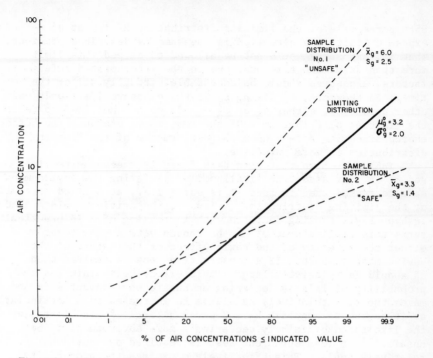

Figure 4.   A limiting distribution for evaluating acute exposures (solid line): sample distribution No. 1 comes from an "unsafe" environment; sample distribution No. 2 comes from a "safe" environment.

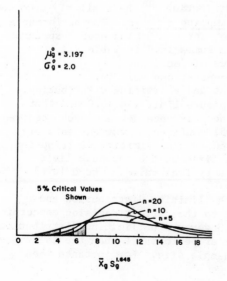

Figure 5.   Distributions, based upon various sample sizes, of the estimated 95th percentile ($\bar{x}_g s_g^{1.645}$) of a limiting distribution ($\sigma_g° = 2.0$; $\mu_g° = 3.2$; $B_o = 10.0$): shaded areas represent critical regions ($\alpha = .05$) for a lower one-sided test of the hypothesis that the sample values were derived from the limiting distribution.

95th percentile of the limiting distribution B_O is set at acute-
exposure limit, the situation is the same as described previously.
That is, exposures would not be allowed to exceed this limit
more than 5% of the time and, due to the incorporated safety
factors, exposures should be acceptable virtually all of the
time. Furthermore, by fixing B_O at the exposure limit only one
other parameter, either μ_g^o or σ_g^o, needs to be specified to define
the limiting distribution. Of the two, σ_g^o is probably the better
choice since its effect upon the performance of the limiting
distribution is more intuitive.

Table VI shows that when B_O is fixed at the exposure limit
and σ_g^o is increased, the likelihood of declaring the environment
safe decreases when in fact it is safe; i.e., at a fixed percentile
of the limiting distribution (α level), the critical values used
in hypothesis testing decrease with increasing σ_g^o. In practical
terms this implies that σ_g^o can be chosen with a view towards
either the severity of the toxic effect or the nature of the
environment or both. If a conservative test is desired then
σ_g^o should be relatively large. This effectively minimizes the
probability of falsely declaring an unsafe environment safe and
makes the test relatively sensitive to increases in B particularly
as it begins to approach B_O. However a large σ_g^o also increases
the likelihood of falsely declaring a safe environment to be
unsafe. If a less stringent test is required σ_g^o should be
relatively small. This effectively minimizes the probability
of falsely declaring a safe environment unsafe and reduces the
sensitivity of the test to increases in B.

The critical values in Table VI are based on lognormal limiting
distributions with values of B_O fixed at 10, σ_g^o values of
1.2, 1.5, 2.0 and 2.5, and computer samples based upon 2500 simulated observations for each case. Tables of percentage points
for any choice of a limiting distribution can be similarly produced.

Performance of Random Air-Sampling Schemes. The performance
of the random air-sampling schemes for evaluating acute exposures
to lethal toxins and irritants is summarized in Table VII.
This table shows the probabilities of declaring
various distributions of air concentrations safe for
schemes based upon a tolerance set and a limiting distribution.
It was assumed that the acute exposure limit was 10.0 and that
10 min samples were required. The tolerance set approach requires
21 samples (10th percentile at 95% confidence) whereas sets of
5, 10 and 20 samples are evaluated for the limiting distribution.
The limiting distribution has B_O fixed at the exposure limit
(10.0) and a σ_g^o of 2.0 which is a typical value. The α level
used for the hypothesis test is .05.

Table VII shows that both the limiting distribution and
the tolerance set generally lead to the correct decision concerning
the safety of the environment. However, the limiting distribution
produces a similar and often better performance than the tolerance
set on the basis of ¼ to ½ the sample size. For instance when

Table VI

Critical Values of Limiting Distributions for
Selected Sample Sizes and Parameters
$(B_O = 10.0)$

Sample Size (n)	σ_g^o	μ_g^o	Percentiles of the Limiting Distribution (α)		
			2.5%	5.0%	10%
5	1.2	7.409	7.72	7.96	8.33
	1.5	5.133	5.62	6.07	6.68
	2.0	3.197	3.64	4.16	5.00
	2.5	2.215	2.68	3.20	4.00
10	1.2	7.409	8.29	8.62	8.86
	1.5	5.133	6.62	7.14	7.66
	2.0	3.197	4.98	5.64	6.40
	2.5	2.215	3.95	4.61	5.45
20	1.2	7.409	8.84	9.00	9.22
	1.5	5.133	7.61	7.94	8.35
	2.0	3.197	6.23	6.68	7.38
	2.5	2.215	5.38	6.00	6.72

Table VII

Performance of Schemes Based upon Tolerance Sets and
Limiting Distributions for Evaluating Acute Exposures to
Lethal Toxins. [Exposure Limit = 10.0]

μ_g^1	σ_g^1	B	$P(\text{Declaring Safe})^2$			
			TS_{21}	LD_5	LD_{10}	LD_{20}
1.0	1.5	1.95	>.999	.996	>.999	>.999
	2.5	4.51	.881	.538	.729	.920
	3.5	7.85	.494	.307	.328	.392
2.0	1.5	3.90	>.999	.681	.967	.998
	2.5	9.03	.430	.159	.161	.186
	3.5	15.7	.112	.087	.049	.027
3.0	1.5	5.84	.969	.177	.455	.890
	2.5	13.5	.125	.046	.032	.010
	3.5	23.6	.021	.039	.015	.008

[1]Parameters of the distribution of air concentrations.
[2]Probability of declaring the environment safe.
Tolerance Set (TS): n = 21, 95% confidence of top 10%.
Limiting Distribution (LD): σ_g° = 2.0, B_0= 10.0,
   α = .05.

the environment is clearly safe (μ_g = 1.0; σ_g = 1.5; B = 1.95) the limiting distribution declares the environment safe more than 99% of the time on the basis of 5 samples whereas 21 samples are required by the tolerance set. A similar result is observed when the environment is clearly unsafe (μ_g = 3.0; σ_g = 3.5; B = 23.6). In this case the probability of declaring the environment unsafe is 96% for the limiting distribution with 5 samples and 98% for the tolerance set. When values of B reach approximately ½ of the exposure limit (e.g. at μ_g = 1.0; σ_g = 2.5; B = 4.51) the performance of the limiting distribution is improved by increasing the sample size from 5 to 10 to 20 at which point the probability of declaring the environment safe is slightly better than that of the tolerance set (.92 vs. .88). However even in this case the limiting distribution would declare the environment safe 54% of the time on the basis of only 5 samples. When B is close to the exposure limit the limiting-distribution approach is somewhat more conservative than that of the tolerance set. Thus the limiting distribution produces a slightly larger probability of falsely declaring a safe environment unsafe (e.g., at μ_g = 2.0; σ_g = 2.5; B = 9.03) but a smaller probability of falsely declaring an unsafe environment safe (e.g., at μ_g = 3.0; σ_g = 2.5; B = 13.5). As previously mentioned this is governed largely by the choice of σ_g^o which in this case is 2.0. If it were desired to reduce the stringency of the test σ_g^o could be reduced thereby reducing the probability of falsely declaring a safe environment unsafe but increasing the probability of falsely declaring an unsafe environment safe.

In short the approach based upon the concept of a limiting distribution offers a viable alternative to that based upon tolerance sets. The stated objective of reducing the number of samples required for making correct decisions has been achieved. Additional refinements in the selection of parameters for the limiting distribution should further enhance its applicability in evaluating acute exposures.

Conclusions

The foregoing discussion has stressed the importance of two factors, i.e., the variability of airborne exposures and the toxic effects posed by chemicals, as determinants in the selection of air sampling schemes. Air sampling schemes which do not adequately consider these factors, such as the one currently used by OSHA, will probably be prone to errors and will either underestimate or overestimate exposures. The consequences in either case will be undesirable.

Concerning the evaluation of acute exposures, two important facts have emerged. First, each environment should be carefully scrutinized to determine whether the maximum exposures can be predicted. When they can, air sampling follows non-statistical lines to ensure, as much as possible, that the highest interval of exposure is measured. When the maximum exposures cannot be predicted, samples must be randomly collected according to statistical schemes to measure or predict the highest exposures. The use of a new statistical approach, the limiting distribution, which is based upon the assumption of lognormality, generally leads to correct conclusions on the basis of few samples relative to the number required by tolerance sets. Limiting distributions may also be indexed to the toxicological and environmental realities so that they can be made as conservative as desired.

Acknowledgement: The authors gratefully acknowledge the assistance of William Popendorf and James Rock in revising the manuscript. This work was supported by the Northern California Occupational Health Center.

List of Symbols

B 95th percentile of the distribution of air concentrations, $\mu_g \sigma_g^{1.645}$

B_o 95th percentile of a limiting distribution, $\mu_g^\circ \sigma_g^{\circ(1.645)}$

$\hat{B}$ 95th percentile of the distribution of sample values, $\bar{x}_g s_g^{1.645}$

s sample standard deviation

s_g sample geometric standard deviation

σ population standard deviation

σ_g population geometric standard deviation

σ_g° geometric standard deviation of a limiting distribution

x a single observation

$\bar{x}$ sample mean

$\bar{x}_g$ sample geometric mean

μ population mean

μ_g population geometric mean

μ_g° geometric mean of a limiting distribution

Literature Cited

1. Oldham, P.D. Brit. J. Industr. Med., 1953, 10, 227-234.
2. Roach, S.A. Brit. J. Industr. Med., 1959, 16, 104-122.
3. LeClare, P.C.; Breslin, A.J.; Ong, L.D.Y. Am. Ind. Hyg. Assoc. J., 1969, 30, 386-393.
4. Sherwood, R.J. Am. Ind. Hyg. Assoc. J., 1966, 27, 98-109.
5. Gale, H.J. Ann. Occup. Hyg., 1967, 10, 39-45.
6. Breslin, A.J.; Ong, L.; Glauberman, H; George, A.C.; LeClare, P. Am. Ind. Hyg. Assoc. J., 1967, 28, 56-61.
7. Sherwood, R.J. Am. Ind. Hyg. Assoc. J.,1971, 32, 840-846.
8. Stevens, D.C. Ann. Occup. Hyg., 1969, 12, 33-40.
9. Gold, A.; Burgess, W.A.; Clougherty, E.V. Am. Ind. Hyg. Assoc. J., 1978, 39, 534-539.
10. Esmen, N.A.; Hammad, Y.Y.; J. Environ. Sci. Health, 1977, A12 (1 and 2), 29-41.
11. Juda, J.; Budzinski, K. Staub. Reinhalt. Luft., 1964, 24, 283-287.
12. Coenen, W. Staub. Reinhalt. Luft., 1966, 26, 39-45 [English Translation].
13. Pearson, E.S.; Hartley, H.O. "Biometrika Tables for Statisticians, Vol. 1," Cambridge University Press, London, 1956, p. 175.
14. Cope, R.F.; Pancamo, R.B.; Rinehart, W.E., TerHaar, G.L. Am. Ind. Hyg. Assoc. J., 1979, 40, 372-379.
15. Leidel, N.A.; Busch, K.A.; Lynch, J.R. "Occupational Exposure Sampling Strategy Manual," USDHEW (NIOSH) Pub. No. 77-173, U.S. Govt. Printing Office, Washington, D.C., 1977.
16. Rock, J.C. "The NIOSH Sampling Strategy... A Useful Industrial Hygiene Tool or an Economic Disaster?," Paper 131, American Industrial Hygiene Conference, Chicago, Illinois, May 31, 1979.
17. Tuggle, R.M. "The NIOSH Decision Scheme - A User's Perspective," Paper 321, American Industrial Hygiene Conference, 23 May 1980.
18. "Criteria for a Recommended Standard... Occupational Exposure to Hydrogen Cyanide and Cyanide Salts," USDHEW (NIOSH) Pub. No. 77-108, U.S. Govt. Printing Office, Washington, D.C., 1976.
19. Nagler, J.; Provoost, R.A., Parizel, G. J. Occ. Med., 1978, 20, 414-416.
20. Corn, M.; Esmen, N.A. Am. Ind. Hyg. Assoc. J., 1979, 40, 47-57.

Appendix

Derivation of Probability that the OSHA/NIOSH Scheme
Will Declare an Environment to be Noncomplying

In terms of probabilities it is possible to investigate the
OSHA/NIOSH sampling scheme and determine its usefulness as a
decision-making tool. Assuming that one worker's exposure was
monitored on a given day the three relevant possibilities can
be represented as: $X > PEL$, $AL \leq X \leq PEL$, and $X < AL$, where X
is the observed concentration. Analysis of the scheme requires
knowledge of the three corresponding probabilities which are:
$p = P(X > PEL)$, $q = P(AL \leq X \leq PEL)$ and $r = P(X < AL)$. Under
the assumption of lognormality with geometric mean, μ_g, and geometric standard deviation, σ_g, the three probabilities can be derived.
They are:

$$p = \phi \left[\frac{\ln(PEL/\mu_g)}{\ln(\sigma_g)} \right]$$

$$r = 1 - \phi \left[\frac{\ln(PEL/2\mu_g)}{\ln(\sigma_g)} \right]$$

$$q = 1 - p - r$$

where the notation $\phi(x)$ represents the probability that a variable
with a standard normal distribution (mean = 0 and variance = 1)
exceeds the value x. The probability of two consecutive samples
falling below the AL after an initial sample between the AL and
the PEL is represented by $1 - P$. That is, the symbol, P, denotes
the probability that a positive result (declaration of noncompliance) would be observed after an intermediate result on the initial
test. The value of P can be derived using the probabilities p,
q and r as follows.

Consider a series of tests where a positive result is observed
after m trials. Two conditions are necessary: (1) the last observed
test $(m + 1)$ is positive and (2) no two consecutive negative results
are observed. The probability of this result, represented by
π_m, is

$$\pi_m = p \left[\sum_{k=0}^{K} \binom{m+1-k}{k} q^{m-k} r^k \right] .$$

The value K is one plus the smallest integer resulting from $(m + 1)/2$ or $[(m + 1)/2] + 1$. For example, if $m = 4$, then $k = 3$
and

$$\pi_4 = \left[p\left(q^4 r^0 + 4q^3 r + 3q^2 r^2 \right) \right] \quad .$$

The factor $\binom{m+1-k}{k}$ is the number of combinations of k decisions to retest and m-k negatives such that two negative tests do not occur consecutively. The product $q^{m-k}r^k$ is the probability that a series with m-k retests and k negative observations occurs. The probability P of ultimately observing a positive result is, therefore,

$$P = \sum_{j=0}^{\infty} \pi_j \quad .$$

For example, if p = .5, q = .2 and r = .3, then π_0 = .5, π_1 = .25, π_2 = .08, π_3 = .031, ... and $\pi_0 + \pi_1 + \pi_2 + ...$ = .878. The probability that a series of samples would ultimately result in observing a concentration greater than the PEL is, therefore, p + qP.

A positive result is not identical to instituting engineering controls. The OSHA/NIOSH sampling scheme allows for repeated testing once a positive result is observed. If two subsequent consecutive negative values are observed after an initial positive value, the system is then considered safe. The assessment described here does not allow for the possibility of additional sampling after a value exceeds the PEL since testing after a positive result will always increase the possibility of falsely declaring an unsafe environment as safe. Furthermore, the rate of falsely declaring unsafe situations as safe depends on the intensity of sampling (i.e., if enough samples are taken two consecutive negative values will undoubtedly occur from random variation).

(The above derivation is based in part upon a conversation with Dr. Chin Long Chiang, University of California, Berkeley.)

RECEIVED October 29, 1980.

Industrial Hygiene Logistics

GARY STOUGH and ALFREDO SALAZAR

SRI International, Menlo Park, CA 94025

Industrial hygiene is both a science and an art that seeks to recognize sources of potentially hazardous occupational exposure. When potentially hazardous agents are identified the industrial hygienist collects appropriate data and observes work routines so the significance of exposure to each agent can be evaluated. Excessive exposure may then be reduced to acceptable levels by application of engineering and administrative controls along with the use of personal protective equipment. Data collection encompasses many difficulties and challenges dependent on the type of operations, material and physical setting that are encountered.

The practice of industrial hygiene often requires the use of sophisticated instrumentation and sample collection techniques to determine exposure to biological, chemical and physical agents. Observations of work practice are collected concurrent with empirical observations. When industrial hygienists must travel to distant locations to collect data, conduct sampling and observe industrial processes the challenges can increase many-fold. While no simple cure-all can ease difficulties associated with field work, many suggestions are possible that may assist industrial hygienists and others who must collect environmental data at field locations.

Preparation for Field Surveys

Anticipation. Because no substitute can satisfactorily replace first hand observation, a preliminary survey of the proposed study area is usually the best preparation for a detailed survey and data collection. If a preliminary survey is not feasible, consultation with knowledgeable colleagues and telephone liaison with personnel at the survey site assumes greater importance.

The success of field operations often depends on the investigator's ability to anticipate conditions at the field location. The information necessary to assist the professional is available from any sources including personal contacts, reviews of previous studies, and consultation with personnel at the study site. The advice of colleagues familiar with facilities similar to the one to be studied can assist the industrial hygienist who has limited experience.

If sample analysis is to be conducted by a commercial analytical laboratory, the investigator and the laboratory personnel should jointly review available sampling techniques to select the best sampling strategy and establish a protocol for the delivery and storage of collected samples.

Professional literature should be reviewed. Manuals of analytical methods (1,2) describe various sampling scenarios and considerations. Overview references (3) discussing processes similar to the one proposed for study may also be available. Professional publications e.g., The Journal of the American Industrial Hygiene Association, Journal of Occupational Medicine, may outline specific aspects of the process to be studied. Toxicological texts (4,5,6,7) provide information about potentially hazardous agents. These sources can provide information about the character of exposure, and their relative expected concentration or intensity, and methods previously used for detection, sampling and analysis.

Plant personnel can assist in determining what parameters and concerns are to be investigated, what conditions are prevalent and what resources are locally available to support field work. Special requirements such as staging area, washing facilities for glassware, refrigeration for samples, calibration gases, reagents and instrumentation should be discussed. Ask for a map of the facilities, a process flow chart, and a list of toxic materials. Request records of previous environmental sampling; these records may include results from area monitoring equipment, personnel sampling, OSHA citations and employee complaints. A plant operating schedule will help predict the best time for the field study. Ask plant personnel about personnel protective equipment requirements at the study site, so the investigators can comply with all regulations.

Preliminary Survey Strategy Development. All available information should be synthesized in preliminary sampling strategy that addresses each concern identified at the study site. The preliminary sampling strategy should consider each potential source, biological, chemical or physical stress as well as ergonomic factors. Determine instrumentation and sampling media requirements. Order supplies and equipment that must be purchased or rented well in advance.

Equipment Selection. Choose direct and indirect reading instruments that address each parameter identified in the preliminary sampling strategy. Select equipment that is accurate, reliable and durable. Air velocity measuring equipment and smoke tubes should be included for the evaluation of exhaust ventilation. Use of equipment in potentially explosive atmospheres requires special consideration. Equipment rental services can bridge gaps in the investigators own instrumentation inventory.

Instrument Preparation. The investigator should collect all the instrumentation, sampling media, supplies and peripherals at a central location. Charge batteries and test each instrument prior to packing. Prepare standard curves that compare the response of each direct reading instrument and air flow measuring device to known standards. Develop a comprehensive equipment inventory that includes battery chargers, instrument manuals, calibration gases, electrical adapters and an assortment of tools, cord, tape and string, etc. Equipment checklists can help ensure against forgetting instrumentation, essential appurtenances or sampling media. One practical way to double check the completeness of the equipment inventory is to arrange all of the components (including collection media) of each instrument system in their normal operating sequence for inspection prior to packing.

Pack equipment carefully into sturdy cases that will protect it during travel. Durable and well protected items are placed outboard, more delicate items inboard in each packing case. Pad gauge and meter faces, orient them inward and away from hard protuberances. Sampling hoses, power cords and other small durable items can be used to fill gaps between larger items. Resilient foam can be placed around the outboard edges of instrumentation and between individual pieces. Pack cases snuggly and tape over all latches with fiber tape to protect against accidental opening during transit. Avoid packing cases that have sharp projections, awkard shapes or sharp edges because they can injure personnel who handle them. Excessively large cases are undesirable since they may become ponderous when fully loaded with equipment.

Many flammable gases and solvents commonly used for instrument calibration or as sorbent media are prohibited from transportation on common carriers by Department of Transportation (DOT) regulations (8). The industrial hygienist should review the equipment inventory for each survey to ensure against potential conflicts with DOT regulations. Express parcel carriers and even commercial airlines can occasionally transport normally restricted materials if prior arrangements are made. Items that are restricted can be transported by some other means or procured at the scene.

Travel and Equipment Transportation. Make travel arrangements that allow adequate time to accomodate unexpected events and allow for the unhurried setup of equipment at the survey site. The effects of jet lag become important when several time zones must be crossed and an intensive sampling schedule is planned. Avoid close connections between flights, especially when journeying to the survey site, since the equipment and luggage may

not make the transfer between aircraft. When rental cars are
reserved, select a car that is large enough to carry all survey
personnel and equipment.

Ground handling of the equipment and the investigators
luggage is a challenging task when large distances to rental
cars or loading areas must be covered. Hand carts and similar
portage devices are useful; however, many available units are
either too frail or too bulky to be useful. Exercise care when
purchasing portage devices to ensure that they are strong enough
to support equipment and luggage. Backpacks allow heavy objects
to be transported while leaving the hands free to carry other
objects. Larger instrument cases can be equipped with wheels at
one end and a handle at the other so that the case doubles as its
own cart.

Implementation

Initial Walk Through Survey. Plant visits usually begin
with a walk-through tour of the facilities to aid in planning
subsequent sampling and data collections. The initial survey is
particularly productive when the investigator has previously
studied the plant layout and process flow charts. Hosey (9)
provides many useful insights that can assist investigators in
surveying industrial operations. The survey team should be alert
for potential exposure to any hazardous agents that may not have
been identified in the preliminary survey strategy. The initial
survey will provide an opportunity to meet supervisors and
operators whose exposures will be monitored during the survey and
identify optimal locations for area sampler deployment.

Direct reading instruments and air movement sensing devices
are used to roughly outline areas of high concentration and to
develop a rough feel for the distribution of contaminants within
the plant environment. Data from direct reading instruments
provides a preliminary estimate of exposure intensity. Sampling
rate or duration can then be adjusted to collect an amount of
contaminant near the center of the analytical range. Several
references produced by the American Conference of Governmental
Industrial Hygienists (10, 11) can aid in the efficient use of
direct reading instruments.

Following the initial survey a detailed schedule for the
succeeding days of sampling and data collection should be
prepared. Frequent rounds to observe sampling equipment can be
made more efficiently if both direct reading and time weighted
data collections are made in one area of large facilities at a
time. Sampling record sheets should be prepared to assist in the
speedy deployment of sampling equipment. Accurate records must
be kept of pump and sample number, the employee's name, location,
job title along with on-off time and flowrate. A careful system
of equipment accounting should be developed to record the where-
abouts of each piece of sampling equipment.

During the initial survey, the investigator should observe the periodicity of operations and the variability of exposure to develop and modify a proposed sampling strategy. Leidel, Busch and Lynch (12) provide statistical procedures that can assist in planning a detailed sampling strategy. We suggest a policy of sampling to excess (i.e., taking as many samples as is feasible) and analyzing selectively, on the theory, that travel, the investigator's time and analytical services are expensive while sampling media is not. Typically our analytical strategy gives priority to personnel and locations we feel are most heavily exposed. Air samples collected on filters can be weighed to establish a relative exposure index; while a representative sub-group are subjected to more rigorous analysis. Lesser priority samples that are not analyzed immediately can often be retained indefinitely and subjected to subsequent analysis as conditions warrant. Consider the retention of collected samples when determining the sampling and analytical strategy.

Equipment Setup. A staging area for charging, calibrating, and occasionally repairing equipment is necessary during most field studies. If no suitable site is available at the plant the industrial hygienist can use the hotel or motel room.

Field investigators can often obtain gases, equipment and standards at the study site. However, when circumstances require, calibration standards may be prepared with a modest array of equipment in the field. A simple vapor or gas mixing system is shown in Figure 1. This system uses a personnel sampling pump and rotameter to supply a metered flow of air. A charcoal tube is used to clean the air stream before it passes through a modified glass "tee" and into an aluminized milar bag. The glass "tee" has a septum on one arm; the opposing arm is bent at a right angle with a length of nichrome wire wrapped and epoxied around the bend to produce a heating element. The heating element which is matched to the output of a personnel sampling pump battery instantly vaporizes aliquots of solvent that are injected through the septum.

Direct reading instruments and air sampling equipment should be thoroughly prepared before being taken into the work area. All sampling media should be logged in, all flow rates adjusted and all hoses and fittings attached. Each sampling train should be dedicated to the particular location, process or occupation it will sample each day. Observe the response of direct reading instruments to prepared standards.

Wheeled carts will assist in deploying large numbers of samplers or large instruments. If carts are unavailable or steps and ladders are encountered, the investigator may consider back-pack, shoulder bags, or other available carrying aids.

Equipment failure can severely hamper data collection. Many equipment failures are caused by simple malfunctions; e.g., blown fuses, dead batteries, loose components or broken wires. The

decision to attempt field repairs pivots on the investigator's
understanding of simple electronic and mechanical systems. Any-
one who considers field repairs should remember that they can
severely damage an expensive instrument if they overestimate their
technical prowess. The instruments' manufacturer may be consulted
by telephone to advise in the correction of instrument failures.
The industrial hygienist who feels competent to undertake field
repairs should include a multimeter and a limited assortment of
hand tools in the equipment inventory.

Deployment of Personnel Samplers. Personnel sampling trains
should be comfortably and securely attached to the workers and
supervisors whose exposures are being monitored. When the
worker's clothing is inappropriate for the attachment of sampling
equipment the industrial hygienist should provide belts or other
supports. Substantial support can be required when workers
engage in strenuous activities (Figure 2). Sampling devices,
hoses, clips, etc. are secured and should not inconvenience or
encumber personnel being sampled. The comfortable and secure
attachment of sampling equipment will prevent discomfort and
help ameliorate one potential source of bias.

The industrial hygienist should explain the purpose of the
survey and its significance to the workers being sampled to
ensure their cooperation and the validity of the data collected.
Workers and supervisors should be asked questions to illuminate
the circumstances of each exposure with an emphasis on
anomalous conditions.

Deployment of Area Samplers. Deploy samplers in positions
that will detect the spread of contaminants from sources of
emanation through the working area. Deployment of area samplers
is often encumbered by the lack of supports and props to support
pumps and media at desired sampling locations. Extendable props
and stands can be placed on the floor or attached to supports to
help answer this requirement (Figure 3). A wide variety of stands
and propping devices are available from photographic equipment
suppliers (Figure 4). Rope, tape, wire and string or materials
available at the survey site can often be used to support sampling
media and equipment (Figure 5). Line power will be required for
instrumentation that is not self powered.

Extreme care should be exercised when large instruments are
handled. Occasionally, heavy and bulky instruments must be
hauled up ladders with ropes and hoisting gear. Acquire hoisting
gear at the study site, if possible and always include an
assortment of cord, wire and tape in the equipment inventory.

Area samples may occasionally be deployed in locations
requiring protection from weather and dirt contamination. In a
recent study of heavy equipment operators' exposure to diesel
exhaust emission; we had to support and protect instrumentation
systems upon heavy earth moving equipment. Exposure to mechanical

Figure 1.   *Calibration standards can be prepared in the field with a simple array of equipment.*

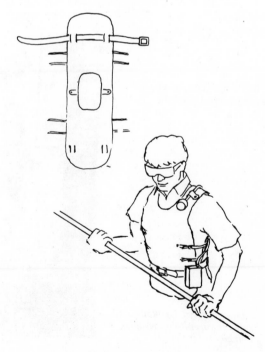

Figure 2.   *Vests provide secure attachment of sampling equipment to workers who are engaged in strenuous activities.*

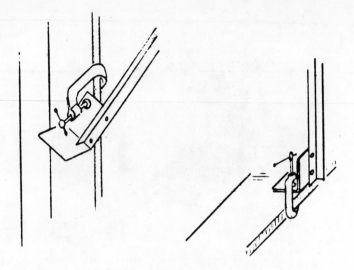

Figure 3.   Props to support sampling devices can be readily fabricated from simple materials.

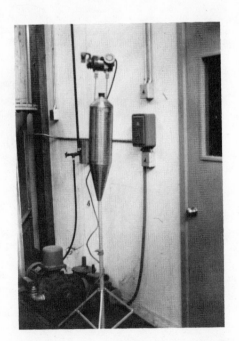

Figure 4.   A wide variety of collapsible stands are available from photographic equipment suppliers.

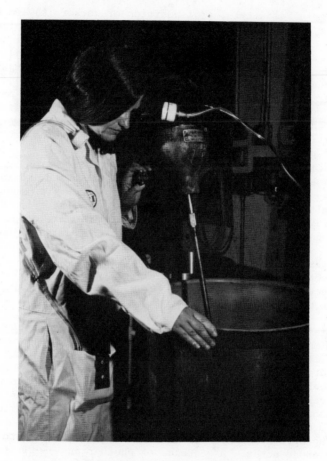

Figure 5.   Here, a coat hanger has been modified to support an aerosol monitoring cassette and sampling hose.

damage, heat and dirt was controlled by shock mounting direct
reading instruments and calibrated air movers in ice chests
(Figure 6). Temperature was controlled by a thermostatically
regulated thermoelectric heat pump. The ice chests were secured
to earth moving equipment with snubbers and straps (Figure 7).

Sample Validity. The disposition of workers toward their
employer can impinge on the validity of collected samples. Employees that are concerned about the employer's fiscal survival
and the survival of their jobs may unconsciously or deliberately
avoid exposure or protect their sampling train from exposure.
Employees who feel animosity toward the employer or feel the
employer is insensitive to occupational health issues may deliberately spike samples. Little information is available to aid
industrial hygienists in evaluating the influence of worker
disposition on sampling validity. Seals and guards provide only
limited protection against sabotage.

Making frequent rounds provides some protection against
sample spiking; however, detection of employees deliberately
avoiding their normal exposure, during sampling, is much more
difficult. Care should be used to avoid confusing innocent
horseplay with deliberate tampering. Workers are often overcome
by the novelty of sampling equipment in their area or on their
person and engage in such activities as shouting into noise
dosimeters or waving solvent soaked rags near the air intake
of direct reading instruments. Typically, employees tire of
these activities after a first few minutes of sampling, and
usually do not exert a significant impact on the values observed.

Field Notes. The collection of accurate field notes is of
vital importance to the success of industrial hygiene surveys.
No estimate of exposure can have any utility unless it can be
associated in the circumstances of its generation. Notes can be
taken both in written form or with miniature tape recorders.
Each method has both strengths and weaknesses; the best strategy
will usually involve a combination of the two. Many miniature
tape recorders require wind screens for use out of doors.

Photography can add an enormous amount of value to written
and recorded notes. A sequence of photographs — first showing
the general area and then the more specific location, and finally,
the actual sampler deployment — can be effective in describing
the circumstances of the sample collected. Photographs of work
operations in progress are also valuable. Written or recorded
notes should be kept to identify the subject, significance, and
orientation of each photograph. Photographs also, assist in
report preparation by refreshing the investigator's memory.
Photographs can provide a permanent record that describes the
exact circumstances of exposure or sample collection.

Many outstanding texts are available that discuss the
intricacies of technical and industrial photography. Prior to

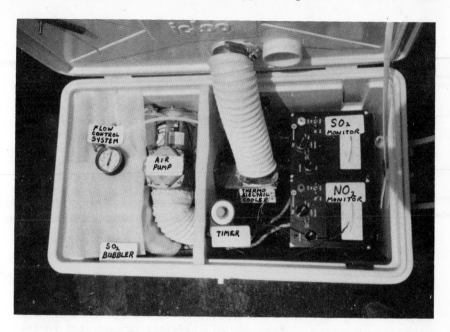

Figure 6. Insulated containers can protect instrumentation against mechanical damage, weather, and dust.

Figure 7. Securing instrumentation can pose problems under severe conditions. Here, webbing straps and rubber snubbers are used to secure instrument packages to earth-moving equipment.

deciding on photographic activities, the plant safety and security requirements must be closely checked.

Transportation of Collected Samples. After the survey is completed, samples and blanks should be transported together. Samples may be carried in the investigator's luggage with the instrumentation, by U.S. mail, or by some commercial parcel carrier; the method used will depend upon Department of Transportation Regulations, analytical protocol and convenience. Some sampling techniques may require fixing procedures or operations prior to transport. Some types of particulate samples must be hand carried. Many sampling methods require collected samples to be refrigerated during transportation. Care should be taken when dry ice is used to avoid freezing some liquid sampling media or denaturing the plastic caps often used to seal charcoal tubes. Ice substitutes can satisfy some short term cooling requirements.

Conclusion

Industrial hygiene logistics provide an infrastructure that supports the collection of data concerning occupational exposures to hazardous agents. The topic of logistics may be divided into two components, software and hardware. Software consists of the procedures for making contacts, collecting information, planning strategy, implementing surveys and conducting surveys.

The software of industrial hygiene logistics used to receive considerable discussion in the professional literature. A wider discussion in the contemporary literature could enhance the spread of solutions to problems investigators encounter conducting field surveys.

The sophistication, precision and sensitivity of industrial hygiene hardware has increased many orders of magnitude since the hand crank pumps, used in the early days. We expect that advances in semiconductor technology and analytical chemistry will allow the quality of industrial hygiene instrumentation to improve more rapidly in the future. If one general suggestion could be applied to the development of hardware, it would be to increase the flow of information between field workers and instrument designers. This exchange could allow the design of instruments that are easier to use and more efficient. Many modern field instruments are awkward to carry on ladders and to use while walking or carrying note pads. Often trivial alterations could substantially ease difficulties of using equipment in the field.

Literature Cited

1. NIOSH, "Manual of Analytical Methods."
   NIOSH: Cincinnati, Ohio, Vol. 1, 1977 - Vol. 5, 1979.

2. NIOSH, "Manual of Sampling Data Sheets."
   NIOSH: Cincinnati, Ohio, 1977.

3. NIOSH, "Occupational Diseases--A Guide to their
   Recognition." NIOSH: Cincinnati, Ohio, 1977.

4. Fassett, David, "Patty's Industrial Hygient and Toxicology,"
   2nd edition. Interscience: New York, 1978.

5. Goodman, L.S. and Gilman, A., "The Pharmacological Basis
   of Therapeutics", 5th edition. Macmillan Publishing
   Company, Inc.: New York, 1975.

6. Gosselin, R.E.; Hodge, H.C.; Smith, R.P.; Gleason, M.N.;
   "Clinical Toxicology of Commercial Products", The Williams
   and Wilkins Company: Baltimore, 1976.

7. Sax, I, "Dangerous Properties of Industrial Materials".
   Van Nostrand Reinhold: New York, 1979.

8. Phemister, T.A., "Hazardous Materials Regulations of the
   Department of Transportation", Association of American
   Railroads, Bureau of Explosives: Washington, D.C., 1980.

9. Hosey, A.D., "General Principle in Evaluating the
   Occupational Environment in the Industrial Environment--
   its Evaluation and Control", U.S. HEW: Washington, D.C.,1973.

10. A.C.G.I.H., "Air Sampling Instruments", A.C.G.I.H.:
    Cincinnati, Ohio, 1978.

11. A.C.G.I.H., "Manual of Recommended Practices for Combustible
    Gas Indicators and Portable Direct Reading Hydrocarbon
    Detectors", A.C.G.I.H: Cincinnati, Ohio, 1980.

12. Leidel, N.A.; Busch, K.A.; Lynch, J.R., "Occupational
    Exposure Sampling Strategy Manual". NIOSH: Cincinnati, Ohio,
    1977.

RECEIVED October 27, 1980.

The NIOSH Action Level

A Closer Look

JAMES C. ROCK

The USAF Occupational and Environmental Health Laboratory,
Brooks Air Force Base, San Antonio TX 78235

Most practicing industrial hygienists in America today must manage their programs with at least two objectives in mind: to limit the physiological risk to employees from their occupational exposures, and to limit the legal risk to employers from OSHA compliance inspections. This paper compares the probability of noncompliance during an OSHA inspection to three different measures of physiological risk: the average long-term exposure; the NIOSH Action Level as originally proposed (1); and the legal action level which is usually set equal to half of the permissible exposure limit (PEL).

The comparison is conducted by identifying those occupational environments which will pass or fail each set of decision criteria. This paper deals only with occupational stresses which pose a chronic health hazard; acute health hazards require different treatment. The most surprising conclusion is that there are some commonly encountered work environments which are very likely to pass the NIOSH Action Level Test even though there is better than a 50% chance that an OSHA compliance officer collecting six samples during one visit would be able to identify a citable violation of an 8-hour PEL.

The Model of the Occupational Environment

The model of occupational exposures which forms the basis for the proposed NIOSH Action Level is introduced in this first section; the NIOSH decision criteria are generalized in the second section; and the decision probabilities are compared in the last section of this paper. A conscious effort has been made to present important results graphically, but the appendix contains equations so that an interested, mathematically-inclined reader can check or extend the results.

Daily Exposures. For the model to be as general as possible, it must represent all likely occupational exposure situations. It is, therefore, useful to define a normalized exposure equal to the actual 8-hour time weighted average exposure (TWAE) divided by the applicable 8-hour standard (PEL).

(1) x = TWAE/PEL

Since the goal of an occupational health program should be to control the exposures for each and every employee, it is convenient to consider the group of daily exposures received by an employee during one year of employment. A typical 50-week work year would include 250 separate daily exposures. Let E represent the number of days during the year that the exposure exceeds the standard. Then the probability of an overexposure on any given day is given by:

(2) $P(x > 1) = e$, and $P(x \leq 1) = (1 - e)$, where $e = E/250$.

Distribution of Daily Exposures. From Equation 1, x represents normalized daily exposures. It is generally recognized that x is a lognormally distributed random variable (1). Therefore, the probability density function for the population of exposures which is experienced by one worker is completely characterized by two parameters: its geometric mean, GM, and its geometric standard deviation, GSD (2). Let $\bar{x}$ represent the true long-term average of all normalized daily exposures and recall from Equation 2 that e is the fraction of those daily exposures which exceeds the standard. GM can be calculated from $\bar{x}$ and GSD using Equation A-7 or from e and GSD using Equations A-2 and A-13 (3, 4). Thus, the lognormal probability density function, pdf(x), is completely characterized by any one of the following pairs of parameters: (GM,GSD), (e,GSD), or ($\bar{x}$,GSD). In each pair of parameters, the first parameter is a measure of the "dirtiness" of the occupational environment, while GSD is a measure of the day-to-day variability in the environment. In this paper, e and GSD are used to identify the various possible occupational environments.

Intuitively, GSD can be considered to be inversely related to the efficacy of installed engineering controls. A GSD of 1.0 means that all exposures are identical, a condition nearly realized in some laminar flow clean rooms. When GSD > 2.5, as may be the case for some kinds of maintenance work, it is likely that there are no functioning engineering controls. The exposure variability for most American workers is characterized by a GSD lying between 1.2 and 2.5 (3).

It is difficult to determine how much "dirtiness" is too much. Nearly everyone would agree that if more than half of the daily exposures exceed the standard ($e > \frac{1}{2}$) or if the long-term average exposure is greater than the standard ($\bar{x} > 1$), then the workplace is too dirty. There is much less agreement on when an environment is clean enough to be considered acceptable.

Long-Term Average Daily Exposure. To shed some light on that question, it is useful to consider how $\bar{x}$ behaves as a function of e and GSD. Figure 1 reveals that even in work environments where exposures are highly variable and where the daily exposure exceeds the standard 20% of the time, the average daily exposure is still less than the standard. Figure 2 is a contour plot of the surface represented by Equation A-7. In Figure 2, each pair of values for (e,GSD) specifies the average daily exposure received by one hypothetical worker and this average is plotted as one point on the surface.

For convenience in the rest of this paper I propose to coin a term which I shall call the Average Exposure Limit (AEL). I have selected for the AEL the contour where the average daily exposure equals 95% of the PEL (AEL = 0.95). Just as the PEL is a standard against which one tests the value of each daily exposure, the AEL is a standard against which one tests the average of many daily exposures. In Figure 2, all those workers whose exposure distributions lie to the left of the AEL contour are deemed to work in an acceptable workplace because, for them, $\bar{x}$ < AEL; those on the right are deemed to work in an unacceptable workplace because, for them, $\bar{x}$ > AEL. By this criterion, six of the workplaces represented in Figure 1 are OK and three of them are NOT OK.

This selection of AEL = 0.95 is a convenient means to enhance the intuitive understanding of the sections which follow. Any readers who disagree with my choice of contour are encouraged to select another contour and to make their own comparisons. Contours not plotted can be easily computed from Equation A-7. Also, please note that the AEL concept does not apply to occupational exposures which have ceiling standards. Further, discussion of the relationship between the AEL, the PEL, and short-term excursion limits is beyond the scope of this paper.

Distribution of Daily Exposure Estimates. It might seem that the mathematical model of the industrial hygiene sampling problem is complete once it has been shown that two parameters are sufficient to uniquely represent the set of daily exposures experienced by an employee during one year. Unfortunately, this is not so. Employee exposures are not directly observable. Each exposure measurement is only an estimate of the true exposure. Let TWAEE be the time weighted average exposure estimate resulting from laboratory analysis of a sample. Then define the normalized exposure estimate as ℓ.

(3) ℓ = TWAEE/PEL

Note that because each sample is likely to be different from the exposure it estimates, the distribution of exposure estimates, $pfd(\ell)$ is different from the distribution of exposures, pdf(x). Leidel, Busch, and Crouse showed that if the uncertainty of the sampling and analytical process is represented by its coefficient

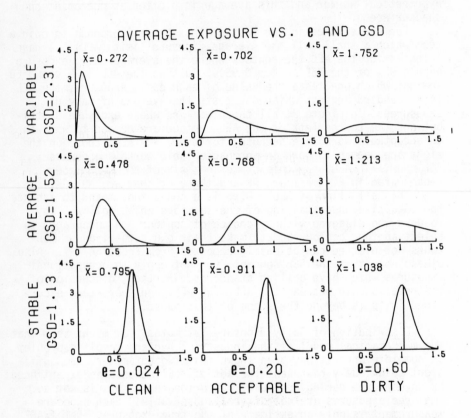

Figure 1.   Nine charts showing how the probability density function, pdf(x), and the long-term average exposure, x, vary as a function of e, the fraction of daily exposures that exceed the standard and GSD, the variability of the work environment.

of variation (CV), and if CV < 0.3, then ℓ can be modelled as a lognormally distributed random variable as shown in Equations A-14 and A-15.

For purposes of illustration in this paper CV = 0.1, a value more or less typical of approved NIOSH sampling and analytical methods for which published values range from 0.01 to 0.25 (5).

Comparing the Population of Exposures to that of Exposure Estimates. The relationship between the distribution of exposures, x, and the distribution of exposure estimates, ℓ, is summarized in the Appendix. The distribution of exposures is characterized completely by two parameters: e and GSD. The distribution of 250 exposure estimates (or samples) which could be observed each year for each employee is broader than the distribution of exposures, and is characterized by three parameters: e, GSD, and CV. Equations A-16, A-17, and A-18 clearly show that although both distributions have the same mean, the median and mode of samples are respectively smaller than the median and mode of exposures.

To put this into practical terms, recall that the only data available to an industrial hygienist are a small fraction of all possible samples; no exposure is directly observable. The average of several industrial hygiene samples is a good estimate of the long-term average exposure, but the median and mode of sample data underestimate the median and mode of the true exposures.

Summary of Key Concepts. The following concepts are so important to understanding the major conclusions of this paper that it is worthwhile to summarize them for repeated reference:

a. The exposures experienced by any one worker over a long period of time are lognormally distributed.

b. The distribution of exposures experienced by each worker can be completely specified by two numbers: e, and GSD.

c. The distribution of sampling and analytical errors is completely specified by the coefficient of variation, CV.

d. The distribution of samples which can be collected from each worker is characterized by three numbers: e, GSD and CV.

A Generalization of the Action Level Decision Criteria

The Decision Rules. The NIOSH action level is a statistical decision threshold designed to help employers attain a high degree of confidence that no more than a small fraction of any employee's daily exposures exceed the standard. After developing the action level, NIOSH provided a set of decision rules to be used by employers who want to determine if their workplaces meet this objective: an exposure estimate smaller than the action level indicates that the employer has probably achieved the objective; an estimate larger than the standard probably indicates a serious

problem; and an estimate between the action level and the standard
means that a decision cannot be made with sufficiently high confidence on the basis of one sample (3).

These decision rules can be generalized to permit easy comparison between OSHA decisions and NIOSH decisions by defining
several new terms. Let UAL represent the upper action level,
which is to be used in place of the standard. Let OK represent
an acceptable workplace and let NOT OK represent an unacceptable
workplace. Let P indicate the probability that the indicated
decision is made. Thus, P(?) represents the probability that a
decision cannot be made with sufficiently high confidence on the
basis of one breathing zone sample. Under these definitions, the
decision rules become:

(4) $P(OK) = P(\hat{x} \leq AL)$

(5) $P(NOT\ OK) = P(\hat{x} > UAL)$

(6) $P(?) = P(AL < \hat{x} \leq UAL)$

The challenge to the statistician is to select values for AL
and UAL such that when the decisions are made, they are made
correctly and with sufficient confidence. It is beyond the scope
of this paper to debate the assumptions underlying the various
competing derivations for values to be assigned to the AL and the
UAL. Instead, I propose to compare the decision probabilities
which result from the various proposed decision thresholds. All
that is required for this comparison is knowledge of $pd_\delta(\hat{x})$ and
repeated applications of Equations A-11 and A-12, with AL = Xq
and UAL = Xg, to compute the decision probabilities.

Once the data are computed in this fashion, there are three
numbers to associate with each point on the surface of Figure 2:
the probability that the environment is OK; the probability that
it is NOT OK; and the probability that no decision can be made on
the basis of one exposure estimate. Therefore, a complete comparison between Figure 2 and the three decision criteria in this
report requires careful consideration of ten 3-dimensional surfaces. This is not practical, so it is necessary to define
contours of significance which can be used to provide a rapid,
easily understood comparison of the different behavior of the
three decision criteria. The first is the contour of unbiased
decisions, and the other three are the contours marking the
boundaries of the regions where one decision is the most likely
outcome of evaluating the meaning of one representative exposure
estimate.

The Unbiased Decision Contour. A workplace which produces
an unbiased decision is defined for purposes of this paper as one
for which P(OK) = P(NOT OK). From Equation 4, P(OK) is represented by the area in the tail of $pd_\delta(\hat{x})$ below the AL. From
Equation 5, P(NOT OK) is represented by the area in the tail of

$pd_\delta(\ell)$ above the UAL. From Equation 6, P(?) is represented by
the area under the $pd_\delta(\ell)$ curve lying between the AL and the UAL.
     There are two important observations to be made concerning
the contour of unbiased decisions. The first is that for workplaces lying to the left of the unbiased decision contour, P(OK)
> P(NOT OK); while for workplaces lying to the right of the
unbiased decision contour, P(NOT OK) > P(OK). The second is that
since our decision criteria have been derived to provide high
confidence in the decisions which are made, it frequently happens
that the most likely outcome from a trial involving one sample
taken from a workplace lying close to the unbiased decision
contour is that a decision cannot be made with sufficient confidence. Because of this, it is important to define additional
decision boundaries.

     The Single Decision Contours. A workplace which is more
likely to produce a decision than to produce no decision and for
which one decision is clearly predominant is defined for purposes
of this paper as lying in a single decision region. There are
two such regions:

(7) P(OK) > P(?) + P(NOT OK)

(8) P(NOT OK) > P(?) + P(OK)

     The boundaries of these regions may be computed from Equa-
tions A-23 and A-24. The region where the predominant decision is
OK lies to the left of the contour defined by Equation A-23. The
region where the predominant decision is NOT OK lies to the right
of the contour defined by Equation A-24. The region between these
two contours is the region where the decision rule makes most of
its errors. Note that there are three types of error: no decision
when one should be made; NOT OK when the decision should be OK;
and OK when the decision should be NOT OK.

     The No Decision Region. A workplace which is more likely to
produce no decision than to produce either or both of the decisions, OK and NOT OK, is defined for purposes of this paper as
lying in the no decision region. There is one such region:

(9) P(?) > P(OK) + P(NOT OK)

     There is no closed form expression for the boundary of this
region. It can only be calculated by numerical methods. This
region is of interest because it identifies those workplaces for
which many samples will be required to insure that a decision can
be made. If workplaces lying in this region need to be sampled,
then it may be wise to develop alternate decision criteria which
exhibit higher decision probabilities, rather than to bear the
unwarranted expense of excess sampling.

LONG TERM AVERAGE EXPOSURE

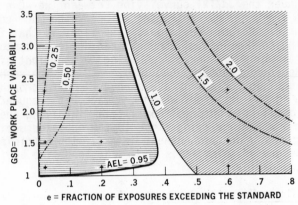

Figure 2. Contour plot showing the long-term average exposure, x, as a function of e, and GSD. The heavy line, where x = 0.95 PEL is the proposed Average Exposure Limit, AEL. Acceptable workplaces lie in the region to the left of the AEL. Each "+" marks the location of one of the charts from Figure 1.

Tabulated Decision Probabilities For Workplaces Shown in Figure 1.

GSD	e = 0.024			e = 0.20			e = 0.60		
	P(OK)	P(?)	P(NOK)	P(OK)	P(?)	P(NOK)	P(OK)	P(?)	P(NOK)

Table I. NIOSH Criteria. AL = f(GSD, CV) and UAL = 1.

GSD	P(OK)	P(?)	P(NOK)	P(OK)	P(?)	P(NOK)	P(OK)	P(?)	P(NOK)
2.306	.0943	.8812	.0245	.0073	.7927	.2000	.0002	.4028	.5970
1.519	.0932	.8803	.0265	.0076	.7892	.2032	.0002	.4069	.5928
1.127	.0824	.8572	.0605	.0118	.7394	.2488	.0010	.4344	.5646

Table II. Legal Action Level Criteria. AL = 0.5 and UAL = 1.

GSD	P(OK)	P(?)	P(NOK)	P(OK)	P(?)	P(NOK)	P(OK)	P(?)	P(NOK)
2.306	.8740	.1016	.0245	.5070	.2930	.2000	.1424	.2606	.5970
1.519	.6261	.3474	.0265	.2167	.5801	.2032	.0323	.3749	.5928
1.127	.0019	.9377	.0605	.0001	.7512	.2488	.0000	.4354	.5646

Table III. OSHA Criteria. One Sample. AL = 0.835 and UAL = 1.165.

GSD	P(OK)	P(?)	P(NOK)	P(OK)	P(?)	P(NOK)	P(OK)	P(?)	P(NOK)
2.306	.9604	.0238	.0158	.7350	.1118	.1532	.3230	.1512	.5258
1.519	.9353	.0537	.0110	.6598	.2222	.1181	.2568	.2908	.4524
1.127	.6544	.3399	.0057	.3173	.6338	.0489	.0940	.6984	.2076

Table IV. OSHA Criteria, Six Samples. AL = 0.835 and UAL = 1.165.

GSD	P(OK)	P(?)	P(NOK)	P(OK)	P(?)	P(NOK)	P(OK)	P(?)	P(NOK)
2.306	.7847	.1241	.0911	.1577	.2110	.6313	.0011	.0102	.9886
1.519	.6694	.2664	.0642	.0825	.3879	.5295	.0003	.0267	.9730
1.127	.0785	.8877	.0337	.0010	.7392	.2598	.0000	.2476	.7524
	P(OK)	P(?)	P(NOK)	P(OK)	P(?)	P(NOK)	P(OK)	P(?)	P(NOK)
GSD	e = 0.024			e = 0.20			e = 0.60		

Decisions Made by Various Decision Criteria.

To gain the maximum amount of insight from the following discussion, it is necessary for the reader to make frequent comparisons between the various figures. To make this as easy as possible, Figures 2, 3, 4, 5, and 6 have been reproduced to the same scale and have been annotated by "+" to mark the points which correspond to the nine charts in Figure 1. The three decision probabilities, P(OK), P(?) and P(NOT OK), are computed for each of these nine sample workplaces using each of the decision criteria. Tables I, II, III, and IV summarize the results of this computation. In the course of the discussion, these sample workplaces are referred to by their coordinates (e,GSD). Thus, the workplace in the lower left corner is (0.024, 1.13), while the middle one in the right-hand column is (0.60, 1.52).

Decisions Made by the NIOSH Action Level Criteria. Figure 3 shows the decision contours for the NIOSH Action Level Decision Criteria, and Table I summarizes the decision probabilities for each of the nine sample workplaces. Recall from Equation A-19 that the AL is computed from GSD to provide 95% confidence that no more than 5% of the daily exposures exceed the standard if one randomly collected sample is less than the AL. In terms of the variables used in this paper, (e > 0.05 with p $\geq$ 0.05 if $\bar{x}$ > AL).

Figure 3 helps to demonstrate that the requirement for 95% confidence makes the NIOSH criteria rather inefficient and conservative. The inefficiency is illustrated by the size of the no decision region, which includes most of the workplaces where less than half of the daily exposures exceed the standard (6). The conservativeness is illustrated by the virtual absence of workplaces for which P(OK) > ½.

In Table I, the NIOSH decision criteria is shown to have poor efficiency by the three dirty workplaces for which e = 0.6. In these cases, a worker would be exposed above the standard three days out of five, and his long-term average exposure would be greater than the 8-hour PEL. Very few people would disagree with the decision to call these workplaces NOT OK. Nevertheless, these workplaces will be declared NOT OK by the NIOSH Action Level decision criteria only about 60% of the time. This inefficiency is further illustrated by the fact that only one of the three average workplaces with e = 0.2 has P(?) < 0.75. That one is (0.2, 1.13) and it also illustrates the conservativeness of the NIOSH criteria since on those infrequent occasions when a decision is made, the odds are 21 to 1 to decide NOT OK. However, Table I most clearly illustrates the conservativeness of the NIOSH criteria by the fact that P(OK) < 0.1 for the three clean workplaces where e = 0.024.

Decisions Made with the Legal Action Level Criteria. Proper application of the NIOSH Action Level (Equation A-19) requires

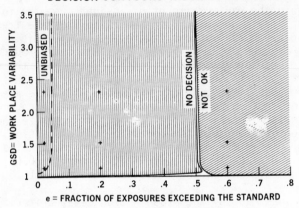

Figure 3. Decision contours resulting from the NIOSH Action Level Criteria: $AL = f(GSD)$, and $UAL = 1$. Each "+" marks the location of one of the charts from Figure 1. Note that the NO DECISION region includes nearly the whole area between the GSD axis and the NOT OK region.

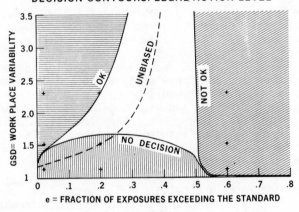

Figure 4. Decision contours resulting from the Legal Action Level Criteria: $AL = \frac{1}{2}$, and $UAL = 1$. Each "+" marks the location of one of the charts from Figure 1.

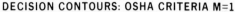

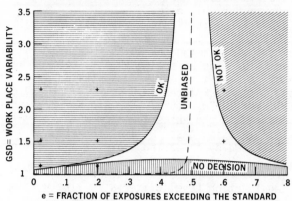

Figure 5. Decision contours resulting from OSHA Compliance Criteria: AL = 0.835, and UAL = 1.165. Each "+" marks the location of one of the charts from Figure 1. Note that P(NOT OK) = probability of a citation based on one sample.

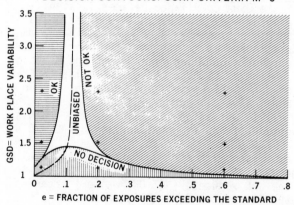

Figure 6. Decision contours resulting from OSHA Compliance Criteria: AL = 0.835, and UAL = 1.165, and assuming that the decision is based on six statistically independent samples. In comparison with Figure 5, many more environments are subject to a citation. This even includes some environments within the acceptable region bounded by the AEL contour in Figure 2.

knowledge of both CV and GSD. It is impossible to estimate CV
and GSD from one sample, and so it is clear that in the real
world the action level criteria have to be modified. A currently
common practice is to set AL = ½ and UAL = 1. Figure 4 shows the
decision contours for this case.

Since UAL = 1, as it did in Figure 3, the single decision
contour for P(NOT OK) does not change. For GSD > 1.22, setting
AL = ½ amounts to increasing the AL as compared with its value
from Equation A-19. The effect of this is to increase P(OK) at
the expense of P(?) as is clearly demonstrated by comparing
Tables I and II for workplaces (0.024, 2.31), (0.20, 2.31),
(0.024, 1.52), and (0.20, 1.52). For GSD < 1.22, setting AL = ½
amounts to decreasing the AL as compared with its value from
Equation A-19. This increases P(?) at the expense of P(OK) as
clearly demonstrated by comparing Tables I and II for workplaces
(0.024, 1.13), (0.20, 1.13), and (0.6, 1.13).

There are two features of Figure 4 which deserve comment.
First, the Legal Action Level Criteria is likely to declare as OK
only three of the six workplaces declared OK by the Average Exposure Limit concept of Figure 2. Second, and in contrast with the
NIOSH Action Level Criteria of Figure 3 where nearly all common
workplaces were in the no decision region, the Legal Action Level
Criteria is slightly more efficient in making decisions; only
those relatively stable workplaces with GSD < 1.7 are in its no
decision region.

Decisions Made with the OSHA Compliance Criteria. The
problem of the OSHA compliance officer is significantly different
than the problem addressed by the NIOSH Action Level. The compliance officer is not nearly so interested in how many exposures
exceed the standard as in whether it can be shown with 95% confidence that the standard was exceeded on the basis of the sample
which was collected. Thus, the OSHA Compliance Criteria is not a
function of workplace variability, GSD, but only of the random
sampling and analytical errors, CV (5). As explained earlier, CV
= 0.1 for all illustrations in this paper. In this case, the
OSHA Decision Criteria is characterized by AL = 0.8355 and UAL =
1.1645.

In comparison with the Legal Action Level Criteria, both the
AL and the UAL for the OSHA Criteria are shifted to the right.
Shifting the UAL reduces P(NOT OK) while increasing P(?).
Shifting the AL reduces P(?) while increasing P(OK). Since the AL
is shifted proportionately further than the UAL, the net result is
that the compliance officer has a higher P(OK) and a lower P(NOT
OK) than does an employer who makes his decisions using the Legal
Action Level Criteria. In addition, for most workplaces the
compliance officer has a much lower P(?) than does the employer.
These comments can be reinforced by careful comparison of Figures
4 and 5 with the aid of Tables II and III.

It is informative to compare the decision regions of the
OSHA Compliance Criteria as shown in Figure 5, with the Average
Exposure Limit Concept of Figure 2. Five of the six workplaces
identified as OK in Figure 2 are likely to be declared OK by an
OSHA compliance officer. However, only one of the three unacceptable workplaces of Figure 2 is likely to be declared NOT OK by
the criteria of Figure 5.

Perhaps the most surprising feature of Figure 5 is the fact
that only those workplaces where more than half of the daily exposures exceed the PEL give the compliance officer better than a 50-
50 chance of collecting a sample which will support a citation!

How, Then, Does OSHA Issue So Many Citations? To support a
citation, a compliance officer must demonstrate with a high
degree of confidence that at least one employee was exposed above
the standard on the day of the inspection. Assuming that day-to-
day variability in exposures is approximately equal to the worker-
to-worker variability on the day of the inspection, then it would
be possible, in principle, for a compliance officer to collect n
statistically independent samples. Under these conditions, the
probability that the workplace will be found in compliance is
equal to the probability that all n samples are less than the AL.
Let Pn(OK) represent the probability of compliance after n samples
are collected and P(OK) represent the probability that a single
sample is less than AL. Then one has:

(10) $P_n(OK) = (P(OK))^n$

Likewise, the probability that at least one sample is greater
than the UAL is given by one minus the probability that none of
the samples is greater than the UAL.

(11) $P_n(NOT\ OK) = (1 - (1 - P(NOT\ OK))^n)$

Figure 6 shows the decision contours for a compliance officer who takes six statistically independent samples during an
inspection. Table IV lists the decision probabilities for each of
our nine sample workplaces. Clearly, collecting more samples
increases the probability of demonstrating noncompliance and
decreases the probability of demonstrating compliance.

By comparing Figure 6 with Figure 2, it is clear that by
taking as few as six samples, a compliance officer can increase
to better than 50-50 the odds of finding evidence to support a
citation, even for some workplaces which lie inside the presumably
acceptable region bounded by the AEL = 0.95 contour.

The principle is very easily generalized. By collecting
enough samples, a compliance officer can move the single decision
contour for a citation as far to the left as he chooses. Given
enough samples, every workplace could be subjected to a citation!
For example, by collecting 14 samples, a compliance officer moves

the single decision contour for issuance of a citation far enough to the left to include five sample workplaces not included in the contour of Figure 5: (0.2, 1.13), (0.2, 1.52), (0.2, 2.31), (0.6, 1.13) and (0.6, 1.52). That is, 14 samples guarantee that P(citation) = P14(NOT OK) is greater than 1/2 for these workplaces.

The most telling comparison is that between Figures 4 and 6, There are many workplaces with GSD > 1.5 which are likely to be declared OK by the Legal Action Level Criteria but which could be cited by a conscientious compliance officer who collects more than five or six samples.

Conclusion.

All of the decision criteria examined in this paper make a decision on the basis of one sample. It should be clear from the discussion that one sample is usually not enough. An employer who wants to document conditions in his workplace would be well advised to collect enough samples to use a more efficient decision rule such as the one-sided tolerance test suggested recently by Tuggle (7). To be truly representative, samples must be drawn from all operating conditions; i.e., from each season of the year, from each day of the week, from each process condition.

In addition, a compliance officer should not expect to be able to sustain a citation based solely on evidence that on the day of his visit one sample exceeded the upper action level defined in the OSHA Industrial Hygiene Field Operations Manual. He should also be required to present corroborative evidence to demonstrate that the sample in question is a representative sample. If it is the only one of a large number of representative samples which exceeds the UAL, a citation is probably not warranted unless a ceiling standard has been violated.

Finally, any decision strategy can be compared to any other on the basis of the decision contours defined in this paper. It is recommended that every candidate decision criteria, whether it is based upon one or several samples, be evaluated to determine which workplaces are accepted and which are rejected before it is adopted. None of the decision criteria examined in this paper are able to fairly determine the quality of all workplaces.

Acknowledgement. My sincere appreciation to Dr David Cohoon of the USAF School of Aerospace Medicine for his continuing support, his constructive criticism, and the use of his computer graphic software. Also to Dr R.M. Tuggle of the US Army Environmental Hygiene Agency for his timely critical review and suggestions for improving my manuscript.

LIST OF MATHEMATICAL SYMBOLS

AEL = Average Exposure Limit: If $\bar{x} \leq$ AEL, Decide OK.

AL = Action Level: If $\hat{x} \geq$ AL, Decide OK.

E = The number of days/year when $x >$ PEL.

e = Fraction of work days when $x >$ PEL.

exp(.) = Raise e, the base of ln(.), to the indicated power.

GM = Geometric Mean = Median of x.

$\hat{GM}$ = Geometric Mean = Median of $\hat{x}$. Note that $\hat{GM} <$ GM.

GSD = Geometric Standard Deviation of x.

$\hat{GSD}$ = Geometric Standard Deviation of $\hat{x}$. $\hat{GSD} >$ GSD.

ln(.) = Take the natural logarithm of the indicated quantity.

NOK/NOT OK = The decision one makes whenever $\hat{x} >$ UAL.

OK = The decision one makes whenever $\hat{x} \leq$ AL.

P(.) = The probability that the indicated event occurs.

p = The (usually small) probability that $e > \theta$.

pdf(x) = The lognormal probability density function of x.

$pd\hat{f}(\hat{x})$ = The lognormal probability density function of $\hat{x}$.

PEL = 8-hour Permissible Exposure Limit.

Pn(OK) = The probability that n statistically independent industrial hygiene samples are all less than AL.

Pn(NOT OK) = The probability that at least one of n statistically independent industrial hygiene samples is $>$ UAL.

UAL = Upper Action Level: If $\hat{x} >$ UAL, Decide NOT OK.

x = The random variable representing true daily exposures.

$\bar{x}$ = The arithmetic average of all true daily exposures.

$\hat{x}$ = The random variable for daily exposure estimates.

Xg = The gth fractile of x: $P(x > Xg) = g$.

Xq = The qth quantile of x: $P(x \leq Xq) = q$.

y = ln(x), a normal random variable.

Za = The ath quantile of y: $P([(y-\mu)/\sigma] \leq [Za]) = a$.

$\mu = ln(GM)$ = The arithmetic average of $y = ln(x)$.

$\sigma = ln(GSD)$ = The standard deviation of $y = ln(x)$.

θ = The allowed fraction of work days for which $x >$ PEL.

Literature Cited.

    1. Leidel, N.A., K.A. Busch, and W.E. Crouse. "Exposure
Measurement Action Level and Occupational Environmental Vari-
ability," Department of Health, Education, and Welfare, National
Institute for Occupational Safety and Health, 4676 Columbia
Parkway, Cincinnati, NIOSH Technical Information, HEW Pub. No.
(NIOSH) 76-131, Cincinnati, Ohio 45226, April 1975.

    2. Aitchison, J., and J.A.C. Brown. "The Lognormal Distri-
bution," Cambridge University Press, New York, 1976.

    3. Liedel, N.A., K.A. Busch, and J.R. Lynch. "Occupational
Exposure Sampling Strategy Manual," Department of Health, Education, and Welfare, National Institute for Occupational Safety and
Health, NIOSH Technical Information, DHEW (NIOSH) Pub. No. 77-
137, 4676 Columbia Parkway, Cincinnati, Ohio 45226, January 1977.

    4. Natrella, M.G. "Experimental Statistics," National
Bureau of Standards Handbook 91, Superintendent of Documents, US
Government Printing Office, Washington DC 20402, October 1966.

    5. OSHA Instruction CPL 2-2.20, Industrial Hygiene Field
Operation Manual, US Department of Labor, OSHA, Washington DC
20210, 30 April 1979.

    6. Rock, J.C. "The NIOSH Sampling Strategy...A Useful
Industrial Hygiene Tool or an Economic Disaster?," Paper 131,
1979 American Industrial Hygiene Conference, Chicago, Illinois,
May 31, 1979.

    7. Tuggle, R.M. "The NIOSH Decision Scheme--A User's
Perspective," Paper 321, 1980 American Industrial Hygiene Conference, 23 May 1980.

Appendix--Key Mathematical Concepts and Equations

Normal Random Variable. The probability density function of a normally distributed random variable, y, is completely characterized by its arithmetic mean, μ, and its standard deviation, σ. This is abbreviated as $y \sim N(\mu, \sigma^2)$ and written as:

(A-1) $pdf(y) = (1/\sqrt{2\pi}\, \sigma)\exp -\frac{1}{2}(((y - \mu)/\sigma)^2)$

The Z-variate can be defined so that there is a one-to-one mapping, $(Za \leftrightarrow a)$ expressed by a definite integral.

(A-2) $a = (1/\sqrt{2\pi})\int_{-\infty}^{Za} \exp(-\tau^2/2)\, d\tau$

Because (a) is the area in the lower tail of the normal distribution, Za is called the ath quantile of the standard normal distribution, (or the (100)(a)th percentile). A useful identity follows directly from the symmetry of the Gaussian distribution in Equation A-2 (4).

(A-3) $Z(1-a) = -Za$

Note that the area (a) also has a more general probabilistic interpretation.

(A-4) $P(y \leq (\sigma Za + \mu)) = a$

Lognormal Random Variable. Every normally distributed random variable, y, is uniquely associated with a lognormally distributed random variable, x, whose probability density function is completely characterized by its geometric mean, GM, and geometric standard deviation, GSD (2).

(A-5)
$$y = \ln(x)$$
$$\mu = \ln(GM)$$
$$\sigma = \ln(GSD)$$

The distribution of x is abbreviated as $\ln x \sim N[\ln(GM), (\ln(GSD))^2]$ and written:

(A-6) $pdf(x) = [1/(x\sqrt{2\pi}\,\ln(GSD))]\exp(-\frac{1}{2}[(\ln(x)-\ln(GM))/\ln(GSD)]^2)$

Since (x) has a skewed distribution, its central measures are distinct with Mean x > Median x > Mode x (2).

(A-7) $\overline{x} = \text{Mean } x = \exp[\ln(GM) + 0.5(\ln(GSD))^2]$

(A-8) Median x = GM

(A-9)   Mode x = exp[ln(GM) - (ln(GSD))^2]

The cumulative distribution function for the lognormal random variable x is derived by combining Equations A-4 and A-5.

(A-10)   P(lnx $\leq$ [Za ln(GSD) + ln(GM)]) = a

Separating Equation A-10 into two equations helps to clarify the procedure needed to quickly calculate the qth quantile, Xq, of pdf(x) when x is a lognormally distributed random variable. Since q and Zq are related by Equation A-2, Equation A-11 can be used to solve for Xq given q or for q given Xq.

(A-11)   ln(Xq) = ln(GM) + Zq ln(GSD), where P(x $\leq$ Xq) = q

Equation A-12 can be derived from Equations A-3 and A-11. It is presented here for easy reference.

(A-12)   ln(Xg) = ln(GM) - Zg ln(GSD),   where P (x > Xg) = g

Recall from Equation 2 that e = P(x > 1). Then the following useful relationship between GM, GSD, and e, follows directly from Equation A-12.

(A-13)   ln(GM) = Ze ln(GSD)

Thus, although a lognormal distribution is usually characterized by GM and GSD, it is equally valid to characterize it by e and GSD, since Equation A-13 permits GM to be calculated uniquely from e and GSD.

The Effect of Imperfect Sampling. Let ℓ be the random variable which estimates the true value of the lognormally distributed random variable (x). The value of (ℓ) is determined by sampling (x) and analyzing with a process which has inherent uncertainty associated with it. The uncertainty is described by the coefficient of variation of the analysis, CV. If CV < 0.3, then (ℓ) can be modelled adequately as a lognormally distributed random varible characterized by GM and GSD as defined below ([1]).

(A-14)   (ln(GSD))^2 = (ln(GSD))^2 + ln(1 + CV^2)

(A-15)   ln(GM) = ln(GM) - ½ ln(1 + CV^2)

It is possible to combine Equations A-7, A-8, A-9, A-14, and A-15 to derive expressions for the mean, median and mode of ℓ.

(A-16)   Mean ℓ = exp(ln(GM) + ½(ln(GSD))^2) = Mean x

(A-17)   Median ℓ = GM = $\dfrac{\text{Median x}}{\exp(\frac{1}{2} \ln(1 + CV^2))}$

(A-18)   Mode $\hat{x} = \exp(\ln(GM) - (\ln(GSD))^2) = \dfrac{\text{Mode } x}{\exp(1.5 \ln(1 + CV^2))}$

Note that since GSD > GSD, the median and Mode of $\hat{x}$ are respectively smaller than the Median and Mode of x, while their means are equal.

The Action Level. By combining the last equation in Ref 2 with Equation A-14, one can derive an explicit equation for the NIOSH Action Level.

(A-19) $\ln(AL) = (Zp)(\ln(GSD)) + (Z\theta)(\ln(GSD)) - \frac{1}{2}(\ln(1 + CV^2))$

$\ln(AL) = (Zp)\sqrt{\ln(1 + CV^2) + (\ln(GSD))^2} + (Z\theta)(\ln(GSD))$
$\quad - \frac{1}{2}(\ln(1 + CV^2))$

In this equation, p & Zp and θ & Zθ are related through Equation A-2. θ is the fraction of daily exposures to be allowed to exceed the standard, and p is the (normally small) probability that at least $(100)(\theta)\%$ of the daily exposures represented by one sample exceeds the standard if that sample exceeds the action level. Numerical examples in this paper assume $p = \theta = 0.05$.

Decision Contours. Let AL = Xq in Equation A-11 and UAL = Xg in Equation A-12. Rearranging these equations:

(A-20) $Zq = (\ln(AL) - \ln(GM))/\ln(GSD)$, where $P(\hat{x} \leq AL) = q$

(A-21) $Zg = (\ln(GM) - \ln(UAL))/\ln(GSD)$, where $P(\hat{x} > UAL) = g$

To derive the unbiased decision contour, note that $P(\hat{x} < AL)$ = $P(\hat{x} > UAL)$ along the contour. This is achieved when $Zq = \overline{Z}g$, which means setting Equations A-20 and A-21 equal. Combine that expression with Equations A-14 and A-15 to derive Equation A-22, which specifies the unbiased contour in terms of e and GSD.

(A-22) $Ze = [\ln(AL) + \ln(UAL) + \ln(1 + CV^2)]/(2 \ln(GSD))$

Note that the single decision contours occur when $P(\hat{x} < AL)$ = 0.5 or $P(\hat{x} > UAL) = 0.5$. These conditions are met respectively when Zq = 0 or Zg = 0. Substitute zero into Equations A-20 and A-21 and combine with Equations A-14 and A-15 to derive equations for the single decision contours:

(A-23) $Ze = (2(\ln(AL)) + \ln(1 + CV^2))/(2 \ln(GSD))$

(A-24) $Ze = (2(\ln(UAL)) + \ln(1 + CV^2))/(2 \ln(GSD))$

Equation A-23 is the contour for OK. Equation A-24 is the contour for NOT OK.

RECEIVED October 27, 1980.

Industrial Hygiene Air Sampling with Constant Flow Pumps

W. B. BAKER, D. G. CLARK, and W. J. LAUTENBERGER

Applied Technology Center, E. I. du Pont de Nemours and Company, Incorporated, Wilmington, DE 19898

In air monitoring program accurate concentration determinations require careful attention to the sample collection and subsequent analysis of the collected sample. The analysis is the most critical step, especially if very low levels (ppb) of contaminants are being determined. However, in many cases, sampling may be the least accurate step since this job is performed in the field with portable equipment under conditions far less favorable than those that can be created and controlled in the lab(1).

NIOSH states that acceptable air monitoring methods must come within ±25% of the true concentration for 95% of the samples taken. The error factor attributed to sampling pumps is a coefficient of variation of 5% (2). If there is no bias in the sampling pump, the accuracy is:

$$\text{Accuracy} \cong 2 \text{ CVp (Assumes no bias)}$$
$$\cong \pm 10\%$$

where: $CV_p = 5\%$

Errors in sampling (sample volume determination) are due to erroneous measurement of time or flow rate. Time can be measured so accurately that flow rate errors make up the majority of the ±10% variation attributed to the average sampling pump.

Sampling Errors for Flow Variations

Flow rate variations cause three types of sampling errors: 1) direct sample size error; 2) time-of-day bias; and 3) respirable sampling errors. A direct error occurs when the

sampler is calibrated to operate at a given flow rate but
changes flow rate. If a sampler is calibrated at 1 liter/min.
but operated at an average flow rate of 0.9 liters/min., this
10 percent flow rate error would cause a 10 percent error in
sample size.

Direct errors can occur in simple positive-displacement
pumps as the result of battery discharge and other effects.
Figure 1 shows the battery discharge effect on flow rate (3).
The initial flow rate decrease is caused by the NiCd battery
discharge characteristic. Direct error can also result from
the effect of inlet pressure changes on flow as shown in
Figure 2(3). Since the pump is working on a compressable fluid,
air, any increase in the pressure drop reduces the flow rate.
In the field, the collection device will accumulate dirt and
dust in addition to the desired sample. This will cause the
pressure drop to increase with time. If the sampling pump does
not control the flow, it will decrease as pressure drop increases.

Some samplers incorporate an accumulated volume indicator
such as a stroke-counter to permit correction for the above
effects. However, time-of-day bias can occur if the exposure
is not constant. A high contaminant level would be erroneously
indicated if a high exposure occurred during the start of the
sampling period rather than at the end.

When performing respirable sampling, the total error will
be greater than the flow rate error. Cyclones used in
respirable sampling are designed to operate at a specific flow
rate. A deviation from this specific flow rate will cause
greater collection efficiency variations depending on the type
of cyclone and flow rate (4).

It becomes evident that accurate sampling requires a
constant flow rate. Many methods have been used in the attept
to provide a constant flow sampler, including: rotometers,
motor-voltage regulators, restricting orifices, pump stroke-
counters and motor-current compensation. These methods attempt
to provide compensation indirectly for the error effects (5,6).
The flow control method described in this discussion provides
direct monitoring and control of sampler flow rate.

Theory of Operation

A simple positive-displacement sampler system is shown in
Figure 3. The basic system contains a battery and motor
connected to a positive-displacement pump mechanism and provides
an efficient means for moving air through the sampler. In order
to provide feedback flow control the system must be expanded to
include the means to monitor air flow through the pump to
compensate for flow variations.

Figure 4 shows a simple version of a feedback flow control
sampler system. An orifice and pressure switch are used to
monitor air flow.

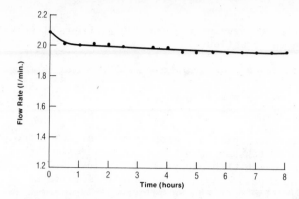

Figure 1.   Flow rate vs. time, simple positive-displacement pump

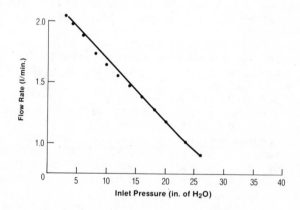

Figure 2.   Flow rate vs. inlet pressure, simple positive-displacement pump

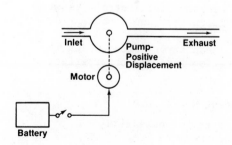

*Figure 3.   Simple positive-displacement
sampler system*

The pressure developed across an orifice varies with flow as shown in the equation([7]):

$$P = \text{Constant} \ x \ \frac{F^2}{\rho}$$

Where: F = flow rate
ρ = fluid density

As flow increases, pressure drop increases. When the pressure exceeds the switch set point, the contacts will close. The flow rate at which this occurs is determined by the orifice size and the pressure switch set point. A variable orifice is used so that flow rate can be adjusted.

The pressure switch is connected to an integrator circuit. The integrator circuit is characterized by an output voltage which changes slowly. This voltage increases when the pressure switch contacts are open indicating that the flow rate is below the set point. When the flow rate is above the set point, the switch contacts are closed and integrator output decreases. The output voltage of the integrator is amplified and applied to the motor.

In this system the motor starts to speed up when the flow rate is low and slows down when the flow rate is high. Thus the control system will always correct in the proper direction for any flow rate error.

Figure 5 further illustrates integrator operation and associated ramping effects. Again it can be seen that with the pressure switch open, the integrator ramps up with time; and conversely when the switch is closed, the integrator ramps down. In a pump with pulsing flow, the switch opens and closes with each pump stroke because it responds to the momentary increases in flow and pressure. During each revolution the pump briefly speeds up and slows down for equal periods of time when operating at constant conditions. Figure 6 shows the operation of the control system when a flow increase is required. Some external influence reduce the flow rate below the set point. This caused the pressure switch to stay open longer during each cycle and the output voltage gradually increased until the flow rate was restored to the set point.

Thus it is shown that the feedback flow control system will act to correct for any condition that would cause flow variation.

With a pulsing flow, there is an error effect which may cause overcompensation. The equation for this error effect is([7]):

$$(\text{Error}) \quad e = \text{Constant} \ x \ f \ x \ \Delta P$$

Where: f is frequency of pulsation
ΔP is the pressure drop

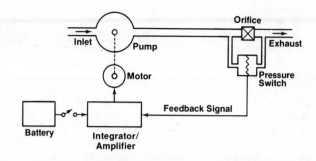

Figure 4. Feedback flow control sampler system

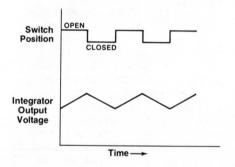

Figure 5. Integrator operation

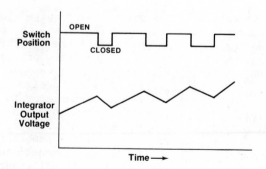

Figure 6. Integrator operation during flow increase

As pump load is increased, this error effect causes the flow
rate to increase more than necessary. This problem is eliminated
by using an accumulator to filter the flow through the orifice.
Figure 7 shows the completed feedback control system which
incorporates an accumulator between the pump and orifice.

Practical Systems

Figure 8 shows the air passages through a typical feedback
control constant flow sampler. The air enters on the left,
proceeds through the valves and diaphragm of the pumping
mechanism into an accumulator. This accumulator is a small
space with a large surface covered by an elastomer sheet to
provide flow-smoothing action. A pressure switch is connected
to taps labeled P.S. High and P.S. Low. The pump mechanism is
operated by a small electric motor and gearhead. Addition of
the control circuit and battery complete this constant flow
system.

Other sampling pump applications require modifications of
this control system for proper performance. When the previous
system was applied to a high flow multi-cylinder pump, the
inherently smooth flow eliminated the need for an accumulator.
The pressure switch experienced rapid wear due to the high
rate of operation, about 100 cycles per second (6000 per minute).
This problem was eliminated by adding a pneumatic filter as
shown in Figure 9. The pneumatic filter consists of a
differential accumulator and four orifices and filters the
pressure signal before it reaches the pressure switch. This
filter had a time delay in passing the signal to the pressure
switch, but by varying the size of the orifices, it was possible
to adjust the time delay so that the pump and control system
cycled at a rate of 8 cycles per second, extending switch life
for several years of normal operation; yet, the cycling
remained rapid enough to eliminate any noticeable variation
in flow.

Another variation of the system was required for a single
cylinder (diaphragm) pump that operated at a high speed and
produced pulsating flow. As shown in Figure 10, this pump
uses a combination of the two control systems previously
described. Because this particular sampler was not required
to fill bags, the pneumatic filter was not designed with
differential accumulators. A larger accumulator in the air
flow passage preceding the orifice provided enough flow-
smoothing for proper operation of the needle valve. An
accumulator with two diaphragms was used to smooth the inlet
flow of this sampler. Thus, it has been demonstrated that
precise flow control is possible with many different pump
mechanisms.

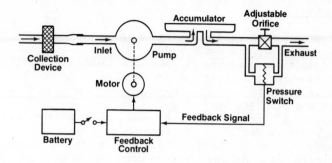

Figure 7. Feedback flow control system for low-speed pump

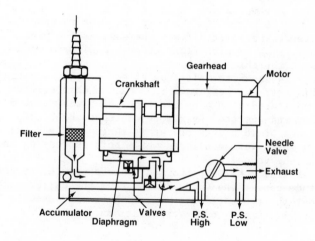

Figure 8. Air flow through low-speed pump

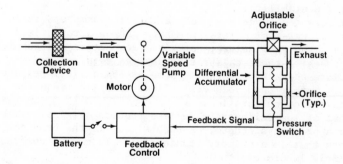

Figure 9. Feedback flow control system for high-speed pump with smooth flow

Performance

The following data illustrates the performance of feedback flow control samplers under test conditions which represent field conditions. Figure 11 shows the flow rate stability versus time for a sampler operated on its battery. The flow control system maintained a constant flow rate even though the battery was discharging(3).

Figure 12 shows the flow rate stability versus load (input pressure drop). A variable orifice connected to the sampler was used to adjust the pressure drop over a range of 0-16 inches water column pressure drop. Flow rate was monitored by a rotometer. The flow rate was constant to within $\pm$ 2 percent(3).

Figure 13 shows the flow rate stability versus ambient air temperature. The samplers were operated in an environmental chamber over a temperature range of 0°C to 70°C. A heat exchanger was connected to the sampler inlet to ensure that the temperature of the inlet air was the same as the ambient air temperature. The sampler flow rate was monitored with a bubble tube outside the chamber.

Figure 14 shows the altitude effect on the sampler flow rate. This test was performed in a vacuum chamber. The flow rate was monitored with a bubble tube which was mounted in the chamber and operated by remote control. Each data point is an average of 18 pieces of data: three samplers and three flow rates which were monitored while both increasing and decreasing the vacuum.

If sampling must be performed under pressure, temperature and altitude conditions which differ from the calibration conditions, the above effects must be known and used as corrections in sample volume calculations.

Calibration

Any sampler, even one which will maintain constant flow, must be calibrated to ensure accuracy. Daily calibration ensures repeatability.

Different methods used for calibrating samplers include: rotometers, wet-test meters, pressure gauges across fixed orifices, mass flow meters, hot wire flow meters and bubble tubes. Each of these calibration devices requires an appropriate correction factor. Some of the devices measure mass flow rather than volumetric flow. Sampling requires volumetric flow calibration.

Most of the calibration devices require original calibration by a primary standard. A bubble tube is the only device mentioned that is a primary standard(8). It can be checked with simple laboratory tools.

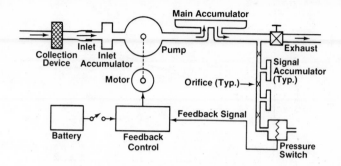

Figure 10. *Feedback flow control system for high-speed pump with pulsating flow*

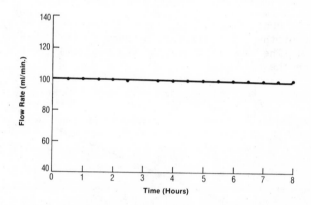

Figure 11. *Flow rate vs. time, pump model P-125*

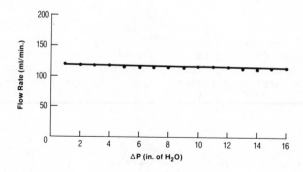

Figure 12. *Flow rate vs. load (pressure differential), pump model P-125*

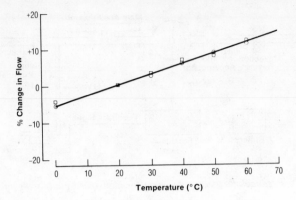

Figure 13.   Temperature effect, pump model P-2500: (○) Pump #1, (□) Pump #2

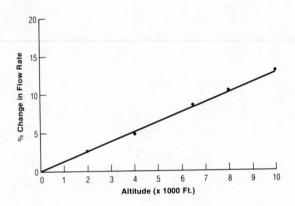

Figure 14.   Altitude effect, pump model P-2500

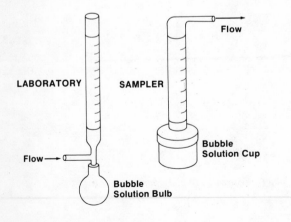

Figure 15.   Bubble tubes

If the bubble tube is connected to the inlet side of the pump, no corrections are required. This requires the sampler bubble tube shown in Figure 15. If a standard laboratory type bubble tube is connected to the exhaust side of the pump, the volume of air which has gone through the sampler pump becomes additionally humidified by the soap solution in the bubble tube causing an increase in the volume. Connecting a bubble tube to the exhaust side of the pump can cause errors in the range of 1 to 3% if the proper correction is omitted(9).

Conclusion

A sampling method is currently available in the field with accuracy that exceeds NIOSH criteria. This accuracy results from a sampler which maintains true constant flow throughout the sampling period.

Acknowledgements

The authors wish to express their appreciation to the many persons who have provided invaluable assistance including: D. O. Conn, III, for early guidance; V. W. Keedy, D. N. Mount and C. J. Lundgren for recent help in design and testing; and A. S. Pollack for editorial assistance.

Literature Cited

1. *The Industrial Environment - Its Evaluation and Control*, NIOSH, 1973, p. 101.
2. Taylor, D. G.; Kupel, R. E.; Bryant, J. M. "Documentation of the NIOSH Validation Tests," NIOSH, Cincinnati, Publication No. 77-185, April 1977.
3. Parker, C. D.; Lee, M. B.; Sharpe, J. C. "An Evaluation of Personal Sampling Pumps in Sub-Zero Temperatures;" Research Triangle Institute Report for NIOSH. Contract No. 210-76-0124, September 1977.
4. "Air Sampling Instruments for Evaluation of Atmospheric Contaminants;" American Conference of Governmental Industrial Hygienists, Fifth Ed., Cincinnati, 1978.
5. Moore, G.; Steinle, S.; LeFebre, H. Am. Ind. Hyg. Assoc. J., 1977, 38, 195.
6. Almich, B. P.; Rubenzahl, M. A.; Carson, G. A. "Electronic Refinements for Improved Operation of Portable Industrial Hygiene for Air Sampling Systems," NIOSH, Cincinnati, Publication No. 75-169; May 1975.
7. Stearns, R. F.; Johnson, R. R.; Jackson, R. M.; Larson, C. A. "Flow Measurements with Orifice Meters," D. Van Nostrand Company, Inc.; New York, 1951.

8. Nelson, G. O. "Controlled Test Atmospheres - Principles and Techniques," Ann Arbor Science Publishers, Inc.; Ann Arbor, 1976.
9. Baker, W. C.; Pouchot, J. F. <u>Flowmeter Survey for Measuring the Flow of Air and Other Gases</u>; Teledyne Hastings - Raydist; Hampton, Va., March 15, 1979, 24-25.

RECEIVED October 17, 1980.

Statistical Protocol for the NIOSH Validation Tests

KENNETH A. BUSH and DAVID G. TAYLOR

National Institute for Occupational Safety and Health, Robert A. Taft Laboratories, 4676 Columbia Parkway, Cincinnati, OH 45226

Early in 1974, the National Institute for Occupational Safety and Health (NIOSH) and the Occupational Safety and Health Administration (OSHA) announced a joint program to complete the existing workroom level standards promulgated by the U.S. Department of Labor in 1972 (29 CFR 1910.1000). At that time, a statistical protocol was developed which has since been used for laboratory validation of over 300 sampling and analytical methods for monitoring employee exposure to the toxic substances in the OSHA regulations. The validations were conducted by Stanford Research Institute (now SRI International) under contracts CDC-99-74-45 and 210-76-0123 with NIOSH. The contractor set up laboratory facilities and air generation-dilution systems to validate methods over a concentration range from one-half to two times the permissible exposure limits (PEL) for the toxic substances shown in 29 CFR 1910.1000, Tables Z-1, Z-2, and Z-3. The OSHA PEL's are occupational health standards for personal exposure limits and may be either an 8-hour time-weighted average (TWA) concentration or a ceiling standard specified for a short time interval (generally 30 minutes or less).

The purpose of the validation program was to assure that accurate personal sampling and analytical methods would be available for use by OSHA in monitoring for non-compliance to the OSHA permissible exposure limits (PEL's). The methods are available to others who may want to use them to determine worker exposure to the substances in the OSHA regulations.

When a standardized sampling/analytical method is used to measure the concentration of a workplace air contaminant, it is certain that there will be some error in the result. But the exact amount of error in a given result is uncertain because quantitative errors occur as if they were random variables, i.e. in a chance manner, even when the method is used correctly. However, for a method which is "in control", what is predictable is the long-term proportion of individual errors which do not exceed a selected limit of error. The

probability that a given error will be less than some selected
limit could be calculated if certain statistical parameters
of the method, were known, namely its coefficient of variation
(CV) and (any) bias. (The CV is referred to as the relative
standard deviation by chemists. It is the ratio of the
standard deviation of replicate concentration measurements
to the mean concentration provided by the method.) Usually,
an approximately normal distribution of errors can be assumed
to exist as a basis for calculating such probabilities. The
CV is assumed to be constant over the four-fold range of
concentrations used in a given method's validation tests.

In this paper, we define an accuracy standard in terms
of its two statistical parameters. However, in order to
evaluate the accuracy of a particular method in terms of its
statistical parameters, we have the problem that estimates
of the method's statistical parameters are themselves subject
to random sampling variations because the estimates must be
calculated from only a finite number of replicate samples.
The high cost of generating and analyzing large numbers of
replicate samples necessitated using only enough samples to
assure that reasonably accurate estimates were obtained of
the CV and bias parameters of a method. Therefore, we also
give statistical decision criteria by which test data for a
method can be evaluated to determine whether there is reasonable confidence that the method meets the accuracy standard.

Several assumptions were made prior to initiating the
actual validation of a given method:

1. The analytical method had to be previously developed
   and tested for items such as sample collection
   efficiency, recovery, and sample stability.
2. Both the air sampling and analytical method were
   to be validated.
3. An independent method was needed to verify the
   laboratory generation atmospheres used to validate
   the method.
4. The accuracy requirement developed for the methods
   had to apply to a single sample analysis, and not
   require an average of the analyses of several
   samples, because OSHA compliance determinations
   may be made on the basis of a single sample.
5. The bias determined in the validation referred to
   the difference between average results of the test
   method and average results of the independent
   reference method. However, it was recognized that
   other sources of bias, e.g. some interferences,
   may increase the true bias of the method in some
   unique field situations.

NIOSH Accuracy Criterion.   Accuracy is determined by
both the precision and bias of the sampling and analytical
method.   Bias was defined under item 5 above as the difference
between average results by the test method and average
results by an independent reference method.   Precision
refers to the distribution of sizes of differences between
results for replicate samples and the mean for the test
method at that concentration.   The accuracy criterion and
its implications with respect to the worst precision and
bias which are allowable are discussed below.   The goal,
however, is to assure that, in the long run, single measurements by the method will come within +25% of corresponding
"true" air concentrations at least 95% of the time.   This
accuracy requirement applies to a concentration range of 0.5
to 2.0 times the environmental PEL.

   In the case of normally distributed sampling and analysis
errors (and no bias) the above requirement implies that the
true coefficient of variation of the total error (i.e. net
precision error of sampling and analysis), denoted by CV_T,
should be no greater than 0.128 derived as follows: CV_T =
0.25/1.96 = 0.128.   The number 0.128 is the largest acceptable
true CV_T for which the net error would not exceed +25% at
the 95% confidence level.   The number 1.96 is the appropiate
Z-statistic (from tables of the standard normal distribution)
at the same confidence level.

   If bias exists, the largest acceptable CV_T would have
to be smaller than 0.128 in order for there to be less than
5% "large errors" (i.e. errors exceeding +25%).   In such
cases, there would not be a 50-50 division of positive and
negative large errors - rather, large errors in the direction
of the bias would occur more often than 2.5% of the time.
Large errors in the other direction would occur correspond-
ingly less often, to keep the total occurrence in both
directions at 5%.

   The solid curve in Figure 1 shows the relationship
between the bias and largest acceptable level of the true
precision parameter (denoted in Figure 1 as the "target
level" of the CV_T of a method).   Note that when the bias is
zero, the largest acceptable true CV_T is 0.128.   Formulae
are given in Appendix I for computing the solid curve giving
the CV_T target level and bias combinations which meet the
NIOSH accuracy standard.

   The dotted curve of Figure 1 gives corresponding maximum
permissible estimates of CV_T (designated $\hat{CV}_T$), based on
laboratory tests performed under the experimental design
described below.   The shaded area indicates the acceptable
$\hat{CV}_T$ region for validation of a method.   The concept of
making allowance for the sampling error in the precision
estimate itself will be developed more fully below under
Statistical Analysis Protocol.   Basically, in the case of an

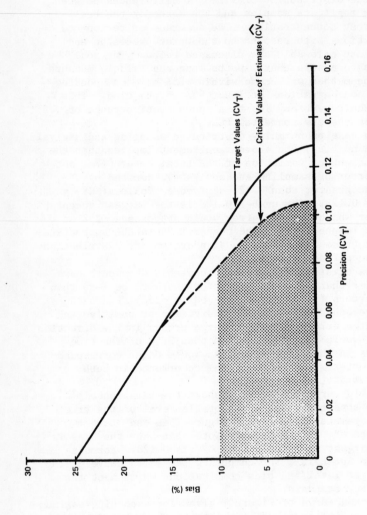

Figure 1.   Combinations of CV_T and bias that meet the NIOSH accuracy standard

unbiased method, the estimate $\widehat{CV}_T$ must be at or below 0.105 in order to be at least 95% confident that the true CV_T is at or below 0.128.

Statistical Experimental Design

Since the accuracy of an air concentration measurement is a function of both sampling and analysis, it is important to evaluate the method by testing both the sampling and analytical portions of the method. The validation program was designed to permit separate evaluation of the levels of error in these two parts of the method, as well as the total (net) error. All validation tests for a given method were carried out in a single laboratory, and although many of the methods had been used in the field previously, no field validations were undertaken.

Initially, the analytical method was tested to assure that it was acceptable for analyte recovery as well as for precision. The sampling medium was spiked with known amounts of the test chemical at three levels corresponding to one-half, one, and two times the occupational PEL for a given air volume. Six spiked samples for each level were analyzed. The success of this portion of the validation assured that the analytical precision was acceptable for the desired concentration range.

The second portion of the validation was to test the net precision due to both the sampling procedure and the analytic method used in sequence. This required the generation of known airborne concentrations of the toxic substance in a laboratory generator-dilution system. Three concentrations, at one-half, one, and two times the PEL, were prepared to test the sampling method. The generated concentrations were verified by a completely independent sampling and analytical method. For some substances this procedure was not possible and calculations based upon known flow and delivery rates, or on the experimentally determined collection efficiency, sample stability, and recovery were necessary to estimate the generated concentration. After selecting the recommended flow rate and sample volume (based upon the sampler capacity), the samples were collected from the laboratory generation-dilution system. Six samples at each of the three concentrations were collected using calibrated critical orifices. The data from these 18 samples, along with the 18 spiked sample results obtained in the analytical validation, were the basic statistical set of data used for the overall method validation. The data were used to determine the precision and bias of the method, which together determine its accuracy. The error of the personal sampling pump was not evaluated experimentally since sample flows in the laboratory tests were controlled by critical orifices in

most cases. However, in the field, sampling pumps are used
and their error was assumed to have a relative standard
deviation of 0.05 (i.e. 5%) based on pump specifications.

Statistical Analysis Protocol

The purpose of the statistical analysis is to estimate
the bias and the precision (measured by the CV_T of the total
precision error of a subject method) and resolve the latter
error into components CV_S due to the sampling method (less
pump error), CV_A due to the analytical method (including
error in the desorption efficiency factor), and CV_P (an
assumed level of pump error). Appendix II gives the definitions and computational formulae for the statistical analysis.

Assuming normally distributed sampling and analysis
errors (and no bias), the NIOSH accuracy standard is met if
the true coefficient of variation of the total error, denoted
by CV_T, is no greater than 0.128. However, estimates of
CV_T (denoted by $\hat{CV}_T$), which were obtained in the laboratory
validations, are themselves subject to appreciable random
errors of estimation. Therefore, a "critical value" for the
$\hat{CV}_T$ was needed (i.e. the value not to be exceeded by an
experimental $\hat{CV}_T$ if the method is to be judged acceptable).
The critical value of $\hat{CV}_T$ has to be lower than the maximum
permissible true value (e.g. lower than $CV_T = 0.128$ when
there is no bias). The maximum permissible value of the
true CV_T will be referred to as its "target level". In
order to have a confidence level of 95% that a subject
method meets this required target level, on the basis of
$\hat{CV}_T$ estimated from laboratory tests, an upper confidence
limit for CV_T is calculated which must satisfy the following
criterion: reject the method (i.e. decide it does not meet
the accuracy standard) if the 95% upper confidence limit for
CV_T exceeds the target level of CV_T. Otherwise, accept the
method. This decision criterion was implemented in the form
of the Decision Rule given below which is based on assumptions
that errors are normally distributed and the method is
unbiased. Biased methods are discussed further below.

For our validations, a $\hat{CV}_T$ is a pooled estimate calculated
from the particular type of statistical data set (36 samples)
described earlier in the Statistical Experimental Design
section of this report. A statistical procedure is given in
Hald[1] for determining an upper confidence limit for the
coefficient of variation. This general theory had to be
adapted appropriately for application to a pooled $\hat{CV}_T$ estimate.
For this design, and under the stated assumptions, there is
a one-to-one correspondence between values of $\hat{CV}_T$ and upper
confidence limits for CV_T. Therefore, the confidence limit
criterion given above is equivalent to another criterion
based on the relationship of $\hat{CV}_T$ and its critical value. The

Decision Rule is as follows:

> Decision Rule: The $\hat{CV}_T$ from lab tests would have to be less than the critical value 0.105 to be 95% confident that the true CV_T is at or below 0.128 (i.e., in order to be 95% confident that future errors by the same method would not exceed ±25% more than 5% of the time).

Figure 1 provides adjustments to critical values for $\hat{CV}_T$ when a method is biased. The dotted curve gives critical values of $\hat{CV}_T$ as a function of bias for a statistical significance test performed at the 5% probability level. Because uniform replicate determinations of the bias were not made in the validation tests, the bias is treated as a known constant rather than an estimated value. The experimental design could be modified to permit determination of the imprecision in the bias by providing for uniform replication of the independent method as well as the method under evaluation. Then the decision chart could be modified to include allowance for variability of replicate bias determinations. In cases where confidence limits can be calculated for the bias, the critical $\hat{CV}_T$ should be read from the dotted curve at a position corresponding to the 95% upper confidence limit for the bias. This is a conservative procedure.

The calculated points through which the curves of Figure 1 were drawn using a French curve are given below.

Bias (%)	Target CV_T(%)	Critical $\hat{CV}_T$(%)
0	12.8	10.5
2.5	12.5	10.3
5.0	11.8	9.8
10.0	9.1	7.9
15.0	6.1	5.8
16.8	5.0	5.0
20.0	3.0	(Unattainable)
25.0	0	(Unattainable)

Operating Characteristics of the Validation Test Program

As would be expected, in order to be able to have at least 95% confidence that the true CV_T does not exceed its target level, we must suffer the penalty of sometimes falsely accepting a "bad" method (i.e. one whose true CV_T is unsatisfactory). Such decision errors, referred to as "type-1 errors", occur randomly but have a controlled long-term frequency of less than 5% of the cases. (The 5% probability of type-1 error is by definition the complement of the confidence level.) The upper confidence limit on CV_T is below the target level when the method is judged acceptable under the Decision Rule.

The validation test program can also have a "type-2 error", which is the mistake of deciding that a method is "bad" ($CV_T > 0.128$) when in fact it is "good" ($CV_T \leq 0.128$). The risk (probability) of making a type-2 decision error is not bounded (as is the case for the type-1 error). Rather, it depends on the true CV_T. In a previous report [2], it was shown that the probability of a type-2 error is large (0.88) for a "borderline" true CV_T (just below 0.128) but decreases to small probabilities of 0.10 for $CV_T = 0.091$, and 0.05 for $CV_T = 0.088$. Thus, more than 95% of methods whose CV_T's are below 0.088 (8.8%) will be accepted on the basis of their test results. "Good" methods whose true CV_T's are in the range 8.8% to 12.8% run a higher risk of not being approved; this risk could be lowered by using more than the now-prescribed 3 sets of 6 samples for the $\hat{CV}_T$ laboratory estimates in (each phase of) this program. However, the rate of improvement, in the precision of the laboratory estimates $\hat{CV}_T$, from using more samples would be small. For example, using 45 samples (15 per each of 3 groups) for each of the two phases instead of 18 (6 per group) only increases the "safe approval level" (0.05 probability of type-2 error) for CV_T from 0.088 (18 samples) to 0.099 (45 samples). The decision was made, therefore, to perform the smaller number (18) of tests for each of the two phases of the program.

Results of Validation Tests

Over 300 methods have been validated using the statistical protocol described above. Histograms have been prepared showing the distributions of precisions and biases obtained in the validation tests. Of 310 methods validated, only 31 (10%) had precision estimates ($\hat{CV}_T$'s) above 9% (See Figure 2). Apparently, only a small number of "good" methods have been tested whose CV_T's are in the borderline range where there is an appreciable chance of rejecting "good" methods. Since the pump error has a CV_P of 5% by itself, no values of $\hat{CV}_T$ fall below this level except for a few cases for which the method does not involve use of a personal sampling pump. It should be noted also that most of the methods have precisions clustering around 6-7% indicating the high quality of analytical methods tested.

The distribution of estimated biases for these methods is shown in Figure 3. Except for a bias of zero, the methods tend to be distributed evenly in the -10% to 10% bias region. The high proportion of zero-bias methods may be explained by the number of filter collection methods which have 100% collection efficiency; many of these methods use low-biased analysis techniques, particularly atomic absorption spectroscopy.

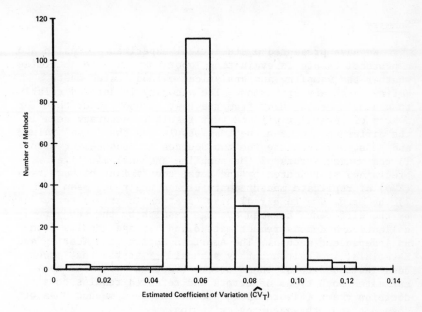

Figure 2.   Histogram of CV_T (estimated coefficient of variation of net error attributable to sampling and analysis) for 310 methods

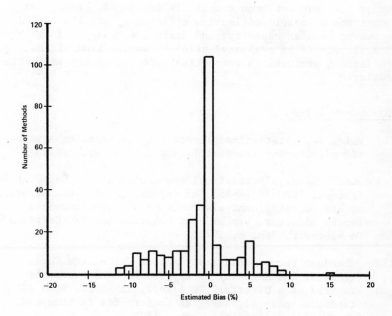

Figure 3.   Estimated biases for 310 test methods

Summary

We have presented a statistical experimental design and a protocol to use in evaluating laboratory data to determine whether the sampling and analytical method tested meets a defined accuracy criterion. The accuracy is defined relative to a single measurement from the test method rather than for a mean of several replicate test results. Accuracy here is the difference between the test result and the "true" value, and thus, must combine the two sources of measurement error: 1) the random errors of the sampling and analysis (i.e. precision) represented by the total coefficient of variation (CV_T) of replicate measurements around their own mean and, 2) the error due to a real bias (systematic error) represented by the difference between average results by the subject collection-and-measurement method and average results from an independent method. The American Society for Testing and Materials, in their accuracy standard[3] states that accuracy does include both of these errors (Section 4.1). We have estimated both types of errors and referred results to a decision chart (Figure 1) to see if the test method does or does not meet the accuracy criterion.

Finally, we would like to point out that the statistical protocol for validation deals mainly with the last step in determining the validity of a monitoring method. The statistical protocol is not appropriate for application to a method that has not been completely developed. Tests for such items as sample collection efficiency, stability, and recovery; sampler capacity; and analytical range and calibration all should be evaluated prior to application of the statistical protocol in connection with laboratory validation testing.

Literature Cited

(1)   Hald, A., "Statistical Theory with Engineering Applications", Chapter 11: part 11.8 and 11.9; Wiley, 1952.

(2)   Busch, K. A., "Statistical Properties of the SRI Contract Protocol (CDC 99-74-45) for Estimation of Total Errors of Air Sampling/Analysis Procedures", memorandum to Deputy Director, Division of Laboratories and Criteria Development, Jan. 6, 1975.

(3)   "Standard Recommended Practice for Use of the Terms Precision and Accuracy as Applied to Measurement of Property of a Material", E 177-71, in Annual Book of Standards, part 41, American Society for testing and Materials: Philadelphia, Pa., 1976.

APPENDIX I

TARGET VALUE OF CV_T FOR A BIASED METHOD

The maximum permissible CV_T (target value) for a biased method can be found by means of the formulae given below.

Let B = Bias ratio for the method

= (mean result by the method)÷(true concentration).

Standard normal deviates for left and right sides of the normal distribution corresponding to large errors (errors beyond ±25%) are given by:

$$Z_L = \frac{0.75-B}{B \cdot CV_T} \quad \text{and} \quad Z_R = \frac{1.25-B}{B \cdot CV_T}$$

For a given B, CV_T is the solution of the equation:

$$\int_{-\infty}^{Z_L} \frac{1}{\sqrt{2\pi}} e^{-(1/2)Z^2} dZ + \int_{Z_R}^{\infty} \frac{1}{\sqrt{2\pi}} e^{(-1/2)Z^2} dZ = 0.05$$

The equation must be solved iteratively. For any selected B, CV_T's are selected by trial and error in order to find the value of CV_T for which the sum of the integrals equals 0.05.

Example: $B = 1.1$, $Z_L = \dfrac{-0.35}{CV_T}$, $Z_R = \dfrac{0.15}{CV_T}$

For $CV_T = 0.09116$, $Z_L = -3.8394$, $Z_R = 1.6455$, and the sum of integrals is $0.0001 + 0.0499 = 0.0500$. Thus a method with B = 1.1 (i.e. 10% bias) has $CV_T = 0.091$ as its target level.

APPENDIX II

COMPUTATIONAL FORMULAE FOR STATISTICAL ANALYSIS

This appendix gives the formulae and definitions used in the protocol to statistically analyze laboratory data from validation tests.

Definitions and symbols are listed below:

Mean - arithmetic mean or average ($\bar{x}$), defined as the sum of the observations divided by the number of observations (n).

Standard Deviation - the positive square root of the variance, which in turn is defined as the sum of squares of the deviations of the observations from their mean ($\bar{x}$) divided by one less than the number of observations (n - 1).

$$\text{Std Dev} = \sqrt{\frac{\sum_{i=1}^{n} (x_i - \bar{x})^2}{n-1}}$$

CV –   coefficient of variation, or relative standard deviation, defined as the standard deviation divided by the mean.

$$CV = \frac{\text{Std Dev}}{\text{Mean}}$$

CV_1 –   coefficient of variation (estimated value) for the six analytical samples at each of the 0.5, 1, and 2X OSHA PEL's for the recommended sample volume.

CV_2 –   coefficient of variation (estimated value) for the six generated samples at each of the 0.5, 1, and 2X OSHA PEL's.

$\overline{CV}$ –   pooled coefficient of variation: the value derived from the coefficients of variation (of a given type, e.g. CV_1 or CV_2) obtained from the analysis of 6 samples at each of the three test levels. The mathematical equation is expressed as:

$$\overline{CV} = \sqrt{\dfrac{\displaystyle\sum_{i=1}^{3} f_i (CV_i)^2}{f}}$$

where:

f_i = degrees of freedom, equal to number of observations minus one ($n_i - 1$), at the ith level.

CV_i = coefficient of variation (CV_1 or CV_2) of the observations at the ith concentration level.

$f = \displaystyle\sum_{i=1}^{3} f_i$

i = index for the 3 concentration levels.

$\overline{CV}_1$ –   pooled coefficient of variation calculated as above based on data for the 18 analytical (spiked) samples (3 groups of 6).

$\overline{CV}_{A+\overline{DE}}$ –   derived correction to $\overline{CV}_1$ including precision error due to the use of the desorption efficiency factor, which is an average of 6 values.

$$\overline{CV}_{A+\overline{DE}} = \overline{CV}_1 \sqrt{7/6} = 1.0801\ \overline{CV}_1$$

$CV_{A+\overline{AMR}}$ –   corrected $\overline{CV}_1$ analogous to use of a desorption efficiency factor noted above except that this notation is used where the factor is associated with analytical method recovery (AMR) other than for solid sorbents.

$$\overline{CV}_{A+\overline{AMR}} = 1.0801\ \overline{CV}_1$$

$\overline{CV}_2$ –   pooled coefficient of variation based on the data for the 18 generated samples (3 groups of 6).

$\overline{CV}_S$ –   coefficient of variation of the sample collection, not including the variability of the personal sampling pump. The value is dependent on the data from the 18 analytical and 18 generated samples.

$$\overline{CV}_S = \sqrt{(\overline{CV}_2)^2 - (\overline{CV}_1)^2} \quad \text{(See "Note" below)}$$

CV_p — coefficient of variation due to the pump error, assumed to be equal to 0.05.

$\hat{CV}_T$ — coefficient of variation of total procedure which consists of the composite variations in sampling and analysis, desorption efficiency, and the pump error.

$$\hat{CV}_T = \sqrt{(\overline{CV}_S)^2 + (\overline{CV}_{A+DE})^2 + (CV_p)^2}$$

or:

$$\hat{CV}_T = \sqrt{(\overline{CV}_2)^2 - (\overline{CV}_1)^2 + 1.1667\,(\overline{CV}_1)^2 + (0.05)^2}$$

or:

$$\hat{CV}_T = \sqrt{(\overline{CV}_2)^2 + 0.1667\,\overline{CV}_1{}^2 + (0.05)^2} \quad \text{(See Note)}$$

NOTE: In case $\overline{CV}_2 < \overline{CV}_1$, take $\overline{CV}_S = 0$. Then replace $\overline{CV}_1$ by a pooled estimate $(\overline{CV}_1{}^*)$ based on $\overline{CV}_1$ and $\overline{CV}_2$,

$$\overline{CV}_1{}^* = \sqrt{\frac{f_1\,\overline{CV}_1{}^2 + f_2\,\overline{CV}_2{}^2}{f_1 + f_2}}$$

where f_1 and f_2 are the respective f-values used in the denominators of $CV_1{}^2$ and $CV_2{}^2$. Thus the equation to be used when $\overline{CV}_2 < \overline{CV}_1$ is:

$$\hat{CV}_T = \sqrt{1.1667(\overline{CV}_1{}^*)^2 + (0.05)^2}$$

GRUBB'S TEST for rejection of an observation is applied in order to determine if one of the observations should be rejected as being an outlier. The following equation was used for the test:

$$B_1{}' = \frac{x - \overline{x}}{s} \quad \text{or} \quad \frac{\overline{x} - x}{s}$$

where:

x = observation being tested (most distant from the mean)

$\overline{x}$ = mean of n observations

s = standard deviation based on n-1 degrees of freedom.

For any 6 observations, a value can be rejected if $B_1' \geq$ 1.944. The B_1' limit is based on a 1% significance level (i.e., a B_1' value calculated from the data can be expected to exceed 1.944 only 1% of the time if the observation is a legitimate one conforming to the underlying theory). For validation testing reject no more than two values in a set of 18 results and the two may not be in any one group of 6 replicates.

BARTLETT'S TEST for homogeneity of CV's is applied in order to test the feasibility of "pooling the coefficients of variation" for any set of 18 generated samples (i.e., 6 at each of the 0.5, 1, and 2X OSHA standard levels). The following equation for the Chi-square, with 2 degrees of freedom, was used:

$$\text{Chi-square} = \frac{f \, \ln (\overline{CV}_2)^2 - \sum_{i=1}^{3} f_i \, \ln (\overline{CV}_{2i})^2}{1 + \dfrac{1}{3(3-1)} \left[\left(\sum_{i=1}^{3} \frac{1}{f_i} \right) - \frac{1}{f} \right]}$$

where:

$\overline{CV}_2$ = pooled coefficient of variation of 18 generated samples

$\overline{CV}_{2i}$ = coefficient of variation of 6 generated samples at the ith level

f_i = degrees of freedom associated with $(CV_{2i})^2$ and equal to number of observations at the ith level minus one.

$f = \sum_{i=1}^{3} f_1$

In order to pass Bartlett's test at the 1% significance level, chi-square must be less than or equal to 9.21 (chi-square has 2 degrees of freedom).

RECEIVED October 21, 1980.

NEW TECHNOLOGIES

The Introduction of Microprocessor-Based Instrumentation for the Measurement of Occupational Exposures to Toxic Substances

R. KRIESEL and H. BROUWERS
MDA Scientific, Incorporated, 1815 Elmdale Avenue, Glenview, IL 60025

K. JANSKY
Compur-Electronic GmbH, Steinerstrasse 15, 8000 München 70, West Germany

Since the advent of OSHA (Occupational Safety and Health Administration), tremendous advances have been made in the degree of sophistication and data gathering ability of toxic substance monitoring instrumentation. Traditionally, exposure standards are often limited by the measurement techniques used to determine exposure. The introduction of new, small computers on a chip, particularly those that have an extensive memory and can be programmed, represents a technology that revolutionizes the measurement of occupational exposures, providing more complete and accurate data. A microprocessor-based dosimeter has been developed with this purpose in mind.

The Chronotox System, as its name implies, measures the concentration of toxic substances over time. It incorporates a programmable, read-only memory and calculating chip and accepts the input from literally any type of sensor with either an analogue or a digital output once the calibration curve for a sensor is determined. The microprocessor, small enough to be worn with a personal sensor, records the sensor output in its memory. At the end of the sampling period, the dosimeter is interrogated and can provide a time versus concentration curve of up to 12 hours. Ninety second signal averaging provides 480 data points over 12 hours. The data output can be connected to an X-Y Recorder, a portable digital printer, a mainframe computer, or, if desired, can be readily transmitted over telephone lines back to a central receiving station. The extensive body of exposure data produced by microprocessor-based instrumentation is vital for compliance purposes, toxicological studies, work practice analysis, and implementation of engineering controls.

Discussion

The recently developed Chronotox is a three component personal monitoring system which provides hard copy documentation of toxic chemical concentrations versus time, as well as 15-minute and 8-hour TWA (Time Weighted Average) data needed for a variety of purposes. The Chronotox System, consisting of a sensor, a microprocessor-based dosimeter, and a readout unit (Figure 1) documents alarm conditions when the permissible exposure level is exceeded, indicating how long the excursion lasted as well as the extent of the excursion. An evaluation of this monitoring data produces information on possible causes of excessive concentrations and the corrective action to be taken.

Data Acquisition and Storage

The concept of a time-based, measuring dosimeter was the baseline design criterion for the development of the Chronotox System. The programmable, microprocessor-controlled profile dosimeter acts as a data acquisition, processing and storage device for an appropriate sensor. The dosimeter is a small, compact, lightweight unit designed to be worn by an individual, and is connected to a detection sensor using a flexible umbilical. By using interchangeable "PROMS" (Programmable Read-Only Memory), the dosimeter is easily programmed to accept the signal from literally any sensing device, whether it has a continuous linear or non-linear, analog or digital output.

A PROM can be programmed for the signal output and calibration curve of up to five different sensors, and inserted into the dosimeter. For example, when used with a continuous output sensor, such as the Monitox, the microprocessor records a concentration value in a temporary memory each second. At the end of 90 seconds, the preceding 90 values are integrated to give an average value which is then stored in the Random Access Memory (RAM). The maximum sampling time is 12 hours. At the end of the sampling period, the microprocessor is disconnected from the sensor and interrogated to provide readout in either of two formats. Since the PROMS are interchangeable, literally any sensor can be accommodated by the dosimeter by inserting an appropriately programmed PROM. The sophisticated software backup includes linearization of non-linear output, data reduction, and communication with the sensor to permit total dose alarms, as well as high level incidental alarms. The only controls for the Chronotox microprocessor include a series of 6 binary switches, as described in Table I, used to call up the appropriate response curve. An LCD readout on the microprocessor indicates the program being used and subsequently provides an instantaneous display in ppm for visual check as the system is being used.

Data Readout

A data processing system incorporating an X-Y Recorder has has been designed to provide a complete graphic display of the data collected in the microprocessor memory during exposure. The display is called a Datagram (Figure 2), and consists of three sections: the first section displays the dosimeter battery voltage to assure that the data stored in the memory is valid; the second section of the Datagram consists of a concentration versus time graph of from 0-12 hours. As the memory is interrogated, a graph is produced showing the ppm concentrations of the measured gas during the sampling period and consists of 480 data points plotted over 12 hours. The graph defines concentrations above and below permissible levels and when they occurred, which is useful for controlling future exposures, i.e., process functions or cycles and/or work practices can be related to times of high exposure, and appropriate action can be taken to preclude or minimize further exposure. The concentration versus time curve also facilitates quick assessment of compliance with allowable ceiling or excursion values. The third section of the Datagram is a graphic readout of the 8-hour TWA (Time Weighted Average), and measurement of the total personal exposure during the monitoring period. The electronics automatically integrate the area under the exposure curve to calculate the total dose. This value is then divided by eight to provide an 8-hour TWA, irrespective of what the total exposure time was. In summary, the graphic display provides a complete profile record of

TABLE I
TYPICAL MICROPROCESSOR PROGRAM SELECTION

Switch Position	Program Selection
All Off	90 second TWA update
1 On	H_2S concentration - 1 second update
2 On	HCN concentration - 1 second update
3 On	$COCl_2$ concentration - 1 second update
4 On	NO_2 concentration - 1 second update
5 On	CO concentration - 1 second update
6 Off	Monitox Alarm off
6 On	Monitox Alarm on

personal exposure suitable for diagnostic procedures, proof of compliance, and permanent recordkeeping. The Datagram Readout System also incorporates up to six charging stations for microprocessor batteries and suitable storage for three microprocessor units.

An alternative method of retrieving the data from the Chronotox microprocessor is the use of a small, portable, battery-operated printer to obtain a digital data format. The dosimeter is inserted into the readout slot of the printer, and, first, a two-digit code displays the sensor program code. Next, the battery voltage of the microprocessor is printed to assure

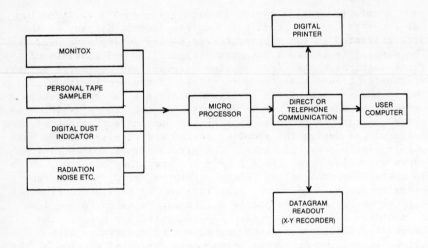

Figure 1.   Chronotox system schematic

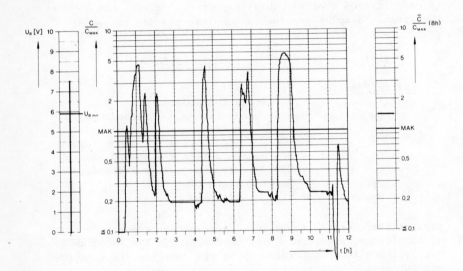

Figure 2.   Carbon monoxide datagram

valid data. The data is then printed out in a series of fifteen-minute TWA (Time Weighted Average) values, expressed in ppm. The cumulative 8-hour TWA is then printed, followed by the number of one-quarter hour segments that the exposure exceeded the program TLV (Threshold Limit Value) level. An example of this printout can be seen in Figure 3. If an asterisk (*) is displayed, this indicates that the sensor was turned off or disconnected sometime during the fifteen-minute sampling period.

The small, compact printer can be utilized where the larger Datagram Readout Unit is not needed or convenient, i.e., for quick readout in field locations, or where a graphic display is not necessary. The choice of two readout systems allows maximum versatility and economy for both the large, multiplant, multi-problem user, as well as the small, single substance installation.

In addition to the Datagram and digital formats for data readout, the Chronotox microprocessor can also be programmed for input into other data storage or communication facilities. Readout of concentrations can be done over telephone lines to a centralized data printer/recorder from remote monitoring sites or directly into a computerized industrial hygiene data storage unit such as is now set up at many industrial and governmental installations.

Sensor Technology

Toxic Gases. Several sensors have been developed as components of the Chronotox System. In addition, sensors may be adapted for use with the system, lending wide ranging versatility to the Chronotox and expanding the capabilities of the selected sensor.

The Monitox personal alarm sensor utilizes a detection principle based on a high technology electrochemical cell. When a toxic gas diffuses through a membrane and impinges on the electrode, a current is generated in the cell which has a linear relationship to the concentration of the toxic gas being detected over a range of 0.1 to 10 times the Threshold Limit Value (TLV). Electronic circuitry amplifies the signal and provides alarm capability at specified levels (Figure 4).

About the size of a package of cigarettes, individual sensors are available for H_2S, phosgene, NO_2, HCN, and CO. In the near future, the series will be expanded to include Cl_2 and hydrazine. When used with the Chronotox microprocessor, the Monitox serves as a personal monitor as well as a gas detection alarm system.

Another sensor system that further expands the capabilities of the Chronotox is the Personal Tape Sampler (PTS). The PTS (Figure 5) is a miniaturized gas monitor designed primarily to be worn by a worker during his entire workshift. It operates on the principle of specific chemical colorimetric reactions occurring on a 'dry' chemically impregnated paper tape and gives

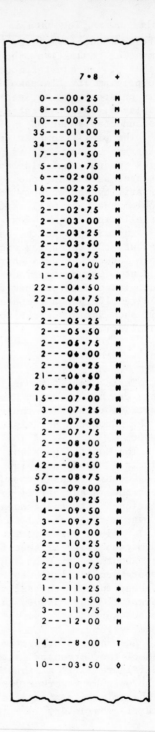

```
                    7•8    +

       0---00•25    M
       8---00•50    M
      10---00•75    M
      35---01•00    M
      34---01•25    M
      17---01•50    M
       5---01•75    M
       6---02•00    M
      16---02•25    M
       2---02•50    M
       2---02•75    M
       2---03•00    M
       2---03•25    M
       2---03•50    M
       2---03•75    M
       2---04•00    M
       1---04•25    M
      22---04•50    M
      22---04•75    M
       3---05•00    M
       2---05•25    M
       2---05•50    M
       2---05•75    M
       2---06•00    M
       2---06•25    M
      21---06•50    M
      26---06•75    M
      15---07•00    M
       3---07•25    M
       2---07•50    M
       2---07•75    M
       2---08•00    M
       2---08•25    M
      42---08•50    M
      57---08•75    M
      50---09•00    M
      14---09•25    M
       4---09•50    M
       3---09•75    M
       2---10•00    M
       2---10•25    M
       2---10•50    M
       2---10•75    M
       2---11•00    M
       1---11•25    •
       6---11•50    •
       3---11•75    M
       2---12•00    M

      14----8•00    T

      10---03•50    ◊
```

Figure 3.   Digital format

valid data. The data is then printed out in a series of fifteen-minute TWA (Time Weighted Average) values, expressed in ppm. The cumulative 8-hour TWA is then printed, followed by the number of one-quarter hour segments that the exposure exceeded the program TLV (Threshold Limit Value) level. An example of this printout can be seen in Figure 3. If an asterisk (*) is displayed, this indicates that the sensor was turned off or disconnected sometime during the fifteen-minute sampling period.

The small, compact printer can be utilized where the larger Datagram Readout Unit is not needed or convenient, i.e., for quick readout in field locations, or where a graphic display is not necessary. The choice of two readout systems allows maximum versatility and economy for both the large, multiplant, multi-problem user, as well as the small, single substance installation.

In addition to the Datagram and digital formats for data readout, the Chronotox microprocessor can also be programmed for input into other data storage or communication facilities. Readout of concentrations can be done over telephone lines to a centralized data printer/recorder from remote monitoring sites or directly into a computerized industrial hygiene data storage unit such as is now set up at many industrial and governmental installations.

Sensor Technology

Toxic Gases. Several sensors have been developed as components of the Chronotox System. In addition, sensors may be adapted for use with the system, lending wide ranging versatility to the Chronotox and expanding the capabilities of the selected sensor.

The Monitox personal alarm sensor utilizes a detection principle based on a high technology electrochemical cell. When a toxic gas diffuses through a membrane and impinges on the electrode, a current is generated in the cell which has a linear relationship to the concentration of the toxic gas being detected over a range of 0.1 to 10 times the Threshold Limit Value (TLV). Electronic circuitry amplifies the signal and provides alarm capability at specified levels (Figure 4).

About the size of a package of cigarettes, individual sensors are available for H_2S, phosgene, NO_2, HCN, and CO. In the near future, the series will be expanded to include Cl_2 and hydrazine. When used with the Chronotox microprocessor, the Monitox serves as a personal monitor as well as a gas detection alarm system.

Another sensor system that further expands the capabilities of the Chronotox is the Personal Tape Sampler (PTS). The PTS (Figure 5) is a miniaturized gas monitor designed primarily to be worn by a worker during his entire workshift. It operates on the principle of specific chemical colorimetric reactions occurring on a 'dry' chemically impregnated paper tape and gives

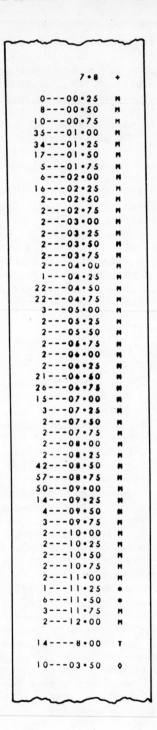

Figure 3.   Digital format

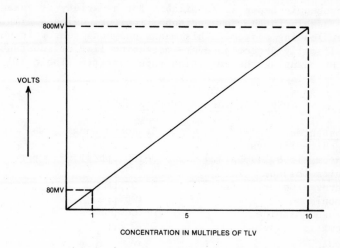

Figure 4. Monitox concentration level vs. output

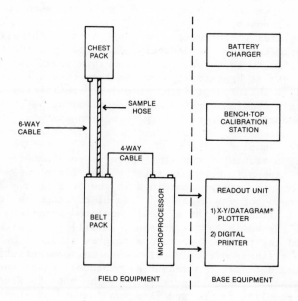

Figure 5. PTS schematic

a continuous concentration record of a particular toxic gas.
Interchangeable tapes are available for a variety of gases and
all are usable in the same personal monitor, using a simple
cassette loading system. This means only one sensor is needed
to specifically detect an ever-increasing list of substances
simply by changing the detection tape cassette (Table II).

TABLE II
PERSONAL TAPE SAMPLER RANGES

Substances	Standard Range
Organic Diisocyanates (Aromatic & Aliphatic)	0-100 ppb or 0-10 ppb
Dinitrotoluene	0-0.5 ppm
Hydrogen Chloride (HCl)	0-50 ppm
Ammonia	0-100 ppm
Toluene Diamine	0-100 ppb
Aromatic Amines	0-10 ppm

The PTS consists of two basic units, the chest pack (worn
near the breathing zone), and the belt pack. The dosimeter
controls the operation of the chest pack and stores the acquired
sampling data. The LC display provides instantaneous update of
the gas concentration being measured for observation or
calibration purposes.

The PTS uses the principle of a specific colorimetric
reaction, each toxic substance being detected by a reaction
system specific to that gas. This principle provides the capability for extreme sensitivity down to less than one part per
billion (10^{-9}) while maintaining excellent specificity. Reactive chemicals are impregnated onto a continuous paper tape by
a special process producing a 'dry' reaction system. The tape
is put into cassettes for convenient use and replacement in the
Personal Tape Sampler (PTS).

In use, a sample of the atmosphere is continuously drawn
through the tape at a constant rate by the microprocessor-
controlled pumping system contained in the belt pack. The
sample is drawn through a defined area of the tape, which is
continuously illuminated by a small regulated tungsten lamp
source. If the toxic gas is present, it will react with the
chemicals on the tape and produce a colored stain. The intensity of the stain is directly related to the concentration of
the gas being detected. A measuring photocell reads the
reflected light from the tape surface. The tape is stationary
during the sampling period and is moved forward in increments
at a preselected time interval, exposing new tape for each
sample. The measuring photocell signal is amplified and made
available as a continuous analog output having a range of 0-1
volt for stains representing the full scale concentration for
each gas.

The PTS is designed to use the Chronotox microprocessor for data storage and retrieval. The flexibility of the micro-processor-based Chronotox dosimeter allows a sophisticated interpretation of the photocell output signal giving excellent accuracy and signal to noise ratio. In operation, the PTS provides a "Data Control Pulse" to the microprocessor which commands it to read the photocell output signal immediately after the tape has incremented - i.e., it takes an initial background reading and stores it in temporary memory. After the preset sampling interval a second "read" command is given by the dosimeter. This final reading represents the stain intensity plus any background color. The first background reading is then taken from memory storage and subtracted from the final reading giving a corrected reading representing only the change that has occurred on the tape during the sampling interval. The corrected reading is then stored as a true value for that sampling period. Typically, 240 such readings will be stored for an 8-hour sampling period. Variable sampling times can be selected to provide maximum sensitivity.

Dusts, Mists, Aerosols and Fumes. The P-5 Digital Dust Indicator is another sensor currently available for use as a component of the Chronotox System. Suitable for the measurement of silica, lead fumes, pharmaceutical powders as well as many other types of particulates found in manufacturing or laboratory situations, the battery-operated P-5 uses the light scattering technique to measure dusts over a range of either 0.01-100 mg/m^3 or 0.001-10 mg/m^3 (Figure 6).

An internal fan aspirates the dust-contaminated air through the inlet separator, allowing only the respirable fraction (sub-micron up to 10 micron particles) to be measured. The geometry of the sampling chamber minimizes dust deposition on the optical surfaces and measuring chamber, eliminating the need for tedious cleaning. The sample is illuminated by a tungsten lamp and scattered light is detected by the photomultiplier tube at a 90° angle, yielding CPM (Counts Per Minute) directly proportional to the concentration. Cumulative dust levels are displayed with an LCD readout over preset intervals of 0.1 to 10 minutes, or con-tinuously if operated in the manual mode. Simultaneously, the ratemeter displays the analog signal. Again, the microprocessor is used to accumulate the sampling data and to produce a con-centration versus time profile. More importantly, the dosimeter is programmed with response factors for specific dust conditions and characteristics, providing actual dust concentration data in mg/m^3 (Figure 7).

Additional Sensors. At this writing, other personal monitoring sensors for noise, heat stress, radiation, etc., are under development for incorporation into the Chronotox System. As previously indicated, there is no limit to the application

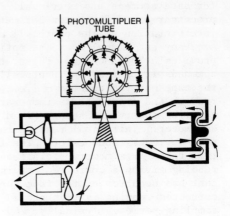

Figure 6.   Digital dust indicator schematic

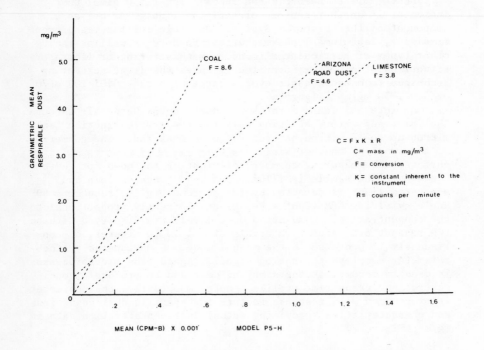

Figure 7.   Typical conversion factors: digital dust indicator vs. gravimetric

versatility of this monitoring approach. It is a relatively simple matter to adapt any type of sensor for continuous concentration/time output and subsequent data processing.

Summary

The use of microprocessor-based instrumentation for the measurement of occupational exposures to toxic substances represents the state-of-the-art in personal monitoring. With the productionizing of theoretical advances in monitoring technology, it is now possible to obtain accurate concentration data on a continuous basis as well as averaged exposure values. The data generated by such instrumentation permits in-depth analysis of production cycles, work practices and engineering controls as related to personnel exposures. In addition, the data can be used for determining compliance with governmental regulations and on an extensive data base for toxicological studies. A typical example is the Chronotox System microprocessor, which incorporates a programmable, read-only memory and calculating chip that can accept either linear or non-linear, analog or digital signals from virtually any type of sensor once its calibration curve is determined. This means the Chronotox is a unique, versatile system permitting use of a wide variety of sensors as well as a choice of data format and storage suitable to a particular monitoring requirement because of the data acquisition, storage and processing capabilities of the microprocessor-based dosimeter.

RECEIVED October 27, 1980.

A Versatile Test Atmosphere Generation and Sampling System

SHUBHENDER KAPILA, RAVINDER K. MALHOTRA, and CORAZON R. VOGT

Environmental Trace Substances Research Center, University of Missouri, Columbia, MO 65201

In recent years, a number of industrially important chemicals have been investigated for their potential carcinogenic, mutagenic and teratogenic properties. In order to protect workers exposed to such chemicals, maximum exposure limits have been set by OSHA. Several methods for the analysis of these chemicals (pollutants) have been reported. These methods can be divided into two catagories, those involving direct monitoring of the atmosphere and those involving trapping of pollutants on or in suitable medium followed by determination with appropriate techniques. Due to their harmful effect (even at very low concentrations), the need for accurate and precise determination can not be overstated. Furthermore, in instances where legal implications are involved, the reliability of the methods is of utmost importance.

The most widely accepted method of evaluating the accuracy and precision of an analytical procedure is to sample known concentrations of contaminants in the atmosphere. Thus an important aspect of analytical method development is the generation of test atmospheres that simulate the conditions (i.e., concentration range, humidity, temperature and interferences) found during the field sampling.

Several test atmosphere generating systems have been reported (1-6). Basic principles and techniques employed have been discussed in detail by Nelson (7). The methods are generally divided into two catagories: static methods and dynamic methods. In static method a known amount of contaminant is introduced into a fixed volume of air in devices such as teflon bags, gas sampling bulbs and gas cylinders, etc. Dynamic methods involve continuous introduction of contaminant (at a controlled rate) into a stream of air. Static methods are generally much simpler to construct and use, however, these suffer from a number of problems. Dynamic methods, while more elaborate and relatively more expensive, offer greater flexibility in concentration range, sample volume and are also less affected by adsorption losses.

The present study describes a dynamic system for generating
known concentrations of analytes in simulated atmoshperes containing organic contaminants, both gases and liquids, at three
different (concentration) levels simultaneously under controlled
humidity. An in-line monitoring device in the form of a gas
chromatograph (GC) was incorporated into the system. Incorporation of the GC with an appropriate detector enhances the versatility of the overall system, making it suitable for most volatile organics. Thus it offers a distinct advantage over the
specific gas analyzers used in some previous generation facilities (5-8).

Description and Working of the Generation Facility

Generation system is divided into a Primary and a Secondary
dilution module as shown in Figure 1.
Primary Dilution Module: The Primary Dilution Module is used for
dilution of the analytes with nitrogen. The volume of contaminant
is regulated by a very fine metering valve V_3 (Nupro Valves). It
is measured by rotameter R_1 and mixed with a known volume of nitrogen controlled by a fine metering valve V_4 and measured by
rotameter R_2. The doped nitrogen stream is split into two streams
by a micro-valve V_{10} (Scientific Glass & Engineering Co.). The
major portion is sent to a burner. The smaller portion is measured through rotameter R_3 and then further diluted with nitrogen
(rotameter R_4). A desired volume of this diluted flow is passed
on to a four-port switching valve V_1 (Valco). The contaminated
analyte-nitrogen stream can then be sent either to the GC
(Bendix Model 2500) through a six-port sampling valve V_2 (Valco)
or to the secondary dilution module. The sampling loop attached
to the six-port valve consists of a stainless steel tube with a
volume of 0.17 mL. The sample in the loop is injected into the
GC by switching the valve, and the concentration of contaminant
in the nitrogen stream is determined by the following equation:

$$C_n = \frac{\text{amt of contaminant in loop(g)}}{\text{size of the loop (mL)}} \times \frac{22.4 \times 10^3}{M} \times \frac{P}{760} \times \frac{T+273}{273} \times 10^6$$

where C_n = concentration of contaminant in N_2 (in ppm)

   M = molecular weight of the contaminant

   T = expt(1) temperature, $^{\circ}C$

   P = expt(1) pressure, mm Hg

The amount of the contaminant in the loop (g) is determined from
a calibration curve established at the time of the analysis.

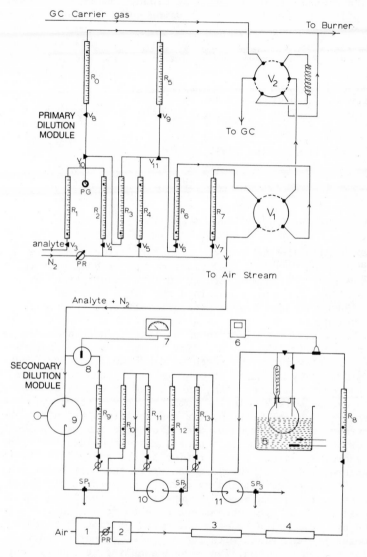

Figure 1. *Schematic of the test atmosphere generation system, various components are as follows: 1, air compressor (Gast model #5HCD); 2, dehumidifier; 3, activated charcoal filter (Alltech 8129); 4, molecular sieve and decimal filter (Alltech 8125); 5, humidifier with thermostated water bath (Neslab model G175); 6, digital mass flow meter (Matheson model 8240); 7, humidity monitor (YSI model 91HC); 8, humidity sensor (YSI double bobbin); 9, primary dilution chamber; 10, 11, secondary dilution chambers; PR, pressure regulators; R_0–R_7, rotameters with range of 5 mL to 150 mL/min (Brooks model AAA); V_1, four-port switching valve (Valco valves); V_2, six-port sampling valve (Valco valves); V_3, double stem, very fine metering valve (Nupro #2SGD); V_4–V_9, fine metering valves (Nupro #25G and 2MG); SP_1–SP_3, sampling ports.*

The concentration of the contaminant can also be calculated from split ratios and flow rates using the following equation:

$$C_n = m \left(\frac{Cf_1}{Nf_1 + Cf_1}\right) \left(\frac{Cf_2}{Nf_2 + Cf_2}\right) \times 10^6$$

where m = mole fraction of contaminant at V_3

Cf_1 = contaminant flow (R_1), mL/min.

Nf_1 = nitrogen flow (R_2), mL/min.

Cf_2 = diluted contaminant flow (R_3), mL/min.

Nf_2 = nitrogen flow (R_4), mL/min.

<u>Secondary Dilution Module:</u> The secondary dilution module is de-signed to produce controlled flows of dry or humid air for the generation of test atmosphere at three concentration levels si-multaneously. The air is obtained through a compressor (Gast Model #5HCD) located approximately 150 ft from the facility. It is de-humidified by a refrigeration unit. Trace organics are re-moved by passing the air through activated charcoal and mol-ecular sieve filters. The air flow is measured by a rotameter R_8 and a mass flow meter (Matheson model #8240). The purified air is humidified by sweeping it over water at constant temperature in a thermostated water bath, and passed through a condensor to remove excess water. Relative humidity is monitored by a hy-grometer (YSI Model 91). The air and analyte-doped nitrogen (from the primary dilution module) are allowed to mix in a dilution chamber. The pressure inside the chamber is maintained at one atmosphere. The exact concentration of the analyte in the test atmosphere, (if 2 to 5 ppm, depending on the particular con-taminant) can be determined by direct analysis of air taken from SP_1. The port SP_1 can also be connected to a sampling manifold for collection of the contaminant on a solid sorbent. A portion of the air is sampled and the unsampled portion is measured by rotameter R_{10} and mixed with the required amount of purified air for second dilution. The second sampling is done at port SP_2; third dilution and subsequent sampling are feasible at port SP_3.

Evaluation of the generation system was done with acetone and ethylene oxide as the test contaminants.

The activated charcoal (Columbia JXC) was selected for col-lection of the analytes. This adsorbent has been reported to give good results for ethylene oxide (<u>9</u>). The sampling tubes con-sisted of borosilicate glass tubes (15 cm x 4 mm i.d.) packed with 400 mg of the charcoal (300 mg in front section and 100 mg

in the back section). The sampling was done for one hour period
with vacuum pump (Gast Model #1031). Methanol and carbon disulfide were used to desorb acetone and ethylene oxide, respectively.
Before proceeding to sampling, desorption efficiencies were carefully determined.

Sampling: A sampling rate of approximately 55 mL/min was maintained by the use of critical orifice constructed in the laboratory. The schematic of the orifice is shown in Figure 2A. The
orifices were made by silver soldering a 0.5 mm o.d. (0.25 mm id)
stainless steel tubing into a drilled out 1/8 inch o.d. stainless
steel rod, which in turn was soldered into a male connector
(1/4 inch swagelok to 1/16 inch NTP). An orifice constructed in
this manner was found to give a maximum flow of 300 mL/min. To
obtain the desired flows (250 and 55 mL/min), stainless steel
wires of various diameters were inserted into the 0.25 mm i.d.
tubings and, in some cases, the tubings were slightly crimped.
The orifices were then connected into the sampling manifold
(Figure 2B).

Each orifice constructed in the above manner was calibrated
with a rotameter and manometer. The set-up used for orifice calibration is shown in Figure 3. The uncorrected flow readings were
obtained from the rotameter calibration curve. The corrected
flows were then calculated using the following equation:

$$F_{cor} = F_{cal} \left(\frac{P_2}{P_1}\right)^{1/2}$$

where: F_{cor} = corrected flow

F_{cal} = flow read from the calibration curve

P_2 = pressure reading from manometer

P_1 = atmospheric pressure during rotameter calibration

The calibration of the orifices was done before and after
each sampling run to ensure accurate flow rates.

Results and Discussion

The simulated test atmosphere with high levels of contaminant (2 to 5ppm) were analyzed by both the in-line GC (in the
primary dilution module) and by direct injection of air taken
from the sampling port (in the secondary dilution module). The
purpose of the latter was to establish the reliability of the
in-line GC. The in-line monitor was particularly important in
test atmosphere with low contaminant concentration because in
this case, direct injection of the air is not feasible. Table I
compares the results obtained by the in-line GC and direct air
injection for ethylene oxide and acetone.

Test atmospheres containing acetone, a relatively innocous
liquid, and ethylene oxide, a mutagenic gas (at room temperature)

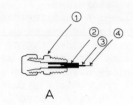

A

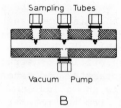

Sampling   Tubes

Vacuum   Pump

B

Figure 2.   A. schematic of the orifices used in sampling: 1, 1/4"–1/16" male connector; 2, 1/8"-o.d. (o. 1/32" i.d.) stainless steel tube; 3, 2.5-mm i.d. stainless steel capillary; 4, stainless steel wire. B. Schematic of the sampling manifold.

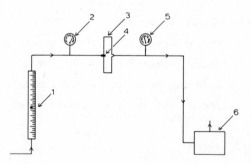

Figure 3.   System used for calibration of the critical orifices: 1, calibrated rotameter; 2, 5, vacuum/pressure gauge; 3, sampling manifold; 4, orifice; 6, vacuum pump.

was generated. One-hour sampling on solid sorbent was done at about 55 mL/min in both humid and dry conditions after two successive dilutions. (No sampling was done at the third level, i.e., at SP_3, because of the lack of a third identical pump).

The results obtained with acetone as the test contaminant are shown in Tables II and III. Since acetone is less hazardous to health than ethylene oxide, the high concentrations shown on the tables were chosen to evaluate the performance of the system at these levels. Recoveries were close to 100% in both dry and humid air.

Table I
Analysis of Generated Test Atmospheres

Ethylene oxide			Acetone		
In-line GC ng/ml air	Direct inj ng/ml air	% error	In-line GC ng/ml air	Direct inj. ng/ml air	% error
162.5	159.8	1.7	1320.0	1330.0	0.07
43.3	42.7	1.4	251.2	246.0	1.7
20.3	20.5	1.0			
5.9	5.8	1.7			

Table II
Sampling of Acetone in Dry Atmosphere, RH < 12%[1]

Acetone 420 ppm			Acetone 230 ppm		
Amount added (mg)[2]	Amount rec. (mg)[3]	Percent Recovery	Amount added (mg)[2]	Amount rec. (mg)[3]	Percent Recovery
3.34	3.97	108.4	1.66	1.75	105.4
3.39	4.03	108.9	1.70	1.70	100.0
3.12	3.71	100.2	1.81	1.75	96.6
2.94	3.50	99.1	1.83	1.94	106.0
2.90	3.45	99.5	1.88	1.99	101.1
$\bar{x}$ = 103.2	σ = 4.9	cv = 0.04	$\bar{x}$ = 101.8	σ = 3.9	cv = 0.03

(1) lowest accurate hygrometer measurement
(2) determined by in-line GC
(3) corrected for a desorption efficiency of 84%

Table III
Sampling of Acetone in Humid Atmosphere, RH = 85%

Acetone 580 ppm			Acetone 285 ppm		
Amount added (mg) [1]	Amount rec. (mg) [2]	Percent Recovery	Amount added (mg) [1]	Amount rec. (mg) [2]	Percent Recovery
4.59	4.47	97.4	2.05	1.91	93.1
4.64	4.37	94.2	2.09	1.86	89.0
4.76	4.73	99.4	2.30	2.37	103.0
4.64	4.38	94.4	2.24	2.24	100.0
4.34	3.80	87.5	2.26	2.33	103.1
$\bar{x}$ = 94.58	σ = 4.5	cv = 0.04	$\bar{x}$ = 98.44	σ = 5.8	cv = 0.05

(1) determined by in-line GC
(2) corrected for a desorption efficiency of 84%

The ethylene oxide experiments were done at four concentration levels starting with a low of about 3 ppm to a high of 44 ppm. (Concentrations greater than 44 ppm ethylene oxide were not attempted because it was felt that the exhaust system in the laboratory at the time of the experiment may not be sufficient to handle safely any higher concentrations of the ethylene oxide.) Sampling at the low concentrations was done only under dry conditions (RH < 12%) whereas, sampling at high concentrations was done both in dry and humid atmospheres. The results are summarized in Tables IV, V and VI. The percent recovery of ethylene oxide is greater than 88% and little or no change in recovery is observed in dry and humid conditions.

Table IV
Sampling of Ethylene Oxide in Dry Atmosphere, RH < 12% [1]

Ethylene oxide 6.0 ppm			Ethylene oxide 3.0 ppm		
Amount added (μg) [2]	Amount rec. (μg) [3]	Percent Recovery	Amount added (μg) [2]	Amount rec. (μg) [3]	Percent Recovery
38.7	40.0	103.4	18.2	16.1	88.4
39.0	38.6	99.0	17.0	15.1	88.4
40.1	35.6	88.8	18.7	16.2	86.6
39.0	36.6	93.6	18.2	16.4	89.9
$\bar{x}$ = 96.2	σ = 2.7	cv = 0.02	$\bar{x}$ = 88.3	σ = 1.3	cv = 0.01

(1) lowest accurate hygrometer measurement
(2) determined by in-line GC
(3) corrected for a desorption efficiency of 94%

Table V
Sampling of Ethylene Oxide in Dry Atmosphere, RH < 12%[1]

Ethylene oxide 40 ppm			Ethylene oxide 20 ppm		
Amount added (µg)[2]	Amount rec. (µg)[3]	Percent Recovery	Amount added (µg)[2]	Amount rec. (µg)[3]	Percent Recovery
234.9	237.1	100.9	128.3	116.6	90.9
237.0	247.0	104.2	134.7	114.1	84.7
237.0	233.0	98.3	131.1	118.4	90.3
226.5	240.5	106.0	132.3	125.3	94.3
222.2	230.9	103.9	135:8	131.2	96.6
$\bar{x}$ - 102.6	σ = 3.0	cv = 0.03	$\bar{x}$ = 92.3	σ = 4.5	cv = 0.04

(1) lowest accurate hygrometer measurement
(2) determined by in-line GC
(3) corrected for a desorption efficiency of 94%

Table VI
Sampling of Ethylene Oxide in Humid Air, RH = 85%

Ethylene oxide 44 ppm			Ethylene oxide 22 ppm		
Amount added (µg)[1]	Amount rec. (µg)[2]	Percent Recovery	Amount added (µg)[1]	Amount rec. (µg)[2]	Percent Recovery
264.6	264.1	99.8	120.4	109.3	91.1
266.0	266.9	99.2	122.7	109.3	89.1
266.9	260.7	97.7	134.7	119.0	88.3
255.0	253.9	99.6	131.1	119.0	99.1
250.3	239.6	95.7	132.3	121.2	91.6
$\bar{x}$ = 98.4	σ = 1.7	cv = 0.01	$\bar{x}$ = 91.89	σ = 4.2	cv = 0.04

(1) determined by in-line GC
(2) corrected for a desorption efficiency of 84%

Tables II-VI further confirmed the accuracy of the in-line GC as a monitor. However, it should be pointed out that for atmospheres of very high concentrations i.e., at levels where the amount of contaminants in the sampling loop exceeds the linear range of the detector, the in-line GC would be of little use.

Conclusion

In this paper, we have presented a system for generating contaminated test air. We have shown that the system can generate test air at two levels (readily expandable to three levels by the use of a third pump) with reasonable accuracy. A broad

range of concentrations (3.0 ppm ethylene oxide and 580 ppm
acetone) have been obtained. It is reasonable to assume that
concentrations higher than what have been demonstrated may be
obtained. However, if the analyte is collected in a solid
sorbent, the collection capacity of the sorbent must not be
exceeded and that once adsorbed, can be desorbed with efficiency
greater than 80%. The in-line GC in the primary dilution module
is shown to give accurate results and therefore the concentration
of the contaminant can be determined at any time during sampling
by means of a four-port switching valve. The system permits
rapid (about 2 min) adjustment of generated concentration levels
by the use of fine metering valves in conjunction with well-cali-
brated rotameters and the in-line GC. However, the equilibration
of the atmosphere in the secondary dilution module takes slightly
longer. The system also permits a wide range of humidity levels
to be set by varying the temperature of the water bath.

Literature Cited

1. Nelson, G.O.; Griggs, K.S Rev. Sci. Instr. 1968, 39, 927.
2. Nelson, G.O.; Hodgkins, D.J. Am. Ind. Hyg. Assoc. J. 1972,
   33, 110.
3. Pella, P.A.; Hughes, E.E.; Taylor, J.K. Am. Ind. Hyg. Assoc. J.
   1975, 36, 755.
4. Ellgehausen, D. Anal. Lett. 1975, 8, 11.
5. Dietrich, M.W.; Chapman, L.A.; Mieure, J.P. Am. Ind. Hyg.
   Assoc. J. 1978, 39, 385.
6. Vincent, W.J.; Hahn, K.J.; Ketcham, N.H. Am. Ind. Hyg.
   Assoc. J. 1979, 40, 512.
7. Nelson, G.O. "Controlled Test Atmospheres: principles and
   techniques"; Ann Arbor Science Publishers, Inc. Ann Arbor, MI
   1971.
8. Nelson, G.O.; Swisher, L.W.; Taylor, R.D.; Bigler, B.E.
   Am. Ind. Hyg. Assoc. J. 1975, 36, 49.
9. Qazi, A.H.; Ketcham, N.H. Am. Ind. Hyg. Assoc. J. 1977,
   38, 635.

RECEIVED September 29, 1980.

Development of Workplace Guidelines for Emerging Energy Technologies

OTTO WHITE, JR.

Brookhaven National Laboratory, Upton, NY 11973

DANIEL LILLIAN

U.S. Department of Energy, Washington, DC 20545

Most of the papers presented at this symposium addressed improving sampling methodologies and new analytical procedures for assessing exposure to compounds and agents for which permissible exposure limits currently exist. It is recognized that there are approximately 500 compounds that the American Conference of Governmental Industrial Hygienist [1] has developed exposure limits and that the NIOSH Registry of Toxic Effects of Chemical Substances [2] list approximately 34,000 individual toxic compounds. The Department of Energy (DOE) has recognized that additional safety and health protection is needed for employees developing new energy technologies and that it is essential to develop in-house interim exposure limits for hazardous agents that may be associated with these technologies. Proposed structural models for base materials associated with new energy technologies such as bituminous coal (Figure 1) and kerogen (Figure 2) from oil shale illustrate the potential capability of producing a myriad of toxic chemical compounds. Therefore, DOE support efforts led to the establishment of the Center for Assessment of Chemical and Physical Hazards (CACPH) at Brookhaven National Laboratory. CACPH is designed to provide broadbase industrial hygiene support functions for the DOE community and has, as its primary functions, the timely development of interim exposure limits.

Safety and health needs for the Department of Energy comes from new energy technologies, but also from new scientific data, unusual occurrences, DOE consultants and inadequate existing standards (Figure 3). A Toxic Material Advisory Committee (TMAC) exists at DOE and consists of members of the DOE headquarter's staff who are particularly qualified to address safety and health issues relating to the uses of hazardous materials. The purpose of this committee is to provide technical and programmatic overview and directions to CACPH and to ensure maximum and effective utilization of this support program. TMAC meets three times a year and functions to review requests for health and safety assistance, to establish priorities for developing interim

0097-6156/81/0149-0543$05.00/0
© 1981 American Chemical Society

FUNCTIONAL GROUP MODEL OF BITUMINOUS COAL (WISER)

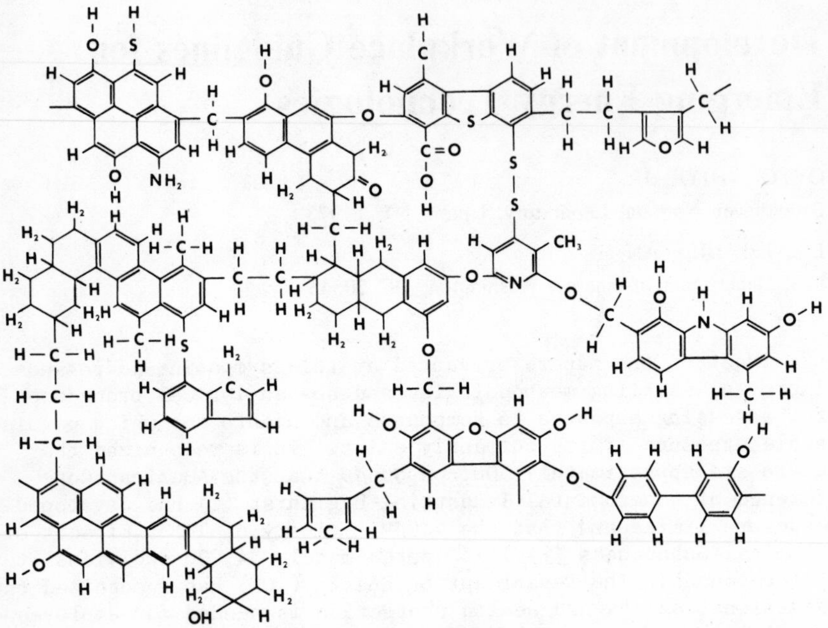

Figure 1.   Proposed structural model for coal

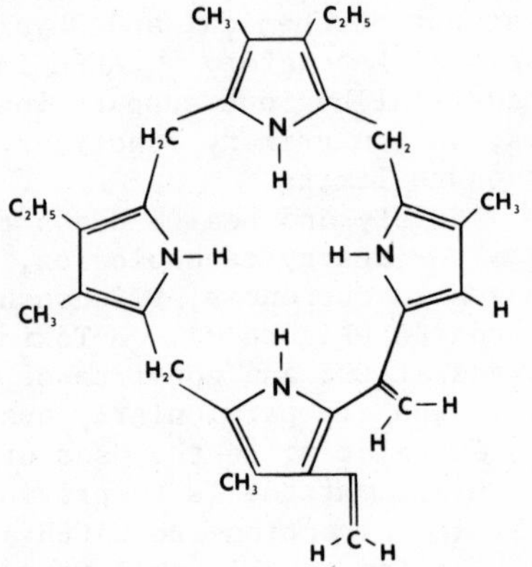

Figure 2.   Proposed structural model for
            kerogen (oil shale)

DECARBOXYLATED HEXAHYDROCHLOROPHYLLIN

guidelines, to evaluate draft interim guidelines for technical quality and impact on the DOE community, to make decisions on acceptability of program documents for use as DOE interim guidelines and to make recommendations to DOE's, the Assistant Secretary for Environment. An additional function for TMAC is to assess the need for research as recommended by CACPH and make appropriate recommendations to the affected Department of Energy program directors.

Immediately after receiving the formal request, CACPH procedes to assess both the published and unpublished literature to facilitate appropriate evaluation of the relevant health and safety data and to develop interim standards for specific agents (Figure 4). Concurrently, CACPH informs the CACPH Advisory Panel of the task.

The CACPH Advisory Panel consists of 10 active members, scientists from Brookhaven and the State University of Stony Brook with expertise in chemistry, biology, medicine, public health, toxicology, pathology, epidemiology and industrial hygiene and three corresponding members including the Associate Director for Chemistry, Biology, Medicine and Safety at the Laboratory. Immediate access exists to these distinquished scientists for CACPH's staff, thereby facilitating meaningful and timely consultation.

In identifying published literature, CACPH's staff not only assesses secondary literature sources such as Chemical Abstracts and Biosis, but also identifies and utilizes computerized data base systems which contain relevant safety and health information such as Medlar, Dialog and DOE-RECON. These secondary literature sources identify the primary literature sources (reports, journals and books) which contain the appropriate safety and health information. The type of information that CACPH's staff retrieves includes chemical properties, physical properties, toxicological data, environmental effects, sampling methods, storage, handling and shipping requirements and medical treatments.

Unpublished literature and information are assessed by contacting DOE and DOE contractor's health and safety personnel, non-DOE safety and health personnel, consultants, industry and union representatives and by performing site visits to users and manufacturers. A Liaison Group has been formed to provide formal contact with a number of agencies, unions, and scientific and technical societies which generate health and safety information. The Liaison Group serves for several purposes: avoids duplications, ensures availability of key personnel from the liaison agency, and assists in the independent review of CACPH's documents. The Department of Energy recognized that publishing safety and health data which describes safe vs. non-safe conditions for materials which do not have current exposure limits would have some impact outside its community. These liaisons have been key elements for ensuring scientific

Figure 3. *Assessment of needs for developing guidelines*

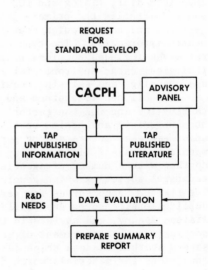

Figure 4. *Search for information*

soundness, minimizing duplicating efforts, but also assuring that
CACPH does not upsurp the activities of other regulatory agencies
which have the specific legislative responsibilities in this
area.

   After assessing the technical literature, a summary document
(Figure 5) of the biological effects data is prepared for review
by CACPH Advisory Panel and TMAC. This review results in a
recommendation to draft a Toxic Material Advisory report which
contains an interim exposure limit in additional pertinent health
and safety requirements. The draft Toxic Material Advisory
report is then reviewed by independent consultants. These
consultants attend a 1-2 day meeting to evaluate the document and
the appropriateness of the interim exposure limit. The
independency of the review meeting is assured by having one of
CACPH Advisory Panel member chair the committee to ensure that
the review process proceeds in an orderly manner. CACPH's staff
participation at this meeting is only to provide overview on the
information that is available for review.

   The independent consultant group can recommend continuation
of the process to develop a revised Toxic Material Advisory Report or recommend research and development needs that could
include the types of toxicological studies that would be required
in order to develop an interim exposure limit. If positive recommendations are obtained at this storage, the draft report is
then subjected to further revision (Figure 6) prior to submittal
for comments to DOE TMAC, DOE safety and health personnel, liaison groups, industry representatives, independent consultants,
and CACPH Advisory Panel. The resulting comments are considered,
and several drafts are generated prior to a pre-final Toxic
Material Advisory report being presented to TMAC. This report
then receives a final review by the Department of Energy's Toxic
Material Advisory Committee, and if appropriate, is recommended
for dissemination (Figure 7). Final dissemination of the report
could include a recommendation to develop a consensus standard,
inclusion in DOE Manual Chapter, issuance as a DOE Order, and
inclusion in DOE contracts.

   Finally a list of potential candidate agents (Table I) has
been prepared for which the Center may be requested to develop
interim exposure limits and Toxic Material Advisory Reports. As
the country becomes more energy independent by developing commercial capabilities in the new energy technology areas, substantial quantities of toxic agents and potential exposure
situations will warrant the attention of regulatory agencies to
develop health and safety requirements. Meanwhile, CACPH will
continue to provide this type of advance health and safety support to the DOE Community.

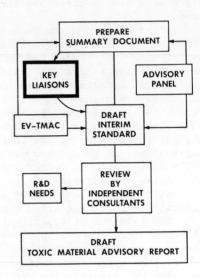

Figure 5. Development of draft Toxic
Material Advisory report

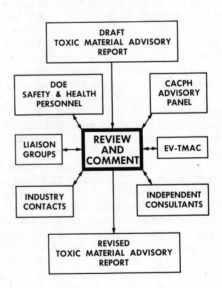

Figure 6. Review and comments pro-
cedure

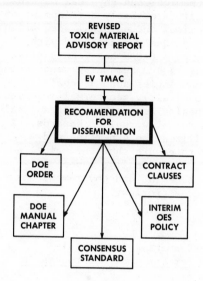

Figure 7. Potential routes of dissemination

Table I. Potential Candidates for Interim Standards

2-MERCAPTOETHANOL
COAL DUST
SHALE OIL EMISSION
UV PLUS POLYCYCLIC AROMATIC
 HYDROCARBONS
POLYCYCLIC AROMATIC HYDROCARBONS
NITROGEN HETEROCYCLIC COMPOUNDS
LOW LEVEL CARCINOGENS
FLYASH
SKIN CONTAMINATION
PHENOLIC COMPOUNDS
FIBERGLASS
CHLOROSILANES
BORON TRICHLORIDE
ORGANIC CD SALTS
TRIMETHYL GALLIUM
ANTIMONY HYDRIDE
CARBON TETRAFLUORIDE
VIBRATIONS
MAGNETIC FIELDS
DIOCTYL PHTHALATE
ELECTROSTATIC FIELDS
ASBESTOS IN PUBLIC BUILDINGS
EXPOSURE STANDARDS FOR FERTILE WOMEN

Abstract

The Center for Assessment of Chemical and Physical Hazards (CACPH) has been established to facilitate timely development of interim exposure guidelines for chemical and physical occupational and environmental hazards which are uniquely associated with activities of the Department of Energy. CACPH has been staffed and equipped to respond to the changing and demanding needs often associated with emerging energy technologies such as coal conversion, oil shale, solar (photovoltaic), geothermal and wind. These guidelines are essential to health and safety professionals responsible for assessment and control of workplace hazards. A system has been developed which utilizes the expertise and experience of reputable scientists to ensure that the guidelines generated represent sound science. Additionally, a pool of consultants from industry, labor, academia, and government has been established to provide independent review.

Literature Cited

1. American Conference of Governmental Industrial Hygienists, Threshold Limit Values for Chemical Substances in Workroom Air, Cincinnati, Ohio, 1979.
2. U.S. Department of Health, Education, and Welfare, Registry of Toxic Effects of Chemical Substances, Cincinnati, Ohio, 1979.

RECEIVED October 27, 1980.

Recent Developments in Electrochemical Solid Polymer Electrolyte Sensor Cells for Measuring Carbon Monoxide and Oxides of Nitrogen

A. B. LA CONTI, M. E. NOLAN, J. A. KOSEK, and J. M. SEDLAK

General Electric Company, Direct Energy Conversion Programs, 50 Fordham Road, Wilmington, MA 01887

Electrochemical gas detection instruments have been developed which use a hydrated solid polymer electrolyte sensor cell to measure the concentration of specific gases, such as CO, in ambient air. These instruments are a spin-off of GE aerospace fuel cell technology. Since no liquid electrolyte is used, time-related problems associated with liquid electrolytes such as corrosion or containment are avoided. This paper describes the technical characteristics of the hydrated SPE cell as well as recent developments made to further improve the performance and extend the scope of applications. These recent advances include development of NO and NO_2 sensor cells, and cells in which the air sample is transported by diffusion rather than a pump mechanism.

The intended use of the SPE detectors is by military, government and industrial personnel involved in air quality measurements. The commercial SPE CO dosimeter and direct-reading detection instruments are being widely used by steel mills, fire departments and various city, state and government regulatory agencies.

0097-6156/81/0149-0551$05.75/0

Experimental

Sensor Cell Operating Mode. The simplest method of sensor operation is as a galvanic cell, whereby the sensor acts as a fuel cell and generates a current proportional to the gas concentration to be detected ([1]). However, when detecting certain species in air, it is difficult to obtain a counter-reference electrode in an acid system that will maintain the sensing electrode at a predetermined potential of approximately 1.0 V, to minimize interference. Counter-reference electrodes such as Pt/air (O_2) or noble metal/noble metal oxide structures have rest potentials in the 1.0 to 1.2 V region. However, they exhibit very low exchange current densities (10^{-10} to $10^{-12} \mu a/cm^2$) and are readily polarized under small current drains.

It has also been demonstrated that the sensing electrode can be maintained at a predetermined voltage of 1.0 V by electrically biasing (constant voltage source) the sensing electrode vs. a stable counter-reference, such as Pt/H_2, H^+ or $PbO_2/PbSO_4$, H^+ ([1], [2,3]). With the former, the sensing electrode is electrically biased 1.0 V above the $Pt/H_2, H^+$ potential (0.0 V) and with the latter 0.7 V below the $PbO_2/PbSO_4$, H^+ potential (1.7 V). Both of these counter-reference electrodes exhibit good reversibility, but reliability and life are not adequate.

The most convenient and reliable electrical biasing method for use with a hydrated SPE cell has been shown to be a three electrode potentiostatic circuit which maintains the sensing electrode at a predetermined potential vs. a stable reference ([1],[3],[4]). The most reversible reference is a Pt/H_2, H^+, static or dynamic. In practical instruments, however, good accuracy and convenience are achieved using a large surface area platinoid metal black/air (O_2). All work reported in this study utilized the air reference which has a potential of approximately +1.05 V vs. a standard hydrogen electrode (SHE). For convenience, all potentials reported are vs. the SHE.

Sensor Cell Characteristics. The membrane and electrode assembly used in the CO sensor cell has three electrodes of identical composition: sensing, counter and reference. These electrodes are fabricated from a platinoid black catalyst composition blended with a Teflon binder. A transition metal screen is embedded into the electrodes to obtain improved mechanical integrity and current collection. The counter electrode is the same size (2.4 cm^2) and configuration as the sensing electrode. The reference electrode has an area of approximately 0.32 cm^2. The electrodes are bonded to the membrane by a proprietary method developed by General Electric. The spatial configuration of the resultant membrane and electrode assembly has been developed to achieve good vapor phase transport and to provide a high output signal.

The membrane and electrode assembly used for the detection of oxides of nitrogen is similar to that used for CO detection, except that the sensing electrode is fabricated from a graphite Teflon mix.

A schematic of the hydrated solid polymer electrolyte sensor cell is depicted in Figure 1. The membrane and electrode assembly is housed in Lexan polycarbonate hardware. Lexan was selected because of its good physical properties (transparency, shock resistance), and chemical inertness/minimal elution. The catalytic sensing and reference electrodes are positioned on one side of the cation exchange membrane; a catalytic counter electrode is positioned on the other side of the membrane opposite the sensing electrode. The counter electrode compartment is flooded with water. Electrolytic contact between the sensing electrode and the platinoid metal/air reference is achieved through a hydrated solid polymer electrolyte membrane bridge. The performance characteristics and other properties of the cell are highly dependent on the morphological structure of the membrane and the method of hydrating the membrane to achieve a fixed water content.

All work has been accomplished using perfluorosulfonate ion exchange membranes manufactured by E.I. duPont and sold under the trade name Nafion. Nafion is a copolymer of polytetrafluoroethylene (PTFE) and polysulfonylfluoride vinyl ether containing pendent sulfonic acid groups. The sulfonic acid groups are chemically bound to the perfluorocarbon backbone.

<pre>
        1100 EW   1500 EW
—(CF₂CF₂)₈—— CFCF₂——   — (CF₂CF₂)₅—— CFCF₂——
             | |
             O   O
             | |
            CF₂   CF₂
             | |
            FC —— CF₃   FC —— CF₃
             | |
             O   O
             | |
            CF₂   CF₂
             | |
            CF₂   CF₂
             | |
            SO₃H   SO₃H
</pre>

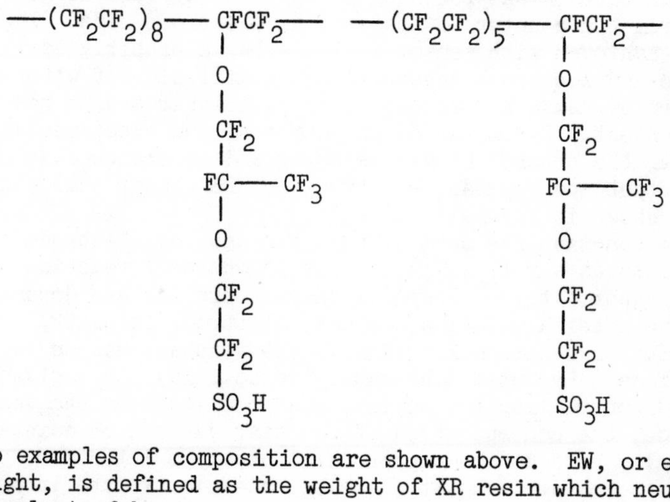

Two examples of composition are shown above. EW, or equivalent weight, is defined as the weight of XR resin which neutralizes one equivalent of base.

Membranes are also characterized by their ion exchange capacity (IEC) - milli-equivalents (meq) of sulfonic acid/dry weight of membrane. The relationship between IEC and EW can be expressed

$$IEC = \frac{1000}{EW} \qquad (1)$$

Generally, IEC is determined by acid/base titration methods and gravimetric weight analysis.

<u>Electrochemical Reaction/Transport.</u> Electrochemical reactions occur at the electrode/electrolyte interface when gas is brought to the electrode surface using a small pump. Gas diffuses through the electrode structure to the electrode/electrolyte interface, where it is electrochemically reacted. Some parasitic chemical reactions can also occur on the electrocatalytic surface between the reactant gas and air. To achieve maximum response and reproducibility, the chemical combination must be minimized and controlled by proper selection of catalyst sensor potential and cell configuration. For CO, water is required to complete the anodic reaction at the sensing electrode according to the following reaction:

$$CO + H_2O = CO_2 + 2H^+ + 2e \qquad (2)$$

Also, approximately 3.5 to 4 millimoles of H_2O are transported with each meq of H^+ from the sensing to the counter electrode side. Water for both these processes is supplied by back-diffusion of water from the counter to the sensing electrode side. This is readily achieved with sensor cells having a properly equilibrated membrane and electrode assembly. Back diffusion of water can be increased by using a membrane which is thinner and/or has higher water content. Hydration of the membrane and electrode assembly, including the channel between sensing and reference electrode, controls output and reproducibility. The SPE sensor cell configuration is shown in Figure 1.

The reactant gas must diffuse through the electrode structure which contains air (O_2, N_2) and any products of reaction (CO_2, NO_2, NO, H_2O vapor, etc.). Response characteristics are dependent on electrode material, Teflon content, electrode porosity, thickness and diffusion/reaction kinetics of the reactant gas on the catalytic surface. By optimizing catalytic activity for a given reaction and controlling the potentiostatic voltage on the sensing electrode, the concentration of reactant gas can be maintained at essentially zero at the electrode/electrolyte interface. All reactant species arriving at the electrode/electrolyte interface will be readily reacted. Under these conditions, the rate of diffusion is proportional to C, where

$$C = C_b - C_e \qquad (3)$$

where C_b and C_e are the reactant concentrations in the bulk gas phase and electrode/electrolyte surface, respectively. However, since C_e is essentially equal to zero, the measured current is proportional to the reactant concentration in the bulk gas phase. The maximum diffusion current, i, can thus be expressed as

$$i = K \, C_b \qquad (4)$$

where K is a proportionally constant.

A detailed review of theories and models for porous gas diffusion catalytic electrodes used in fuel cell-type applications is presented by Austin (5). Two mechanisms of mass transport are possible. There may be a total pressure gradient across the electrode so that forced flow (permeability) occurs. Alternately, there may be other components (N_2, O_2, water vapor, CO_2, etc.) that build up within the electrode structure to give an almost constant total pressure. Mass transfer is then accomplished by diffusion of the reactant gas through a stagnant film of gas in the pores. This process is much slower than forced flow and leads to large partial pressure gradients through the electrode structure. It is this mechanism of mass transfer which likely predominates. Mass transfer relationships for porous gas diffusion electrodes have been solved using modifications of Fick's law. The gas diffusion electrodes used in the sensor cell contain a mixture of large and fine pores. When the diameter of the pores in the electrode is less than 0.1 micron, mass transfer is by Knudsden diffusion.

Solutions of the combined equations of mass transfer, kinetics and electrochemical transport expressed in terms of the limiting current, i_L generally are of the form

$$i_L = \frac{nFDC_bA}{\delta} \qquad (5)$$

where

  n = No. of electrons/molecule
  F = Faraday's constant
  D = Diffusion coefficient of the reactant gas
  A = Active electrode area
  δ = Diffusion layer thickness
thus a linear relationship is predicted where

$$i_L = KC_b \qquad (6)$$

which is in agreement with the empirical equation previously discussed. For the concentration range studied, good linearity was observed when detecting CO, NO and NO_2 in air.

Potentiostatic Circuit. The electrical circuit used for breadboard testing of three-electrode sensor cells is shown in Figure 2. Amplifier U1 sensed the voltage between the reference and

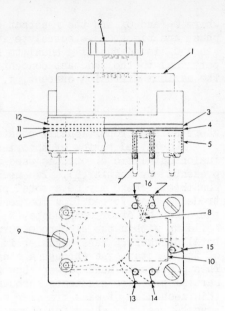

Figure 1. Sensor cell assembly: 1, reservoir housing; 2, cap; 3, support plate; 4, M & E assembly; 5, base plate; 6, gasket; 7, contact pin; 8, thermistor; 9, nylon screw; 10, Teflon tape; 11, gasket; 12, gasket; 13, counter electrode; 14, sensor electrode; 15, reference electrode; 16, thermistor.

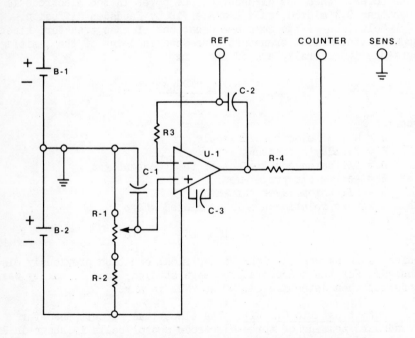

Figure 2.   Schematic of simple potentiostatic circuit for sensor tests

sensing electrodes (common) and compared it with a preset reference voltage at R1. Any differences in signal were used to generate a current between the counter and sensing electrode, which acted to eliminate the difference voltage producing it. The current required could be sensed across resistor R4 or measured directly with an ammeter in series with the counter electrode.

The circuitry used for the breadboard testing of NO and NO_2 sensor cells was very similar to that shown in Figure 2; only the applied potential was changed. An applied potential of +1.30 V versus the SHE reference electrode was used for NO oxidation while a potential of 0.75 V versus the same reference electrode was used for NO_2 reduction. Current measurements were again made by measuring the voltage drop across resistor R-4. Three electrode systems were used for both gases.

Gases. Mixtures of carbon monoxide in air or nitrogen were obtained from either Matheson, Scott or Airco Gas Products. All gases were factory-analyzed to ± 5% and were used without further analysis.

Mixtures of nitric oxide in nitrogen were obtained from Matheson Gas Products. High concentrations (250 ppm), factory analyzed to ± 5%, were diluted with air using standard gas mixing techniques. NO concentrations were calculated from NO and air flow rates.

Nitrogen dioxide permeation tubes were obtained from Metronics, Inc. These tubes were factory calibrated for use at 30.0°C. The permeation tubes were thermostated in a water bath and air passed over at a known flow rate (of at least 150 cc/min). Gas samples were drawn off at 80 cc/min.

Results and Discussion

Current-Voltage Characteristics of Platinoid and Graphite Electrodes in Air. The SPE sensor cells have been optimized for detection of certain oxidizable or reducible species in air. To achieve this, the sensing electrode is maintained at a voltage when there is minimal interference by the O_2 in the air. The two electrodes studied in detail have been platinoid and graphite sensing electrodes.

The anodic and cathodic behavior of a platinoid electrode structure is shown in Figure 3. All voltage measurements are referred to the SHE. The rest potential of the Pt/air (O_2) electrode is approximately 1.05 V and background current is in the order of micro-amperes. The most useful range for detection of certain oxidizable species such as CO, H_2, ethylene, acetylene and ethanol is 1.05 to 1.20 V. Above 1.2 V the anodic background current rises due to formation of some oxygen. Below 1.05 V a cathodic current due to oxygen reduction is observed. A limiting current for oxygen reduction is generally observed in the range 0.4 to 0.8 V (2). This voltage range is optimum for oxygen detection using these

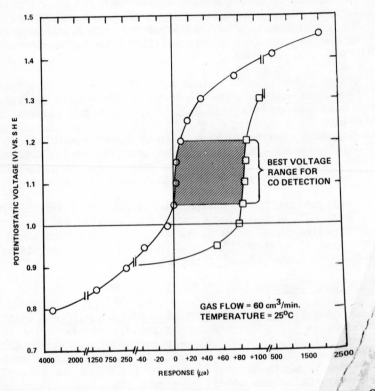

Figure 3.   CO sensor cell voltage–current relationships: (○) air, (□) 40 ppm CO in air

sensing electrode structures. In this paper, only the detection
of CO in air on platinoid structures is addressed. The corresponding curve for oxidation of CO in air is also shown in Figure 3.
The cell response is essentially constant for the range 1.05 to
1.20 V. The optimum voltage selected for oxidation of CO in air
was 1.15 V.

The anodic and cathodic behavior of a graphite electrode
structure is shown in Figure 4. With graphite electrode structures
the useful range for detecting certain oxidizable (NO, H_2S) or reducible (NO_2, Cl_2) species is extended from 0.7 to 1.3 V before
interference due to air reduction or oxygenation of graphite is observed, respectively. In this paper, only the detection of NO and
NO_2 on graphite electrode structures is addressed. The optimum
voltages selected for NO oxidation and NO_2 reduction in air were
1.25 and 0.75 V respectively.

<u>Linearity of Response and Reaction Products</u>. The response vs.
concentration curves obtained for CO, NO and NO_2 gas sensor cells
are depicted in Figures 5, 6 and 7 respectively. In all instances
good linearity over the range studied was observed between current
and partial pressure of each of the above gases (as depicted by
equation (4)). The proportionality constants, K, with standard deviation for the gases are as follows:

<u>Gas</u>	<u>Range Investigated (ppm)</u>	<u>K (μa/ppm)</u>
CO	0 to 1500	2.21 ± 0.05
NO	0 to 200	2.60 ± 0.13
NO_2	0 to 18	2.91 ± 0.04

The observed sensor cell response linearity with standard deviation makes a one-point calibration of a detector possible for
the concentration ranges indicated.

When measuring CO the only reaction product found was CO_2 and
mass balance measurements indicate the reaction does proceed according to equation (2).

Experiments were also conducted to measure NO oxidation products and aid in identification of the reaction occurring at the
electrode surface. Using an NO_2 sensor cell in series with an NO
sensor cell, 70-80 ppm NO_2 were detected in the exhaust of an NO
sensor cell which had been exposed to 240 ppm NO.

Two NO sensor cells in series were used to measure n, the
number of electrons involved in the oxidation reaction. The following relationship was derived from Fick's Law, and was used to
calculate n:

$$n = \frac{2.24 \times 10^7 \, i}{FfC \, (i^\circ_2 - i_2^\circ)} \frac{i^\circ_2}{1} \qquad (7)$$

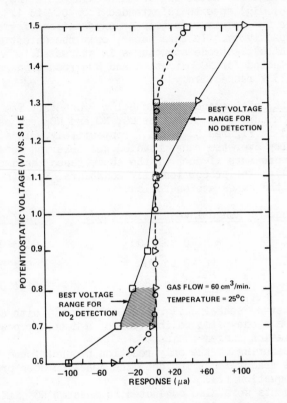

Figure 4.   NO and NO₂ sensor cells voltage–current relationships: (○) air or 100 ppm CO in air, (△) 15 ppm NO in air, (□) 10 ppm NO₂ in air

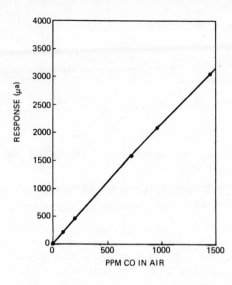

Figure 5. Response vs. concentration for CO sensor cell: potentiostatic voltage, 1.15 V; gas flow, 60 cm³/min; temperature, 25°C; TLV, 50 ppm; STEL, 400 ppm; OSHA PEL, 50 ppm.

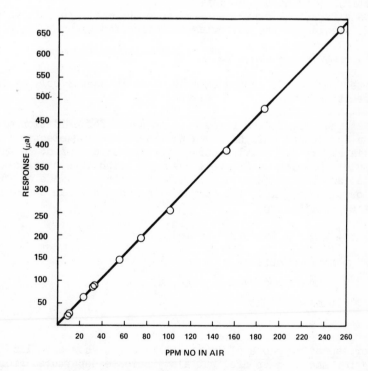

Figure 6. Response vs. NO concentration: potentiostatic voltage, 1.3 V; gas flow, 60 cm³/min; temperature, 25°C; TLV, 25 ppm; STEL, 35 ppm.

where F   = the value of Faraday's constant (96,500 coulombs/
                mole)

      f   = the sensor cell flow rate (liters/second)

      C   = the NO concentration (ppm)

2.24 x 10^7 = a conversion factor to change from ppm to molar
                volume

      i_1 = the current at the first sensor cell (amps)

      i_2 = the current at the second sensor cell measuring the
                exhaust of the first (amps)

      i°_2 = the current obtained when flowing NO directly into
                the second sensor cell (amps)

Flow rates were varied from 60 to 5 cc/min. The NO concentrations which were used varied from 243 ppm at low flow rates to 11.9 ppm at higher flow rates, Values of n at higher (40 - 60 cc/min) flow rates were found to be independent of NO concentration. Due to the low sensor cell currents measured at low flow rates, NO concentrations were held constant.

Data from this experiment indicate the most probable value for n is 2, leading to the following oxidation reaction:

$$NO + H_2O \longrightarrow NO_2 + 2H^+ + 2e^- \qquad (8)$$

The detection of NO_2 in the exhaust of an NO sensor cell verifies this reaction.

Response Time. The response-time curve for oxidation of CO with an SPE sensor cell having a platinoid sensing electrode is shown in Figure 8. Similar curves for the oxidation of NO and reduction of NO_2 with an SPE cell having a graphite sensing electrode are also shown in Figure 8. All measurements were made at 25°C at gas flow of 60 cc/min. The current-time response can be estimated from the relationship

$$1 - I/i = \exp(-t/\tau) \qquad (9)$$

where i = maximum current

      I = current observed at time, t, sec.

      τ = time constant

      t - time

The corresponding plots of ln (1 - I/i) vs. t show excellent linearity for the NO response, but also indicate departures from linearity for CO and NO_2 after 10-12 seconds. The response time characteristics of these cells are sensitive to cell geometry,

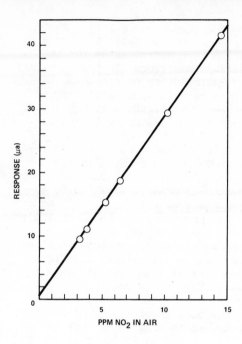

Figure 7.   Response vs. NO₂ concentration: potentiostatic voltage, 0.75 V; gas flow, 80 cm³/min; temperature, 25°C; TLV, 5 ppm.

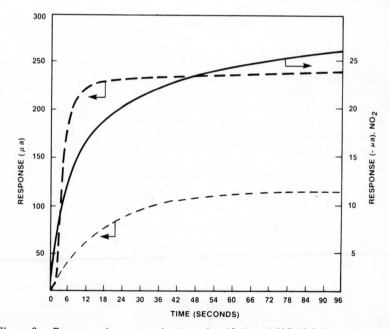

Figure 8.   Response–time curves for (– – –) oxidation of NO (1.3 V) and (———) reduction of NO₂ (0.75 V) on a graphite electrode and (– – –) oxidation of CO (1.15 V) on a platinum electrode. Test gases: 50 ppm CO in air; 51.4 ppm NO in air; 7.4 ppm NO₂ in air.

electrode composition, and reactant species. In general, response times are lengthened for strongly adsorbed reactants and for electrodes with a large surface area. Active surface area for the platinoid black and graphite catalytic electrodes is approximately 40 and 1 meter^2/g, respectively. Optimization of diffusion layer thickness (δ), decreases in electrode surface area, and improved water management are all under active investigation as methods of increasing the speed of sensor response to the subject gases.

Effect of Feed Flow. The effect of feed flow on sensor response for the CO sensor cell is shown in Figure 9. Similar curves were observed for NO and NO$_2$. Generally, there is a marked increase in sensor cell response with increasing flow (10 to 40 cc/min), followed by a slow rising slope as flow exceeds 50 cc/min. A flow of 60 cc/min was selected for practical use based on trade-off studies of water management and flow dependence.

Studies by Sedlak (6) have shown a similar response-flow relationship for liquid electrolyte cells which utilize a teflon-bonded diffusion electrode. The empirical equations and relationships derived generally apply to the SPE sensor cells.

Effect of Temperature. The response level as a function of temperature (1 to 40°C) at a fixed feed rate of 60 cm^3/min of 100 ppm CO in air is shown in Figure 10. The data follow an Arrhenius relationship of the type

$$i = K \exp(-E/RT) \qquad (10)$$

where E = activation energy, k cal/mole

A similar relationship was found for the NO and NO$_2$ sensor cells. The background current observed over the same temperature range with zero air (air containing none of the test gases) remained essentially constant $0 \pm 2\ \mu a$. A least square fit of the temperature data for CO in air (0 to 1000 ppm), NO in air (0 to 200 ppm) and NO$_2$ (0 to 14 ppm) yielded activation energies as follows:

Test Gas	E, k cal/mole
CO in air	7.7
NO in air	7.1
NO$_2$ in air	7.6

The values of E are indicative of a diffusion controlled process.

The compensated span response over the temperature range 1 to 40°C is shown in Figure 11. The circuitry used for compensation is depicted in the same figure. The optimum parallel resistor to achieve accuracy within ± 5% of the 25°C calibrated value is 2.2K ohm.

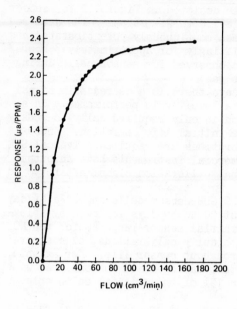

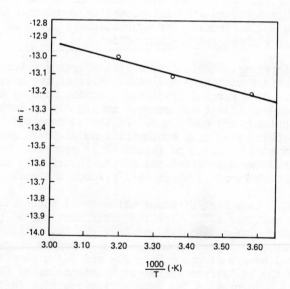

Figure 9. Effect of gas flow on CO response level: potentiostatic voltage, 1.15 V; temperature, 25°C; 100 ppm CO in air.

Figure 10. Effect of temperature on CO span response: potentiostatic voltage, 1.15 V; gas flow, 60 cm³/min; 100 ppm CO in air.

Life Characteristics of the Sensor Cells. Typical life behavior of the SPE CO sensor cells is depicted in Figure 12 for sensor cells with responses in the range 1.8 to 2.4 μa/ppm. The data reported are for cells that had been continuously potentiostated and operated on the test gas intermittingly for approximately 8 hours/day. Similar stability has been observed for sensor cells during continuous operation with a test gas.

During the first 15 to 20 days there is a decrease in signal amounting to 2%/day followed by a leveling in performance. In a practical instrument, calibration is only required daily to weekly to achieve ± 10% accuracy during initial life operation. Subsequently, weekly to monthly calibrations are required. Typical SPE CO sensor cells operated in commercial instruments have an output that is within 40% of the original calibration signal after two to three years of operation.

Life testing conducted on NO SPE sensor cells showed a 1%/day increase in signal over the first 20 to 30 days followed by a leveling in performance. Typical initial sensor response for the NO sensor cell is 3 μa/ppm NO. With daily calibrations, high accuracy levels can be maintained throughout sensor life. For example, a prototype NO detection instrument has operated for one year with the response to NO-in-air within 15% of the original calibration value.

The NO_2 sensor cells demonstrate a decrease in signal ranging from 0 to 0.5%/day. Typical initial sensor response is 2.5 to 3.0 μa/ppm NO_2. This instrument also demonstrates the capability to remain highly accurate with infrequent calibration. Operation of a prototype NO_2 instrument for approximately 200 days has verified the need for minimal calibration effort.

Interferences. Table I is a list of gases which could potentially interfere with a CO analysis. Most of the gases are removed by use of a Purafil (potassium permanganate on alumina) scrubber column. Gases not removed completely by Purafil are hydrogen, ethylene, and acetylene. If the CO detector must be used in an atmosphere with high concentrations of these gases, special precautions must be taken to insure their removal. Column 3 of Table I lists the concentrations of the interfering species, after passing through Purafil, which will produce a signal equivalent to 1 ppm CO.

Table II is a list of gases which could potentially interfere with an NO analysis along with the concentrations of these gases which produce a signal equivalent to 1 ppm NO. Only H_2S had an effect on sensor cell performance and was found to decrease the response level by 0.2 μa/ppm. H_2S, SO_2 and NO_2 were effectively filtered from the gas stream by use of triethanolamine (TEA) as shown in Table II. To prevent TEA vapors from reaching (and thus poisoning) the sensing electrode surface, a short column of a cation exchange bead was placed after the filter.

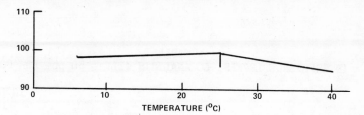

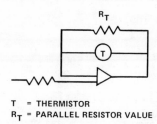

T   = THERMISTOR
R_T = PARALLEL RESISTOR VALUE

Figure 11.   Temperature compensated output for CO sensor cells

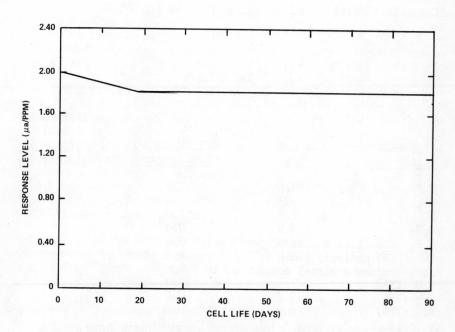

Figure 12.   CO sensor cell response as a function of time: potentiostatic voltage, 1.15 V; gas flow, 60 cm³/min; temperature, 25°C.

TABLE I
INTERFERENCE LEVELS, CO ANALYSIS (TLV = 50 ppm FOR CO)

INTERFERENT* GAS	INTERFERENT GAS CONC. IN AIR	TLV	EQUIVALENT CONC. (ppm) OF CO	INTERFERENT RATIO (ppm GAS) (ppm CO)
H_2O VAPOR	50 TO 100% RH		0	-
O_2	16 TO 20%		0	-
CH_4	3%		0	-
CO_2	1%	5000	0	-
NO	50 ppm	25	0	-
NO_2	10 ppm	5	0	-
SO_2	25 ppm	5	0	-
H_2	100 ppm		2	50.0
C_2H_2	100 ppm		16	6.0
$HCHO$	5 ppm	2	0	-
H_2S	10 ppm	10	0	-
C_2H_4	100 ppm		18	5.4

*$KMnO_4$ ON ALUMINA (PURAFIL) FILTER USED IN ALL TESTS

TABLE II
INTERFERENCE LEVELS, NO ANALYSIS (TLV = 25 ppm FOR NO)

INTERFERENT GAS*	GAS CONC. (ppm)	EQUIVALENT NO CONC. (ppm)	INTERFERENT RATIO (ppm GAS) (ppm NO)
H_2S	7.8	2.8	2.6
SO_2	9.3	0.9	10.0
NO_2	6.6	17.0	1/2.5
C_2H_2	201.0	0.0	-
C_2H_4	55.0	0.0	-
CO	98.0	0.0	-
H_2S (TEA)	7.8	0.0	-
SO_2 (TEA)	9.3	0.0	-
NO_2 (TEA)	6.6	0.0	-

*ALL TESTS RUN WITHOUT A TEA FILTER EXCEPT WHERE INDICATED

Table III is a list of gases which might interfere with an NO_2 analysis. Of all the gases tested, only H_2S and SO_2 had a significant response; oxidation currents were observed for both species. Neither gas gave a signal which was constant with time. After an initial response, the response to both gases slowly started to drift back to the background value. Exposure of the sensor cell to these gases did not have any effect on the response level to NO_2. Both gases may be filtered out by use of $HgCl_2$ immobilized on an inert support as shown in Table III.

TABLE III
INTERFERENCE LEVELS, NO_2 ANALYSIS (TLV = 5 ppm FOR NO_2)

INTERFERENT GAS*	INTERFERENT GAS CONC. (ppm)	EQUIVALENT NO_2 CONC. (ppm)	INTERFERENT RATIO (ppm INTERFERENT) (ppm NO_2)
H_2S	7.8	36	1/4.6
SO_2	6.9	18	1/2.6
C_2H_2	201.0	0	-
C_2H_4	55.0	0	-
NO	20.8	0	-
CO	167.0	0	-
H_2S (0.1$\underline{M}$ $HgCl_2$)	5.7	0	-
SO_2 (0.1$\underline{M}$ $HgCl_2$)	6.9	38	1/1.6

*ALL TESTS RUN WITHOUT A $Hg Cl_2$ FILTER EXCEPT WHERE INDICATED

Development of a Diffusion Head Sensor Cell. The use of air sampling pumps in portable electrochemical gas detection apparatus introduces potential problems into the instrument. First, the sensor cell response is dependent on gas flow rate. The sample flow rate, therefore, must be accurately controlled to obtain reproducible results, or the sample flow rate must be set high enough to insure a flow independent response. Secondly, failure of the pump itself could prevent a sample from reaching the sensor cell. Thirdly, the pumps are usually one of the largest users of current in a portable instrument and thereby limit usable battery life. The need for an air sampling pump can be eliminated by use of a diffusion tube having a set length to diameter (L/d) ratio in its geometry for introduction of a gas sample. Proper selection of the geometry and L/d ratio of the diffusion tube results in an electrochemical cell with a response which is independent of external gas flow rate. A schematic of a solid polymer electrolyte diffusion head sensor cell is shown in Figure 13.

Equations were derived to predict response times and the expected steady state current for a pure diffusion system.

For response time, the steady state current occurs at

$$t = L^2/4D \qquad (11)$$

and the steady state current is given by

$$i = \frac{2nFAC_b D}{L} \exp(-1) \qquad (12)$$

In these equations,

   D = diffusion coefficient (0.175 cm^2/sec for CO)

   C_b = ambient CO concentration (moles/cm^3)

   t = time, sec (from time of admission of sample to base of tube)

   L = tube length

   A = electrode area

First, an estimate can be made of expected response time from Equation (11). Tube length is $L = 10$ cm and $D = 0.175$ cm^2/sec. Using these values, $t = 143$ seconds was computed. Secondly, the steady state current was estimated for 1 ppm CO by means of Equation (12). Note that tube area $A = 2.85$ cm^2, $nF = 2 \times 96,500$ coul/mole and 1 ppm CO is equivalent to 4.08×10^{-11} moles/cm^3. Using these numbers, a sensor response of 0.29 μA/ppm CO was calculated. The simplified diffusion model provided predictions of operating characteristics which were sufficiently promising to proceed with an experimental study.

A pre-prototype "pumpless" CO detector fabricated for test purposes included an interference gas filter cartridge which could be replaced easily without disassembly of the sensor/diffusion tube assembly. The signal response in this arrangement was 0.35 μA/ppm CO - in good agreement with theoretical predictions. Temperature characteristics of the diffusion head unit were defined over 4-37°C. The results are summarized in Table IV. Zero stability was excellent over this temperature range; the variability in zero reading was only to the equivalent of a few ppm CO. Signal response was essentially constant between 4-37°C at $0.31 - 0.34$ μA/ppm CO. This relative invariance of signal may be a consequence of a compensation effect. That is, as temperature increases, the test gas becomes slightly less dense on a moles CO/cm^3 basis, but the diffusion coefficient increases slightly.

The effect of air flow past the diffusion tube inlet was studied to simulate ambient convection. The effect of external air flows perpendicular to the diffusion tube for two different CO concentrations is shown in Figure 14. Little variation was seen in the response over the range 200 - 600 ft/min. Response times to

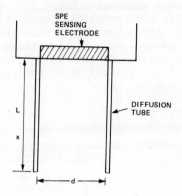

Figure 13. *Schematic of SPE diffusion head gas sensor*

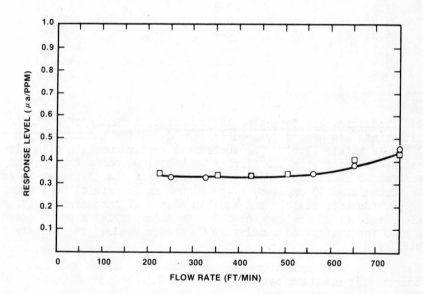

Figure 14. *Response level vs. flow rate: (○) ~ 100 ppm CO in air; (□) ~ 200 ppm CO in air*

ca 90% of steady state signal were within 2-3 minutes. This also
is in reasonable conformance to theoretical predictions.

Recent work has extended the use of a diffusion based instrument to one in which the diffusion tube has been rotated 90° with
respect to that of Figure 13. In this configuration, there is a
possibility of flow directly into the diffusion tube. This tube
has dimensions L = 1.6 cm and d = 0.87 cm. Substituting these
values into Equation (2) gives a predicted diffusion cell response
of 1.5 μa/ppm CO. Observed values of 1.1-1.2 μa/ppm CO were again
in good agreement with theoretical predictions.

TABLE IV
TEMPERATURE DEPENDENCE, DIFFUSION HEAD SENSOR CELL

TEMPERATURE °C	BACKGROUND (μa)	CO RESPONSE (μa)
23.3	1-3	137-138
4.0	1-3	133-134
24.0	3-4	131-132
37.0	2-4	141-142
27.0	2-3	131-132
22.4	0-2	127-128
36.0	2-3	145-146
24.5	1-3	132-134
3.9	0-2	141-143
23.8	2-3	121-122

TEST GAS 417 ppm CO IN AIR

Production Gas Detection Instruments. A family of portable
instruments has been developed for the detection and monitoring of
CO levels in air (7,8). The instrument family consists of a direct
reading detector with LCD display of actual CO concentration and a
personal CO dosimeter. Both the detector and the dosimeter measure
the accumulated CO exposure of personnel in industrial environments
and provide both visible and audible alarms if instantaneously unsafe levels of CO are encountered. An accompanying support console
is used for integrated cumulative CO dosage readout and battery
charging.

Summary

Gas detection sensor cell systems highly specific for CO, NO
and NO_2 have been developed. All sensor cells utilize a solid
polymer electrolyte with integrally-bonded fuel cell electrodes.
Current-voltage characteristics of various catalyst mixtures were
studied to determine optimum sensor cell operating voltages. Activation energies indicative of diffusion-controlled processes were
calculated for CO and NO oxidation and NO_2 reduction. Linear responses over wide concentration ranges were also observed. Additionally, a CO sensor cell system utilizing a diffusion tube in-

stead of an air sampling pump was constructed. Good agreement was obtained between theoretical and calculated responses.

Acknowledgements

This work was partially supported by the U.S. Department of Interior, Bureau of Mines. The authors would like to acknowledge the technical guidance of Drs. Emery Chilton and George Schnakenberg of the Bureau of Mines, whose many helpful suggestions and comments contributed significantly to the technology developments described in this publication.

Literature Cited

1. LaConti, A.B.; Dantowitz, P.; Kegan, R.; Maget, H.J.R TIS Report No. 66, ASD 3, General Electric Company: Lynn, Mass., 1966

2. LaConti, A.B.; Maget, H.J.R. J. Electrochem. Soc., 1971, 118, p. 506

3. Dempsey, R.M.; LaConti, A.B.; Nolan, M.E. "Gas Detection Instrumentation Using Electrically-Biased Fuel Cell Sensor Technology", PB254823, NTIS, Springfield, Va., 1978

4. Schnakenberg, G.H.; LaConti, A.B., "Improved CO Detection Systems", Proceedings, 17th International Conference, Bulgaria, 1977

5. Austin, L.G. "Fuel Cells", NASA SP-120 Office of Technology Utilization, NASA, Washington, D.C., 1967

6. Bay, H.W.; Blurton, K.F.; Sedlak, J.M.; Valentine, A.M.; Anal. Chem., 1974, 46, p. 1837

7. Jones, R.C.; LaConti, A.B.; Nuttall, L.J. "Carbon Monoxide Monitor Features Hydrated Solid Polymer Electrolyte Cell", Proceedings, American Industrial Hygiene Conference, New Orleans, Louisiana, May, 1977

8. Gruber, A.H.; Goldstein, A.G.; LaConti, A.B. "A New Family of Miniaturized Self-Contained CO Dosimeters and Direct Reading Detectors", Paper Presented at U.S. Environmental Agency Symposium on Personal Air Pollution Monitors, Chapel Hill, N.C., Jan., 1979

RECEIVED October 6, 1980.

A New Passive Organic Vapor Badge with Backup Capability

W. J. LAUTENBERGER and E. V. KRING

Applied Technology Division, E. I. du Pont de Nemours and Company, Incorporated, Wilmington, DE 19898

J. A. MORELLO

Engineering Test Center, E. I. du Pont de Nemours and Company, Incorporated, Wilmington, DE 19898

Recently, interest has been growing in sampling organic vapors with a passive sampling device (1-8). In general these devices collect organic vapors by means of molecular diffusion and adsorption onto an activated charcoal collection element. After exposure, the activated charcoal is desorbed with a measured volume of desorbing solvent. The desorbing solutions are then analyzed using gas chromatographic techniques outlined in NIOSH Analytical Method, P&CAM 127.

A new passive organic vapor badge has been developed with the capability of determining when breakthrough or saturation of the charcoal adsorbent has occurred. The Du Pont Pro-Tek Organic Vapor Air Monitoring Badge with Back-Up (G-BB) contains two 300 milligram charcoal strips, one in the front section and another in the backup section. (See Figure 1.) The front section of the badge normally collects all of the contaminant. The backup section serves two purposes: 1) it can indicate when saturation of the front section has occurred, and 2) it can extend the total sampling time up to the point of its saturation.

Experimental

The dynamic contaminant generation system used has been previously described by Du Pont for the purpose of laboratory validation of sampling methods where an accurate measure of the true contaminant concentration could be determined (9,10).

The badge exposure chamber used in the laboratory testing consists of a miniaturized wind tunnel made of glass rectangular tubing jacketed with a water condenser which permits temperature control. (See Figure 2.) G-BB badges were placed in the exposure channel chamber and held in position so that the flow of air remained parallel to the face of the badges. Exposure tests involved the placement of six badges in the chamber and

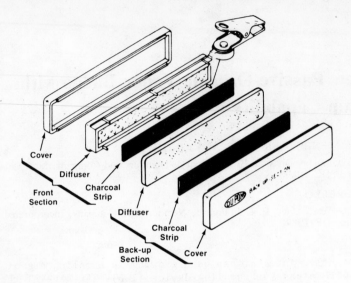

Figure 1.   Pro-Tek G-BB badge components

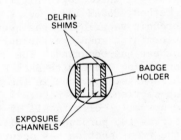

END VIEW

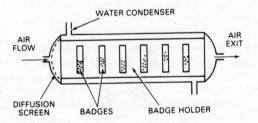

SIDE VIEW

Figure 2.   Badge exposure chamber

the removal of one every hour. Exposure dose levels were,
therefore, generated over a range of conditions. Test vapors
chosen were toluene, acetone, methyl chloroform, and
trichloroethylene. Generated concentrations were approximately
one to two times the TLV . All organic chemicals used were
Fisher Scientific Company spectro quality grade. Face velocity
conditions were maintained at 100 ft/min. Temperature and
pressure conditions were 298°K and 760 mm Hg respectively.
Relative humidity was either 5 or 80 percent.

For storage stability tests, a sufficient number of badges
were exposed to an exposure dose level of approximately one-half
the charcoal capacity at 80 percent relative humidity. Half the
samples were refrigerated (40-45°F) and half were left at
ambient temperatures (75-80°F). Badges were then analyzed at
intervals ranging from one day to three weeks.

Analysis

After exposure, the activated charcoal strips are removed
from both front and backup sections for analysis. Each strip
is desorbed separately with 1.5 ml of spectro quality carbon
disulfide and agitated for 30 minutes with a shaker (SKC Model
Charcoal Developer). The desorbing solution is then analyzed
by removing a $0.5\mu \cdot \ell$ aliquot and injecting it into a Hewlett-
Packard Model 5840A gas chromatograph equipped with an automatic
sampler (Model 7672A) and flame ionization detector. The
analytical column used was a 6-foot by 1/8-inch glass tube
packed with 10 percent Carbowax 20M on 80/100 mesh
Chromosorb W. The unit was operated isothermally with column
temperature ranging between 65°C and 100°C depending on the
contaminant being measured. The carrier was nitrogen at a
flow rate of 30 cm^3/min. Calibration standards for the analysis
were prepared daily so that known quantities of the contaminant
were contained in the desorbing solvent. The weight of the
desorbed contaminant is determined by comparison with peak
areas of known calibration standards. The total mass collected
was determined by correcting the desorbed mass for desorption
efficiency. Previously determined desorption efficiencies were
found to be: 0.67, acetone; 0.97, toluene; 0.98,
trichloroethylene; and 1.01, methyl chloroform (10).

Results and Discussion

Theory

For diffusional badges, according to Fick's First Law
of Diffusion, the amount of mass, M (ng), adsorbed on the
charcoal is a function of the badge sampling rate, (D A/L)
(cm^3/min), times the ambient concentration, C (mg/m^3) and
the sampling time, t (min) (3,5,7,11-14).

$$M \text{ (ng)} = D\left(\frac{A}{L}\right)\left(\frac{cm^3}{min}\right) \times C\left(\frac{mg}{m^3}\right) \times t \text{ (min)} \qquad (1)$$

The badge sampling rate is a direct function of the diffusion coefficient (D) of the organic vapor(s) being sampled and the total cross-sectional area (A) of the badge cavities. The rate is an inverse function of the diffusion path or length (L) of the cavities.

$$\text{Sampling Rate,}\left(\frac{cm^3}{min}\right) = D\left(\frac{A}{L}\right) \qquad (2)$$

where:

$D, \frac{cm^2}{min}$ = Diffusion coefficient at 25°C, 760 mm Hg

A, cm^2 = Total cross-sectional area of the cavities

L, cm = Length of cavities

According to equation (1), if the badge geometry is held constant, the mass collected will be a linear function of the exposure dose, C x t (See Figure 3) up to the point of saturation of the charcoal. In a two-stage badge such as the G-BB badge the front section of the badge samples at a rate of approximately 50 cm^3/min. up to the point of saturation. (See Figure 4.) No sampling should be taking place at this time in the backup section. After saturation of the front section, however, the backup section samples additional material at a reduced rate. (See Figure 5.) The sampling rate of the backup section is a function of the total area and length through which the contaminant must travel from the front face of the badge to the backup section charcoal strip. This sampling rate can be determined by the use of equation (3) (14).

$$\left(\frac{L}{A}\right)_{Total} = \left(\frac{L}{A}\right)_1 + \left(\frac{L}{A}\right)_2 + \left(\frac{L}{A}\right)_3 \qquad (3)$$

where:

$\left(\frac{L}{A}\right)_1$ = diffusional resistance through diffuser element 1

$\left(\frac{L}{A}\right)_2$ = diffusional resistance through charcoal strip

$\left(\frac{L}{A}\right)_3$ = diffusional resistance through diffuser element 2

For the G-BB badge:

$\left(\frac{L}{A}\right)_{Total}$ = 0.135 cm + .023 cm + 0.135 cm = 0.293 cm

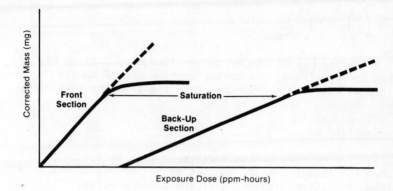

Figure 3. Backup badge theory

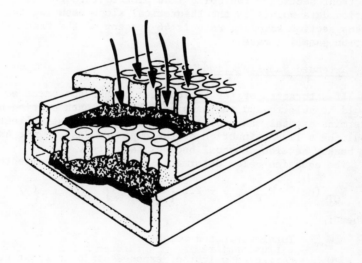

Figure 4. Front section sampling

$$\left(\frac{A}{L}\right)_{Total} = \left(\frac{A}{L}\right)_{Backup} = 3.41$$

$$\left(\frac{A}{L}\right)_1 = \left(\frac{A}{L}\right)_{Front} = 7.4$$

Since the $\left(\frac{A}{L}\right)$ of the backup section is 2.2 times the $\left(\frac{A}{L}\right)$ of the front section, the backup section should sample additional material at 46 percent of the sampling rate of the front section. (See Figure 3.)

Sampling Data

Studies of four contaminants were made to determine whether the backup section samples at 46 percent of the front section rate after saturation of the front section. The four contaminants chosen had varying vapors pressures and adsorption properties on charcoal. (See Figures 6-9.) In all four cases, breakthrough was detected after saturation of the front section charcoal and the sampling rate in the backup section was approximately 46 percent of the sampling rate of the front section. The solid line plotted through the backup section data points is the theoretical slope assuming the backup section sampling rate is 46 percent of the front section sampling rate.

Extended Sampling Time

If saturation of the backup section has not occurred, one should be able to add the mass collected in each section and determine a total mass collected. This total mass collected should be a linear function of the exposure dose, C x t until the backup section charcoal becomes saturated.

The equation below determines the total mass collected.

$$W_{CT} \text{ (ng)} = W_{CF} \text{ (ng)} + 2.2 \left[W_{CB} \text{ (ng)} \right] \qquad (4)$$

Where:

W_{CT} = Total corrected weight

W_{CF} = Corrected weight of exposed strip in front section.

W_{CB} = Corrected weight of exposed strip in backup section.

2.2 = Correction factor based on the fact that the sampling rate of the backup section is 46 percent of the sampling rate of the front section.

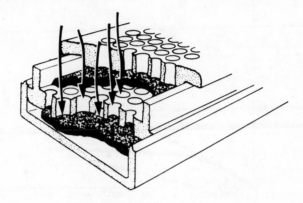

Figure 5. Backup section sampling

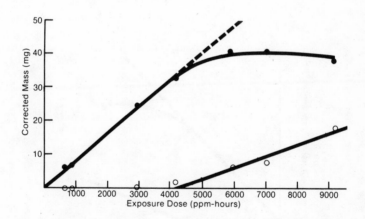

Figure 6. Toluene sampling data: T = 298 K; RH = < 5%; P = 760 mm Hg; concentration = 125 ppm; (●) front section; (○) backup section

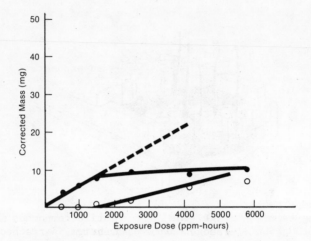

Figure 7. *Acetone sampling data: T = 298 K; RH = 90%; P = 760 mm Hg;*
concentration = 1100 ppm; (●) front section; (○) backup section

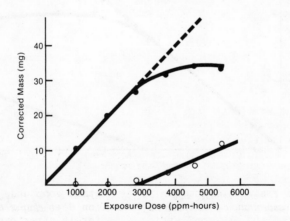

Figure 8. *Trichloroethylene sampling data: T = 298 K; RH = 80%; P = 760*
mm Hg; concentration = 990 ppm; (●) front section; (○) backup section

Weights have been corrected for any correction from a blank (i.e., charcoal strip from an unexposed badge).

Figure 10 represents the total corrected mass as a function of exposure dose for toluene from the data collected in Figure 6. A linear relationship exists out to exposure dose levels of at least 9,000 ppm-hours versus 4,000 ppm-hours for the front section of the badge. This extended linearity allows one to increase the sampling time for measuring an exposure.

Storage Stability

After a test exposure, the cover was replaced on the badge and the badge resealed in a pouch. Then, half of the samples were refrigerated and half were stored at ambient temperatures. Figure 11 shows that the storage stability of the total mass collected is approximately two weeks for acetone, trichloro- ethylene, and toluene whether stored refrigerated or at ambient temperatures.

During sample storage, the more volatile compounds may migrate throughout the badge until equilibrium is reached. Figure 12 shows that for volatile compounds such as acetone and trichloroethylene significant migration occurs within one to two days. This migration process is somewhat reduced by refrigerating the samples. For compounds in which migration is expected, separation of the charcoal strips immediately after exposure appears to be necessary.

Conclusion

To date, laboratory sampling tests for acetone, methyl- chloroform, trichloroethylene and toluene have confirmed the Pro-Tek   G-BB Organic Vapor Air Monitoring Badge's ability to:

1. Indicate when saturation of the front section of charcoal has occurred.
2. Extend the total sampling time for measuring an exposure.

Storage stability studies indicate that badge samples can be stored up to two weeks either refrigerated or at ambient temperatures. However, for volatile compounds such as acetone and trichloroethylene migration does occur and separation of the charcoal strips immediately after exposure is recommended.

Acknowledgements

The authors express their gratitude to the following for their assistance: Vince Keedy and Dave Mount for assistance in badge design and construction; John Pratt, Juanita Reeves,

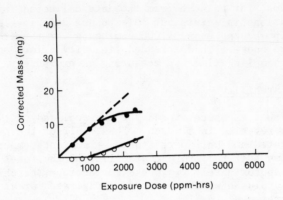

Figure 9.   Methyl chloroform sampling data: $T = 298$ K; $RH = 80\%$; $P = 760$ mm Hg; concentration = 400 ppm; (●) front section; (○) backup section

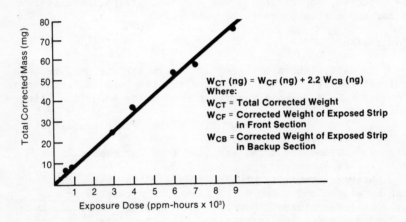

Figure 10.   Total corrected mass (weight); $T = 298$ K; $RH = < 5\%$; $P = 760$ mm Hg; organic vapor = toluene; concentration = 125 ppm

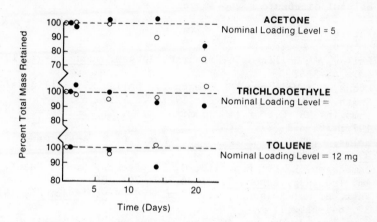

Figure 11. G-BB storage stability—percent total mass retained vs. time; data points, average of two determinations: (○) ambient (75–80°F); (●) refrigerated (40–45°F)

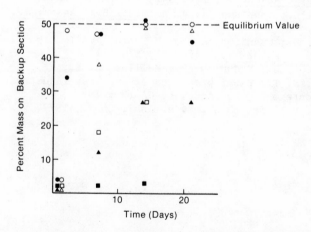

Figure 12. G-BB storage stability—percent mass on backup section vs. time; data points, average of two determinations: (△, ▲) acetone, (△, ▲) trichloroethylene, (□, ■) toluene; open symbols, ambient temperature; closed symbols, refrigerated

and Lois Lin for performing many of the sampling and analytical procedures; Anita Pollack for editorial assistance; and Mary Lynn Woebkenberg of NIOSH's Measurements Systems Section for helpful discussions and suggestions.

Literature Cited

1. Bailey,A.; Hollingdale-Smith, P. A. Ann. Occup. Hyg., 1977, 20, 345-356.
2. Evans, M.; Molyneux, M.; Sharp T.; Bailey A.; Hollingdale-Smith, P. Ann. Occup. Hyg., 1977, 20, 357-363.
3. Tompkins, F. C., Jr.; Goldsmith, R. L. Am. Ind. Hyg. Assoc., 1977, 38, 371-377.
4. Nelms, L. H.; Reiszner, K. D.; West, P. W. Anal. Chem., 1977, 49, 994-998.
5. Bamberger, R. L.; Esposito, G. G.; Jacobs, B. W.; Mazur, J. F. Am. Ind. Hyg. Assoc. J., 1978, 39, 701-708.
6. West, P. W.; Reiszner, K. D. Am. Ind. Hyg. Assoc. J., 1978, 39, 645-650.
7. Montalvo, J. G., Jr. Am. Ind. Hyg. Assoc. J., 1979, 40, 1046-1054.
8. Toshiko, H.; Ikeda,M. Am. Ind. Hyg. Assoc. J.,1979, 40, 1091-1096.
9. Freeland, L. T. Am. Ind. Hyg. Assoc. J., 1977, 38, 712-720.
10. Lautenberger, W. J.; Kring, E. V.; Morello, J. A., Am. Ind. Hyg. Assoc. J., 1980, 41, 737-747.
11. Palmes, E. D.; Gunnison, A. F. Am. Ind. Hyg. Assoc. J., 1973, 34, 78-81.
12. Palmes, E. D.; Gunnison, A. F.; DiMatto, J.; Tomczk, C. Am. Ind. Hyg. Assoc. J., 1976, 37, 570-577.
13. Gossilink, D. W.; Braun, D. L.; Mullins, H. E.; Rodriguez, S. T.; paper presented at American Industrial Hygiene Conference, Los Angeles, CA (May 1978).
14. Palmes, E. D.; Lindenboom, R. H. Anal. Chem., 1979, 51, 2400-2401.

RECEIVED October 15, 1980.

New Technology for Personal Sampling of NO_2 and NO_x in the Workplace

R. McMAHON and T. KLINGNER

MDA Scientific, Incorporated, 1815 Elmdale Avenue, Glenview, IL 60025

B. FERBER and G. SCHNAKENBERG

U.S. Bureau of Mines, P.O. Box 18070, Pittsburgh, PA 15236

Employee health problems, such as respiratory effects of exposure to nitrogen dioxide (NO_2) and nitric oxide (NO), have prompted the development of new technology for personal monitoring of toxic gases. Prevalent in underground mines where diesel equipment is used, NO_2 and NO levels can be analyzed by chemilumenescence, detector tubes, impingers or area monitors. These analytical methods, however, do not reflect <u>individual</u> employee exposures during an eight-hour workshift, expressed as Time Weighted Average (TWA) measurements. To obtain accurate and complete employee histories of exposure to toxic gases, personal monitoring techniques are required. For the detection of NO_2 and NO (NO_x) in mining and other applications, such techniques must be easily implemented, readily portable and must differentiate NO exposures from NO_2 exposures. Combining the principles of a passive diffusion personal sampler and the technology of a dedicated colorimeter, a monitoring system has been developed which meets the above parameters and which is the subject of this report.

NO_2/NO_x Passive Sampler

Based on the research of Dr. E. D. Palmes (1) at New York University's Institute of Environmental Medicine and on additional study supported by a contract from the United States Bureau of Mines, a unique personal sampler has been designed to passively collect NO_x. This is accomplished by way of molecular diffusion and subsequent trapping of the molecules onto a matrix coated with triethanolamine (TEA) at the closed end of the sampler. Constructed of polypropylene, the tubular sampler shown in Figure 1 is small, lightweight, unbreakable and can be easily worn in the breathing zone of the employee whose exposure is to be monitored. No pumping mechanism is required. The components of NO_x diffuse at constant, known rates towards the sealed end of the sampler and are efficiently collected onto the TEA coated medium. (2) These rates have been calculated using Fick's first law of diffusion, the estimation of coefficients of diffusion of gases in air, the effects of temperature and pressure on diffusion, as well as the dimensions of the sampling tube.

The development of a passive sampling device designed for this application is based on Palmes' (3) work as given below.

Derived from Fick's first law, the equation for unidirectional, isothermal diffusion of gas 1 (NO_2) through a constant pressure mixture of gas 1 and gas 2 (air) is:

$$J_1 = D_{1\ 2}\left(\frac{dc_1}{dz}\right) \qquad \text{Eqn. 1}$$

where

J_1 = diffusion flux of gas 1 (NO_2), in moles/cm^2/sec.

$D_{1\ 2}$ = diffusion coefficient of gas 1 through gas 2, in cm^2/sec.

c_1 = concentration of diffusing gas 1, in moles/cm^3

z = distance in direction of diffusion (length of sampling tube), in cm.

The concentration of gas 1 (NO_2) is maintained at zero by a highly efficient sorbent so that the flux of molecules will occur toward the closed end of the sampler. The gradient then becomes numerically equal to the ambient concentration of the gas to be measured, c_1.

When Eqn. 1 is used to calculate the quantity of gas 1 (Q_1, in moles) transported by molecular diffusion through a tube in a given time, one must also consider the cross-sectional area of the tube (A, in cm^2) and the time (t, in sec.) for which the flux persists. The modified equation becomes:

$$Q_1 = J_1At = D_{1\ 2}\left(\frac{c_1}{z}\right)At \qquad \text{Eqn. 2}$$

The coefficient of diffusion of NO_2 (gas 1) in air (gas 2), $D_{1\ 2}$, was determined by an equation which incorporates temperature, pressure, molecular weights of gases 1 and 2, force constants and

collision integrals. The best estimate of $D_{1\ 2}$ for
NO_2 in air at 70°F was found to be 0.154 cm^2/sec.

The molar quantity of NO_2 transferred, Q_{NO_2},
can be calculated using the following values for
the constants $D_{1\ 2}$, A and z and the variables c_1
and t in Eqn. 2:

$D_{1\ 2}$ = 0.154 cm^2/sec.

c_1 (ppm) = 0.0414 x 10^{-9} g moles/cc.

z = 7.1 cm.

A = 0.71 cm^2. $\rangle$ actual dimensions of sampler

t (hrs) = 3600 sec.

Q_{NO_2} (in nanomoles) = 2.3×(ppm NO_2)×(hours of exposure)

$$\text{Eqn. 3}$$

Exposing the sampler to 1 ppm NO_2 for 1 hour
results in the collection of 2.3 nanomoles of NO_2.
Exposure of a sampler to 1 ppm NO, however, for 1
hour results in the collection of 3.0 nanomoles of
NO (NO actually collected as NO_2 due to oxidative
conversion) since the $D_{1\ 2}$ value for NO in air is
0.199 cm^2/sec. (4).

Palmes has further calculated that correction for change in
temperature at which the sampling takes place amounts to only
1% per 10°F deviation from 70°F. Changes in pressure may be
disregarded, as well, since they cancel each other out in the
equations used to derive Q_{NO_2}.

Sampler Preparation

The samplers are prepared by loading each with three TEA-
treated stainless steel screens made of 40 x 40 per inch mesh
and 0.010 inch diameter wire. The screens are coated with TEA
by a simple process in which the cleaned screens are dipped into
a 25% vol/vol solution of TEA/acetone and allowed to air-dry.
This solution coats the screens with enough TEA to trap roughly
20,000 nanomoles of NO_2 and yet does not block the interstices
of the screens, thereby allowing unhindered diffusion of NO_x
through the screens. Once loaded with these treated screens,
both ends of the sampler are sealed. Now ready for use, samplers
can be stored for at least one month prior to TWA sampling.

Preparation of NO_x samplers involves the insertion of a
chromic acid disc behind the TEA-treated screens held in the
sampler. The disc, made by soaking fiberglass filters in an
aqueous solution of $Na_2Cr_2O_7$ and H_2SO_4, oxidizes NO to NO_2.
This NO_2 then back-diffuses and is trapped onto the TEA coated

screens. Analysis proceeds as for NO_2, but, in actuality, is for NO_x ($NO + NO_2$). By analyzing an NO_2 sampler that has been exposed simultaneously with an NO_x sampler, the NO concentration can be determined by subtracting the NO_2 value from the NO_x value.

While the chromic acid discs are stable for at least one month, they are unstable in the presence of TEA and, therefore, must not be inserted behind the TEA-treated screens until just before the sampler is to be used. Likewise, the disc must be removed as soon as possible after the sampler has been exposed. Color coded indicators are used to differentiate NO_x samplers from NO_2 samplers.

Sample Analysis

Samples may be analyzed up to one month after exposure. This is accomplished by adding a 2.0 milliliter (ml) aliquot of azo dye reagent to the sampler, transferring the sample to a cuvette and reading the transmittance of the resulting colored solution with a specially designed colorimeter. The developing solution contains a diazotizing reagent (sulfanilamide), phosphoric acid, and a coupling reagent (N-1-Naphthylethylenediamine dihydrochloride).

As shown in Figure 2, the reaction product of the analysis reagent and NO_2 has nine conjugated double bonds, rendering the solution a violet color which is optimally read at 540 nm between ten and thirty minutes after the reaction takes place.

The sampler is designed to be leak-proof so that the analysis reagent can be conveniently added directly to it. Since the samplers are worn with the exposed, open end facing downward, dust or particulate matter cannot enter and, therefore, cannot affect the transmittance of the developed solution. The TEA on the stainless steel screens does not interfere with the color reaction. Due to the selectivity of this system, that is, the specificity of TEA for collecting nitrites (NO_2^-) and the characteristic response of the diazotization reaction, other gases will not interfere in the determination of NO_x. Other oxides of nitrogen, such as N_2O_3 or N_2O_5, are either nonexistent or unstable in most industrial environments, presenting no potential interference.

The components of the passive sampler are inert, protecting the integrity of the analyte. Once used, the sampler can be recycled by washing and reloading with fresh TEA coated screens (Figure 3).

A colorimeter (Figure 4) has been designed specifically for analyzing NO_2/NO_x passive samplers. It is lightweight, portable and simple to use. Features incorporated into the unit include a ten-minute timer with audible alarm to ensure complete color development of the samples before analysis, a cuvette slot which

ITEM	DESCRIPTION
1	POLYPROPYLENE SAMPLING TUBE
2	TEA COATED SCREENS
3	POLYETHYLENE SPACER
4	SCREW IN CAP
5	VITON O-RING
6	PROTECTIVE CAP
7	CLIP ASSEMBLY
8	LABEL
9	CHROMIC ACID DISC FOR NO, SAMPLER
10	UNCOATED SCREEN FOR NO, SAMPLER
11	COLOR COATED INDICATOR FOR NO, SAMPLER

Figure 1. NO_2/NO_x sampler assembly

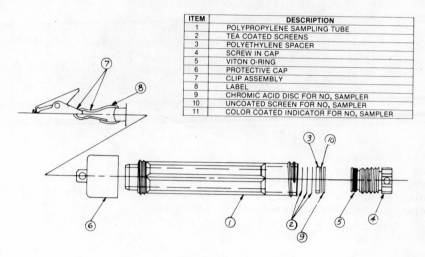

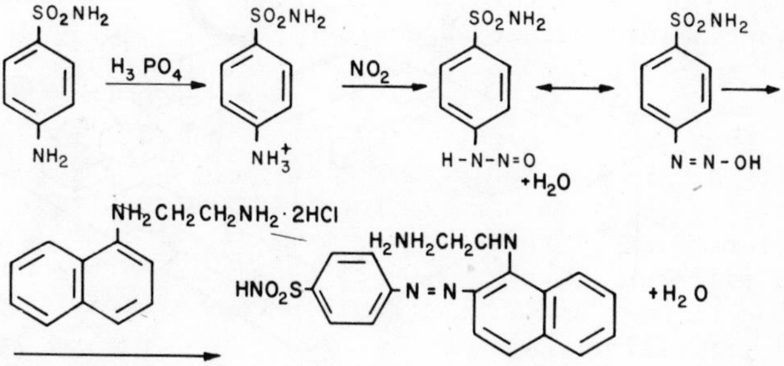

Figure 2.   Diazotization reaction mechanism

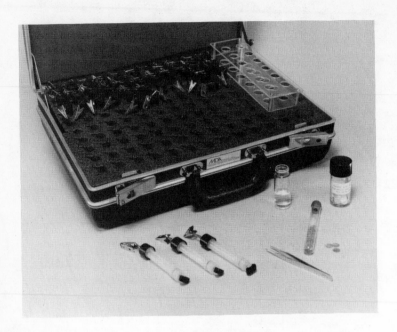

Figure 3.   NO₂/NOₓ passive sampler kit

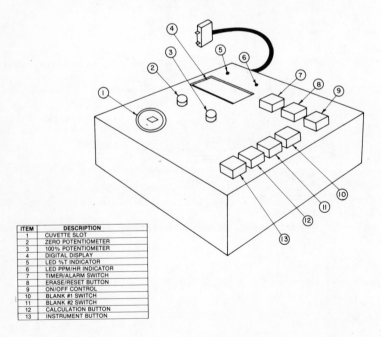

ITEM	DESCRIPTION
1	CUVETTE SLOT
2	ZERO POTENTIOMETER
3	100% POTENTIOMETER
4	DIGITAL DISPLAY
5	LED %T INDICATOR
6	LED PPM/HR INDICATOR
7	TIMER/ALARM SWITCH
8	ERASE/RESET BUTTON
9	ON/OFF CONTROL
10	BLANK #1 SWITCH
11	BLANK #2 SWITCH
12	CALCULATION BUTTON
13	INSTRUMENT BUTTON

Figure 4.   NO₂/NOₓ colorimeter

accommodates 1-cm path length cuvettes and blocks stray outside light, a memory which averages two "blank" values and automatically subtracts them from subsequent readings, potentiometer knobs for adjusting 0% and 100% transmittance, and a clearly visible digital readout for display of pertinent data.

The colorimeter is designed to convert % transmittance (% T) to absorbance (Abs) (Figure 5). It is also programmed with the slope from the standard curve of absorbance vs. nanomole NO_2 and thus derives a nanomole NO_2 value for each sampler analyzed. Additionally, the factor of 2.3 nanomoles per ppm·hour exposure is incorporated so that the colorimeter can calculate and display a ppm·hour value for each sampler by performing the following functions:

1. Average % T at 540 nm from two blanks is converted to absorbance (Abs) by the following:
    Abs = 2-log % T ⟶ Stored in memory
2. % T at 540 nm read from sample cuvette containing color developed solution. % T displayed in digital format.
3. Technician depresses designated pushbutton to obtain the ppm·hour NO_2 value.
    a. Colorimeter converts % T to Abs for sample.
    b. Blank Abs from memory subtracted from sample Abs.
    c. Colorimeter calculates:
        $$\text{ppm·hour } NO_2 = \frac{\text{Abs}}{\text{Slope}} \div 2.3$$
    d. Colorimeter displays ppm·hour NO_2 value until cuvette is removed.

For each exposed and analyzed sampler then, one can readily calculate the TWA by considering the time in hours during which a sampler was exposed. A dose of 20 ppm·hour NO_2, for example, divided by 8 hours, gives a TWA of 2.5 ppm NO_2.

The sensitivity range for the colorimeter, following the method above, is 0-20 ppm·hours NO_2. If a sampler exceeds 20 ppm·hours, it can be diluted with an additional, known aliquot of azo dye reagent and analyzed again. This value is multiplied by the dilution factor to arrive at the accurate ppm·hour value. By performing a one fold dilution (dilution factor = 2), the sensitivity range of the colorimeter can be extended to 40 ppm·hours NO_2. A dilution flask is supplied in case of the need for further dilutions.

Samplers were exposed in the laboratory to incrementally increasing NO_2 doses from 0 to 55 ppm·hours and analyzed one day later. In the first series, they were exposed to 6.0 ppm NO_2 ($\pm$ 10%) for 0, 1, 2, 3, 4, 5 and 6 hours. In the second series, samplers were exposed to 9.0 ppm NO_2 ($\pm$ 10%) for 0, 1, 2, 3, 4, 5 and 6 hours. Stable NO_2 atmospheres were generated by flow dilution of 100 ppm NO_2 into a glass chamber. Concentrations were verified by a continuous monitor, the Model 7030 for NO_2

available from MDA Scientific, Inc., and by the NIOSH wet
method for the detection of NO_2. (5)   Both verification
methods are reliable to $\pm$ 10%. The data are shown in Tables I&II.

TABLE I - SERIES I
NO_2 DOSE VS. RESPONSE - 6 PPM

NO_2 Dose (ppm·hr)	NO_2 Response (ppm·hr)		Average Deviation of Response from Dose (%)
	Sampler 1	Sampler 2	
0	0	0	0
6	5.7	6.4	+0.8%
12	10.5	11.7	-7.5%
18	17.1	16.5	-6.7%
24	22.0	22.6	-7.1%
30	28.0	29.1	-4.8%
36	33.4	33.3	-7.4%

TABLE II- SERIES II
NO_2 DOSE VS. RESPONSE - 9 PPM

NO_2 Dose (ppm·hr)	NO_2 Response (ppm·hr)		Average Deviation of Response from Dose (%)
	Sampler 1	Sampler 2	
0	0	0	0
9	8.7	8.2	-6.1%
18	16.6	17.0	-6.7%
27	24.3	24.3	-10.0%
36	30.9	32.0	-12.6%
45	40.2	40.2	-10.7%
54	48.0	47.1	-11.9%

It can be seen that the NO_2 collection efficiency is linear
and has an accuracy of approximately 10% for any given dose,
regardless of the concentration level of NO_2 present.   Even
though the higher level (9 ppm) results are 5% lower than are the
6 ppm level results, one cannot infer that this is significant
since the 5% drop is not cumulative with increasing exposure
times.

For determining NO exposures, the technician analyzes an
NO_2 sampler and an NO_x sampler which have been exposed simultaneously. He determines the ppm·hour value for each sampler as
described previously and performs the following calculation:

$$\frac{\text{ppm·hour } NO_x - \text{ppm·hour } NO_2}{1.304} = \text{ppm·hour NO}$$

It is necessary to divide the difference by 1.304 due to the
faster diffusion rate of NO

$$\frac{3.0 \text{ nanomoles NO/ppm·hour}}{2.3 \text{ nanomoles } NO_2/\text{ppm·hour}} = 1.304 \frac{\text{nanomoles NO}}{\text{nanomoles } NO_2}$$

since the colorimeter is programmed with the factor 2.3 for NO_2.

Exposure data for NO_2 and NO is then converted to TWA (ppm) measurements by dividing the total dose (ppm·hours) by the personnel exposure period (hours).

Subsequent to this developmental work, the U. S. Bureau of Mines ran parallel evaluation studies of the NO_2/NO_x passive sampler and colorimeter and determined their suitability for field operation.

NO_2/NO_x Field Kit

Personal samplers and the designated colorimeter are contained in a complete kit (Figure 6) designed for use in the field. Housed in two attache cases, the portable kit contains all necessary materials for exposure and analysis of one hundred samplers. These components include prepared azo dye reagent, a dispenser bottle for delivering 2.0 ml aliquots of this reagent to exposed samplers, disposable polystyrene cuvettes, cuvette adapter units for direct attachment of the sampler to the cuvette, sampler labels, and clips for securing a sampler to an employee. Also included are a four-function calculator, instructions and recordkeeping sheets for collection of employee exposure data, and plastic bags for disposing or recycling used items. Once the field technician has exposed and analyzed his supply of samplers, his kit can be replenished in the home lab or replacement items can be sent to him. Likewise, the data compiled is sent to the home office where it is analyzed for exposure patterns as a function of operational procedures before being incorporated into employee health history files.

Conclusion

Passive personal samplers for NO_2 and NO are an efficient collection device for these toxic gases and are accurate to $\pm$ 10% for determining individual employee exposures to NO_2 and/or NO during an entire workshift. It is speculated that other toxic gases, such as SO_2 or CO, could also be collected by passive samplers, given that suitable adsorbents or absorbents could be identified and incorporated into the system.

As the colorimeter described here has been designated specifically for NO_2 and NO evaluation, so could a colorimeter be dedicated to analyzing other contaminants, given a reliable colorimetric wet analysis for determination of exposed sampler contents.

As presently designed, this personal monitoring system for NO_x is best used for measuring concentrations at or below the Threshold Limit Value (TLV): 5 ppm for NO_2 and 25 ppm for NO, equivalent to 40 and 200 ppm·hours, respectively, of total dose. Should monitoring of higher NO_x concentrations be desired, sampling parameters could be changed by modifying the length

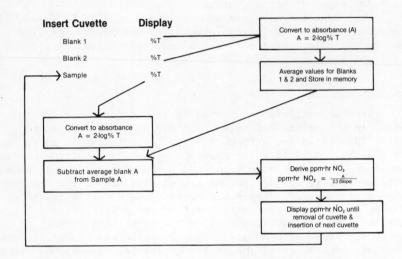

Figure 5.   Colorimeter functions

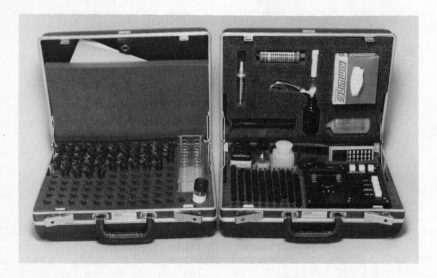

Figure 6. NO_2/NO_x field kit: passive samplers and colorimeter

or cross-sectional area of the sampler tube. Examination of Eqn. 2 shows that the diffusion rate is inversely proportional to the sampler size and could be manipulated by altering the length to area ratio of the sampling tube. A longer tube, for instance, would collect less NO_x per unit time and would effectively extend the range of NO_x detection.

By combining passive samplers and a designated colorimeter into a versatile field kit, employee exposures to toxic gases can be more efficiently and accurately examined than is possible with other methods requiring skilled personnel and cumbersome equipment. The cost/benefit ratio of maintaining worker health is improved, as is the quality of such maintenance. With the current advances being made in industrial hygiene research, it is hoped that this new system of personal exposure monitoring can soon be expanded to include other gaseous contaminants affecting worker safety.

Literature Cited

1. Palmes, E. D.; Gunnison, A. F. Am. Ind. Hyg. Assoc. J., 1973, 34, 78.
2. Blacker, J. H. Am. Ind. Hyg. Assoc. J., 1973, 34, 390.
3. Palmes, E. D.; Gunnison, A. F.; DiMattio, J.; Tomczyk, C. Am. Ind. Hyg. Assoc. J., 1976, 37, 570.
4. Palmes, E. D.; Tomczyk, C. Am. Ind. Hyg. Assoc. J., 1979, 40, 588.
5. NIOSH Analytical Methods, 1977, 1, 108, DHEW NIOSH Publication No. 78-175.

RECEIVED October 27, 1980.

Ion Chromatographic Analysis of Formic Acid in Diesel Exhaust and Mine Air

ITAMAR BODEK and KENNETH T. MENZIES

Arthur D. Little, Incorporated, Cambridge, MA 02140

Formaldehyde, a suspected carcinogen (1) and known irritant (2), is found at low concentrations in diesel engine exhaust and in environments subjected to diesel emissions (3). A typical concentration in diesel exhaust is 5-25 ppm and about one hundred times lower in a mine atmosphere subjected to such exhaust (3). The fate of formaldehyde in mines is of interest to mine workers and operators due to its potential health hazard. It has been suggested that oxidation of formaldehyde to formic acid may occur in the mine environment and thus reduce its concentration (4). Previous attempts to measure concentrations of formic acid at ppm levels have been thwarted by inadequate detection limits. Generally, analytical methods for formic acid employ collection in aqueous solutions and reaction with oxidizing or reducing agents. Measurement of formic acid at high concentrations can be made by adding an excess of oxidizing agent and titrating the remaining excess of oxidant with reducing agents (5). Another method (6) relies on the distillation of a chloroform-formic acid azeotrope to separate formic acid from higher carboxylic acids and analysis of the formic acid by potentiometric titration with sodium hydroxide. Detection limits for this method are generally in the 0.1% range (6). Gas chromatography has been used to measure formic acid both directly and after derivitization. Direct analysis poses problems of corrosion of metal surfaces and of high detection limits with a flame ionization detector. Derivitization (7) eliminates these problems and achieves a detection limit of about 25 µg/mL.

Ion chromatography has recently been successfully used for analysis of formaldehyde after oxidation to formic acid (8) and thus can be used for direct analysis of formic acid.

In order to prevent interference with such inorganic anions as fluoride, chloride, and nitrate which occurs with $Na_2CO_3/NaHCO_3$ eluents, a weak aqueous eluent, i.e., $Na_2B_4O_7$, was used to achieve adequate separation. The detection limit in such an ion chromatographic analysis is limited by the conductance of the suppressed eluents, but the development of ion chromatography exclusion (ICE),

which permits separation of weak acids in a lower conductance
background, has circumvented this problem. The determination of
weak acids in complex media (9) has been reported with this technique.

This paper describes the analysis of formic acid in diesel
engine exhaust and mine air using ion chromatography (IC) and ion
chromatography exclusion (ICE).

Experimental

Collection. Formic acid in diluted (1:10) diesel exhaust or
mine air was collected by drawing the sample atmosphere through
two fritted bubblers (10) in series, each containing 15 mL of
10^{-3} $\underline{M}$ Na_2CO_3. A flow rate of 1.0 liter per minute and collection
time of 60 minutes for diluted diesel exhaust or 240 minutes for
mine air was used. A 37 mm glass fiber filter (Gelman Type A/E)
was placed before the bubblers to remove particulates.

Analysis. The solution in each bubbler was transferred to a
25 mL volumetric flask and diluted to volume with the collection
medium (10^{-3} $\underline{M}$ Na_2CO_3). In the case of samples collected from
diluted diesel exhaust, an excess (3 mL) of this solution was
flushed through a 100 µL sample loop of Dionex Model 14 ion chromatograph. The sample was analyzed on the anion system with instrumental conditions presented in Table I.

Table I

Ion Chromatographic Conditions

Instrument:	Dionex Model 14 Ion Chromatograph
Eluent:	0.005 $\underline{M}$ $Na_2B_4O_7$
Flow Rate:	2.3 mL/min.
Detector:	Conductivity
Sensitivity:	30 µmho full scale
Anion Columns:	3 x 125 mm Dionex Anion Pre-column
	3 x 500 mm Dionex Separator (Borate Form)
	6 x 250 mm Dionex Suppressor (H^+ Form)
Sample Volume:	100 µL Sample Loop
Recorder:	HP 7133A
Chart Speed:	0.5 cm/min.

In the case of samples collected from mine air, the bubbler
solution was concentrated by freeze drying. Samples were transferred to wide mouth jars, frozen to -25°C and freeze-dried under
0.5 cm Hg vacuum in a Vacudyne, Inc. Pilot Freeze Dryer. Once
reduced to dryness, 1 mL of distilled/deionized water was added
to the samples. After shaking to ensure complete dissolution,
300 µL aliquots were placed in micro vials available for use in a
Waters Associates Autoinjector Model 710 A. One hundred µL sample

volumes were automatically injected into the ion chromatograph and
analyzed on the ion chromatography exclusion (ICE) system with
instrumental conditions presented in Table II.

TABLE II

Ion Chromatographic Exclusion (ICE) Conditions

Instrument:	Dionex Model 14 Ion Chromatograph
Eluent:	0.0001 $\underline{M}$ HCl
Flow Rate:	0.7 mL/min.
Detector:	Conductivity
Sensitivity:	3 µmho full scale
ICE Columns:	9 x 250 mm Dionex Exclusion
	3 x 500 mm Dionex Halide Suppressor
	(Ag^+ Form)
Sample Volume:	100 µL, Waters Associates Autoinjector
	(Model 710 A)
Recorder:	HP 7133A
Chart Speed:	0.5 cm/min.

Results And Discussion

Analytical Method. Initial experiments were conducted on
the Dionex Model 14 ion chromatograph to determine the feasibility
and sensitivity of ion chromatographic analysis of formic acid.
The conditions of IC analysis (Table I) chosen were based on the
need for separation of formate ion from common atmospheric contaminants, such as chloride ion, and other organic anions, such
as acetate. The conditions were also optimized to obtain a suit-
able detection limit. Thus, a weak borate eluent (0.005 $\underline{M}$ $Na_2B_4O_7$)
combined with a 500 mm anion separator column (and a 150 mm pre-
column) was chosen. Conversion of the pre-column and separator
column to their borate forms (from the normal carbonate form) was
necessary. The process of continually passing the borate eluent
through the columns until a stable baseline was obtained required
several hours.

Standard stock solutions of formate anion were prepared from
reagent grade sodium formate. Standard solutions of other organic anions were prepared for assessment of potential interferences.
Injections of 3 mL were made to fill a sample loop of 100 µL volume.

A typical ion chromatogram of formate and acetate is shown in
Figure 1. Peak identifications and concentrations are reported as
the free acids to facilitate sample analysis. The retention times
for acetate and formate under these conditions are 6.6 and 8.8
minutes, respectively. The preceding negative peak grouping seen
in the 2-5 minute region is due to water whose conductivity is
lower than boric acid. The peak height is linear with formic acid
concentration over the range from 0.1 to 35 µg/mL. The corre-

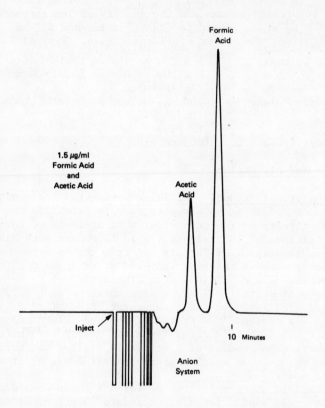

Figure 1.   Ion chromatogram of formic acid and acetic acid

sponding linear range for acetic acid is about 0.2 to 70 μg/mL. The presence of fluoride or chloride in the sample does not interfere with the formic acid analysis. Retention time for these ions under the stated conditions are F^- (5.4 minutes) and Cl^- (29 minutes). Sulfate ion is retained for a longer time. Ultimately, these anions are eluted and may interfere with subsequent samples. Generally, these highly retained anions show very broad peaks, which are easily distinguished from the organic acid peaks.

Due to this interference, the need for frequent regeneration of the suppressor column and the ability of ion chromatography exclusion to easily separate organic acids and Cl^- and $SO_4^=$, an ICE system was tested for similar analyses. Typical ICE conditions (Table II) provide easy separation of strongly ionized species (e.g., H_2SO_4, HCl, HNO_3) from weakly ionized species (e.g., organic acids) due to the greater retention of uncharged species in the interstitial fluid of the packing material (9). Acidic eluents reduce the ionization of weak organic acids and thus increase their retention time. A 10^{-4} M hydrochloric acid eluent offers adequate separation of the strong acids and the organic acids of interest with a reasonable retention period. Standard stock solutions were prepared as before and injected onto the ion chromatograph. In order to utilize very small sample volumes, the normal sample loop, which requires excess sample, was by-passed and a Waters Associates Autoinjector (Model 710 A) installed. One hundred μL injections were made directly into the flowing eluent by the autoinjector and analyzed on the ICE system. A typical ion chromatogram of formic acid and sulfuric acid is shown in Figure 2. The retention time for the sulfuric acid and formic acid are 9.8 and 15.4 minutes, respectively. The peak height is linear with formic acid concentration over the range of 0.3 to 10 μg/mL. The corresponding linear range for acetic acid is 3 to 100 μg/mL (Figure 3). As well as providing separation of these organic acids and precluding interference from strong acids, the ICE system permits continuous analysis of samples for up to 30 hours without regeneration of the suppressor column.

The precision of the ICE method was determined by analyzing six replicates of two standard solutions containing strong acids (i.e., H_2SO_4) and several weak acids (i.e., formic acid, acetic acid and carbonic acid). Carbonic acid is present as a result of the use of Na_2CO_3 in the standard solution matrix (as in the collection medium) and dissolution of atmospheric carbon dioxide (Figure 4). At formic acid concentrations of 5.0 and 10 mg/L, the measured mean concentrations (Table III) were 5.08 and 10.0 mg/L, respectively. The relative standard deviation (CV) was 0.025 and 0.016, respectively.

Collection Method. The goal of the initial phase of our work was to determine the concentration of formic acid in engine exhaust subject to different forms of control, e.g., catalytic oxidation. Initially, samples were collected from diesel engine

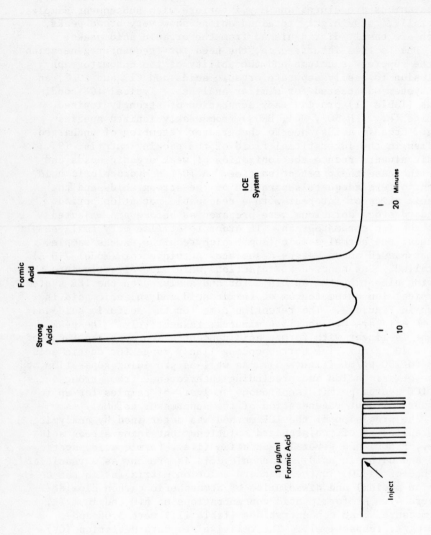

Figure 2. ICE chromatogram of formic acid and other strong acids

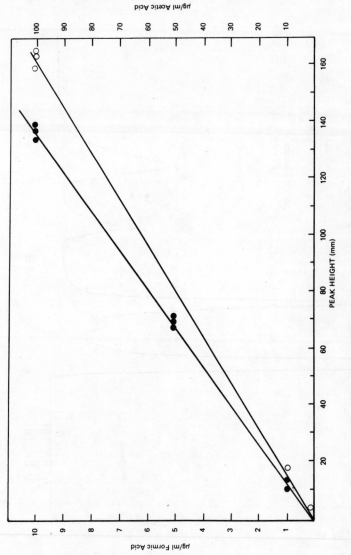

Figure 3.   Standard curve for ICE system: (●) formic acid, (○) acetic acid

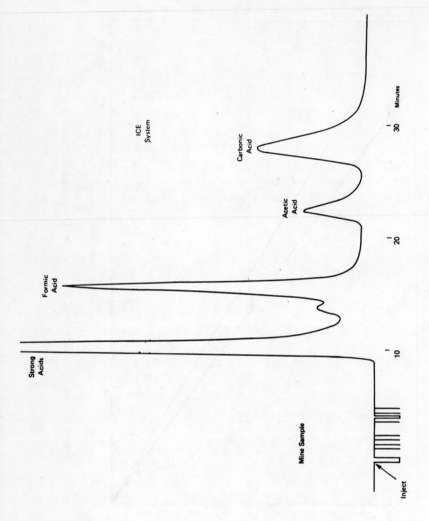

Figure 4. Chromatogram of mine sample

TABLE III

Formic Acid Analytical Data

Calculated Concentration (µg/mL)		Observed Concentration (µg/mL)
5.00		5.23
		5.18
		5.18
		4.95
		4.95
		5.01
	Mean	5.08
	Std Dev	0.127
	CV	0.025
10.00		10.2
		10.2
		10.1
		9.9
		10.1
		9.8
	Mean	10.0
	Std Dev	0.163
	CV	0.016

exhaust diluted by a factor of 10 in a stainless steel dilution
tunnel. A collection medium of 15 mL of 10^{-3} $\underline{M}$ sodium carbonate
solution was utilized for two reasons. First, such an aqueous
solution (7) is reported to provide good collection efficiency of
soluble organic acids at flow rates of 0.5 to 5 L/min. Second,
due to the volatility of free formic acid, it was felt that a
basic solution would provide improved stability of the samples
over periods up to seven days. The collection efficiency of this
medium was determined by measuring the amount of formic acid collected at 1.0 L/min in two bubblers connected in series. The
amount of formic acid found in the front and back bubbler at three
challenge concentrations produced in the dilution tunnel is reported in Table IV and shows that the collection efficiency is
greater than 92%.

TABLE IV

Collection Efficiency

Challenge Concentration (mg/m^3)	µg Formic Acid Found		% Collection in Front
	Front	Back	
12	153	<2.5	>98.4
6.7	90.0	<1.0	>98.9
2.5	32.5	<2.5	>92.3
0.06	10.3	0.8	92.2

In order to provide at least 99% collection of formic acid
in diluted exhaust, two bubblers were routinely used in series.
In a later phase of work, the formic acid concentration in mine
air subject to diesel emissions was measured. The expected concentrations were about one hundred times lower than those found
in engine exhaust. The efficiency of the collection scheme was
again measured under these conditions of challenge concentration
$(0.06\ mg/m^3)$. The collection efficiency was found to be 92.2%
at this level (Table IV).

Sample Analysis. For sample collection in both diluted
diesel exhaust and a mine atmosphere, the collection technique
described previously was used, and the sampling periods were 60
minutes and 240 minutes, respectively. In the mine samples, the
amount of formic acid collected was too small to be analyzed reliably after dilution to 25 mL. Larger amounts could have been
collected by utilizing a higher flow rate (e.g., 5 L/min) or
longer sampling period. However, in order to utilize the samples
as collected, they were concentrated by freeze-drying. The pH
of samples was checked prior to freeze-drying to ensure that the
solutions were slightly basic. Strongly acidic species, e.g.,
H_2SO_4, formed in the bubblers by oxidation and hydration of SO_2
during mine air sampling, slowly depleted the Na_2CO_3 in the col-

lection medium. Weakly ionized organic acids may be lost if
a sufficiently low pH is achieved. Recovery of formic acid during
the freeze-drying process was checked by carrying standard solutions through the entire analytical procedure. As is shown in
Table V, recovery of triplicate samples at two concentrations
averaged about 0.88.

TABLE V

Formic Acid Recovery
After Freeze-Drying

Initial Sample Concentration (µg/mL)	Freeze-Dried Sample Concentration (µg/mL)	Recovery
4.3	3.9	0.91
	3.7	0.86
	3.9	0.91
9.7	9.0	0.93
	8.3	0.86
	7.8	0.80

Storage stability of these samples was checked by replicate analysis after a period of seven days. Losses of less than 2% were
observed after storage for this period of time.

The accuracy of the ICE method was assessed by using the
standard addition technique. Four mine samples containing measured concentrations of formic acid of about 0.8 to 1.1 µg/mL were
spiked with a known volume for formate standard solution sufficient to double the sample concentration. The observed concentration indicated agreement of better than ±12% with an average
agreement within ±5% (Table VI).

TABLE VI

Standard Addition of Formic Acid

Initial Concentration (µg/mL)	With Spike Calculated	With Spike Observed	Agreement
0.8	1.6	1.8	1.12
1.0	2.0	2.1	1.05
1.1	2.2	2.3	1.04
1.1	2.2	2.2	1.00

Standard addition was also used to confirm the identification
of the chromatographic peaks based on comparison of retention
time. No interference was observed due to chloride, sulfate or
carbonate ion. Both formate and acetate were adequately resolved.

A small unidentified peak did precede formic acid but did not significantly affect quantitation.

Samples of diluted diesel engine (i.e., Ford 2401 E, 50 horsepower) exhaust were collected under several conditions of engine speed and load and applicable control technology to determine their impact on the emission of formic acid. Two to six replicate samples were collected under each test condition, and the mean concentration is reported in Table VII. It is generally apparent that at a higher speed and load (higher engine temperature), the concentration of formic acid is lower presumably due to more complete combustion. The catalyst has little impact except at the higher speed and load condition where the catalyst is at an elevated temperature and thus where its oxidation efficiency is greater. The water conditioner, which is normally installed on diesel engines in underground coal mines, shows a significant reduction of formic acid concentration which is probably due to its solubility in water.

To determine the fate of formaldehyde and formic acid in a coal mine, an unused shaft about 120 m long and 6 m^2 in cross sectional area was selected for study. With a ventilation air flow of 190 m^3/min and an engine exhaust flow of 1.5 m^3/min, complete exhaust dispersion and dilution was observed in about 10 m. Samples collected in the mine air downstream of the diesel engine indicate no significant change in formic acid concentration at increasing distances from the engine (Table VIII). This is certainly not consistent with the loss of formaldehyde in the same interval. The mechanism for loss of formaldehyde is apparently not a gas phase oxidation to formic acid. Interaction with surfaces may be a more suitable explanation of the observed reduction in formaldehyde concentrations.

Conclusions

Two ion chromatographic techniques were utilized to quantify formic acid in both diesel engine exhaust and mine air subjected to diesel emissions. A commonly used anion separation system utilizing a weak borate eluent adequately separated the acids of interest in diesel exhaust. It was, however, affected by the presence of strong acids during subsequent consecutive analyses.

In order to preclude this problem and the necessary frequent regeneration of the anion system's suppressor column, an ion chromatography exclusion scheme was utilized. Samples collected in a mine environment were reliably concentrated by freeze-drying and then analyzed on an ICE system with dilute hydrochloric acid eluent. The precision of the ICE method was experimentally determined to be ±2.5% in a concentration range of 1 to 10 µg/mL. The accuracy was not independently determined but good precision and recovery yield confidence that measured values are within ±5% of the true value. No interferences were observed in the ICE system due to strong acids, carbonic acid or other water soluble species present in mine air subject to diesel emissions.

TABLE VII

Formic Acid in Engine Exhaust

Control Device	Engine Condition (RPM)	Percent Load (%)	Formic Acid Concentration (mg/m^3)	Relative Standard Deviation
None	1,000	0	13.0	0.09
	1,800	16	9.8	0.05
	2,650	34	5.6	0.10
Monolithic Catalyst (Exhaust Controls, Inc.)	1,000	0	12.0	0.15
	1,800	16	9.4	0.01
	2,650	34	1.7	0.14
Catalyst and Water Conditioner (MSA, Inc.)	1,000	0	3.0	0.14
	1,800	16	2.1	0.10
	2,650	34	0.5	0.10

TABLE VIII

Fate of Formaldehyde and Formic Acid in Mine

Distance from Engine (m)	Formaldehyde Concentration[b] (10^{-3} mg/m^3)	Formic Acid Concentration (10^{-3} mg/m^3)
0[a]	4.6	8.5
15	62	62
46	31	62
77	18	61
108	20	66

[a] Background concentration upstream of engine exhaust.

[b] Measured by chromotropic acid method (11).

Results of analysis of formic acid in diesel engine exhaust subjected to various forms of post-combustion control, i.e., catalytic oxidation and water conditioning, indicate both a reduction of formic acid due to oxidation in the catalyst and dissolution in the water scrubber. In-mine analysis of formic acid at increasing distances from a source of diesel exhaust indicates that no significant change in concentration occurs. This finding contradicts a hypothesis that formaldehyde concentration decreases with increasing distance due to gas phase oxidation to formic acid. Surface reactions may, however, be important sinks for formaldehyde.

Abstract

Low molecular weight carboxylic acids are among the highly water soluble compounds that are difficult to quantify by conventional organic analytical procedures. The relatively new technique of ion chromatography has the potential for analyzing these acids in complex matrices. Concentrations of formic acid found in samples of diluted diesel exhaust and coal mine air were determined using two ion chromatographic procedures.

Samples were collected in midget bubblers containing dilute sodium carbonate solutions. Analysis of formate ion in these solutions was performed using a 500 mm anion separator column and sodium borate eluent. Analysis of formate in these solutions was also performed using the ion chromatography exclusion mode (ICE) using the Dionex IE-C-1 column with dilute hydrochloric acid eluent.

Retention time for formate ion using the borate system is 11 minutes at a flow rate of 2.3 mL/min. Fluoride, chloride, sulfate and acetate do not interfere in the analysis. Linear response is obtained over the concentration range of 0.1-4 μg/mL. The detection limit is estimated at 0.05 μg/mL for an injection volume of 100 μL at 3 μmhos full scale sensitivity.

The ICE system was used to confirm the identity of the formate detected in the samples and determine its concentration. The retention time for formate using the ICE system at an eluent flow rate of 0.7 mL/min is 16 minutes. No interference is observed from chloride, sulfate and acetate. Linear response is obtained for formate in the concentration range of 1-20 μg/mL. The detection limit for a 100 μL sample injection volume at 30 μmhos full scale sensitivity is estimated at 0.5 μg/mL. Concentration of samples by freeze-drying affords better detection limits with minimal loss of formic acid. Acetic and carbonic acids are also analyzable under these conditions.

Literature Cited

1. Billings, C. E. "Industrial Pollution"; Sax, N. I., Ed.: Van Nostrand Rheinhold Co.: New York, NY, 1974; p. 120.

2. Patty, F. A., Ed. "Industrial Hygiene and Toxicology", Second Revised Edition, Volume II, Toxicology; Interscience Publishers: New York, NY, 1963; p. 1959.

3. Lawter, J. R.; Kendall, D. A. "Effects of Diesel Emissions on Coal Mine Air Quality"; Final Report, U.S. Bureau of Mines Contract J0166009, 1977.

4. Menzies, K. T. "Fate of Reactive Diesel Exhaust Contaminants"; Draft Final Report, U.S. Bureau of Mines Contract J0188061, 1980.

5. Peters, D. G.; Hayes, J. M.; Hieftze, G. M. "Chemical Separations and Measurements"; Saunders Co.: New York, NY; 1974; p. 180.

6. Warner, B. R.; Raptis, L. Z. Anal. Chem., 1955, 27, 1978.

7. Smallwood, A. W. Amer. Ind. Hyg. Assoc. J., 1978, 39, 151.

8. Kim, W. S.; Geraci, C. L.; Kupel, R. E. Amer. Ind. Hyg. Assoc. J., 1980, 41, 334.

9. Rich, W.; Smith, F. C.; McNeil, L.; Sidebottom, T. "Determination of Strong and Weak Acids and Their Salts by Ion Chromatography Coupled with Ion Exclusion"; unpublished, available from Dionex, Inc.

10. U.S. Public Health Service "Selection Methods for the Measurement of Air Pollutants"; Publication No. 999-AP-11, May 1965.

11. Katz, M., Ed. "Methods of Air Sampling and Analysis", Second Edition; American Public Health Association: Washington, D.C., 1977; p. 303.

RECEIVED October 27, 1980.

INDEX

INDEX

A

Absorbance ratios, peak height 407*t*
Accuracy criterion, NIOSH 505–507
Acetaldehyde 14
Acetic acid 184
  ion chromatogram 602
Acetone 184, 536, 539*t*, 577, 583
  sampling of 540*t*, 582*f*
Acetylene tetrabromide 186
Acid mist generation 140*f*
Acid mist sampling methods 137
Acrolein 14
Acrylonitrile 436
  OSHA standards 439*t*
Activated
  charcoal tubes 536
  coconut base carbon, collection
    efficiency 168*f*
  coconut charcoal 184
  recovery of diphenyl from 190, 192*t*
Action level 436, 489
  decision criteria 475–478
  NIOSH 471–489
Acute
  hazards, air sampling for 440–451
  hazards, OSHA/NIOSH scheme
    and 440
  toxin 434
Adsorbents, carbonaceous 167
Aerosol(s)
  characteristics, metal 97–100*t*
  collection, sorbent tube 144–145
  generation 111
    validation by 89–92
  generator, lead chloride 98*f*
  packed bed collection 138
  sampling studies 96–97
  sensors for 529
  test atmospheres 15
Air concentrations, cumulative
  distribution 435*f*
Air sampling
  for acute hazards 440–451
  with constant flow pumps 491–502
  scheme, OSHA/NIOSH 436–440
Aldehydes 14
Alkanolamines 169, 173
  gas chromatographic separation of 171*f*
Allyl glycidyl ether 187
Amberlite XAD-2 ....164, 165*f*

Ames *Salmonella* test 294
Amines, aromatic 371, 377, 379*t*
  analysis 374
  in body fluids 413–427
  in commercial dyes 413–427
  gas chromatogram 378*f*
  GC separation of 378*f*
p-Aminophenylarsonic acid
  (*p*-APA) 383–400
2-Aminopyridine 187
Ammonia 169, 188
  and methylamines, ion chromato-
    gram 170*f*
Amperometric titration of NaDCC 131*f*
Analysis for formic acid 600–601
Analyte stability 58
  sorbed 60–61
Aniline(s) 186, 416
  and benzidine, chromatogram 27*f*
  HPLC detection limits 418*t*
Animal skin, fluorescence
  characteristics 276
Anisidine 187
Anthracene 273, 358
ANTU 14
Area samplers 462, 464*f*, 466
Aroclor 1254 235–267
  protective garment permeation by .. 258
Aroclor 1260 328
Arsine 434
  derivatives 385
Arsenic 227, 436
  OSHA standards for 439*t*
Atomic absorption spectrometry
  (AAS) 96, 100–103, 110, 111
Atomization, test atmospheres by 15
Average daily exposure, long-term 473
Average Exposure Limit (AEL) 473

B

Badge with backup, organic vapor 575–586
Badges, passive charcoal 179–180
Bartlett's test 517
Beer's law 405
Benzene 199, 436
  OSHA standards for 439*t*
Benzaldehyde 14
Benzidine 415
  and aniline, chromatogram 27*f*

Benzidine (*continued*)
-based dyes 21–35
   chromatogram of hair dye 422*f*
   visible spectrum studies 26*t*
  electrochemical reaction 414*f*
  from filter samples, recovery 28*t*, 30*t*–32*t*
  in hair dyes 419–420
  and *o*-tolidine, chromatogram 27*f*
Benzo[*a*]anthracene 358, 363, 374
Benzo[*b*]fluoranthene 358
Benzo[*b,j&k*]fluoranthene 367
Benzo[*ghi*]fluoranthene 367
Benzo[*k*]fluoranthene 358, 374
Benzo[*c*]phenanthrene 374
Benzo[*a*]pyrene 273, 358, 367,
               370, 374, 375
Benzo[*e*]pyrene 358, 367
Bioabsorption of toxicants 228–229
Bioassay 223
Biological monitoring surveillance 223–233
Biphenyl, method development 190–193
Biphenyl, polychlorinated 235–267
Bituminous coal 543
Blood
  analysis for nitrosamines 286
  cholinesterase levels 227
  PCCDs and PCDFs in samples
    from
     leather workers 338*t*
     saw mill workers 337*t*
     textile workers 338*t*
Body uptake 223
Boric acid 87–94
Breakthrough
  determination(s) 237, 244
  inorganic acid 148
  organotin 113, 116
  test 308, 393
    chlordane 309*f*
    demeton 313*f*
    endrin 311*f*
    heptachlor 309*f*
    ronnel 311*f*
  time 247, 249
  and protective garment weight
    change 258–261
  and protective garment volume
    change 261
  volume 183
Bubble tubes 500*f*
Bubblers, fritted 600
Bubblers, midget 13
1,3-Butadiene 200
2-Butanone 371
Butyl rubber 241–267
  weight change 243*f*
n-Butyl mercaptan 187–188
n-Butylamine 188
Byssinosis 66, 71

C

Cadmium 229
  urinary 226
Calibration
  chromatographic 23–25
  of constant flow pumps 498, 501
  curve(s)
    aromatic diamines 406*f*
    inorganic acid IC 142, 143*f*
    iodometric titration 129–130
    NaDCC 131*f*
    TCCA 132*f*
  monitor 199–200
  standards 463*f*
Capacity ratio for CN-column 404*f*
Capacity ratio for NO$_2$-column 404*f*
Carbon disulfide 184, 198, 537
Carbon monoxide 525
  datagram 524*f*
  sensor cell 551–573
Carbonaceous adsorbents 167
Carbosieve B 12
Cassette rack, pump and 112*f*
Ceiling limits for airborne chemicals .. 444*t*
Cellulose ester filters 103, 384, 387
Cellulose ester, mixed (MCE) .12, 303–306
Center for Assessment of Chemical and
  Physical Hazards (CACPH) ..543–550
Charcoal
  104, recovery of diphenyl from 192
  sorbents 184–186
  tube(s) 179
    vs. diffusion sampler 215*f*, 216*f*
    vs. gasbadge 214*f*
    vs. OVM 214*f*
    vs. pair difference 218*f*–220*f*
    spiked 41
Chemical properties for sorption 181
Chlordane 13, 312
  breakthrough test 309*f*
Chlorinated hydrocarbons 370, 375
  analysis 371, 374
  gas chromatogram 376*f*
Chlorinated isocyanuric acids 123–135
Chloride 603
Chlorine gas 525
1-Chloro-1-nitropropane 187
Chloroacetaldehyde 14, 186
Chlorocarbons, collection media for .. 59*t*
Chlorocarbons, GC analytical
  procedure for 53*t*
Chloromethylmethyl ether (CMME) .. 173
  chromatograms 172*f*
  derivative collection 172*f*
Chlorophenate 327–328
  pyrolytic dimerization 321*f*
Chlorophenols, PCDDs and PCDPs
  in 326–328, 329*t*

Chlorophenols in urine 336t
Chloroprene 184
Chloropyrifos, recovery 167
Cholinesterase levels, blood 227
Chromosorb 102 ....187, 306, 308
Chromosorb 103 187
Chromosorb 104 ....187, 188
Chromosorb 106 187
Chromosorb 108 187
Chromosorb P 188
Chronic toxins 434, 436
Chrysene 358, 363, 367
Clophen A-60 328
CMME (see Chloromethylmethyl ether)
Coal(s)
    bituminous 543
    gasifier, monitoring skin con-
        tamination at 278, 279t
    on paper, detection 273–275
    structural model 544f
    tars, skin contamination by 269–281
Collection
    of airborne organoarsenicals 384
    of CMME, derivative 172f
    of diesel emissions 358
    devices 10, 11f
    efficiency 12, 82, 149t, 312, 393,
                394t, 507, 510, 594, 608t
        of activated coconut base carbon 168f
        particle 79
    electrodes 82
    filter sample 98f
    of formic acid 600, 603, 608
    media 54–58
        for chlorocarbons 59t
    operations 96
    ratio, silica gel mesh size and 147f
    studies, metal fume 100
    of toluene 204f–205f
    tube geometry 145–146
    tube, silica gel 143f
Collimating optics 270
Colorimeter
    functions 596f
    for NO₂/NOₓ passive samplers ..590–593
Colorimetric titrations 88
Condensation technique for aerosol
    generation 97
Condensation, test atmospheres by 15–17
Constant flow pumps, air sampling
    with 491–502
Contact coupling 270
Continuum mechanics 67
Coronene 358
Cotton dust 65–85
Cresol 186
Critical values of limiting
    distributions 449t
Crotonaldehyde 14

Cyclohexane 371
Cyclopentadiene 188

D

2,4-D 14
DCB (see 3,3'-Dichlorobenzene)
n-Decane 273
Decision
    contour(s) 480f, 481f
        unbiased 476–477
    criteria, statistical 504
    rule 509
Demeton 13, 308, 312
    breakthrough test 313f
Density function, probability 474f
Derivative collection of CMME 172f
Derivatization after desorption 169, 173
Derivatization techniques 155–177
Derivatizing agent 13, 169, 173
Desorption
    from activated charcoal 577
    derivatization after 169, 173
    efficiency(ies) 58, 61, 156, 158,
                210, 537, 577
        of Freon 113 41
    inorganic acid 141
    and recovery 156–158
    recovery for hydrocarbons 41
    solvent 37
    techniques 155–177
    two-phase 161
Detection limits, AAS 106t
Detection limits for anilines, HPLC .. 418t
Diamines in hair dyes, aromatic 401–411
2,4-Diaminoanisole 402–411
    metabolites of 379t
2,4-Diaminotoluene 401–411
Diarylide yellow, DCB in 420
    detector potential/response curve
        for 422f
Diazotization mechanisms 591f
Diazotizing reagent 590
Dibenz[a,h]anthracene 369–370
Dibenzofurans, polychlorinated
    (PCDF) 319–342
Dibenzopyrene 374
Dibromochloropropane 230
3,3'-Dichlorobenzidine (DCB) 415
    in diarylide yellow 420
        detector potential/response
            curve 422f
    in rat urine 243, 425f
1,2-Dichloroethane 51, 239, 241
    permeation rate 250t
    protective garment permeation
        by 249–258
Dichloromethane 286, 358
1,2-Dichloropropane (1,2-DCP) 50–64

Diels-Alder adduct 188
Diesel emissions, PAHs in
    particulate 357–368
Diesel exhaust and mine air, formic
    acid in 599–613
Diethylamine 186
Diethylaminoethanol 186
O,O-Diethylphosphorochlorido-
    thioate (DEPCT) 164
Diffusion
    coefficient(s) 197, 247, 249, 578, 588
    head gas sensor 571f
    head sensor cell 569–572
    rate 554–555
    sampler(s) 209–221
    vs. charcoal tube 215f, 216f
Diffusional monitoring 195–207
1,3-Dihydroxybenzene 402–411
1,4-Dihydroxybenzene 402–411
Dilution, test atmospheres by 17
Dilution tunnel 358, 608
Dimethylacetamide 186
Dimethylamine 169, 186, 343, 351
    sulfate 346, 351, 353
Dimethylarsenic acid 383–400
7,12-Dimethylbenzanthracene 273
Dimethylformamide 186
N,N-Dimethylformamide 371
1,1-Dimethylhydrazine 343
Dimethylphenanthrene/anthracene 363
O,O-Dimethylphosphorochlorido-
    thioate (DMPCT) 164
1,2-Dinitrobenzene 403
Dioctylphthalate 67
Diorganotin compounds 109
Dioxins, polychlorinated (PCDD) ..319–342
Diphenyl 187
    method development for 190–193
    properties of 191f
Direct Blue 6
    analysis for 419–420
    chromatogram of hair dye with 412f
    extract, chromatogram 421f
Direct reading instruments 460
Discharge electrodes 82
Distribution(s)
    of air concentrations, cumulative .. 435f
    of daily exposures 472–475
    of exposures to inorganic lead,
        cumulative 437f
    limiting 445–448
    critical values 449t
    lognormal 438, 472
Dosimeter, microprocessor-based ..521–531
Dosimeters, organic vapor passive 209–221
Drift time, average 197
Dust(s)
    dispersion, test atmospheres by 17
    fibrogenic 434

Dust(s) (continued)
    indicator, digital 530f
    respirable 78
    sensors for 529
    and trash in cotton 71–78
Dye(s)
    aromatic amines in commercial 413–427
    without benzidine-based dyes,
        chromatogram 422f
    benzidine in hair 419–420
    with Direct Blue 6, chromatogram
        of hair 421f

E

EGDN 187
Electrochemical
    detection, HPLC with 413–427
    reaction for benzidine 414f
    sensor cell(s) 525
        reaction in 554–555
    SPE 551–573
Electrometric titration(s) 89
    curves 91f
Electron capture detector 50
End-point detection, potentiometric .. 128
Endrin 13, 312
    breakthrough test 311f
Energy technologies, workplace
    guidelines for 543–550
Engine exhaust, formic acid in 611f
Environmental surveillance 223–233
2,3-Epoxides, metabolism via 322
Equilibrium
    vapor concentration 303
    volume change of protective
        material 262f–263f
    weight change of protective
        material 259f–260f
Ethanes
    chlorinated 235
    GC conditions for halogenated 240t
    liquid halogenated 235–267
Ethyl acetate 200, 286
Ethyl salicate 187
Ethylamine 186
Ethylene glycol 304, 325
Ethylene oxide 536, 539t
    sampling of 540t, 541t
n-Ethylmorpholine 186
Ethylphenanthrene/anthracene 363
Exposure(s)
    to chlorinated phenols 334–339
    daily 471–475
    estimates and population of
        exposures 475
    index of 223
    to inorganic Pb, cumulative
        distribution 437f

Exposure(s) (*continued*)
  level, time-weighted 197–198
  Limit, Average (AEL) 473
  long-term average 474*f*, 478*f*
  to nitrosamines, leather workers' 346–354
  permissible
    level 522
    limit(s) 436, 471
      for inorganic acids, OSHA 144*t*
  to PCDDs and PCDFs 333–334
  variability of airborne 431–434
Extraction of filter samples 389

F

Feces analysis for nitrosamines 286–287
Feed flow and sensor response 564
Feedback control constant flow
  sampler 496
Feedback flow control
  system 494, 497*f*, 499*f*
  sampler 495*f*
Fiberoptics skin contamination
  monitor 269–281
Fick's law 196, 555, 559,
                          577–578, 588–589
Field
  kit, NO₂/NO_x 595, 596*f*
  studies, hydrocarbon 42–47
  surveys, preparation for 457–460
  testing of the OVM 200–203
Filter(s)
  cassette 13
  cellulose ester 103
  extraction 127
  glass fiber 97, 113, 115*t*
  membrane 384
  pneumatic 496
  PVC copolymer membrane 124
  sample collection 98*f*
  samples, recovery of benzidine
    from 28*t*, 30*t*–32*t*
  /sorbent sampling
    methods 303–306, 308–312
Flame ionization detector 51
Fluid electrode precipitator 81–84, 83*f*
Fluids, aromatic amines in body 413–427
Fluoranthene 363, 367
Fluorescence characteristics on
  animal skin 276
Fluorescence spectra of oil blend 275*f*
Fluoride 603
Fluoropore filters 387
Fly ash, PCDDs and PCDFs in 328, 333
Formaldehyde 14, 167, 599
  and formic acid in mine 611*f*
Formic acid 187
  ICE chromatogram 604*f*
  in diesel exhaust and mine air 599–613

Formic acid (*continued*)
  and formaldehyde, in mine 611*f*
  ion chromatogram 602*f*
Freon 113 37
  desorption efficiency 41
  hydrocarbon recovery with 43*t*, 44*t*
Fumes, sensors for 529
Furfural 14
  alcohol 187

G

Gas
  chromatogram
    of aromatic amines 378*f*
    of chlorinated aromatics in Mg
      plant effluent 376*f*
    of chlorinated hydrocarbons 376*f*
    of PAH(s) 361*f*, 362*f*, 372*f*–373*f*
    standards 360*f*
  chromatographic separation of
    alkanolamines 171*f*
  chromatography (GC) 14, 37–38,
                  50–54, 198, 306, 537, 577
  analytical procedure for
    chlorocarbons 53*t*
  detection of volatile nitrosamines 285
  glass capillary 357–368, 369–381
  interference testing 52*t*
  and IR hydrocarbon analysis ...46*t* 47*t*
  flow and CO response level 565*f*
Gases and vapors, test atmospheres ...17–18
Gasbadge vs. charcoal tube 214*f*
Generation, test atmosphere 138–139,
                                   533–542
Girard-T derivatives of aldehydes 14
Glass capillary gas
  chromatography 357–368, 369–381
Glass fiber 12
  filter(s) 97, 113, 115*t*, 303–306
Gradient elution 401
Graphite electrodes, characteristics
  of 557–559
Graphite furnace AAS 110
Graphitized carbons 184, 186
Grubb's test 516
Guidelines for energy technologies 543–550

H

Hair dyes, aromatic diamines in 401–411
Hall electrolytic conductivity detector 51
Hamster skin, signal intensities for
  materials on 276*t*
Health hazard, defining 434–436
Heat stress sensors 529
Heptachlor 13, 187, 308
  breakthrough test 309*f*
Heptafluorobutyryl imidazole 173

n-Heptane 240
Herbicide Orange 325–326
Hexachlorobutadiene (HCBD) 50–64
Hexachlorocyclopentadiene (HCCP) .. 50
Hexachlorophene, PCDDs and
    PCDFs in 326
n-Hexane 239
Hide processing drum 351, 353
High performance liquid chromatog-
    raphy (HPLC) 358, 413–427
Histogram of CV 511*f*
Humidity and organoarsenicals
    collection 395*t*
Hydrazine 525
Hydrocarbons 37–48
    collected on charcoal tubes, IR
        determination 37–48
    polycyclic aromatic (PAH) 271
Hydrogen cyanide 434, 440, 525
Hydrogen sulfide 434, 525
Hydroquinone 402–411

                        I

ICE *(see* Ion chromatography
    exclusion)
Imidazole, heptafluorobutyryl 173
Industrial chemicals, PCDDs and
    PCDFs in 325–333
Industrial hygiene logistics 457–469
Infrared
    determination of hydrocarbons col-
        lected on charcoal tubes 37–48
    and GC hydrocarbon analysis 46*t*, 47*t*
    hydrocarbon analysis, statistical
        data on 44*t*, 45*t*
Inorganic acids, airborne 137–152
Integrator operation 495*f*
Interference(s)
    in sensor cells 566–569
    in sorption 181
    testing, GC 52*t*
Iodometric titration 124–125
Ion
    chromatogram
        of acetic acid 602*f*
        of ammonia and methylamines .. 170*f*
        of formic acid 602*f*
        of PAHs 366*f*
        of sulfuryl fluoride 170*f*
    chromatographic separation of
        spiked solvents 163*f*
    chromatography (IC) 137, 141–142,
                    167–169, 304, 599–613
    exclusion (ICE) 600–613
    exchange membranes, perfluoro-
        sulfonate 553
Isobutyl alcohol 203*t*
Isocyanuric acids, chlorinated 123–125

Isomers
    PCDD and PCDF 324*f*
    of PCDDs and PCDFs, positional .. 320*t*
    diaminotoluene 401–411
Isomeric PAHs by GC–MS,
    identification of 363–367

                        K

Kerogen 543
    structural model 544*f*
Knudsden diffusion 555

                        L

LAQL *(see* Lowest analytically
    quantifiable level)
Latex 241
Lead 227–228, 436
    aerosol(s) 97, 100
        particle size distribution 99*f*, 101*f*
    chloride aerosol generator 98*f*
    cumulative distribution of expo-
        sures to inorganic 437*f*
    dissolution and measurement 103
    fumes 529
    OSHA standards for 439*t*
    recovery 104*t*
Leather workers' exposure to
    nitrosamines 346–354
Leather workers, PCDDs and PCDFs
    in blood samples from 338*t*
Legal action level criteria 479, 482
Limiting distribution(s) 445–448, 447*f*
    critical values 449*t*
    for evaluating acute exposures 447*f*, 450*t*
Limits of optical detection (LOD) 274*t*
Lint, spiked cotton 72
Liquid chromatogram of hair dye 408*f*
Liquid chromatography 404–411
    high performance
        (HPLC) 13–14, 22, 358, 413–427
Lognormal distribution(s) 472
    of air concentrations 432, 434
Lognormal random variable 487–488
Lowest analytically quantifiable
    level (LAQL) 50
    determination of 58–60
Luminescence of SRC-I recycle
    solvent 275*f*
Luminoscope 269–271, 272*f*

                        M

Magnesium plant effluent, gas chro-
    matogram of chlorinated
    aromatics in 376*f*
Magnesium, production of 370
Mass spectrometry 323, 363–367

MBOCA in human urine 424, 426f
MCE (*see* Mixed cellulose ester)
Medical–legal aspects of biological
   monitoring 230
Medical surveillance 223–233
Membrane filters 384
Mercaptan 14–15
Mercury 188
   vapor 188
Mesh size and collection 180
Metabolism of PCDDs and
   PCDFs 322–323
Metabolism of TCDD 322
Metal fumes 95–107
Metal oxide fumes 17
Methanol 537
Method development 188–193
Method validation 183–184
4-Methoxy-*m*-phenylenediamine
   (MMPDA) 415
   in human urine 423–424, 425f
Methyl
   alcohol 186
   chloride 167
   chloroform 577
      sampling 584f
   formate 186
   methacrylate 187
   -orange test for free chlorine 128
Methylamines 167, 169, 186
   and ammonia, ion chromatogram .. 170f
Methylchohexanone 187
Methylene chloride 402, 420
4,4′-Methylenebis(2-chloroaniline)
   (MBOCA) 415
4,4′-Methylenedianiline (MDA) 415
Methylethylphenanthrene/anthracene 363
Methylphenanthrene/anthracene 363
β-2-Microglobulin 229
Microprocessor-based instrumen-
   tation 521–531
Midget impinger 56, 88, 111
Mine
   air, formic acid in 599–613
   formaldehyde and formic acid in .. 611f
   sample, chromatogram of 606f
Mists, sensors for 529
Mitex filters 384, 387
Mixed cellulose ester 12, 13
Mixed solvents 160–161
MMPDA (*see* 4-Methoxy-*m*-
   phenylenediamine)
Model of the occupational
   environment 471–475
Monomethylamine 169
Monomethylarsonic acid (MMA) ..383–400
Monoethylphosphorodichlorodo-
   thioate (MEPCT) 164
Morpholine 186

N

NaDCC (sodium dichloroisocyanurate
   dihydrate) 123–135
NaOH aerosol generation, validation
   by 89–92
α-Naphthol 402–411
α-Naphthyl–thiorea 304
N-1-Naphthylethylenediamine
   dihydrochloride 590
Nebulizer 139
   pneumatic 15
Neoprene latex 241–267
   rubber, weight change of 245f
Neutron activation 395, 398
Nicotine 187
NIOSH
   accuracy criterion 505–507
   action level 471–489
   validation tests, statistical
      protocol 503–517
Nitric acid digestion 103
   impingers 113–116
Nitric–hydrochloric acid digestion 105
p-Nitrochlorobenzene 186
Nitrile latex 241–267
Nitrobenzene 186
Nitroethane 187
Nitrogen
   dilution of analytes 534
   dioxide 188, 434, 525
      permeation tubes 557
   oxides, personal sampling of 587–597
   oxides sensor cell 551–573
Nitroglycerin 187
Nitromethane 187
Nitrosamines in industrial
   atmospheres 343–356
Nitrosamines in tire manufacturing 283–299
N-Nitrosamine(s)
   air samples 295f
   airborne 292f
   volatilized 284f
Nitrosation capacity of tannery air 350
N-Nitrosodiethanolamine (NDEIA) .. 286
N-Nitrosodimethylamine (NDMA) 283, 343
   in tanneries 345t
N-Nitrosodiphenylamine (NDPhA) .. 283
N-Nitrosomorpholine
   (NMOR) 283, 344, 353
Nitrosopiperidine (NPiP) 286
Nitrotoluene 186
Noise sensors 529
Normal random variable 487

O

1-Octanesulfonic acid 169
Optical detection, limits of (LOD) 274t

Organic Vapor
  Monitor (OVM) 198, 200
    vs. charcoal tube 214f
    badge with backup 575–586
    passive dosimeters 209–221
Organics peremation through pro-
  tective garment 247–249
Organo-thiophosphates 164
  chromatogram 165f
Organoarsenicals, airborne 383–400
Organoarsenicals sampling 384
Organochlorine compounds 49–64
Organochlorine sorbent evaluation 55t
Organolead compounds 14
Organotin compounds 109–121
OSHA
  compliance criteria 482–483
  /NIOSH air–sampling scheme ..436–440
  permissible exposure limits for
    inorganic acids 144t
OVM (see Organic Vapor Monitor)

P

Packed bed collection of aerosols 138
Packed bed length and
  collection 146–148
PAH (see Polynuclear aromatic
  hydrocarbons)
Paraquat 14
Particle
  collection efficiency 79
  size distribution
    of Pb aerosols 99f, 101f
    of Se aerosols 101f
    of Te aerosols 99f
  transport 96
Particulate(s)
  burden, cotton 77t
  collection of 12
  /vapor sampler 11f
Passive
  charcoal badges 179–180
  dosimeters, organic vapor 209–221
  monitors 18
  sampler kit, NO_2/NO_x 587–589, 592f
  sampling device 575–586
Pathophysiologic change 229
PCB (see Polychlorinated biphenyl)
PCP (pentachlorophenol) 326–327, 334
Peak
  exposures, predictable 442
  exposures, unpredictable 442–443
  identity confirmation 405–407
  potentials for aromatic amines 417t
Pentachlorophenol (PCP) 326–327, 334
Perchloroethylene 37
Perfluorosulfonate ion exchange
  membranes 553

Permeation
  cell(s) 237, 238f, 239
  protective garment 235–267
  test atmospheres by 18
  tubes, NO_2 557
Permissible exposure level 522
Permissible exposure limit 436, 471
Personal
  alarm sensor 525
  monitoring 54, 195–207
  sampler 11f
  sampling
    of nitrogen oxides 587–597
    pump(s) 10, 18
    trains 462
  sampling equipment, attachment
    of 463f
  tape sampler (PTS) 525, 527f, 528
Pesticides, airborne 13–14, 301–315
Petroleum-based charcoal 184
Phase equilibrium 158–160
Phenanthrene 358, 363, 367
Phenoclor DP-6 328
Phenols, chlorinated
  exposure to 334–339
  PCDD isomers in 330f
  PCDF isomers in 330f
Phenoxy acids, TCDD in 325–326
Phenyl ester-biphenyl vapor mix 186
2-Phenylnaphthalene 358
Phosdrin 13, 187
Phosgene 434, 525
Phosphine 188
Phosphoric acid 590
  collection of 147f
Phosphorus yellow 187
Photosensitizing chemicals 273
Phototoxic agents 271
Physical properties for sorption 181
Platinoid electrodes, characteristics 557–559
Platinum dissolution and
  measurement 105–106
Pleural plaques, calcified 229
Plictran (tricyclohexyltin
  hydroxide) 109–121
Pneumatic filter 496
Poisons, systemic 434
Polychlorinated biphenyl (PCB) 235–267
  PCDDs and PCDFs in 328, 331t
  PCDF isomers 332f
  permeation rate 251t
Polychlorinated dibenzofurans
  (PCDF) 319–342
Polychlorinated dioxins (PCDD) 319–342
Polyethylene weight change 245f
Polynuclear aromatic hydrocarbon(s)
  (PAH) 271, 357–370, 374–375
  analysis 371
  in particulate hazards diesel emissions 357–368

Polytetrafluoroethylene (PTFE) 12
    filters 303–306
Poly(vinyl alcohol) 241–267
Polyvinyl chloride (PVC) 12
    copolymer membrane filter 124
Polyvinylidene chloride 167
Porapak Q 187
    recovery of diphenyl from 192
Pore size and collection 180
Porous polymer sorbents 186–188
Potentiostatic circuit in sensor cells 555–557
Precipitator, fluid electrode 81–84, 83f
Precipitator, wet wall electro-
    inertial 79–81, 80f
Precision sampling and analytical 149t
Precutter 71
    sampler 70f
Pressure gradient, total 555
Probability density function 474f
Programmable read-only memory 522
Propane 350–353
Propionaldehyde 14
Protective garment material 243f
Protective garment permeation 235–267
PTFE (polytetrafluoroethylene) 12
PTS (personal tape sampler) 525, 527f, 528
Pump
    altitude effect on 500f
    and cassette rack 112f
    flow rate
        and inlet pressure 493f
        and load 499f
        and time 493f, 499f
    temperature effect on 500f
PVC (see Polyvinyl chloride)
Pyrene 358, 363, 367
Pyrethrum 14
Pyrolytic dimerization of
    chlorophenate 321f

Q

Quinone 187

R

Reactivity and sorption 181
Radiation sensors 529
Random air-sampling 448–451
Recovery(ies) 507
    desorption and 156–158
    formic acid 609t
    improving 160–161
    inorganic acid 144
    of NaDCC 130, 133t, 134t
    NaOH 90f
    of organoarsenicals 392t
Resins, carbonized 167
Resorcinol 402

Respirable dust 78
Respiratory system, effect of
    metal fumes 95
Response
    vs. concentration, sensor
        cells 559, 561f, 563f
    level vs. flow rate 571f
    time for sensor cells 562–564
Ronnel 13, 312
    breakthrough test 311f
Rotenone 14

S

Safety factors for airborne chemicals 444t
Sample
    distribution 211f, 212f
    sizes for OSHA/NIOSH air
        sampling 441t
    stability 507
        inorganic acid 148–149
    validity 466
Sampler(s)
    area 462, 466
    preparation 589–590
    ranges, personal tape 528t
    system, positive-displacement 492–496
Sampling
    and analytical equipment setup ..461–462
    and analytical methods validation .. 4
    device criteria, pesticide 301–302
    effect of imperfect 488–489
    efficiency of VE, isokinetic 66–67, 68t
    errors for flow variation 491–492
    field 41
    method development 190
    methods, acid mist 137
    methodology 18
    NaDCC and TCCA 124
    pump, personal 10
    rate, backup 578, 580
    statistical schemes for air 431–455
    studies, aerosol 96–97
    studies, organotin 115t, 117t–120t
    tannery air 346, 351, 353
    train(s) 301, 308
        organoarsenical vapor 387
        organotin stationary 110–111
        personal 462
    tube, silica gel 138–141
    velocity and collection 146
Saw mill workers, PCDDs and
    PCDFs in blood samples from 337t
Sawdust, chlorinated contaminants in 335t
Selenium
    aerosol(s) 100
        particle size distribution of 101f
    dissolution and measurement 103, 105
    recovery 104t

Sensor
  cell characteristics 558f, 560f
  cells, electrochemical SPE 551–573
  technology 525–531
Separation
  of alkanolamines, GC 171f
  of aromatic diamines 406f
  of spiked solvents, IC 163f
Signal intensities for materials on
  hamster skin 276t
Silica 529
Silica gel
  collection tube 143f
  mesh size and collection ratio ...146, 147f
  sampling tube 138–141
  sorbent 186
Silver membrane 12
Single decision contours 477
Single-photon counting (SPC)
  technique 270
Skin contamination monitor,
  fiberoptics 269–281
Smiles rearrangement 327, 328
Sodium
  dichloroisocyanurate dihydrate
    (NaDCC) 123–135
  fluoroacetate 12, 304, 306
  hydroxide 402
    airborne 87–94
  thiosulfate standardization 125–126
Solvent(s)
  IC separation of spiked 163f
  mixed 160–161
  permeation through protective
    garment 237–241
  recycle
    luminescence of SRC-I 275f
    -refined coal 274
    solvent-refined coal 274
  selection 183
Sorbent(s)
  bonded 164–167, 168f
  coated 188
  evaluation, organochlorine 55t
  reactant-coated solid 173
  sampling methods 306–308
  selection 180–183
  solid 155, 161–167, 179–193
  tube aerosol collection 144–145
  techniques, specialized 155–177
Spiked filters 127, 389
Spiking procedure 56–58
Spray drying, test atmospheres by 15
Stability, sensor cell 566
Standard(s)
  addition of formic acid 609t
  calibration 463f
  curve for organotins, AAS 114f

Standard(s) (continued)
  gas chromatogram of PAH 360f
  hydrocarbon 38
  interim 545
  method, native cotton 72
  model of native cotton 75f
Standardization, sodium thiosulfate 125–126
Statistical
  data on IR hydrocarbon analysis 44t, 45t
  evaluation of data 215f, 217f
  protocol for NIOSH validation
    tests 503–517
  schemes for air sampling 431–455
Steric effects and recovery 158
Stibine 188
Storage stability 184, 187, 302–303,
                        388t, 390t, 585f, 590, 609
  of dyes 29–32
  of passive badges 583
Styrene 210
Sulfanilamide 590
Sulfate ion 603
Sulfur dioxide 167
  and organoarsenicals collection 369t
Sulfuric acid, collection of 147f
Sulfuryl fluoride 169
  ion chromatogram 170f
Surface groups and collection 180–181
Surgical rubber latex 241–267
  volume change 248f
Synfuel plants 269–281
Synthetic charcoals 184, 186
Syringe injection, test atmospheres
  by 17
Systox 312

T

2,4,5-T 14
TEPP 13
Tanneries, N-nitrosodimethylamine in 345t
Tannery air sampling 346, 351, 353
Tars on paper, detection of 273–275
TCCA (trichloroisocyanuric acid) ..123–125
Teflon gloves 241–267
Tellurium
  aerosol(s) 100
    particle size distribution of 99f
  dissolution and measurement 105
  recovery 104t
Temperature
  and CO span response 565f
  dependence, diffusion head sensor
    cell 572t
  and organoarsenicals collection 395t
  and sensor response 564
Tenax-GC 12, 187
  recovery of diphenyl from 192

Test atmosphere generation 12, 15–18, 16f, 96–97, 533–542
Tetraamine copper(II) chloride coating 111
Tetrachloro-1,1-difluoroethane 51
1,2,4,5-Tetrachlorobenzene 325
2,3,7,8-Tetrachlorodibenzo-p-dioxin .. 320
n-Tetradecane 273
Tetraethyl lead 14, 187
Tetramethyl lead 14, 187
Tetramethyl thiuram disulfide 296
Textile workers, PCDDs and PCDFs in blood samples from 338t
Thermal decomposition, test atmospheres by 17
Thiram 14, 304
Threshold limit value (TLV) 88, 273, 525
Time
  and CO sensor cell 567f
  weighted average (TWA) 88, 432
  -weighted exposure level 197–198
Tire manufacturing, nitrosamines in 283–299
α-Tocopherol 286
Tolerance set(s) 443–445, 448
  for evaluating acute exposures 450t
o-Tolidine and benzidine, chromatogram 27f
Toluene 203t, 306, 308, 577, 583
  collection 204f
  sampling 581f
o-Toluidine 186
Toxic
  gases, sensors for 525–529
  Material Advisory Committee (TMAC) 543–550
  properties of inorganic acids 137
Toxicants, bioabsorption 228–229
Toxicity, index of 223
Toxicity of PCDDs and PCDFs 320–322
Transistor-Transistor Logic (TTL) signals 270
Transnitrosation 283, 293, 296
Trapping agent 87
Trash and dust in cotton 71–78
1,1,2-Trichloro-1,2,2-trifluorethane .. 37
1,1,1-Trichloroethane 239
  permeation rate 250t
  protective garment permeation by 249–258
1,1,2-Trichloroethane 239, 247, 249
  permeation rate 251t
  protective garment permeation by 249–258
Trichloroethylene 577, 583
  sampling 582f
Trichloroisocyanuric acid (TCCA) 123–135
2,4,5-Trichlorophenol 325

Tricyclohexyltin hydroxide (Plictran) 109–121
Triethanolamine 566, 587
Triiodide ion 129
Trimethylamine 169
Trimethylphenanthrene/anthracene .. 363
Triorganotin compounds 109
Triphenylene 363, 367
TTL (Transistor-Transistor-Logic) signals 270
Two-phase desorption 161

U

Ultraviolet radiation for skin illumination 273
Urinary cadmium 226
Urine analysis, samples for nitrosamines 286
Urine, aromatic amines in 423–426

V

Validation
  method 58–62
  by NaOH aerosol generation 89–92
  sampling and analytical methods 4
  by spiking 89
  test(s) 190
    program 509–510
    statistical protocol for 503–517
Validity, sample 466
van der Waals attraction 164
Vapor(s)
  monitors, spiked 41
  particulate ratio 12
  test atmospheres, gases and 17–18
Vaporization, test atmospheres by 17–18
Variation, coefficient of (CV) ...26, 42, 90f, 304, 475, 491, 504–517
Ventilatory capacity 229
Vertical elutriator (VE) 66–69, 70f
Vinyl chloride 167, 436
  OSHA standards for 439t
Visible spectrum studies of benzidine-based dyes 26t
Viton elastomer 241–267
  volume change 248f
Volume change
  protective garment 236–237, 244–247
    and breakthrough time 261
  of surgical rubber latex 248f
  of Viton elastomer 248f

W

Warfarin 14
Weight change
  of butyl rubber 243f

Weight change (*continued*)
  of neoprene rubber latex 245*f*
  of polyethylene 245*f*
  of protective garment 242*t*, 244
    and breakthrough time 258–261
    experiments 236
Wet wall electroinertial
  precipitator 79–81, 80*f*
Wood preservation 334
Workplace guidelines for energy
  technologies 543–550

X

X-ray fluorescence 395, 398
XAD-2 ....12, 187, 306, 308, 312
  Amberlite 164, 165*f*
  recovery of diphenyl from 192
Xylidine 186

Y

Yusho disease 322
Yusho patients, PCDF isomers
  retained by 324*f*

DATE DUE

NOV 0 3			
OCT 1 7 REC'D			
GAYLORD			PRINTED IN U.S.A.